GLACIERS OF ASIA

I0027520

F–1.   GLACIERS OF THE FORMER SOVIET UNION
By VLADIMIR M. KOTLYAKOV, *senior author, with contributions from* A.M. DYAKOVA (Siberia), V.S. KORYAKIN (Russian Arctic Islands), V.I. KRAVTSOVA (Caucasus, Altay), G.B. OSIPOVA (Tien Shan), G.M. VARNAKOVA (Pamirs and Alai Range), V.N. VINOGRADOV (Kamchatka), O.N. VINOGRADOV (Caucasus), *and* N.M. ZVERKOVA (Ural Mountains and Taymyr Peninsula)
With *sections on* FLUCTUATIONS OF GLACIERS OF THE CENTRAL CAUCASUS AND GORA EL'BRUS (*With a subsection on* THE GLACIOLOGICAL DISASTER IN NORTH OSETIYA, *by* VLADIMIR M. KOTLYAKOV, O.V. ROTOTAEVA, *and* G.A. NOSENKO, INVESTIGATIONS OF THE FLUCTUATIONS OF SURGE-TYPE GLACIERS IN THE PAMIRS BASED ON OBSERVATIONS FROM SPACE, *by* VLADIMIR M. KOTLYAKOV, G.B. OSIPOVA, *and* D.G. TSVETKOV, *and* THE GLACIOLOGY OF THE RUSSIAN HIGH ARCTIC FROM LANDSAT IMAGERY, *by* J.A. DOWDESWELL, E.K. DOWDESWELL, M. WILLIAMS, *and* A.F. GLAZOVSKII
F–2.   GLACIERS OF CHINA
By SHI YAFENG, MI DESHENG, YAO TANDONG, ZENG QUNZHU, *and* LIU CHAOHAI
F–3   GLACIERS OF AFGHANISTAN
By JOHN E. SHRODER, JR., *and* MICHAEL P. BISHOP
F–4   GLACIERS OF PAKISTAN
By JOHN E. SHRODER, JR., *and* MICHAEL P. BISHOP
F–5   GLACIERS OF INDIA
By CHANDER P. VOHRA
With an updated supplement on A STUDY OF SELECTED GLACIERS UNDER THE CHANGING CLIMATE REGIME
By SYED IQBAL HASNAIN, RAJESH KUMAR, SAFARAZ AHMAD, *and* SHRESTH TAYAL
F–6   GLACIERS OF NEPAL — GLACIER DISTRIBUTION IN THE NEPAL HIMALAYA WITH COMPARISON TO THE KARAKORAM RANGE
By KEIJI HIGUCHI, OKITSUGU WATANABE, HIROJI FUSHIMI, SHUHEI TAKENAKA, *and* AKIO NAGOSHI
With a supplement by YUTAKA AGETA
F–7   GLACIERS OF BHUTAN
By SHUJI IWATA
F–8   THE PALEOENVIRONMENTAL RECORD PRESERVED IN MIDDLE-LATITUDE, HIGH-MOUNTAIN GLACIERS: AN OVERVIEW OF U.S. GEOLOGICAL SURVEY RESEARCH IN CENTRAL ASIA AND THE UNITED STATES
By L. DeWAYNE CECIL, DAVID L. NAFTZ, PAUL F. SCHUSTER, DAVID D. SUSONG, *and* JAROMY R. GREEN

SATELLITE IMAGE ATLAS OF GLACIERS OF THE WORLD

Edited by RICHARD S. WILLIAMS, JR., *and* JANE G. FERRIGNO

U.S. GEOLOGICAL SURVEY PROFESSIONAL PAPER 1386–F

Landsat images, other satellite images, and other data are used to discuss the geographic distribution of ice caps, outlet glaciers, valley glaciers and other glaciers in Asia, and glaciological hazards in the Himalaya

UNITED STATES GOVERNMENT PRINTING OFFICE, WASHINGTON: 2010

U.S. DEPARTMENT OF THE INTERIOR
KEN SALAZAR, *Secretary*

U.S. GEOLOGICAL SURVEY
Marcia K. McNutt, Director

Technical editing by Jane R. Eggleston

Illustrations by Ronald S. Spencer and James Tomberlin

Design and layout by Twila Darden Wilson

Cover: A Landsat MSS false-color composite image of the ice caps on
Northern Savernaya Zemlya, Russian High Arctic, acquired 22 July 1975.
The 5,575 square kilometer Lednik Akademii Nauk (upper right) is the
largest ice cap in the archipelago. See page F101.

Foreword

On 23 July 1972, the first Earth Resources Technology Satellite (ERTS 1 or Landsat 1) was successfully placed in orbit. The success of Landsat inaugurated a new era in satisfying mankind's desire to better understand the dynamic world upon which we live. Space-based observations have now become an essential means for monitoring changes in local, regional, and global environments.

The short- and long-term cumulative effects of processes that cause significant changes on the Earth's surface can be documented and studied by repetitive Landsat and other satellite images. Such images provide a permanent historical record of the surface of the planet on a specific date; they also make possible comparative two- and three-dimensional measurements of change over time. This Professional Paper demonstrates the importance of the application of Landsat images to global studies by using them to determine the 1970s distribution of glaciers on our planet. As images become available from future satellites, the new data will be used to document global changes in glacier extent by reference to the baseline Landsat image record of the 10-year period 1972 to 1981.

Although many geological processes take centuries or even millenia to produce obvious changes on the Earth's surface, other geological phenomena, such as glaciers and volcanoes, cause noticeable changes over shorter periods. Some of these phenomena can have a worldwide impact and often are interrelated. Explosive volcanic eruptions, such as the 1991 Mount Pinatubo, Philippines, eruption can produce dramatic short-term effects on the global climate. Natural or culturally induced processes can cause global climatic cooling or warming. Glaciers respond to such warming or cooling periods by decreasing or increasing in size, which in turn causes sea level to rise or fall.

As our understanding of the interrelationship of global processes improves and our ability to assess changes caused by these processes develops further, we will learn how to use indicators of global change, such as glacier variation, to manage more wisely the use of our finite land and water resources. This Professional Paper is an excellent example of the way in which we can use technology to provide needed earth-science information about our planet. The international collaboration represented by this report is also an excellent model for the kind of cooperation that scientists will increasingly find necessary in the future in order to solve important Earth-System-science problems on a global basis.

Marcia K. McNutt
Director
U.S. Geological Survey

Preface

This chapter is the ninth to be released in U.S. Geological Survey Professional Paper 1386, Satellite Image Atlas of Glaciers of the World, a series of 11 chapters. In each of the geographic area chapters, remotely sensed images, primarily from the Landsat 1, 2, and 3 series of spacecraft, are used to analyze the specific glacierized region of our planet under consideration and to monitor glacier changes. Landsat images, acquired primarily during the middle to late 1970s and early 1980s, were used by an international team of glaciologists and other scientists to study various geographic regions and (or) to discuss related glaciological topics. In each glacierized geographic region, the present areal distribution of glaciers is compared, wherever possible, with historical information about their past extent. The atlas provides an accurate regional inventory of the areal extent of glacier ice on our planet during the 1970s as part of a growing international scientific effort to measure global environmental change on the Earth's surface.

The chapter is divided into seven geographic parts and one topical part: Glaciers of the Former Soviet Union (F–1), Glaciers of China (F–2), Glaciers of Afghanistan (F–3), Glaciers of Pakistan (F–4), Glaciers of India (F–5), Glaciers of Nepal (F–6), Glaciers of Bhutan (F–7), and the Paleoenvironmental Record Preserved in Middle-Latitude, High-Mountain Glaciers (F–8). Each geographic section describes the glacier extent during the 1970s and 1980s, the benchmark time period (1972–1981) of this volume, but has been updated to include more recent information.

Glaciers of the Former Soviet Union are located in the Russian Arctic and various mountain ranges of Russia and the Republics of Georgia, Kyrgyzstan, Tajikistan, and Kazakstun. The Glacier Inventory of the USSR and the World Atlas of Ice and Snow Resources recorded a total of 28,881 glaciers covering an area of 78,938 square kilometers (km^2).

China includes many of the mountain-glacier systems of the world including the Himalaya, Karakorum, Tien Shan and Altay mountain ranges. The glaciers are widely scattered and cover an area of about 59,425 km^2. The mountain glaciers may be classified as maritime, subcontinental or extreme continental.

In Afghanistan, more than 3,000 small glaciers occur in the Hindu Kush and Pamir mountains. Most glaciers occur on north-facing slopes shaded by mountain peaks and on east and southeast slopes that are shaded by monsoon clouds. The glaciers provide vital water resources to the region and cover an area of about 2,700 km^2.

Glaciers of northern Pakistan are some of the largest and longest mid-latitude glaciers on Earth. They are located in the Hindu Kush, Himalaya, and Karakoram mountains and cover an area of about 15,000 km^2. Glaciers here are important for their role in providing water resources and their hazard potential.

The glaciers in India are located in the Himalaya and cover about 8,500 km^2. The Himalaya contains one of the largest reservoirs of snow and ice outside the polar regions. The glaciers are a major source of fresh water and supply meltwater to all the rivers in northern India, thereby affecting the quality of life of millions of people.

In Nepal, the glaciers are located in the Himalaya as individual glaciers; the glacierized area covers about 5,324 km^2. The region is the highest mountainous region on Earth and includes the Mt. Everest region.

Glaciers in the Bhutan Himalaya have a total area of about 1,317 km^2. Many recent glacier studies are focused on glacier lakes that have the potential of generating dangerous glacier lake outburst floods.

Research on the glaciers of the middle-latitude, high-mountain glaciers of Asia has also focused on the information contained in the ice cores from the glaciers. This information helps in the reconstruction of paleoclimatic records, and the computer modeling of global climate change.

Richard S. Williams, Jr.
Jane G. Ferrigno
Editors

About this Volume

U.S. Geological Survey Professional Paper 1386, Satellite Image Atlas of Glaciers of the World, contains 11 chapters designated by the letters A through K. Chapter A provides a comprehensive, yet concise, review of the "State of the Earth's Cryosphere at the Beginning of the 21st Century: Glaciers, Global Snow Cover, Floating Ice, and Permafrost and Periglacial Environments," and a "Map/Poster of the Earth's Dynamic Cryosphere," and a set of eight "Supplemental Cryosphere Notes" about the Earth's Dynamic Cryosphere and the Earth System. The next 10 chapters, B through K, are arranged geographically and present glaciological information from Landsat and other sources of historic and modern data on each of the geographic areas. Chapter B covers Antarctica; Chapter C, Greenland; Chapter D, Iceland; Chapter E, Continental Europe (except for the European part of the former Soviet Union), including the Alps, the Pyrenees, Norway, Sweden, Svalbard (Norway), and Jan Mayen (Norway); Chapter F, Asia, including the European part of the former Soviet Union, China, Afghanistan, Pakistan, India, Nepal, and Bhutan; Chapter G, Turkey, Iran, and Africa; Chapter H, Irian Jaya (Indonesia) and New Zealand; Chapter I, South America; Chapter J, North America (excluding Alaska); and Chapter K, Alaska. Chapters A–D each include map plates.

The realization that one element of the Earth's cryosphere, its glaciers, was amenable to global inventorying and monitoring with Landsat images led to the decision, in late 1979, to prepare this Professional Paper, in which Landsat 1, 2, and 3 multispectral scanner (MSS) and Landsat 2 and 3 return beam vidicon (RBV) images would be used to inventory the areal occurrence of glacier ice on our planet within the boundaries of the spacecraft's coverage (between about 81° north and south latitudes). Through identification and analysis of optimum Landsat images of the glacierized areas of the Earth during the first decade of the Landsat era, a global benchmark could be established for determining the areal extent of glaciers during a relatively narrow time interval (1972 to 1981). This global "snapshot" of glacier extent could then be used for comparative analysis with previously published maps and aerial photographs and with new maps, satellite images, and aerial photographs to determine the areal fluctuation of glaciers in response to natural or culturally induced changes in the Earth's climate.

To accomplish this objective, the editors selected optimum Landsat images of each of the glacierized regions of our planet from the Landsat image data base at the EROS Data Center in Sioux Falls, South Dakota, although some images were also obtained from the Landsat image archives maintained by the Canada Centre for Remote Sensing, Ottawa, Ontario, Canada, and by the European Space Agency in Kiruna, Sweden, and Fucino, Italy. Between 1979 and 1981, these optimum images were distributed to an international team of more than 50 scientists who agreed to write a section of the Professional Paper concerning either a geographic area or a glaciological topic. The total number of scientists who made contributions eventually reached 109. In addition to analyzing images of a specific geographic area, each author was also asked to summarize up-to-date information about the glaciers within the area and to compare their present areal distribution with historical information (for example, from published maps, reports, and photographs) about their past extent.

The atlas provides a regional inventory of the areal extent of glaciers on the planet during the 1970s, a period of time, in restrospect (2010), that was cooler climatically than the increasingly warmer period that followed at the end of the 20th century and has continued into the first decade of the 21st century.

Richard S. Williams, Jr.
Jane G. Ferrigno
Editors

Acknowledgments

The editors and authors want to thank the following individuals for reviewing parts of this report, USGS Professional Paper 1386-F, Glaciers of Asia. Reviewers are listed with their affiliation at the time of the review: **Yutaka Ageta**, Nagoya University, Graduate School of Environmental Studies; **Vladimir B. Aizen**, University of Idaho, Department of Geography; **John T. Andrews**, University of Colorado, Institute for Arctic and Alpine Research (INSTAAR); **Richard L. Armstrong**, University of Colorado, National Snow and Ice Data Center (NSIDC); **Yves Arnaud**, Laboratoire de Glaciologie et Géophysique de l'Environment (LGGE), Institute de Recherche pour le Developpement (IRD), Grenoble; **L. DeWayne Cecil**, U.S. Geological Survey, Water Resources Discipline, **Mark Dyurgerov**, University of Colorado, Institute for Arctic and Alpine Research (INSTAAR); **D.R. Gurung**, Bhutan Department of Geology and Mines; **Jon Øve Hagen**, University of Oslo, Department of Geosciences; **L. Flint Hall**, Idaho Department of Environmental Quality; **Gordon Hamilton**, University of Maine, Climate Change Institute; **Herman Häusler**, University of Vienna, Center for Earth Sciences; **Kenneth Hewitt**, Wilfrid Laurier University, Department of Geography and Environmental Studies; **Jianli Song**, USGS, Geologic Discipline, Woods Hole Science Center; **Andreas Kääb**, University of Oslo, Department of Geosciences; **Ulrich Kamp**, University of Montana, Department of Geography; **V.I. Kravtsova**, Moscow State University, Faculty of Geography; **Robert M. Krimmel**, USGS, Water Resources Discipline; **Karl Kreutz**, University of Maine, Climate Change Institute; **Li Jun**, National Aeronautics and Space Administration, Goddard Space Flight Center; **Nozomu Naito**, Hiroshima Institute of Technology, Department of Environmental Information Studies; **Masayoshi Nakawo**, Research Institute for Humanity and Nature, Kyoto; **Mitchell A. Plummer**, Idaho National Laboratory; **Charles F. Raymond**, University of Washington, Department of Earth and Space Sciences; **John F. Shroder**, University of Nebraska at Omaha, Department of Geography and Geology; **Patrick Wagnon**, Laboratoire de Glaciologie et Géophysique de l'Environment (LGGE), Institute de Recherche pour le Developpement (IRD), Grenoble.

The editors are also thankful for the generous assistance of **Nora Tamberg**, U.S. Geological Survey, who translated several Russian articles.

CONTENTS: Major Sections

Foreword --- **III**

Preface --- **V**

About this Volume -- **VI**

Acknowledgments -- **VII**

F–1. Glaciers of the Former Soviet Union, by Vladimir M. Kotlyakov,
        senior author, with contributions from A.M. Dyakova (Siberia),
        V.S. Koryakin (Russian Arctic Islands), V.I. Kravtsova (Caucasus, Altay),
        G.B. Osipova (Tien Shan), G.M. Varnakova (Pamirs and Alai Range),
        V.N. Vinogradov (Kamchatka), O.N. Vinogradov (Caucasus),
        and N.M. Zverkova (Ural Mountains and Taymyr Peninsula) ---------------- **F1**
          With sections on Fluctuations of Glaciers of the Central
            Caucasus and Gora El'Brus (with a subsection on
            The Glaciological Disaster in North Osetiya), by Vladrim M.
            Kotlyakov, O.V. Rototaeva, and G.A. Nosenko, Investigations
            of the Fluctuations of Surge-Type Glaciers in the Pamirs Based
            on Observations from Space, by Vladimir M. Kotlyakov,
            G.B. Osipova and D.G. Tsvetkov, and the Glaciology of the Russian
            High Arctic from Landsat Imagery, by J.A. Dowdeswell,
            E.K. Dowdeswell, M. Williams, and A.F. Glazovskii -------------------- **F59**

F–2. Glaciers of China, by Shi Yafeng, Mi Desheng, Yao Tandong, Zeng
        Qunzhu, and Liu Chaohai --- **F127**

F–3. Glaciers of Afghanistan, by John E. Shroder, Jr., and
        Michael P. Bishop --- **F167**

F–4. Glaciers of Pakistan, by John E. Shroder, Jr., and Michael P. Bishop --------- **F201**

F–5. Glaciers of India, by Chander P. Vohra ------------------------------------ **F259**
        With an updated supplement on A Study of Selected Glaciers
          Under the Changing Climate Regime, by Syed Iqbal Hasnain,
          Rajesh Kumar, Safaraz Ahmad, and Shresth Tayal -------------------- **F275**

F–6. Glaciers of Nepal — Glacier Distribution in the Nepal Himalaya
        with Comparison to the Karakoram Range, by Keiji Higuchi,
        Okitsugu Watanabe, Hiroji Fushimi, Shuhei Takenaka, and
        Akio Nagoshi -- **F293**
          With a supplement by Yutaka Ageta ------------------------------------ **F317**

F–7. Glaciers of Bhutan, by Shuji Iwata -- **F321**

F–8. The Paleoenvironmental Record Preserved in Middle-Latitude,
        High-Mountain Glaciers: An Overview of U.S. Geological Survey
        Research in Central Asia and the United States, by L. DeWayne
        Cecil, David L. Naftz, Paul F. Schuster, David D. Susong,
        and Jaromy R. Green -- **F335**

Glaciers of Asia—

GLACIERS OF THE FORMER SOVIET UNION

By V.M. KOTLYAKOV, *senior author, with contributions from* A.M. DYAKOVA (Siberia), V.S. KORYAKIN (Russian Arctic Islands), V.I. KRAVTSOVA (Caucasus, Altay), G.B. OSIPOVA (Tien Shan), G.M. VARNAKOVA (Pamirs and Alai Range), V.N. VINOGRADOV (Kamchatka), O.N. VINOGRADOV (Caucasus), *and* N.M. ZVERKOVA (Ural Mountains and Taymyr Peninsula)

SATELLITE IMAGE ATLAS OF GLACIERS OF THE WORLD

Edited by RICHARD S. WILLIAMS, JR., *and* JANE G. FERRIGNO

U.S. GEOLOGICAL SURVEY PROFESSIONAL PAPER 1386–F–1

CONTENTS

Glaciers of the Former Soviet Union — by V.M. Kotlyakov, senior author, with contributions from A.M. Dyakova (Siberia), V.S. Koryakin (Russian Arctic Islands), V.I. Kravtsova (Caucasus, Altay), G.B.Osipova (Tien Shan), G.M. Varnakova (Pamirs and Alai Range), V.N. Vinogradov (Kamchatka), O.N. Vinogradov (Caucasus), and N.M. Zverkova (Ural Mountains and Taymyr Peninsula) --- 1

Abstract -- 1

Introduction -- 1

Discussion of Glaciers by Region ----------------------------------- 2

FIGURE 1. Glacierized regions of the Former Soviet Union according to the Glacier Inventory of the USSR -------------------- 3

TABLE 1. Data on the glaciers of the Former Soviet Union ------------------- 3

FIGURE 2. Morphological types of mountain glaciers adopted for the Glacier Inventory of the USSR ------------------------------ 4

Caucasus (in Russia and the Republic of Georgia) ------------------------ 4

FIGURE 3. Map of the glacierization of a large portion of the Caucasus region, compiled for the World Atlas of Snow and Ice Resources --- 5

TABLE 2. Distribution of glaciers in the main regions of the Caucasus --------------- 5

TABLE 3. Glacier complexes and large glaciers of the Caucasus ------------------------ 6

TABLE 4. The distribution of glaciers and glacier complexes in the Caucasus based on their area ------------------------------------ 6

TABLE 5. The distribution of glaciers of the Caucasus according to morphological types -------------------------------------- 7

FIGURE 4. Part of an enlarged Landsat 2 MSS image taken on 9 July 1976, and map of the morphologial types of glaciers of the same part of the Western Caucasus, including Gora El'brus------------------------- 7

FIGURE 5. Aerial photograph of the Lednik Marukhskiy taken on 15 September 1955, and a series of glaciological maps of this glacier based on detailed studies done during the IHD---------------- 8–9

FIGURE 6. Enlargement of part of a Landsat 2 MSS image taken on 9 July 1976, depicting the El'brus massif ------------------------------- 10

FIGURE 7. Panoramic photographic mosaic of the northern and southern slopes of Gora El'brus--- 10

FIGURE 8. Part of an enlargement of a Landsat 2 MSS image of the Central Caucasus, taken on 13 September 1977, and map of the morphological types of glaciers of the Central Caucasus ---------------- 11

FIGURE 9. Two enlargements of the part of the Central Caucasus around Lednik Bezengi depicted on Landsat 2 MSS images taken on 9 July 1976 and 13 September 1977------------------------------ 12

FIGURE 10. Illustrations of the surging *Lednik Kolka*, Gora Kazbek: Aerial photograph, sketch map of the advance, and morphology of the glacier basin--- 13

FIGURE 11. Part of the Eastern Caucasus with the *Djul'tydag* range depicted on a Landsat 2 MSS image taken on 28 August 1976--------------------- 14

Pamirs, Alai (Alayskiy), and Gissar (Hisor) Ranges (in the Republics of Tajikistan, Kyrgyzstan, and Uzbekistan)----------------14

TABLE 6. Largest glaciers of the Pamirs and the Gissar (Hisor) and Alai (Alayskiy) Ranges--- 15

TABLE 7. Distribution of glaciers based on their area ------------------------------- 15

FIGURE 12. Landsat 2 MSS image of the Central Pamirs acquired on 28 September 1977, and the distribution pattern of the area and number of glaciers related to glacier orientation in the Hisor and Alayskiy Ranges and the Pamirs ----------------------------- 16

FIGURE 13. The Lednik Bivachnyy surging glacier, tributary of Lednik Fedchenko with loops of medial moraines ----------------------------- 17

FIGURE 14. Enlargement of part of the Landsat image, showing the Lednik Fedchenko basin, and sketch map showing the distribution pattern of morphological types of glaciers in this area --------------------**18**

FIGURE 15. Enlargement of part of the Landsat image of 28 September 1977, showing the eastern part of Zaalayskiy Khrebet (Qatorkŭhi Pasi Oloy) and part of the Eastern Pamirs, and sketch map showing the pattern of morphological types of glaciers in this region --**19**

FIGURE 16. Debris-free slope glaciers of the Eastern Pamirs-----------------------**20**

FIGURE 17. Enlargement of part of the Landsat image, showing the upper reaches of the Saukdara river, and sketch map of surges of the Saukdara group of glaciers from 1973 to 1977----------------------**20**

FIGURE 18. Photograph taken August 1976 of the rapid advance of ice of *Lednik Valy* caused by its surge ---------------------------------------**21**

FIGURE 19. Photograph showing the western part of the Khrebet Alayskiy on 18 September 1976---**22**

Tien Shan (in the Republics of Kyrgyzstan and Kazakhstan)------------**22**

FIGURE 20. Morphological types of glaciers in the western part of Khrebet Alayskiy delineated from space photographs and images ----------------**23**

TABLE 8. Distribution of Tien Shan glaciers in various river or lake basins----------**23**

Piks Pobedy and *Khan-Tengri* in Central Tien Shan -------------**24**

FIGURE 21. Part of a Landsat 2 MSS image of the Pobedy and *Khan Tägiri/ Kan-Too* peak region taken on 16 October 1976 -------------------------**24**

TABLE 9. Largest glaciers in the Tien Shan --------------------------------------**25**

TABLE 10. Distribution of glaciers in two regions of the Central Tien Shan based on their area ---**25**

FIGURE 22. Enlargement of part of a Landsat 1 MSS image taken on 18 September 1973, and aerial photograph and map of the same area --------**26**

Khrebet Ak-Shyyrak in Central Tien Shan --------------------------**27**

FIGURE 23. Enlargement of part of a Landsat 1 MSS image of Khrebet Ak-Shyyrak taken on 18 September 1973, map of the morphological types of glaciers, and distribution pattern of the area and number of glaciers --**28**

FIGURE 24. Aerial photograph of *Lednik Severniy Karasay* in Khrebet Ak-Shyyrak --**28**

Altay (Russia and Kazakhstan) --**29**

FIGURE 25. Glaciers in the main mountain-glacier areas of the Altay — on the Katunskiy, Severo-Chuyskiy, and Yuzhno-Chuyskiy Khrebet, and the distribution pattern of the area and number of glaciers--**29**

TABLE 11. Largest glaciers in the Russian Altay----------------------------------**30**

FIGURE 26. Image of the central Altay with the Gora Belukha glacier group taken from the Salyut–1 space station on 14 June 1971; and comparison of the delineation of the glacier boundaries shown on topographic maps and derived from the space image -----------------**31**

FIGURE 27. Part of a Landsat 2 MSS image of Katunskiy Khrebet taken on 11 July 1977, map of surface morphology, and morphological types of glaciers in this region---**32**

FIGURE 28. Part of the Landsat 2 MSS image taken on 11 July 1977 of the *Bish-Irdu* glacier complex in the Severo-Chuyskiy Khrebet, enlarged portion of the image, and depiction of this region on a map ---**33**

Ural Mountains (Ural'skiye Gory) ---**34**

FIGURE 29. Enlargement of a Landsat 1 MSS image of the Northern Urals acquired on 25 June 1973--**34**

FIGURE 30. Distribution pattern of glaciers in the Polar Urals, and by area and number related to orientation ----------------------------------**35**

FIGURE 31. Part of an enlargement of a Landsat 1 MSS image of part of the Polar Urals acquired on 6 March 1973: *Bol'shoy Khadyta* river-valley, and location of the *Lednik Obrucheva* -------------------------**36**

FIGURE 32. Aerial photograph taken on 30 July 1953 of *Lednik Obrucheva*; morphology of the surface; and changes of the surface elevation of the *Lednik Obrucheva* in the Polar Urals ----------------------**36**

The Byrranga Mountains (Gory Byrranga) in Northern Russia --------**37**

FIGURE 33. Landsat 1 MSS image acquired 26 July 1973 showing the distribution of glaciers in the Gory Byrranga in the region near lat. 76°N., long. 108°E.; sketch of glacier location; and distribution pattern of glacier orientation in the Gory Byrranga -----------**37**

Northeastern Siberia --**38**

FIGURE 34. Distribution of glaciers over Khrebet Suntar-Khayata; distribution patterns of the number and area of glaciers relating to the orientation in Khrebet Suntar-Khayata; and Khrebet Cherskogo --**38**

TABLE 12. Distribution of glaciers in the Khrebet Suntar-Khayata based on their size --**39**

FIGURE 35. Enlargements of two Landsat 2 MSS images of the Koryakskoye highland, northeastern Siberia, taken on 2 July 1975 and 26 August 1975; and distribution of glaciers in the highest part of the Koryakskoye highland------------------------**39**

FIGURE 36. Distribution of aufeis areas in northeastern Siberia-------------------**40**

Kamchatka ---**40**

FIGURE 37. Morphological types of present-day glaciers in the volcanic regions of Kamchatka-------------------------------------**41**

FIGURE 38. Aerial photograph of glaciers in the crater and on the slopes of Vulkan Mutnovskaya Sopka-------------------------**41**

Russian Arctic --**42**

Franz Josef Land (Zemlya Frantsa-Iosifa) --------------------------**42**

FIGURE 39. Glacierized regions of Franz Josef Land: western region; central region; and eastern region------------------------**43**

TABLE 13. The area of glaciers in Franz Josef Land ----------------------------**43**

FIGURE 40. Landsat 3 MSS image of the western part of Franz Josef Land acquired on 24 March 1978 ------------------------------**44**

Severnaya Zemlya ---**44**

FIGURE 41. Glacierization of Severnaya Zemlya ---------------------------------**45**

TABLE 14. The area of glaciers on Severnaya Zemlya ---------------------------**45**

FIGURE 42. Landsat 1 MSS image of the northern part of Severnaya Zemlya taken on 28 July 1973 --------------------------------------**46**

Novaya Zemlya ---**47**

Conclusions--**47**

FIGURE 43. Glacierization of Novaya Zemlya -----------------------------------**48**

Acknowledgments---**48**

TABLE 15. The glaciers on Novaya Zemlya -----------------------------------**49**

FIGURE 44. Landsat 1 MSS image of the central part of *Novozemel'skiy ice cap* taken on 23 July 1973 ------------------------------------**49**

TABLE 16. Optimum Landsat 1, 2, and 3 MSS images of glaciers of the Former Soviet Union------------------------------------**50–55**

FIGURE 45. Index map of the optimum Landsat 1, 2, 3 MSS images of the Former Soviet Union --------------------------------**56–58**

Fluctuations of Glaciers of the Central Caucasus and Gora El'brus, with a Section on the Glaciological Disaster in North Osetiya — by V.M. Kotlyakov, O.V. Rototaeva, and G.A. Nosenko --------------------------------------**59**

Introduction --**59**

FIGURE 46. Image of Gora El'brus and its glaciers on 25 August 2002 ------------------**60**

FIGURE 47. ASTER image of the glaciers on the southern slope of Gora El'brus taken on 15 September 2001 ---------------------------------**60**

FIGURE 48. Map of the glaciers of Gora El'brus------------------------------------**61**

TABLE 17. Changes in the position of the termini of El'brus glaciers during
          different periods of observation --------------------------------**61**

FIGURE 49. Part of the 25 August 2002 International Space Station image
          enlarged to show the lake-moraine complex near the terminus
          of the retreating *Lednik Birdzhalytchiran, Gora El'brus* --------------------**63**

FIGURE 50. Aerial and space photographs of *Lednik Bol'shoy Azau,*
          Gora El'brus, showing changes in the terminus shape and
          position --**64**

FIGURE 51. Aerial photograph (1987) of the terminus of *Lednik
          Ullumalienderku,* Gora El'brus, during a surge event ----------------------**65**

Glaciological Disaster in North Osetiya--**66**

FIGURE 52. Photographs (1981 and 2004) of the Lednik Kukurtli,
          Gora El'brus--**67**

FIGURE 53. ASTER image taken on 15 September 2001, showing glaciers
          of the central Caucasus --**68**

TABLE 18. Changes in the position of glacier termini from 1987 to 2001/04
          in the Tcherek River basin --**68**

FIGURE 54. Image taken from the International Space Station in 2002,
          showing the retreating Lednik Karaugom ----------------------------------**69**

FIGURE 55. Aerial photograph taken in July 2004 of the terminus of *Lednik
          Ailama* and the stagnant terminus of *Lednik Dykh-su* covered
          with debris --**69**

FIGURE 56. Photograph taken in August 2003 of the advancing front of
          Lednik Mizhirgitchiran--**70**

FIGURE 57. Oblique aerial photograph of avalanche debris in the Karmadon
          basin caused by the 20 September 2002 catastrophic collapse
          of the *Lednik Kolka* --**70**

FIGURE 58. Two ASTER images of the Genaldon river valley, scene of the
          Lednik Kolka disaster--**71**

FIGURE 59. Oblique aerial photograph taken on 21 September 2002, looking
          southwest at the avalanche debris saturated with water -------------------**72**

FIGURE 60. Oblique aerial photograph taken on 24 September 2002,
          looking south at the mountain crest in the source region of
          Lednik Kolka--**73**

FIGURE 61. International Space Station image taken on 13 August 2002 of
          Lednik Maili and *Lednik Kolka* --**74**

FIGURE 62. Oblique aerial photograph taken September 2002, looking south
          at the empty cirque of *Lednik Kolka* after the disaster--------------------**74**

FIGURE 63. International Space Station image taken on 19 October 2002,
          looking south at Gora Kasbek volcanic massif and the cirques
          of Lednik Maili and *Kolka*--**75**

FIGURE 64. Enlarged section of 27 September 2002 ASTER image ---------------------**75**

FIGURE 65. Terrestrial photograph taken on 28 June 2003 of one of the lakes
          in the *Kolka* bed dammed by a combination of ice and debris-----------**76**

FIGURE 66. Oblique aerial photograph taken on September 2003, looking
          southeast at the collapsing avalanche debris in Karmadon
          basin and the dammed lake near Saniba settlement one year
          after the disaster --**76**

Investigations of the Fluctuations of Surge-Type Glaciers in the Pamirs based
on Observations from Space — by V.M. Kotlyakov, G.B. Osipova, and
D.G. Tsvetkov--**77**

Introduction --**77**

Compilation of an Inventory of Surging Glaciers of the Pamirs ----------------**77**

TABLE 19. The number of surge-type glaciers in the large river basins of
          the Pamirs--**78**

TABLE 20. The largest observed surges of the glaciers of the Pamirs from
          1960 to 2003 --**78**

FIGURE 67. KFA-1000 photograph taken in 1973 from a Resurs-F1 satellite
          showing surge-type glaciers in the Central Pamirs-----------------------**79**

FIGURE 68. KFA-1000 photograph taken in 1985 from a Resurs-F1 satellite showing surge-type glaciers in the central part of Zaalayskiy Khrebet (Qatorkŭhi Pasi Oloy in Tajikistan, Chong Alay Kyrka Toosu in Kyrgyzstan) --80

Investigation of the Fluctuation Regime of Compound Glaciers----------------80

FIGURE 69. Morphology of the lower part of Lednik Oktyabr'skiye from 1946 to 1990 --81

FIGURE 70. Photographs from space of Lednik Oktyabr'skiye -------------------82

FIGURE 71. Graph of the velocity of the ice movement along longitudinal profile of Lednik Oktyabr'skiye, showing the stage of recovery (1972–1985) and the surge (1985–1990) ------------------------82

FIGURE 72. Photographs of *Lednik Bivachniy* and Lednik Moscow State University: 1973, pre-surge; 1980, post-surge; and 2001 soon after the successive surge ------------------------------------83

FIGURE 73. Images from space of the Gando-*Dorofeeva* glacier system before, during, and after the 1989–1990 surge ------------------------84

The State of Surging Glaciers — Ledniki Medvezhiy and Geograficheskogo Obshchestva --85

FIGURE 74. Aerial photographs showing the tongue of Lednik Medvezhiy during the surge of 1988 to 1989, and in 2001------------------86

FIGURE 75. Sketch map of the Lednik Geograficheskogo Obschestva (RGO Glacier) basin showing glaciers, the reconstructed outline of the glacier's terminus in 1916, a reconstructed dammed lake in 1916, locations of velocity profiles, and a mudflow on the surface of the glacier in September 2002 ------------------88

FIGURE 76. Aerial photograph taken on 13 October 2002 of the terminus of Lednik RGO, looking northeast ------------------------------89

FIGURE 77. International Space Station images of Lednik RGO taken on 25 June 2001 and 30 September 2002 ------------------------------90

Investigations of the Fluctuations of "Normal" Non-Surging Glaciers in the Eastern Pamirs--90

FIGURE 78. Landsat image of the Eastern Pamirs where glacier-retreat measurements listed in table 21 were made -------------------------91

TABLE 21. Retreat velocity of the termini of glaciers of the Eastern Pamirs from 1972 to 2001 --92

FIGURE 79. Sketch map showing the retreat of the terminus of *Lednik Severniy Zulumart* (Eastern Pamirs) based on interpretation of a 1946 aerial photograph and 1972, 1978, 1990, and 2001 space images and photographs------------------------------------92

TABLE 22. Retreat of the terminus and decrease in the area of the tongue of *Lednik Severniy Zulumart* from 1946 to 2001 --------------------93

The Glaciology of the Russian High Arctic from Landsat Imagery — by J.A. Dowdeswell, E.K. Dowdeswell, M. Williams, and A.F. Glazovskii--------94

Abstract --94

Introduction --95

Glaciological Background --95

FIGURE 80. Map of the Russian High Arctic showing the major ice-covered archipelagos of Franz Josef Land, Novaya Zemlya, and Severnaya Zemlya, together with the smaller glacierized islands further east--96

FIGURE 81. Diagrams showing the effects of cloud cover on Landsat image quality for selected Path–Rows in the Russian High Arctic, including Franz Josef Land, Novaya Zemlya, and Severnaya Zemlya: April to September, May, July, and September -------------------97

Franz Josef Land--98

Severnaya Zemlya--98

FIGURE 82. Map of the Franz Josef Land archipelago showing the distribution of ice masses --99

FIGURE 83. Map of the islands of Severnaya Zemlya showing the distribution of ice masses -- 100

FIGURE 84. Landsat false-color Multispectral Scanner image of the ice caps in northern Severnaya Zemlya------------------------------------- 101

Novaya Zemlya---102

Other Glacierized Islands in the Russian High Arctic ------------------ 102

FIGURE 85. Map of Novaya Zemlya showing the distribution of ice masses -------- 103

Glaciological Investigation of Russian High Arctic Ice Masses from Landsat Imagery-- 104

Ice-Surface Topography ---104

FIGURE 86. Map of the ice divides on the ice caps of Franz Josef Land, shown on a georeferenced mosaic of Landsat TM images acquired on 25 July 1986 -------------------------------------- 105

Ice Dynamics ---105

Outlet Glaciers and Icebergs--105

FIGURE 87. Landsat TM image sub-scene showing an example of the major ice divides on the Ostrov Gallya ice cap in Franz Josef Land --- 106

FIGURE 88. Digitally enhanced Landsat TM image sub-scene of the major ice divide running along the northeast-southwest trending spine of Novaya Zemlya -------------------------------------- 107

FIGURE 89. Landsat TM images of iceberg-producing outlet glaciers draining ice caps in the Russian High Arctic ---------------------- 108

FIGURE 90. Four georeferenced and edge-enhanced sub-scenes of Landsat images of the Lednik Znamenityy on Zemlya Vil'cheka, Franz Josef Land, showing large numbers of tabular icebergs --------- 109

Floating Ice Shelves or Ice Tongues --------------------------------110

FIGURE 91. Annotated Landsat TM band 3 image sub-scene of the *Lednik Matusevicha* on Ostrov Oktyabr'skoy Revolyutsii, Severnaya Zemlya --- 110

FIGURE 92. Graph showing ice-surface profiles from two ice shelves and their parent drainage basins on Severnaya Zemlya, derived from analysis of Russian aerial photographs from the 1950s ---------- 111

FIGURE 93. Landsat TM image of three flat ice-marginal glacier termini on the ice cap on Ostrov Zemlya Georga, western Franz Josef Land--- 112

FIGURE 94. Map of the ice caps on Franz Josef Land, with marginal areas of low surface gradient --- 113

Surge-Type Glaciers --114

FIGURE 95. Digitally enhanced Landsat TM image of the outlet glaciers entering Zaliv Nordenshel'da, Novaya Zemlya -------------------------- 115

Snow-and-Ice Facies and Glacier Mass Balance-------------------------- 115

FIGURE 96. Digitally enhanced Landsat TM sub-scene showing the snow-and-ice facies on a summer image of the Lednik Kropotkina on Ostrov Zemlya Aleksandry, Franz Josef Land -------- 116

FIGURE 97. Landsat TM sub-scene of Ostrov Ketlitsa, Franz Josef Land ----------- 117

Conclusions--- 118

Acknowledgments--- 119

References Cited--- 119

GLACIERS OF ASIA—

GLACIERS OF THE FORMER SOVIET UNION[1]

By V.M. KOTLYAKOV[2], senior author, with contributions from
A.M. DYAKOVA (Siberia), V.S. KORYAKIN (Russian Arctic
Islands), V.I. KRAVTSOVA (Caucasus, Altay), G.B. OSIPOVA
(Tien Shan), G.M. VARNAKOVA (Pamirs and Alai Range),
V.N. VINOGRADOV(Kamchatka), O.N. VINOGRADOV (Caucasus),
and N.M. ZVERKOVA (Ural Mountains and Taymyr Peninsula)

Abstract

Glaciers in the Former Soviet Union cover a total area of 78,938 km^2; 72 percent (56,894 km^2) are in the Russian Arctic (Franz Josef Land, Severnaya Zemlya, Novaya Zemlya, and Wrangel Island) and 28 percent (22,044 km^2) in various mountain ranges (mainly the Caucasus, Pamirs, Alai Range (Alayskiy Khrebet), Tien Shan, Altay, Ural Mountains, and those of Northeastern Siberia) in Russia and in the Republics of Georgia, Kazakhstan, Kyrgyzstan, Tajikistan, and Uzbekistan. The Glacier Inventory of the USSR and the World Atlas of Snow and Ice Resources records a total of 26,881 glaciers, 1,983 in the Russian Arctic, and 24,898 in Russia and in the four independent republics. Information from the Glacier Inventory, World Atlas, Landsat 1, 2, and 3 MSS images from the 10-year Landsat Baseline Period (1972–1981), Soviet space imagery, aerial photographs, and maps were used to provide information on the glacierized areas. Russian glaciologists recognize 20 morphological types of mountain glaciers and 7 types of ice caps, ice fields, and ice sheets from a global perspective (World Atlas). Glaciers in the Former Soviet Union represent 11 of the 20 types of mountain glaciers and all of the types of ice caps and ice fields. The early (1972–1981) Landsat MSS images were found to be useful in delineating the margins and ice divides on ice caps and the termini and some margins of larger mountain glaciers, if the images were acquired under conditions of minimum cloud cover and late season, residual snow pack, but before new snowfall. Analysis of changes in smaller glaciers was not possible with the limitations of pixel resolution (80 m) of MSS images. Later Landsat TM and more recent Landsat ETM+ and ASTER images have overcome the spatial resolution problem. The Resurs-F1 KFA-1000 photographs and Landsat MSS images were especially useful for identifying surge-type glaciers and for determining the position of termini of glaciers.

Introduction

The information presented in this part of the chapter is the result of interpretation of Landsat images, Soviet space images, and aerial photographs. The data presented in the Soviet Glacier Inventory (USSR Academy of Sciences, 1965–1983) are also widely used in this chapter, as are some of the sketches and sections of maps compiled for the World Atlas of Snow and Ice Resources (Kotlyakov, 1997).

---

[1]Editors' Note: This manuscript was originally written in 1981 to describe the glaciers of the Former Soviet Union in the late 1970s and early 1980s, the "benchmark" time period for the Satellite Image Atlas of Glaciers of the World, U.S. Geological Survey Professional Paper 1386-A–K. Because there were delays in publishing, the authors and editors updated this manuscript by adding references to more recent work while still retaining the original benchmark information. In addition, three supplemental sections were added to give a fuller description of more recent glaciological work — "Fluctuations of Glaciers of the Central Caucasus and Gora (Mount) El'brus with a section on the Glaciological Disaster in North Osetiya," "Investigations of the Fluctuations of Surge-type Glaciers in the Pamirs Based on Observations from Space," and "The Glaciology of the Russian High Arctic from Landsat Imagery."

[2]Institute of Geography, Russian Academy of Sciences, Staromonetny 29, Moscow, Russia.

The Glacier Inventory of the USSR was published between 1965 and 1983 as a series of publications that comprise 108 issues. The Inventory provides quantitative information on all of the main characteristics of glaciers — dimensions, shape, position, and regime. These data were obtained from interpretation of aerial photographs and cartographic materials, sometimes supplemented by field and aerial studies of the glaciers. In some cases, the Inventory has made it possible to correctly interpret space images; and in some areas such as the Altay, interpretations using the Soviet space images have provided additional data to the Glacier Inventory.

According to the Glacier Inventory of the USSR, glaciers are situated in 26 regions of the Former Soviet Union (fig. 1). In addition to the glacierized regions shown on the index map (fig. 1), isolated glaciers and perennial snow patches are situated in the Kharaulakhskiy[3] Khrebet (Mountains) to the north of the Khrebet Orulgan, and on the Khrebet Pekul'ney (Range) on the Chukotka peninsula north of the Koryakskoye Nagor'ye (Highland). [Recent work includes Sedov (1997a, b) and Dolgushin and Osipova (1989).]

Based on the data in the Glacier Inventory of the Former Soviet Union, glaciers had an area of 78,938 km^2, of which 56,894 km^2 is in the Arctic and 22,044 km^2 is in mountain regions. The total number of individual glaciers is 28,881, of which 1,983 are in the Arctic and 24,898 are in the mountains. Table 1 lists the number of glaciers and the total glacierized area for each of the 26 glacierized regions of the Former Soviet Union.

The Former Soviet Union had previously established its own morphological classification of glaciers, which was standardized with due regard to the inventorying of glaciers. The Guide for the Compilation of the Glacier Inventory of the USSR (USSR Academy of Sciences, 1966) distinguishes 20 morphological types of mountain glaciers and 7 types of ice caps/ice fields/ice sheets. Figure 2 shows examples of the 11 major morphological types of mountain glaciers described in the Former Soviet Union. They are shown on the maps of glacier morphology contained in this chapter. The colors assigned to each glacier type are consistent on all of the maps.

Discussion of Glaciers by Region

Glaciers within seven of the major regions of the Former Soviet Union are discussed in this section. These regions include the Caucasus, the Pamirs and surrounding ranges, Tien Shan, Altay, the Ural Mountains, Kamchatka, and the Russian Arctic. Within the Tien Shan, the Piks Pobedy, *Khan-Tengri*, and Khrebet Ak-Shyyrak areas are discussed in more detail. After the Ural Mountain region, the Byrranga Mountains and Northeast Siberia are discussed. And within the Russian Arctic, Franz Josef Land, Severnaya Zemlya, and Novaya Zemlya are described in more detail.

---

[3]Official geographic place-names for foreign countries are required to be used in U.S. Government publications to the greatest extent possible. In this section, the use of geographic place-names is based on the U.S. Board on Geographic Names (BGN) as listed on the GEOnet Names Server (GNS) website: *http://earth-info.nga.mil/gns/html/index.html*. The website lists some conventional names and those will be used when available, for example Caucasus, Franz Josef Land, and Kara Sea. All other names will be the Russian names introduced by or with the English equivalent in parenthesis if needed for understanding. Names not listed in the BGN website will be shown in italics. Commonly used Russian generic terms are lednik, ledniki (glacier, glaciers); khrebet (mountain range); Gora, Gory (mount, mountains); pik, piks (peak, peaks); ozera (lake); poluostrov (peninsula); proliv (strait); and ostrov (island).

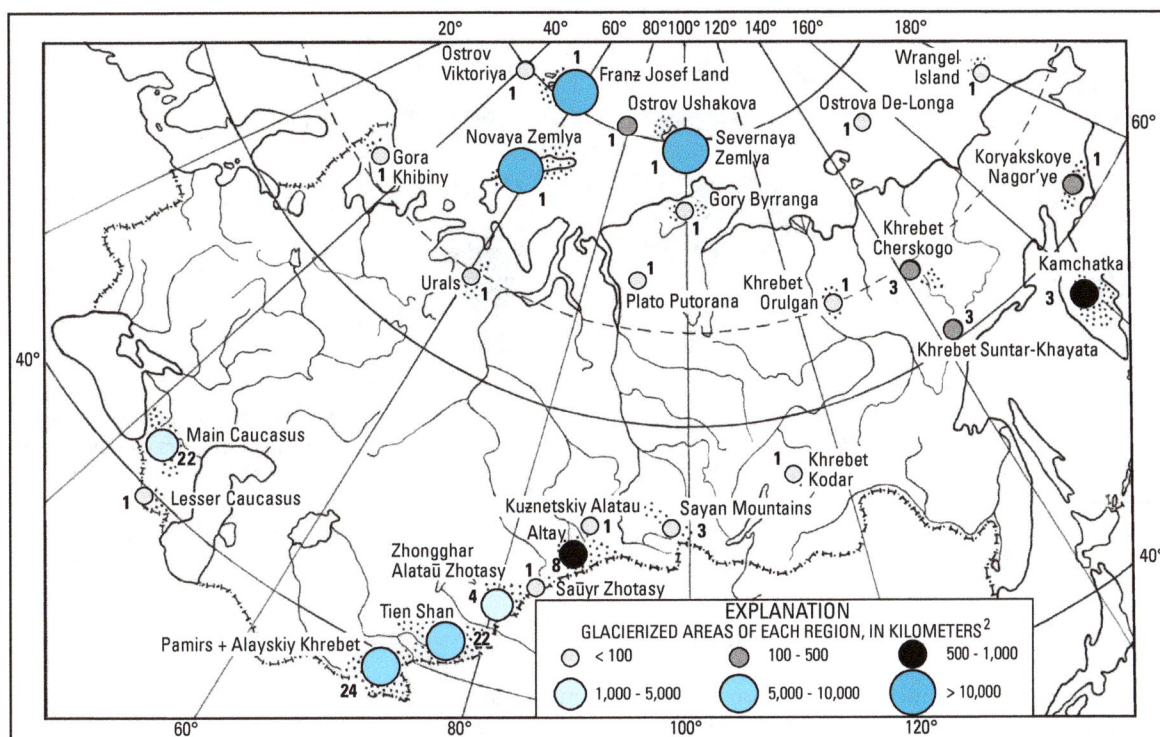

Figure 1.—*Glacierized regions of the Former Soviet Union according to the Glacier Inventory of the USSR. The number near the symbol for each region denotes the section of the Glacier Inventory that describes that region.*

TABLE 1.—*Data on the glaciers of the Former Soviet Union*

[Data source: USSR Academy of Sciences (1965–1983). Unit: km², square kilometer]

Glacierized area	Number of glaciers	Total area of glacieriza-tion (km²)	Date of area calculation	Glacierized area	Number of glaciers	Total area of glacieriza-tion (km²)	Date of area calculation
ICE CAPS AND ASSOCIATED OUTLET GLACIERS				MOUNTAIN GLACIERS (Continued)			
Novaya Zemlya	685	24,413	1960–80	Kamchatka	405	874.1	1960–80
Severnaya Zemlya	285	18,325	1960–80	Koryakskoye Nagor'ye (upland)	1,302	240.6	1960–80
Franz Josef Land (Zemlya Frantsa-Iosifa)	995	13,738.8	1960–80	Khrebet Suntar-Khayata	208	201.6	1960–80
Ostrov Ushakova (Ushakov Island)	2	325.4	1960–80	Khrebet Cherskogo	372	156.2	1960–80
Ostrova De-longa (De Long Islands)	15	80.6	1960–80	Gory Byrranga	96	30.5	1960–80
Ostrov Viktoriya (Victoria Island)	1	10.7	1960–80	Sayany	105	30.3	1960–80
Total	1,983	56,893.5	1960–80	Ural Mountains	143	28.7	1960–80
				Kodar	30	18.8	1960–80
MOUNTAIN GLACIERS				Orulgan	74	18.4	1960–80
Qatorkŭhi Hisor/Hisor Tizmasi and Alayskiy Khrebet	3,893	2,335.8	1960–80	Saŭyr	18	16.6	1960–80
				Kuznetskiy Alatau	91	6.8	1960–80
Pamirs	6,730	7,493.4	1960–80	Lesser Caucasus	42	3.8	1960–80
Subtotal	10,623	9,829.2	1960–80	Wrangel Island (Ostrov Vrangelya)	101	3.5	1960–80
Tien Shan	6,347	7,251.7	1960–80	Plato Putorana	22	2.5	1960–80
Main Caucasus	2,047	1,424.4	1960–80	Khibiny	4	0.1	1960–80
Zhongghar Alatau-Zhostasy	1,369	1,000	1960–80	**Grand Total**	24,898	22,044.3	1960–80
Altay	1,499	906.5	1960–80				

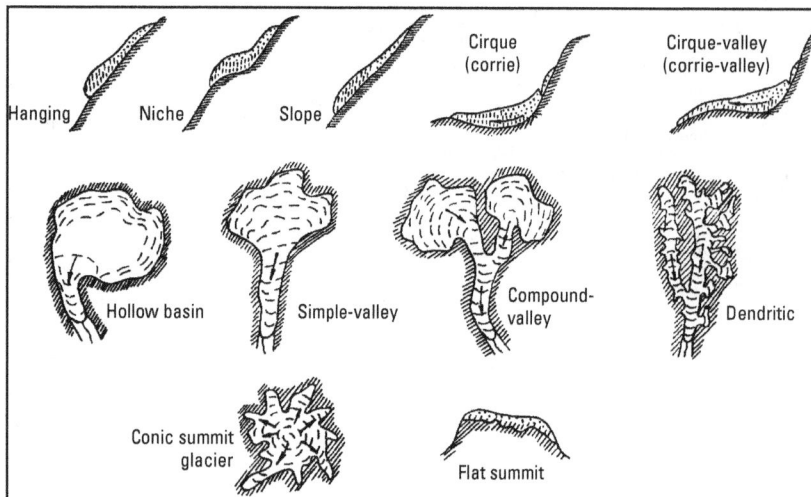

Figure 2.—*Morphological types of mountain glaciers adopted for the Glacier Inventory of the USSR and used in this chapter.*

Caucasus (in Russia and the Republic of Georgia)

The Caucasus mountain system consists of two major mountain ranges: the southern Lesser Caucasus, and the northern Greater, or Main, Caucasus. Precipitation in the Lesser Caucasus is much lower, and glaciers are concentrated in the Main Caucasus, which is the subject of this section and referred to herein as the Caucasus. The study of glaciers in the Caucasus began in the first quarter of the 18th century. The first data on glaciers can be found in the works of Vakhushti Bagrationes (1997). In the middle and the second half of the 19th century, studies of glaciers were carried out together with geological investigations, geographical travels, mountaineering, and topographic studies. At the beginning of the 20th century, active glaciological investigations of the Caucasus were undertaken by the Russian Geographical Society.

The compilation of the first reliable map of the Caucasus, at a scale of 1:420,000 and depicting the largest glaciers, was completed by 1862. Topographic surveys of the Caucasus at a scale of 1:42,000 were accomplished 30 years later. Having analyzed these maps, K.I. Podozerskiy (1911) published the first inventory of Caucasus glaciers, compiling data on the lengths and areas of all of the glaciers.

In 1911, the German explorer H. Burmester (1913) performed the first phototheodolite surveys of the *Bol'shoy Azau* and *Shkelda* Glaciers. Tachymetric surveys of the termini of many glaciers were undertaken in 1932 during the Second International Polar Year (IPY, 1932-1933) (Russia, Committee of the Second International Polar Year, 1936b). Aerial photographs acquired during the 1940s and topographic maps compiled and published during the 1950s provided the basis for more modern studies and inventories of glaciers.

During the International Geophysical Year (IGY) (1957–1959) and for the next ten years, the glaciers of the Caucasus were studied during expeditions by Kharkov University (Kharkov University, 1960–1961), Moscow University (Tushinskiy, 1968), Institutes of Geography of the USSR Academy of Sciences (Various authors, 1964) and Georgian Academy of Sciences. Repeated aerial photography and phototheodolite surveys of some glacier areas were also undertaken. The comprehensive surveys performed by Moscow University in the El'brus area resulted in the compilation of the Atlas of El'brus Glaciers (Various authors, 1967).

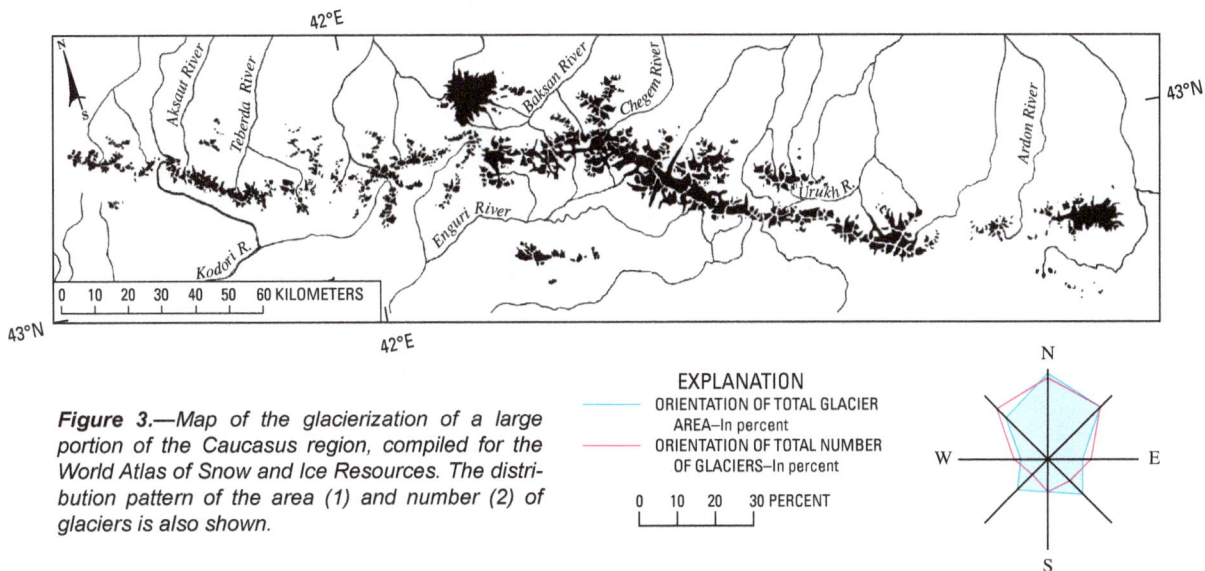

Figure 3.—Map of the glacierization of a large portion of the Caucasus region, compiled for the World Atlas of Snow and Ice Resources. The distribution pattern of the area (1) and number (2) of glaciers is also shown.

Comprehensive Caucasus studies used to compile the Glacier Inventory of the USSR, as well as water, ice, and heat-balance studies, were carried out in the representative Marukhskiy, Dzhankuat, and Gergetskiy glacier basins under the International Hydrological Decade (IHD) program from 1965 to 1974 (Golubev and others, 1978; Krenke and others, 1988; Panov, 1993). Results of these studies provided the basis for the Caucasus maps in the World Atlas of Snow and Ice Resources (Kotlyakov, 1997). Cartographic sources, aerial photographs, and space images were used to compile later maps of the glaciers of the Caucasus (Dolgushin and Osipova (1989)). Figure 3 is an example.

The glacierized area of the Caucasus extends for about 1,000 km from the northwest to the southeast. It is traditionally subdivided into three main regions: Western, Central, and Eastern Caucasus. The Western Caucasus region lies generally between the Black Sea and Mount El'brus, the Central Caucasus region, between Mount El'brus and Mount Kazbek, and the Eastern Caucasus region, between Mount Kazbek and the Caspian Sea. The northern and southern slopes are also considered separately. The distribution of glaciers in these areas is shown in table 2.

The Caucasus contains two large glacier complexes — one includes 23 glaciers on Gora (Mount) El'brus, at the west end of the Central Caucasus region, and the other includes 14 glaciers on Gora Kazbek, at the eastern end of the Central Caucasus region. Additionally, there are 11 valley and compound valley glaciers, each with an area of more than 10 km^2 (table 3). However, small glaciers with areas less than or equal to 1 km^2 prevail in the Caucasus (table 4). The number of these small glaciers is 85 percent of the total number of glaciers, but the area covered by them is 29 percent of the total area of glacierization. More than 60 percent of glaciers (57 percent of the glacierized area) have a northern exposure: northwest, north, or northeast (see fig. 3). However, the glaciers facing south are, in general, 1.5 times larger than the glaciers that have a northern exposure (Vinogradov and others, 1976).

Most of the morphological types of glaciers can be found in the Caucasus. Compound-valley glaciers and two glacier complexes on conic summits spread along the main ridge of the Central Caucasus, and the extent of glacierization in that area is more than 50 percent. Valley glaciers are situated at the edges of compound-valley glaciers. Cirque glaciers occur most frequently on the

TABLE 2.—Distribution of glaciers in the main regions of the Caucasus

[Unit: km^2, square kilometer]

Region of glacierization	Number of glaciers	Total area (km^2)
Western Caucasus	565	277.7
Central Caucasus	1,110	1,033.3
Eastern Caucasus	372	113.4
Total	2,047	1,424.4
Northern Slope	1,471	992.4
Southern Slope	576	432.0
Total	2,047	1,424.4

TABLE 3.—*Glacier complexes and large glaciers of the Caucasus*

[Abbreviations: km, kilometer; m, meter; AP, aerial photography; SI, surface investigation]

Glacier name	Name of the river basin	Area (km²)	Length (km)	Elevation of the tongues (m)	Elevation of the firn line (m)	Firn line observations	
						Methodology	Dates
Gora El'brus glacier complex	Kuban' Malka, Baksan	122.6	—	2,480–3,697	3,550–4,000	AP	22 Aug 1957
The glacier complex of Gora Kazbek	Terek, Gizel'don	64.8	—	2,260–3,450	3,260–3,720	AP	1960–1965
Bezengi	Cherek	36.2	17.6	2,080	3,400–3,750	AP	15 Aug 1957
Dykh-Su	Cherek	34.0	13.3	2,070	2,900–3.750	AP	16 Sep 1957
Lekziry	Enguri	33.7	11.8	2,020	3,090	SI	3 Sep 1959
Tsaneri	Enguri	28.8	10.1	2,390	3,190	SI	25 Aug 1960
Karaugom	Urukh	26.6	13.3	1,830	3,440	AP	15 Sep 1957
Tviberi	Enguri	20.1	5.8	2,200	3,150	SI	15 Aug 1960
Agashtan	Cherek	15.9	9.8	2,250	3,140–3,380	AP	10 Sep 1957
Kvitlodi	Enguri	11.9	7.8	2,320	3,160	SI	30 Aug 1960
Chalaati	Enguri	11.3	7.3	1,850	2,980	SI	5 Sep 1959
Khalde	Enguri	11.0	8.4	2,450	3,440	SI	2 Sep 1962
Shaurtu	Chegem	10.1	9.1	2,220	3,220	AP	15 Aug 1957

TABLE 4.—*The distribution of glaciers and glacier complexes in the Caucasus based on their area*

[Unit: km², square kilometer]

Dimensions of glaciers and glacier complexes	Number of glaciers	Total area (km²)
Largest (>100 km²)	1 complex (El'brus)	122.6
Large (10.1–100 km²)	1 complex (Kazbek); 11 other glaciers listed in table 3	304.4
Middle-sized (1.1–10.0 km²)	250	581.9
Small (≤1 km²)	1,749	415.5

Western Caucasus and on the northern slope of the Central Caucasus. Hanging glaciers are found everywhere, but they are most numerous in the Central Caucasus. Table 5 summarizes data on the distribution of the main morphological types of glaciers in the Caucasus. In the table, the category of valley glaciers also includes cirque-valley glaciers; the category of cirque glaciers incorporates hanging cirque glaciers, and the category of hanging glaciers also includes spill-over glaciers, glaciers lying near the crest, and niche glaciers. Table 5 only contains data on glaciers whose area is ≥0.1 km²; the number of glaciers included in glacier complexes is shown in parenthesis.

Morphological characteristics of certain glacierized areas of the Caucasus are well depicted on Landsat images. The image of the Western Caucasus and Mount El'brus (fig. 4A) shows the spectacularly scattered and dispersed nature of the glacierization, with prevailing cirque and cirque-valley glaciers extending along and abutting the mountain crests. The relationship between morphological types of glaciers of this area is well represented on the map of the same area compiled for the World Atlas of Snow and Ice Resources (fig. 4B).

[Values in parenthesis indicate number of individual glaciers. Abbreviations: N, number of glaciers; A, glacierized area, in square kilometers]

Glacierized area	Glacier complexes on conic summits		Compound-valley		Valley (includes cirque-valley)		Cirque (Corrie) (includes hanging cirque)		Hanging (includes spill-over and niche)		Flat summits' glaciers		Total	
	N	A	N	A	N	A	N	A	N	A	N	A	N	A
Western Caucasus	—	—	—	—	61	121.6	338	135.6	53	13.7	—	—	452	270.9
Central Caucasus	2 (37)	187.4	14	230.1	188	416.8	413	155.0	129	26.3	—	—	781	1,015.6
Eastern Caucasus	—	—	—	—	33	39.5	182	55.1	64	12.4	2	1.8	281	108.8
Northern Slope	2 (37)	187.4	11	165.0	143	306.0	668	267.7	212	43.5	2	1.8	1,073	971.4
Southern Slope	—	—	3	65.1	139	271.9	265	78.0	34	8.9	—	—	441	423.9
Total Caucasus	2 (37)	187.4	14	230.1	282	577.9	933	345.7	246	52.4	2	1.8	1,514	1,395.3

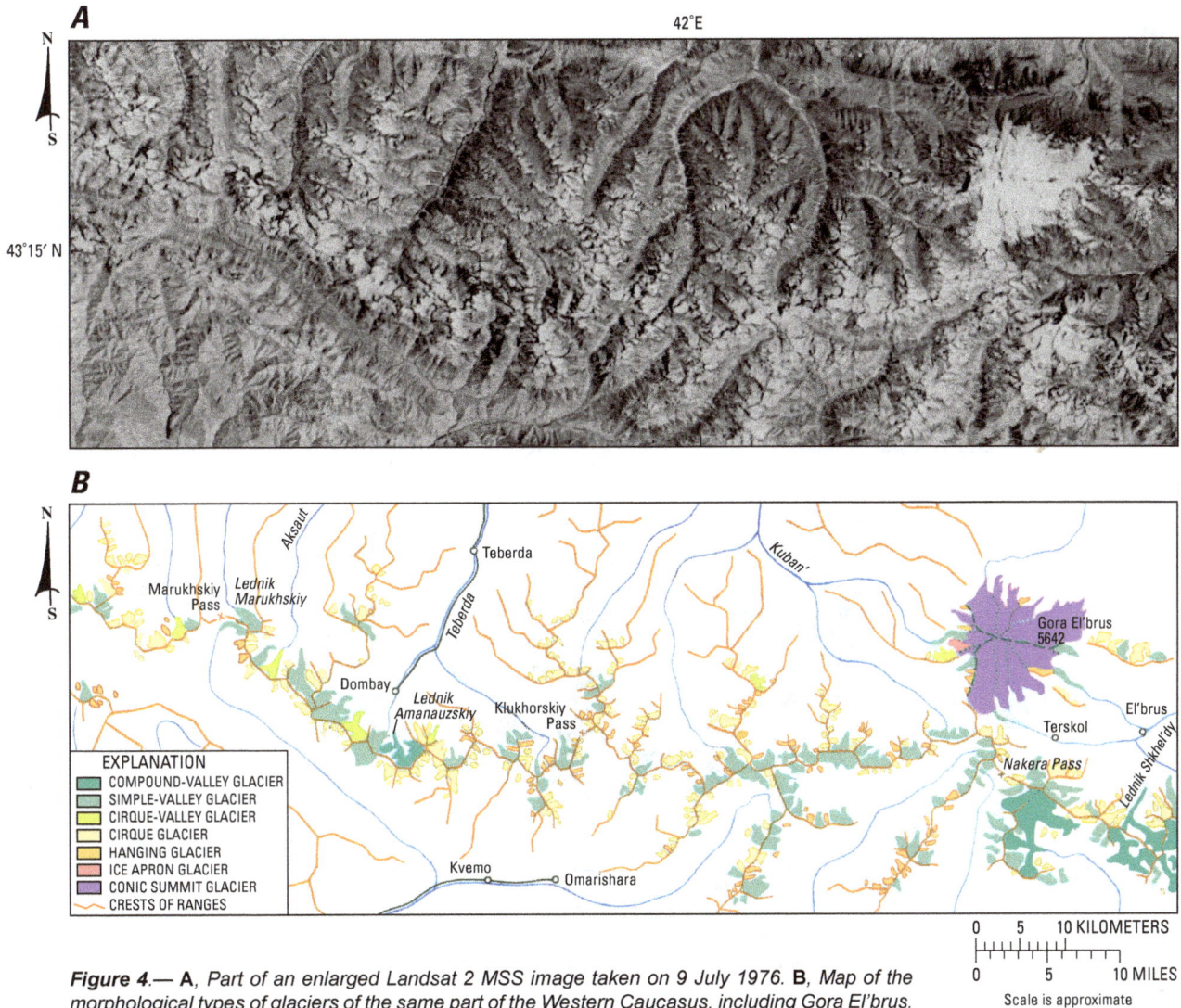

Figure 4.— A, *Part of an enlarged Landsat 2 MSS image taken on 9 July 1976.* **B**, *Map of the morphological types of glaciers of the same part of the Western Caucasus, including Gora El'brus. Landsat 2 MSS image 2185030007619190, Path 185–Row 30 is from the U.S. Geological Survey, EROS Data Center, Sioux Falls, S. Dak. 57198.*

During the IHD, detailed studies were undertaken in the Western Caucasus, in the representative Marukh mountain-glacier basin (fig. 4). These studies resulted in a series of regime maps of the Marukh Glacier (Lednik Marukhskiy) (fig. 5), which served as a model for similar maps in the World Atlas of Snow and Ice Resources.

The Landsat image makes it possible to distinguish the star-like shape of the Gora El'brus glacierization (fig. 6) with its complex of conic summit glaciers. Despite the small scale of this image, the termini of 14 large glaciers can be recognized. The glaciers can be seen in detail on the phototheodolite panoramas (photographic mosaic) of the northern (fig. 7A) and southern (fig. 7B) slopes of Gora El'brus.

Glacierization of the Central Caucasus has a fairly continuous nature, as can be seen on part of the Landsat image shown in figure 8A. It is well represented by the map of morphological types of glaciers, compiled for the World Atlas of Snow and Ice Resources (fig. 8B). Unlike the mostly debris-free glaciers of Western Caucasus, the Central Caucasus glaciers are characterized by well-developed moraines, which can be seen in space images on the surface of large valley glaciers.

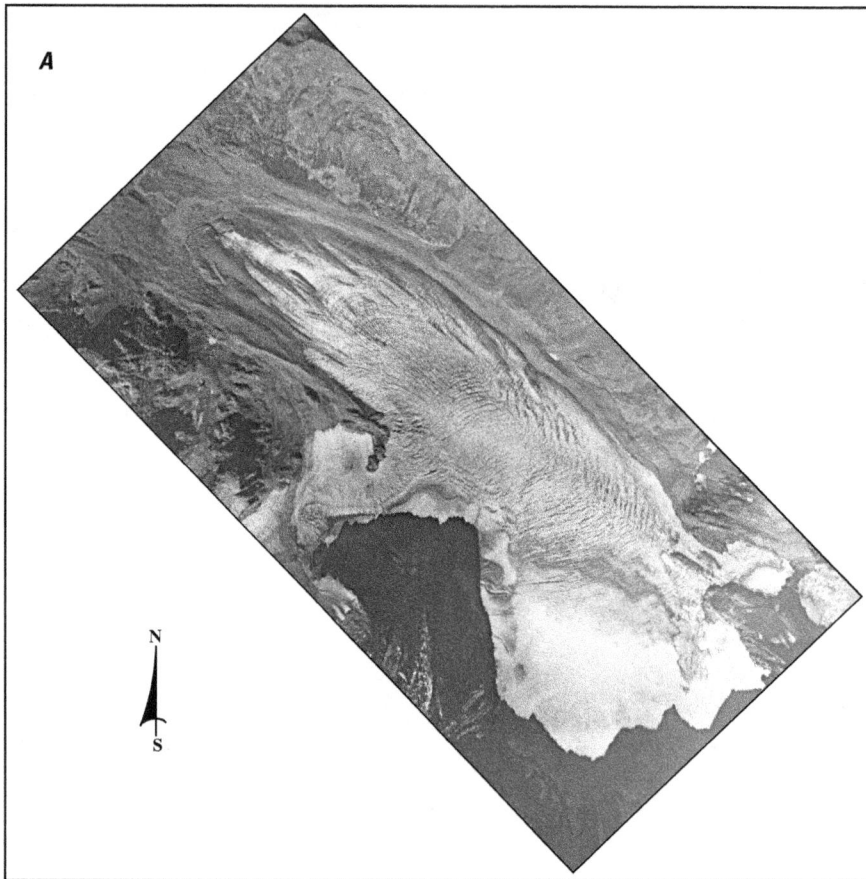

Figure 5.—(above and continued at right) **A**, Aerial photograph of the Lednik Marukhskiy (see fig. 4 for location) taken on 15 September 1955, and **B**, a series of glaciological maps of this glacier based on detailed studies done during the IHD.

B

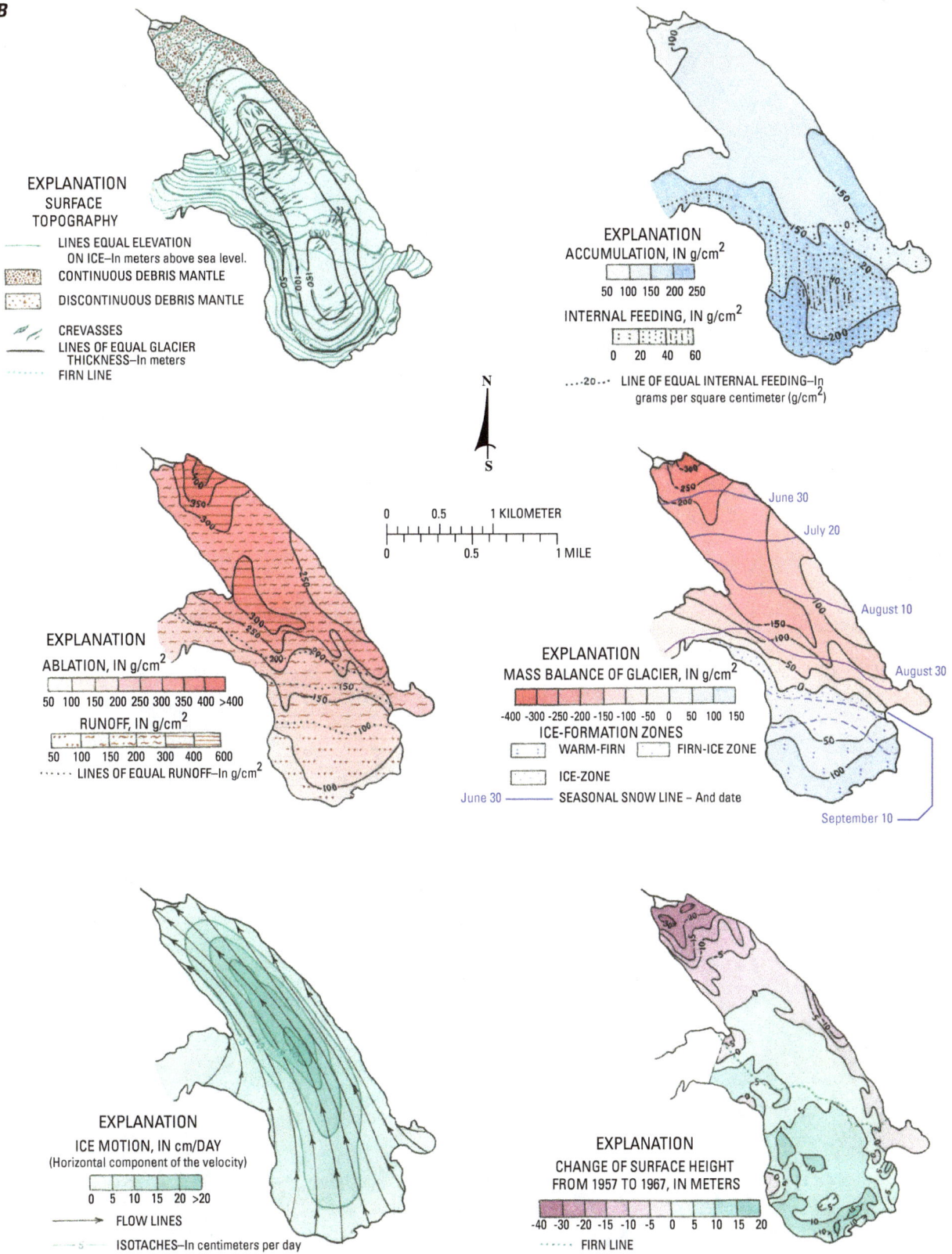

EXPLANATION
SURFACE TOPOGRAPHY

LINES EQUAL ELEVATION ON ICE–In meters above sea level.
CONTINUOUS DEBRIS MANTLE
DISCONTINUOUS DEBRIS MANTLE
CREVASSES
LINES OF EQUAL GLACIER THICKNESS–In meters
FIRN LINE

EXPLANATION
ACCUMULATION, IN g/cm^2

50 100 150 200 250

INTERNAL FEEDING, IN g/cm^2

0   20   40   60

....-20-.. LINE OF EQUAL INTERNAL FEEDING–In grams per square centimeter (g/cm^2)

N
S

0   0.5   1 KILOMETER

0   0.5   1 MILE

EXPLANATION
ABLATION, IN g/cm^2

50 100 150 200 250 300 350 400 >400

RUNOFF, IN g/cm^2

50 100 150 200 300 400 600

.... LINES OF EQUAL RUNOFF–In g/cm^2

EXPLANATION
MASS BALANCE OF GLACIER, IN g/cm^2

-400 -300 -250 -200 -150 -100 -50 0 50 100 150

ICE-FORMATION ZONES
WARM-FIRN   FIRN-ICE ZONE
ICE-ZONE

June 30 ——— SEASONAL SNOW LINE – And date

June 30
July 20
August 10
August 30
September 10

EXPLANATION
ICE MOTION, IN cm/DAY
(Horizontal component of the velocity)

0   5   10   15   20   >20

——► FLOW LINES

- - - 5 - - - ISOTACHES–In centimeters per day

EXPLANATION
CHANGE OF SURFACE HEIGHT FROM 1957 TO 1967, IN METERS

-40 -30 -20 -15 -10 -5   0   5   10   15   20

.... FIRN LINE

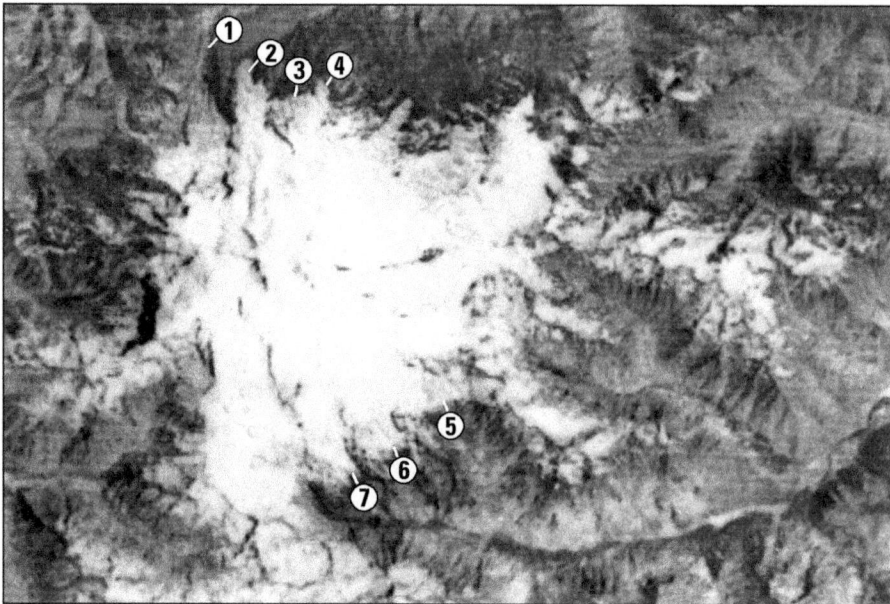

A

B

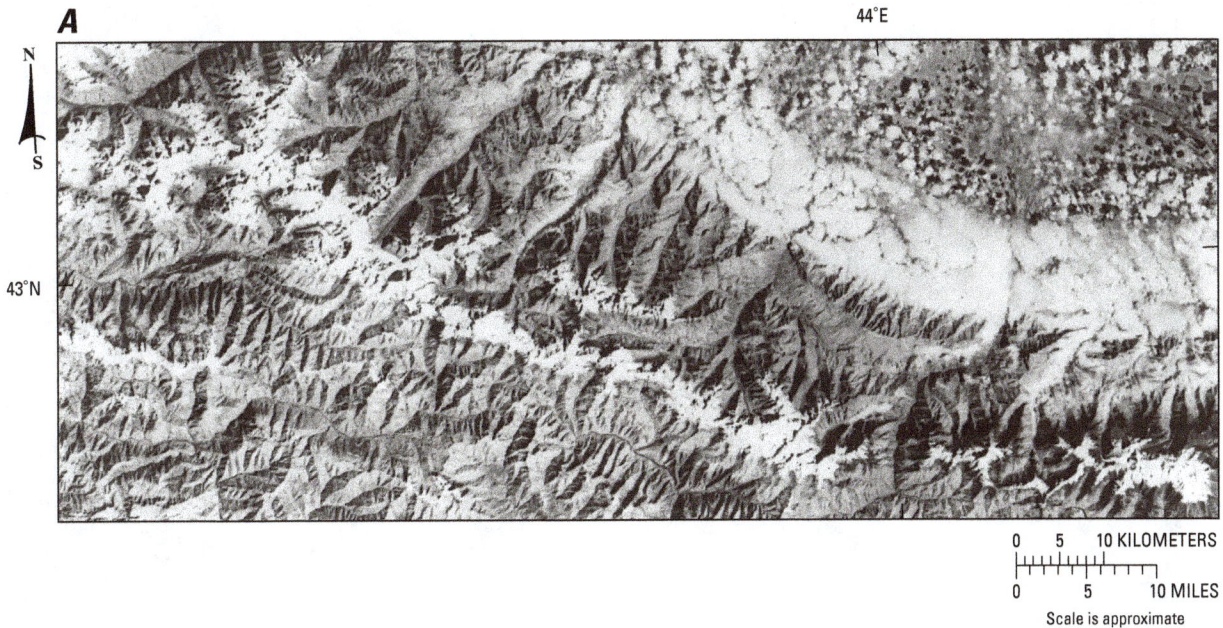

A   44°E

43°N

| 0 | 5 | 10 KILOMETERS |
| 0 | 5 | 10 MILES |

Scale is approximate

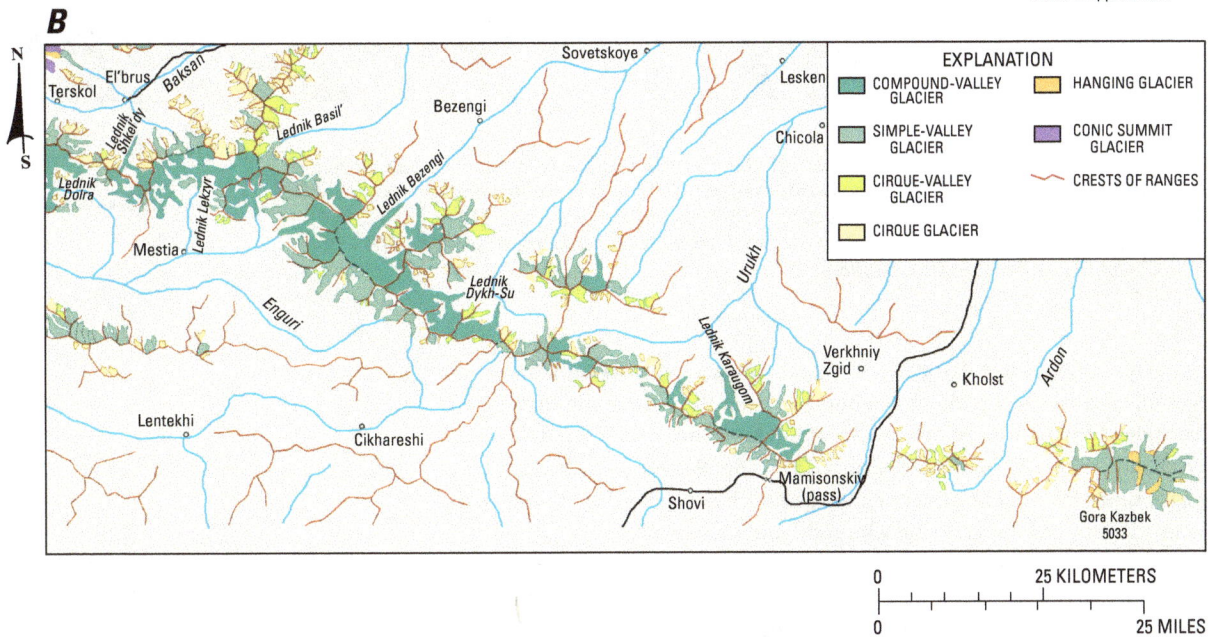

B

EXPLANATION

▨ COMPOUND-VALLEY GLACIER	▨ HANGING GLACIER
▨ SIMPLE-VALLEY GLACIER	▨ CONIC SUMMIT GLACIER
▨ CIRQUE-VALLEY GLACIER	⟋ CRESTS OF RANGES
▨ CIRQUE GLACIER	

| 0 | | 25 KILOMETERS |
| 0 | | 25 MILES |

Figure 6.— (at left, top) *Enlargement of part of a Landsat 2 MSS image taken on 9 July 1976, depicting the El'brus massif (lat 43°21'N., long 42°26'E.). Landsat 2 MSS 2185030007619190, Path 185–Row 30, is from the U.S. Geological Survey, EROS Data Center, Sioux Falls, S. Dak. 57198.*

Figure 7.— (at left, bottom) *Panoramic photographic mosaic of the northern (**A**) and southern (**B**) slopes of Gora El'brus. Phototheodolite surveys were done by F.N. Nikulin in 1957 and A.V. Brukhanov in 1958, respectively.*

Figure 8.— (this page) **A**, *Part of an enlargement of a Landsat 2 MSS image of the Central Caucasus, taken on 13 September 1977.* **B**, *Map of the morphological types of glaciers of the Central Caucasus. Landsat 2 MSS image 2184030007725690, Path 184–Row 30, is from the U.S. Geological Survey, EROS Data Center, Sioux Falls, S. Dak. 57198.*

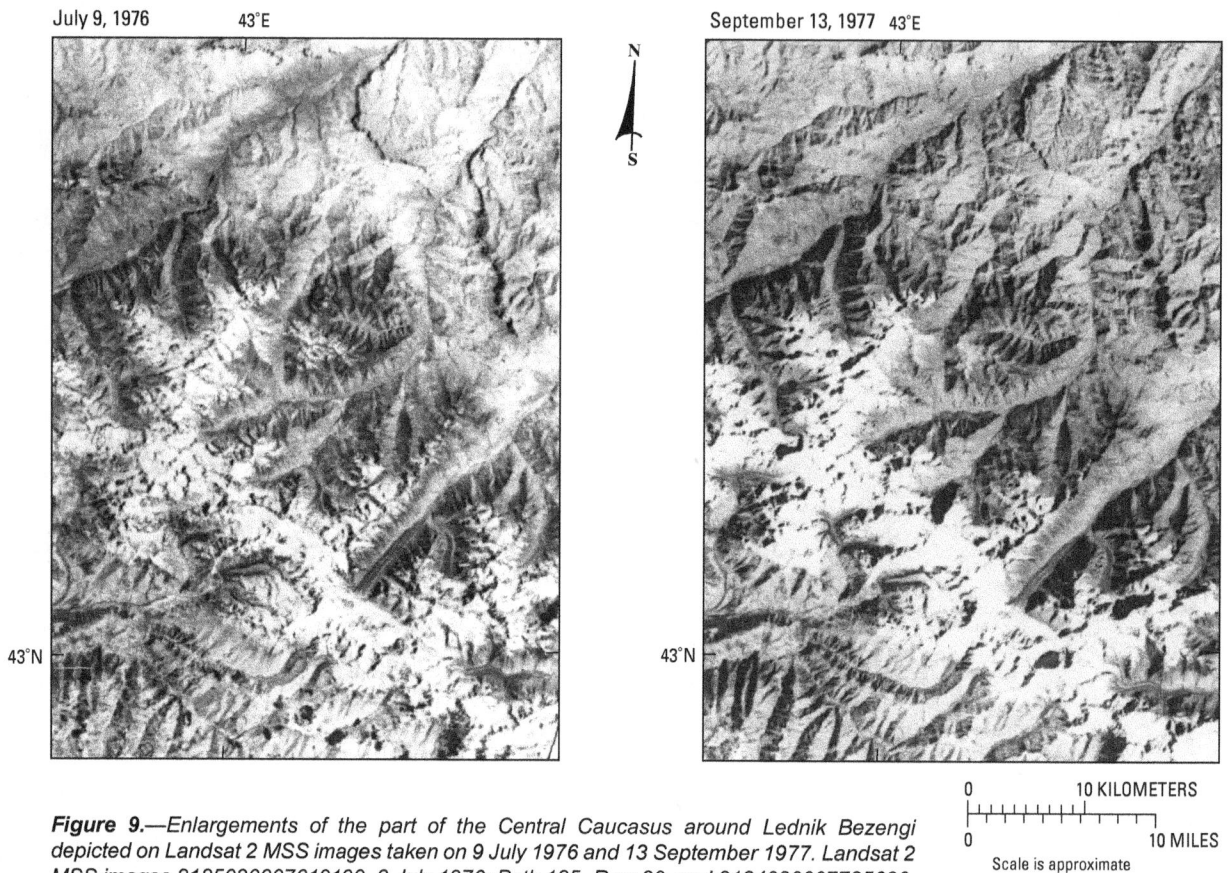

N
S

43°N

43°N

0   10 KILOMETERS
0   10 MILES
Scale is approximate

Figure 9.—Enlargements of the part of the Central Caucasus around Lednik Bezengi depicted on Landsat 2 MSS images taken on 9 July 1976 and 13 September 1977. Landsat 2 MSS images 2185030007619190, 9 July 1976; Path 185–Row 30, and 2184030007725690, 13 September 1977; Path 184–Row 30, are from the U.S. Geological Survey, EROS Data Center, Sioux Falls, S. Dak. 57198.

The ability to identify glaciers on space images depends naturally on the season, residual snow pack, or new snow cover at the time of the surveys. For example, a comparison of two images of the same area of the Caucasus taken on 9 July 1976 and on 13 September 1977 (fig. 9), suggests that the latter image is not as useful for observing the glaciers because the upper parts of the ranges already had been covered with early autumn snowfalls.

The Kazbek massif in the eastern part of the Central Caucasus can be seen very well in figure 8. The surging *Lednik Kolka* is situated here, and a surge was observed in the winter of 1969/70. Continued investigations undertaken in this area by the Institute of Geography, the USSR Academy of Sciences, made it possible to compile the sketches shown in figures 10*B* and *C*. They indicated that the overloading of this glacier by debris was one of the main causes of its surge, which is clearly seen in the aerial photograph (fig. 10*A*) (see also companion section in this part of the chapter which provides a discussion of the 2002 surge of the *Lednik Kolka*).

Glacierization of the Eastern Caucasus has a disparate nature. Separate groups of glaciers are confined to the high parts of crests that have cirques. On the northern slopes of these ranges, small cirque glaciers are situated in the near-crest zone. One such area of glacierization situated on the *Djul'tydag* range in the upper reaches of the Karakoysu river is seen in the part of the Landsat image shown in figure 11. This image, together with two other parts of Landsat images of the Western and Central Caucasus, testifies to the differences in the general nature of the mountain glaciers depicted on space images.

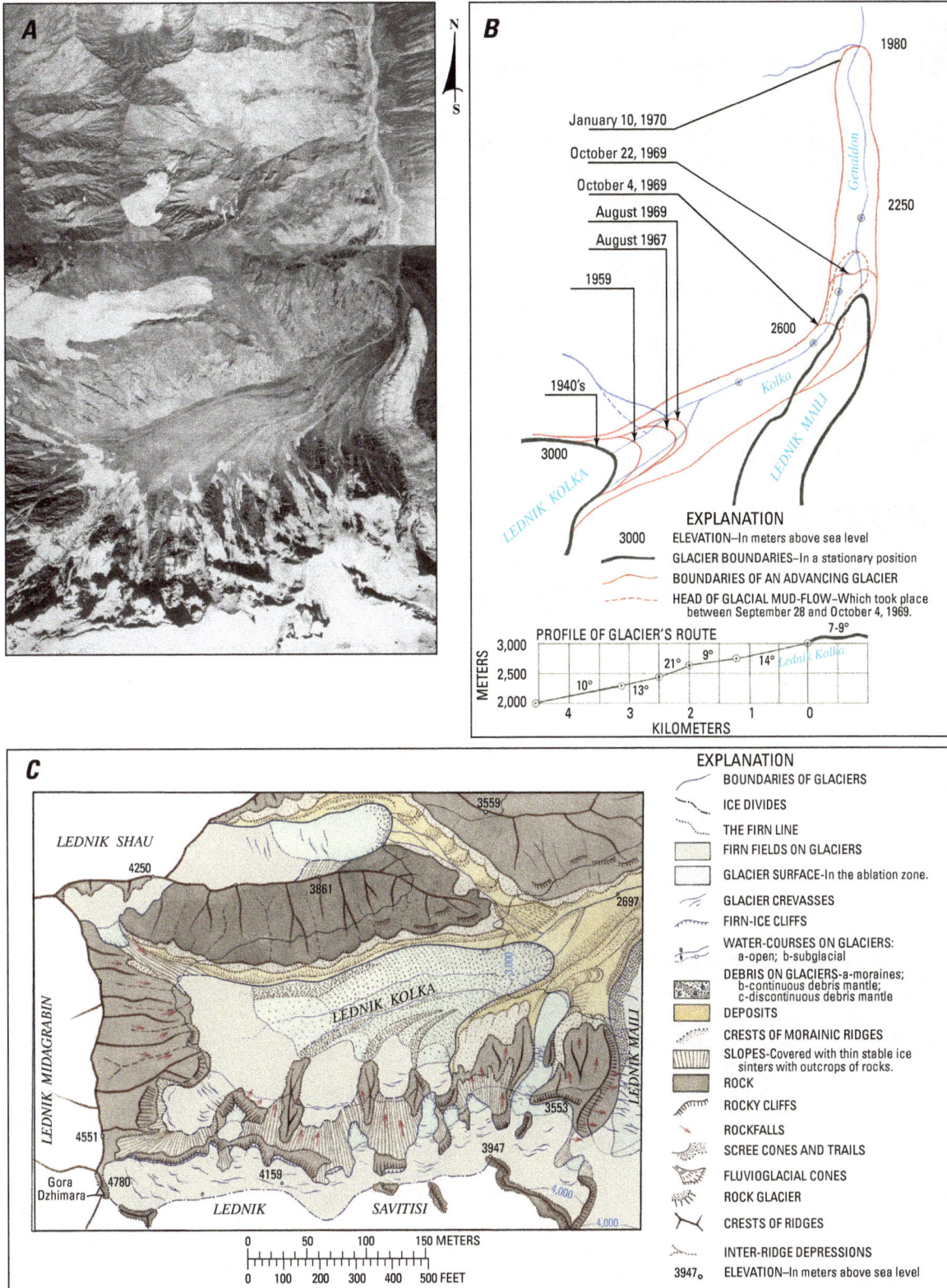

Figure 10.— Illustrations of the surging Lednik Kolka (approximate lat 42°50'N., long 44°30'E.), Gora Kazbek: **A**, Aerial photograph, **B**, sketch map of the advance, and **C**, morphology of the glacier basin.

47°E

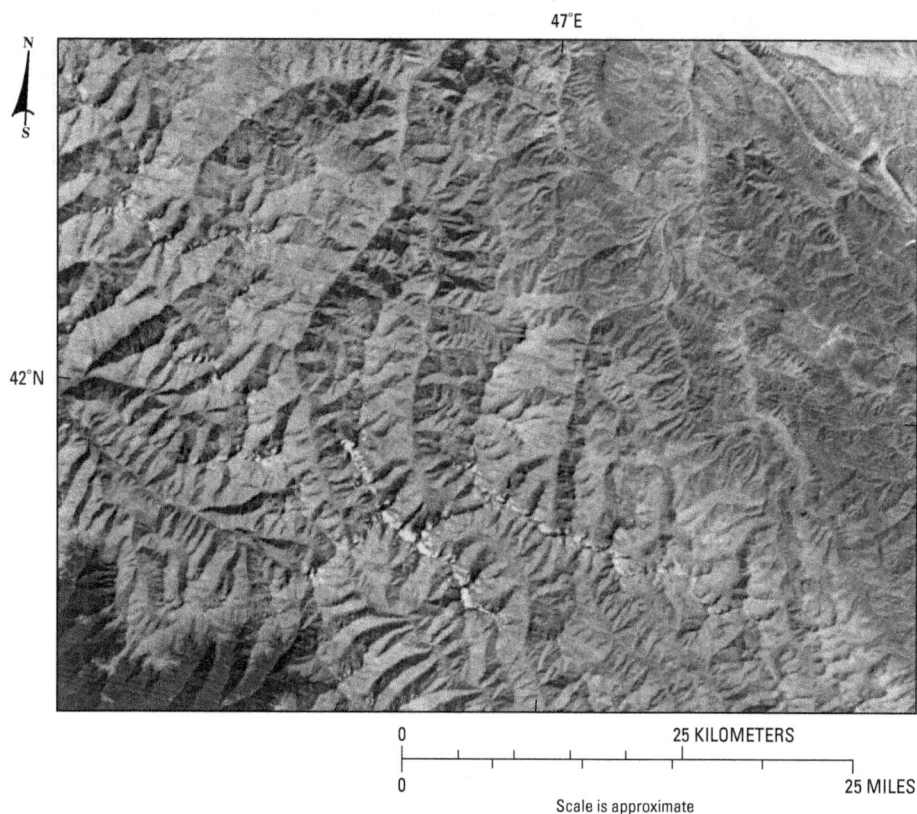

42°N

Figure 11.—*Part of the Eastern Caucasus with the Djul'tydag range depicted on a Landsat 2 MSS image taken on 28 August 1976. The Landsat 2 MSS image 2181031007624190, Path 181–Row 31, is from the U.S. Geological Survey, EROS Data Center, Sioux Falls, S. Dak. 57198.*

```
0                          25 KILOMETERS
├──┼──┼──┼──┼──┼──┼──┤
0                                    25 MILES
             Scale is approximate
```

Pamirs, Alai (Alayskiy), and Gissar (Hisor) Ranges (in the Republics of Tajikistan, Kyrgyzstan, and Uzbekistan)

Studies of the glaciers of the Pamirs, located in eastern Tajikistan (fig. 1), were begun during the second half of the 19th century and were initially connected with the names of well-known geographers, geologists, and botanists: A.P. and B.A. Fedchenko, I.V. Mushketov, V.F. Oshanin, G.E. Grum-Grzhimailo, and V.I. Lipskiy. The first systematic data on the glaciers of the Pamirs were obtained by N.L. Korzhenevskii, who began studying these glaciers in 1903, and in 1930, published the first inventory of Central Asian glaciers (Korzhenevskii, 1930).

Field studies conducted during the Soviet-German expedition of 1928 and later during the Tajik-Pamirian expedition of the USSR Academy of Sciences (1929-1932) resulted in the compilation of a topographic map of the alpine area of the Central Pamirs, a region previously unknown. The expeditions also resulted in the discovery of the highest peak (7,495 m) in the USSR — Communism Peak (Pik Kommunizma, now named Pik Imeni Ismail Samani) and the describing of numerous glaciers on many Pamirian ranges.

During the Second International Polar Year (1932–1933), the primary glaciological studies were concentrated around the area of Fedchenko Glacier (Lednik Fedchenko). Here, at an elevation of 4,200 m a.s.l., an observatory was constructed and repeated phototheodolite surveys of the Tanymas group of glaciers and Lednik Fedchenko were carried out (Russia, Committee of the Second International Polar Year, 1936a).

During the post-World War II years, complete aerial photography of the area was acquired, and compilation of a large-scale topographic map was completed. Using these data, R.D. Zabirov compiled an inventory of glaciers of the Pamirs and published it as a monograph (Zabirov, 1955). This inventory lists 1,085 glaciers that were each more than 1.5 km in length, covering a total area of about 8,041 km^2.

Since 1962, the Institute of Geography, the USSR Academy of Sciences (now the Russian Academy of Sciences) has been responsible for investigations of Pamirian glaciers. From 1968 to 1978, the Institute carried out a series of comprehensive field studies connected to the inventorying of glaciers. The new glacier inventory of the Pamirs and the Alai Range (Alayskiy Khrebet) was compiled on the basis of interpretation of aerial photographs complemented by analysis of space images.

Glaciers are located on nearly all of the Pamirian ranges and occupy more than 10 percent of the total surface area at higher elevations. Their sizes are varied — from scattered small glaciers to large glacier complexes. Table 6 contains a list of the largest glaciers of the Pamirs and the Alayskiy Khrebet; table 7 presents their distribution based on size. Medium-sized glaciers (more than 1 km^2 but less than 10 km^2) are most common, but large glaciers (more than 10 km^2 but less than 100 km^2) are also numerous, a fact that differentiates this area from the glacierization of the Caucasus.

TABLE 6.—*Largest glaciers of the Pamirs and the Gissar (Hisor) and Alai (Alayskiy Khrebet) Ranges*

[Units: km, kilometer; m, meter]

Glacier name	River basin	Area (km^2)	Length (km)	Elevation of the tongue (m)	Elevation of the firn line (m)
Fedchenko	Muksu	651.7	77.0	2,910	4,700
Grumm-Grzhemaylo	Bartang	142.9	37.0	3,610	5,080
Zeravshanskiy	Zeravshan	132.6	27.8	2,810	4,000
Garmo	Obikhingob	114.6	30.4	2,970	4,900
Oktyabr'skiye	Karakul'	88.2	19.0	4,440	5,200
Korzhenevskogo	Kyzylsu	73.0	21.5	3,890	5,100
Geograficheskogo Obshchestva	Vanj	64.4	24.2	2,580	4,200
Fortambek	Muksu	57.3	27.2	2,850	4,170
Lenina	Kyzylsu	55.3	13.5	3,760	4,500
Severniy Tanymas	Bartang	55.0	18.0	3,790	4,900
Nura	Kyzylsu Vostochnaya	54.2	13.8	3,520	4,550
Preobrazhenskogo-Rama	Zeravshan	54.0	12.8	2,960	4,080
Bol'shoya Sauk-Dara	Muksu	53.0	20.6	4,240	5,470
Uisu	Markansu	49.9	13.8	4,390	5,200

TABLE 7.—*Distribution of glaciers based on their area*

[Abbreviations: km^2, square kilometer; N, number of glaciers; A, area of glaciers, in square kilometers]

Glaciers	Small ≤1 km^2		Medium ≤10 km^2		Large ≤100 km^2		Largest >100 km^2		Totals	
	N	A	N	A	N	A	N	A	N	A
Pamirs	5,488	1,559.6	1,151	2,988.5	88	2,036.1	3	909.2	6,730	7,493.4
Qatorkŭhi Hisor/ Hisor Tizmasi and Alayskiy Khrebet	3,449	945.6	431	1,008.5	12	249.1	1	132.6	3,893	2,335.8
Total	8,937	2,505.2	1,582	3,997.0	100	2,285.2	4	1,041.8	10,623	9,829.2

The nature of the Pamirian glacierization changes from west to east and is also affected by elevation. Scattered cirque and small valley glaciers prevail in the western part of the mountains. They are gradually replaced eastward by larger valley, compound-valley, and lastly, dendritic glaciers. A Landsat image (fig. 12) demonstrates well the narrow valleys, surrounded by sharp crests, where valley glaciers occur. Compound-valley glaciers are situated in the upper heads of the valleys, and many of the glaciers' tongues have thick mantles of debris (fig. 13).

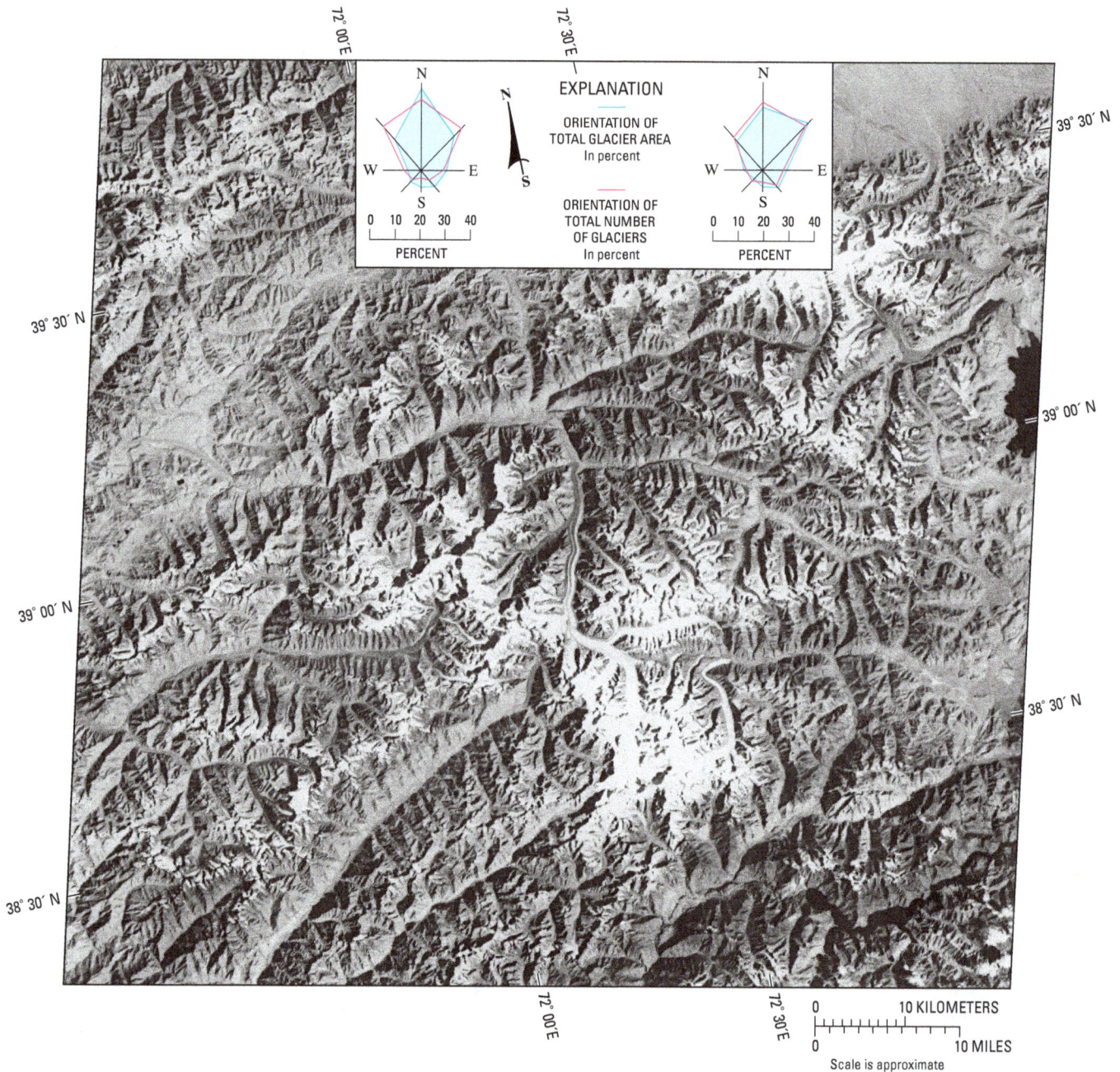

Figure 12.—Landsat 2 MSS image of the Central Pamirs acquired on 28 September 1977 and the distribution pattern of the area (in blue) and number (in red) of glaciers related to glacier orientation in the Hisor and Alayskiy Ranges (left) and the Pamirs. Landsat 2 MSS image 2980-04445 (band 6), Path 163–Row 33, is from the U.S. Geological Survey, EROS Data Center, Sioux Falls, S. Dak. 57198.

Figure 13.—The Lednik Bivachnyy surging glacier, tributary of Lednik Fedchenko with loops of medial moraines.

In the Central Pamirs, compound glacier complexes consisting of adjoining compound-valley and dendritic glaciers are present. The largest is the Lednik Fedchenko complex. It has 26 tributaries and is illustrated in the enlarged part of a Landsat image (fig. 14A) and on the map prepared for the World Atlas of Snow and Ice Resources (fig. 14B). Within the area of the satellite image presented in figure 14A, the extent of glacierization in different areas is very large — from 60 to 85 percent.

The *Zaalayskiy Khrebet* lies in northeastern Tajikistan on the border with Kyrgyzstan, in the northern portion of the Pamirs. Glacierization of the central part of the *Zaalayskiy Khrebet* (Qatorkŭhi Pasi Oloy (in Tajikistan)) or Chong Alay Kyrka Toosu (in Kyrgyzstan), is similar in nature. The similarity in glacierization can be seen in figure 15, where large compound-valley glaciers flow from the second highest summit of the Pamirs, Lenin Peak (Pik Lenina).

However, in some areas, the typical glacierization of the Eastern Pamirs is quite different from that of the Central Pamirs. Eighty percent of the Eastern Pamirs (both numerically and in area) is occupied by small valley glaciers, and especially niche glaciers, which were first identified here. They cover the wide and gently sloping mountains with a relatively thin layer (some tens of meters) of glacier ice (fig. 15); the ice surface is debris free (fig. 16). In general, East Pamirian glacierization has a scattered nature; individual groups of glaciers are separated by dozens of kilometers.

The good quality early Landsat 1, 2, and 3 MSS images of the Pamirs that are cloud-free in glacier areas, have minimal snow cover, and have a pixel resolution of 80 m, make it possible to easily interpret the boundaries of Pamirian glaciers that have an area >1 km^2. Big valley, compound-valley, and dendritic glaciers depicted on space images convey a great deal of information. Upper boundaries of such glaciers can be well delineated when they are surrounded by rock. Debris mantle of varying compactness is clearly seen on the surface of glacier tongues, and the linear outlines of medial moraines are highly visible, making it possible to determine the tributaries of compound-valley glaciers and their relative importance. The most informative images are those taken at the beginning of September for the Western and Central Pamirs and those taken at the end of August for the Eastern Pamirs.

The Pamirs is a classic area of surge-type glaciers, similar to Alaska (Post, 1969). About 100 surge-type glaciers are known in the Pamirs, and many of them are easily identified on space images. This work was pioneered by R. Krimmel of the U.S. Geological Survey on the basis of Landsat imagery (Krimmel and others, 1976; Krimmel, 1978), and, soon thereafter, interpretation of Soviet space images made it possible to detect a number of glaciers at different stages of a surge cycle (Desinov and others, 1978). The diagnostic features of surge-type glaciers — looped medial moraines (fig. 13), active crevassing of the surface, the reduction of surface albedo, and rapid advance of

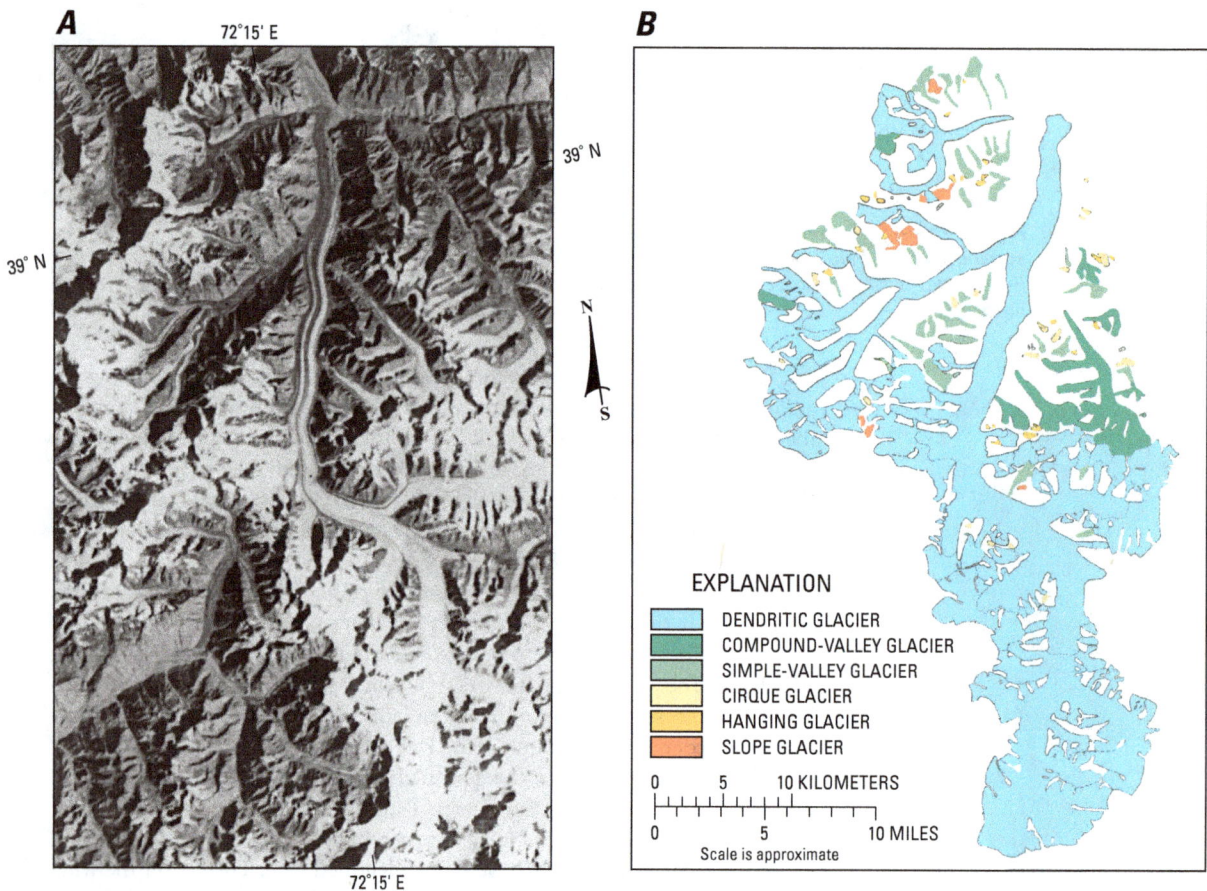

Figure 14.—A, Enlargement of part of the Landsat image (fig. 12), showing the Lednik Fedchenko basin; and B, the sketch map showing the distribution pattern of morphological types of glaciers in this area.

A

0   25 KILOMETERS

0   25 MILES

Scale is approximate

B

EXPLANATION

DENDRITIC GLACIER	CIRQUE GLACIER
COMPOUND-VALLEY GLACIER	HANGING GLACIER
SIMPLE-VALLEY GLACIER	MORAINE MATERIAL
	SNOW AND FIRN

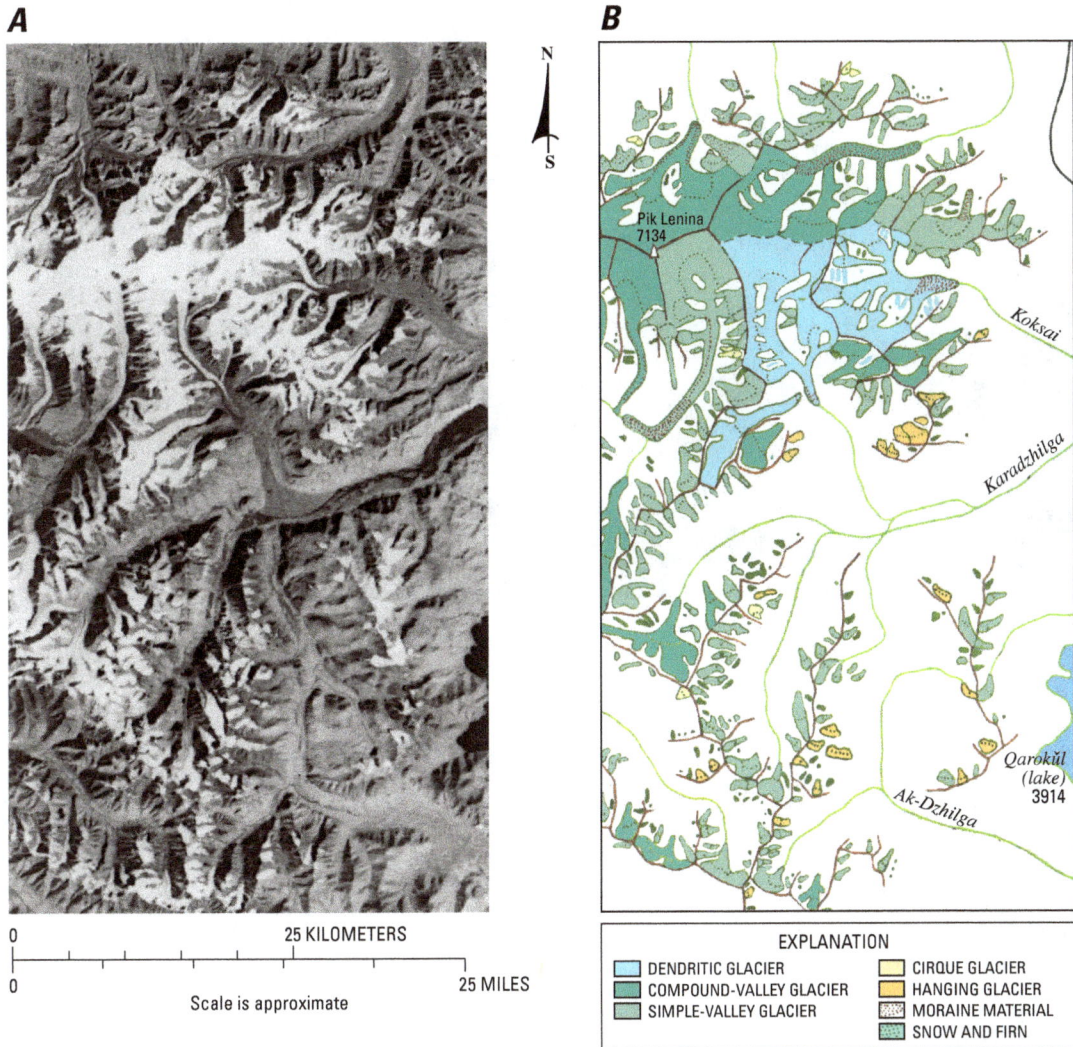

Figure 15.—A, *Enlargement of part of the Landsat image of 28 September 1977 (fig. 12) showing the eastern part of Zaalayskiy Khrebet (at the top) (Qatorkŭhi Pasi Oloy) and part of the Eastern Pamirs (at the bottom).* **B**, *Sketch map showing the pattern of morphological types of glaciers in this region.*

a large volume of ice beyond the previous position of the terminus (figs. 17 and 18) — permit one to reliably detect surge-type glaciers on space images and to survey them from space. The Salyut–6 astronauts made such observations, which permitted them not only to identify a number of new surge-type glaciers in the Pamirs, but also to predict the beginning of their next surge.

Very interesting results were obtained by observing a group of glaciers in the upper reaches of the Saukdara river on the southern slope of Qatorkŭhi Pasi Oloy (fig. 17*A*), where three large glaciers — *Valy, Dzerzhinskogo*, and *Maliy Saukdara* — advanced abruptly from 1973 to 1977 (fig. 17*B*). The advance was first seen in 1973 on *Ledniki Dzerzhinskogo* and *Maliy Saukdara*. Their surges stopped in 1976 and 1975, respectively. *Lednik Valy* started advancing in 1975 and continued into 1977. Changes in the positions of these glaciers were clearly visible on Landsat images taken in July 1973 and in September 1977, and on some Soviet space images taken annually of this area.

Figure 16.—*Debris-free slope glaciers of the Eastern Pamirs.*

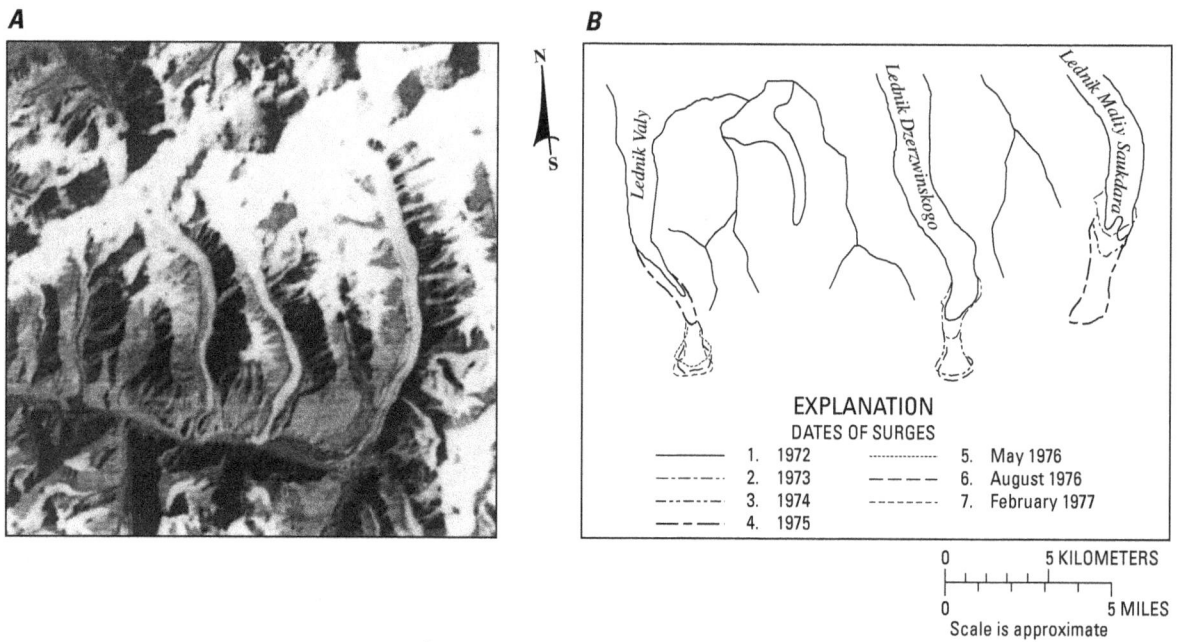

A

B

EXPLANATION
DATES OF SURGES

———	1. 1972	············	5. May 1976
—·—·—	2. 1973	— — —	6. August 1976
—··—··—	3. 1974	--------	7. February 1977
—·—·—	4. 1975		

0   5 KILOMETERS

0   5 MILES
Scale is approximate

Figure 17.—**A**, *Enlargement of part of the Landsat image (fig. 12), showing the upper reaches of the Saukdara river (lat ~39°20'N., long 72°50'E.); and* **B**, *A sketch map of surges of the Saukdara group of glaciers from 1973 to 1977.*

The Gissar Range lies northwest of the Pamirs, in the northwest portion of Tajikistan (there called Qatorkŭhi Hisor) and in southeastern Uzbekistan (there called Hisor Tizmasi). This mountain range has very few glaciers, and will not be discussed further in this chapter.

Figure 18.—*Photograph taken August 1976 of the rapid advance of ice of* **Lednik Valy** *caused by its surge.*

The Alai Range (Alayskiy Khrebet) lies to the north of the Pamirs, mainly in southern Kyrgyzstan. On the Alayskiy Khrebet, cirque and cirque-valley glaciers with areas of 1 to 3 km^2 prevail in the mountains; and the upper reaches of big valleys contain valley and compound-valley glaciers 5 to 10 km in length with areas of 10 to 15 km^2. Among them is the *Abramov Glacier*, situated in a representative mountain glacier basin, where detailed glaciological observations were initiated during the IHD. Glacier tongues are debris-mantled for one third of their area, which hampers the delineation of their boundaries.

For the interpretation of glaciers in the Alayskiy Khrebet, high resolution (20–30 m) multispectral photographs were taken from the Soyuz-22 spacecraft on 18 September 1976 and analyzed (Kravtsova and Chaikina, 1980). Figure 19 is one of these photographs, showing the western part of the Alayskiy Khrebet. For comparison, Landsat images of the same region, obtained on 10 September 1972 and on 12 July 1973, were also studied. By analyzing these images in several spectral bands, it was possible to draw the boundaries of glacier tongues and firn basins, distinguish snow-and-firn covered areas of glaciers from the bare ice surface, map medial moraines on the glaciers and terminal and lateral moraines in the glacier valleys, and to see the largest icefalls. Morphological types of glaciers were determined, and a map was compiled for the World Atlas of Snow and Ice Resources. Figure 20 shows the map of the same area photographed in figure 19. From the space photograph taken from the Soyuz-22 spacecraft, it became possible to detect and map approximately four times more glaciers than are shown on conventional topographic maps of this region (fig. 20*A, B*).

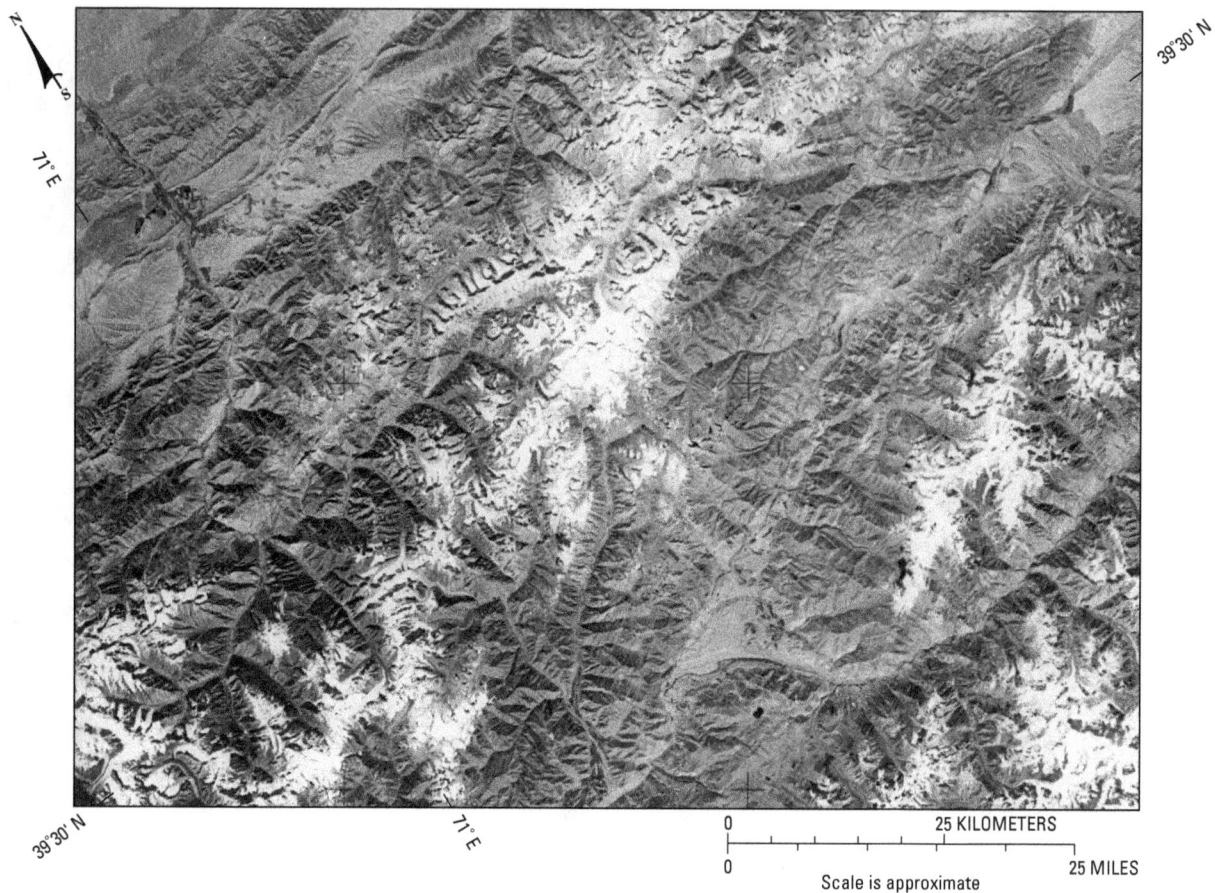

Figure 19.—*Part of the photograph taken by a cosmonaut on the Soyuz-22 spacecraft, showing the western part of the Khrebet Alayskiy on 18 September 1976.*

Based on these studies in regions of wide-spread glacierization, space images can serve as a reliable source of information for (1) identification of glaciers that have an area larger than 1 km², and for (2) detection of surge-type glaciers. Multispectral images may also reveal morphological features of certain types of glaciers.

Tien Shan (in the Republics of Kyrgyzstan and Kazakhstan)

The Tien Shan trends west to east for nearly 2,500 km, 1,200 km of which lies in the Republics of Kyrgyzstan and Kazakhstan (both Soviet Socialist Republics within the Former Soviet Union). The orographic features of the Tien Shan make it possible to subdivide the range into 4 parts: Northern, Western, Central, and Eastern. The Eastern part is within China (see the Glaciers of China portion of this chapter). Distribution of glaciers within the Former Soviet Tien Shan is shown in table 8; it was compiled from data in the Glacier Inventory of the USSR and used in the World Atlas of Snow and Ice Resources. Glacierization of the Central Tien Shan is discussed here with special reference to the two glacier groups, Piks Pobedy and *Khan-Tengri* (Khan Tägiri in Kazakhstan, Kan-Too in Kyrgystan), and also Khrebet Ak-shyyrak, because these regions are covered by good Landsat images.

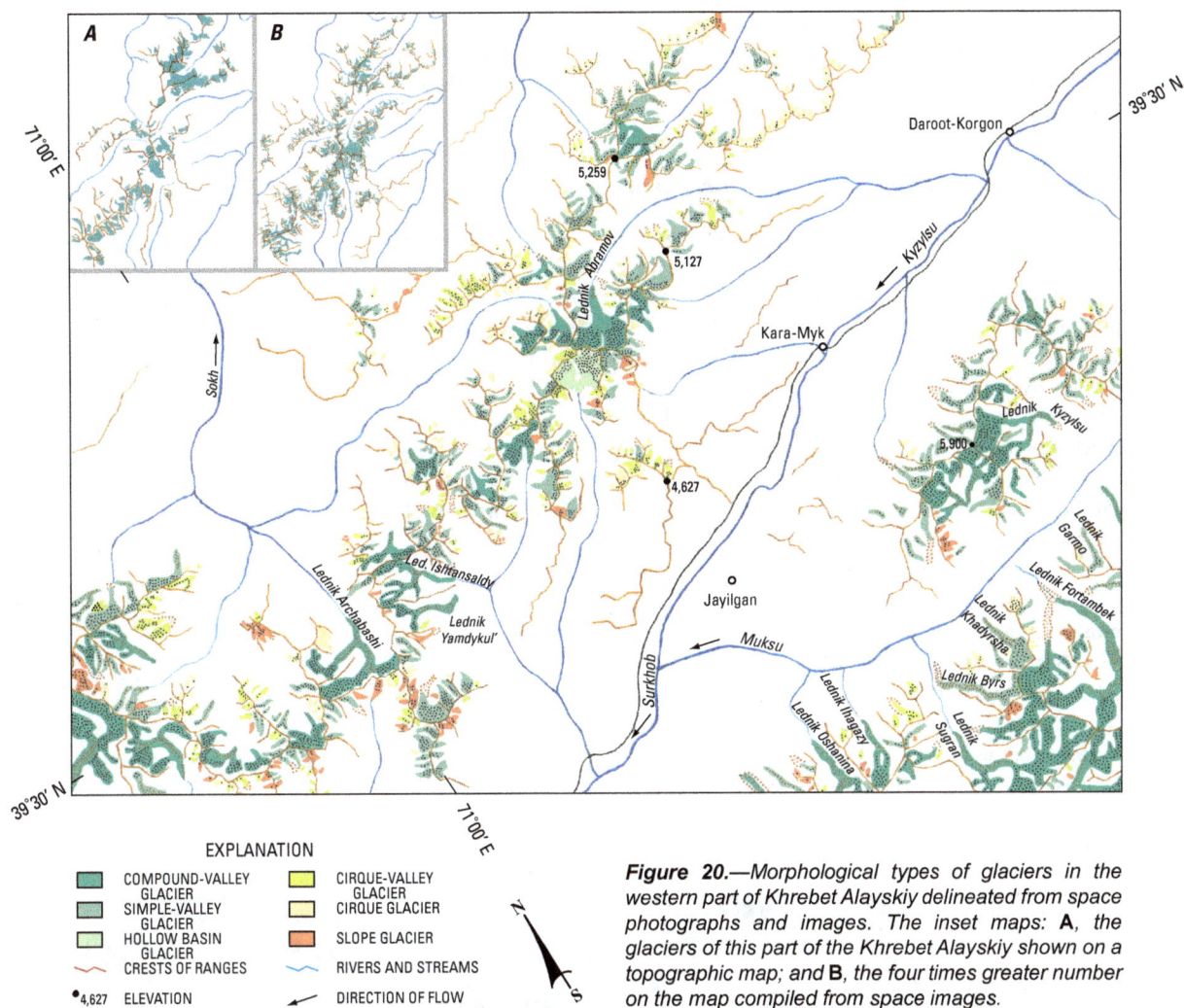

EXPLANATION

▇ COMPOUND-VALLEY GLACIER	▇ CIRQUE-VALLEY GLACIER
▇ SIMPLE-VALLEY GLACIER	▇ CIRQUE GLACIER
▇ HOLLOW BASIN GLACIER	▇ SLOPE GLACIER
⌐⌐ CRESTS OF RANGES	⌐⌐ RIVERS AND STREAMS
•4,627 ELEVATION	⌐ DIRECTION OF FLOW

Figure 20.—*Morphological types of glaciers in the western part of Khrebet Alayskiy delineated from space photographs and images. The inset maps: **A**, the glaciers of this part of the Khrebet Alayskiy shown on a topographic map; and **B**, the four times greater number on the map compiled from space images.*

TABLE 8.—*Distribution of Tien Shan glaciers in various river or lake basins*

[Data from Vinogradov and others (1980). Unit: km², square kilometer]

Name of river or lake basin	Number of glaciers	Glazierized area (km²)	Average area of individual glaciers (km²)
Syr Darya	2,441	1,658.0	0.7
Assa	48	42.0	0.9
Talas	162	117.3	0.7
Chu	790	722.9	0.9
Ili	583	751.6	1.3
Ozero Issyk-Kul' (Lake)	631	636.4	1.0
Ozero Chatyr-Kël' (Lake)	3	2.8	0.9
Taushkan Darya	372	674.7	1.8
Aksu	1,317	2,646.0	2.0
Total Tien Shan	6,347	7,251.7	1.14

[Recent work includes Dyurgerov and others, 1995; Kuzmichenok and Chaohai, 1995; Vilesov and Uvarov, 2001; Solomina and others, 2004; Aizen and others, 2006, 2007; Kotlyakov, 2006; Narama and others, 2006; Aizen and Kuzmichenok, 2007; and Bolch, 2007.]

Piks Pobedy and *Khan-Tengri* in Central Tien Shan

The largest glacier complex in the Tien Shan is situated in the area of Piks Pobedy and *Khan-Tengri*. P.P. Semenov was the first scientist to travel there in 1856 to 1857 (Semenov, 1858), and he discovered the large glaciers. The first glaciological map of this region was compiled by G. Merzbacher (1905) in the 1900s. During the Second International Polar Year, a glaciological expedition headed by S.V. Kalesnik worked in the Central Tien Shan (Kalesnik, 1937). From 1929 to 1934, a topographical survey of this region was undertaken and surveys of the largest Tien Shan glaciers were carried out. During the course of topographical surveys from 1941 to 1943, the highest summit of the Tien Shan, Pik Pobedy was discovered. Subsequent aerial photography of the Tien Shan has revealed the dynamics of its glacierization.

The Landsat image in figure 21 shows the general nature of Tien Shan glacierization: latitudinal extension of ranges and valleys bounded by the high mountain range, *Khrebet Meridional'niy*. The Former Soviet Union part of the Tien Shan region contains 704 glaciers which cover an area of 1,697.5 km^2. Large compound-valley and dendritic glaciers with numerous tributaries are widespread; their termini are often debris-mantled (table 9). As can be seen in table 10, the major part of the area is covered by large valley glaciers, all of which can be easily seen on space images.

Figure 21.—Part of a Landsat 2 MSS image of the Pobedy and Khan Tägiri/Kan-Too peak region taken on 16 October 1976. The Landsat image, 2158031007629090, Path 158–Row 31, is from the U.S. Geological Survey, EROS Data Center, Sioux Falls, S. Dak. 57198.

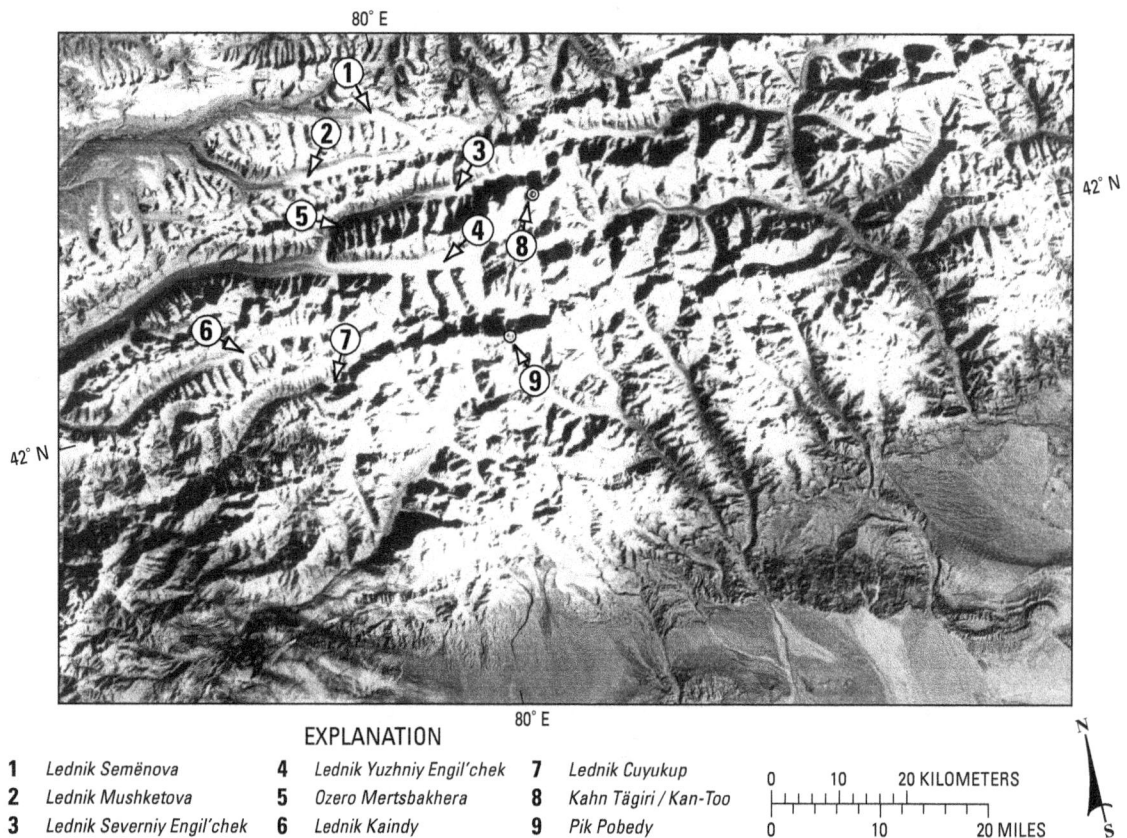

EXPLANATION

1	Lednik Semënova	**4**	Lednik Yuzhniy Engil'chek	**7**	Lednik Cuyukup
2	Lednik Mushketova	**5**	Ozero Mertsbakhera	**8**	Kahn Tägiri / Kan-Too
3	Lednik Severniy Engil'chek	**6**	Lednik Kaindy	**9**	Pik Pobedy

0   10   20 KILOMETERS

0   10   20 MILES

TABLE 9.—*Largest glaciers in the Tien Shan*

[Units: km, kilometer; m, meter]

Glacier name	River Basin	Length (km)	Total area, including tributaries (km²)	Debris-covered area (km²)	Elevation of the terminus (m)	Elevation of the firn line (m)
GLACIERIZED AREA OF POBEDY AND KHAN TÄGIRI/KAN-TOO PEAKS						
Yuzhniy Engil'chek	Engil'chek	60.5	567.2	25.2	2,800	4,500
Severniy Engil'chek	Engil'chek	32.8	181.2	37.6	3,400	4,550
Kaindy	Sarydzhaz	19.0	84.1	11.0	3,400	4,440
Mushketova	Adyr Ter	20.5	68.7	—	3,440	4,150
Semënova	Sarydzhaz	20.2	56.6	0.9	3,340	4,200
KHREBET AK-SHYYRAK						
Petrova	*Kumtor*	11.3	69.8	1.9	3,730	4,270
Severniy Karasay	Karasay	12.2	49.2	1.2	3,800	4,160
Kaindy	Kaindy	8.0	24.3	1.1	3,850	4,200

TABLE 10.—*Distribution of glaciers in two regions of the Central Tien Shan based on their area*

[Abbreviations: km², square kilometer; N, number of glaciers; A, total area, in square kilometers]

Region	<1 km²		1.0–4.9 km²		5.0–9.9 km²		10.0–49.9 km²		>50 km²		Total	
	N	A	N	A	N	A	N	A	N	A	N	A
Pobedy and *Khan Tägiri/Kan-Too* peaks	429	176.5	209	445.6	46	308.0	16	339.2	4	428.2	704	1,697.5
Khrebet Ak-Shyyrak	78	28.1	52	113.2	13	94.2	6	130.1	1	69.8	150	435.4

Similar to the Pamirs, one can easily distinguish on space images the large, primarily linear features on the surface of the large glaciers — medial moraines, zones of large crevasses, and other features. Accuracy in the delineation of glacier boundaries depends on the time of the survey. For example, it is easy to detect the features on the Landsat image of 18 September 1973 (fig. 22), because the image was obtained at the end of the ablation period. However, it is difficult to distinguish these features on the Landsat image of 16 October 1976 (fig. 21), because the upper heads of glaciers and abutting slopes were already covered by a fresh snowfall.

Surging glaciers can be distinguished by the contorted loops of medial moraines. Such loops were visible on the Lednik Mushketova, whose surge was observed in 1956 and 1957. By August 1957, the glacier tongue had advanced nearly 5 km in one year, and its surface became covered by a dense network of crevasses for several kilometers. Since then, the glacier has been retreating. Lednik Kaindy advanced 1,300 m between 1943 and 1960. Variations on this scale for relatively small time intervals can be measured on repeated space images.

Figure 22 presents part of a Landsat image (fig. 22A) an aerial photograph (fig. 22B), and a map (fig. 22C) compiled from them of *Ledniki Severniy and Yuzhniy Engil'chek,* and Ozero Mertsbakhera (lake). The lake is situated at an elevation of about 3,400 m a.s.l., and at high water level it has an area of 4.5 km², a length of about 4 km, an average width of 1 km, and an approximate volume of 200×10⁶ m³. The bottom of the lake is composed of lacustrine sediments. The lake was created by the retreat of *Lednik Severniy Engil'chek* and the filling of the mouth of this glacier valley with melting water and ice flowing down *Lednik Yuzhniy Engil'chek* (Zabirov and Bakov, 1981). The flow of this ice, with its distinctive medial moraine, can be clearly seen on the Landsat image as it turns into the valley of *Lednik Severniy Engil'chek*. Ozero Mertsbakhera reaches its maximum dimensions by the end of the ablation period. Glacier outburst floods (jökulhlaups) from the lake occur annually, sometimes twice a year, and usually in the summer (Kotlyakov, 1997), similar to Alaska (Post and Mayo, 1971). The Landsat image in figure 22*A* shows the lake after the jökulhlaup.

Figure 22.—A, *Enlargement of part of a Landsat 1 MSS image taken on 18 September 1973. Compare with figure 21 for location.* **B**, *Aerial photograph; and* **C**, *Map of the* Ledniki Severniy *and* Yuzhniy Engil'chek *basin and the Ozero Mertsbakhera. The map was compiled from the Landsat image and the aerial photograph. The Landsat 1 MSS image 1159031007326190, Path 159–Row 31, is from the U.S. Geological Survey, EROS Data Center, Sioux Falls, S. Dak. 57198.*

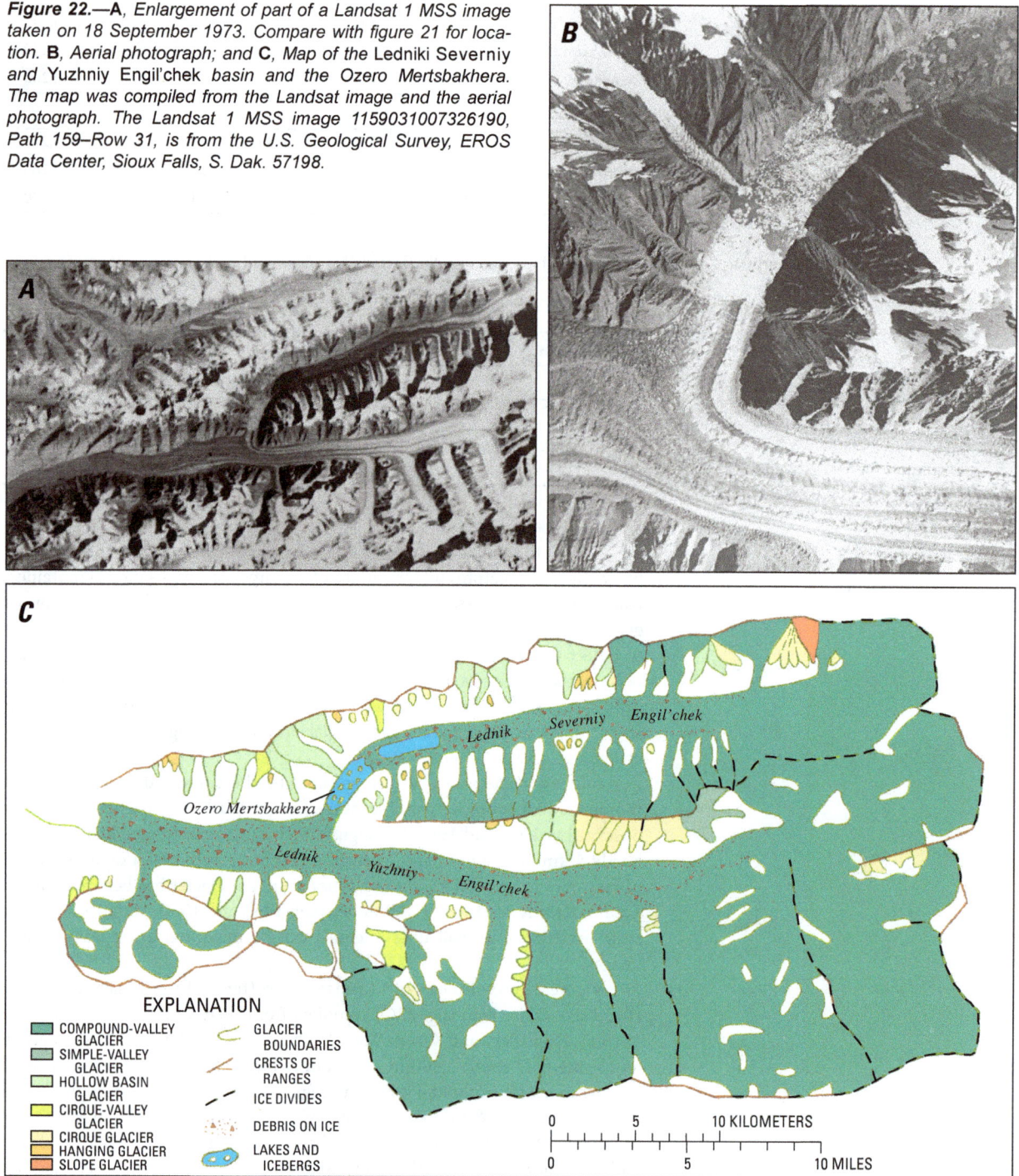

Khrebet Ak-Shyyrak in Central Tien Shan

Khrebet Ak-Shyyrak consists of three parallel mountain ranges trending northeast-southwest. One hundred and fifty glaciers, covering an area of 435.4 km^2 are situated here. The extent of the glacierization of the massif is very high — 46 percent. [Editors' note: Recent work by Aizen and others (2006) suggests that this massif has 178 glaciers with a total area of 371.6 km^2.]

The first information on this region was obtained by A.V. Kaulbars in 1869 (Kaulbars, 1874). Instrumental topographical surveys were undertaken here from 1911 to 1913. The inventory by N.L. Korzhenevskiy (1930) contains data on 29 glaciers of Khrebet Ak-Shyyrak. During the Second IPY, the entire region encompassed by the massif was explored, and the termini of some glaciers were surveyed by phototheodolite. In the second half of the 1940s, the glaciers of this region were studied by G.A. Avsiuk (1952), and, in the 1950s, a comprehensive investigation of the massif were undertaken by the Tien Shan geographical station. These efforts served as the basis for a monograph on the glaciers of Khrebet Ak-Shyyrak (Bondarev, 1963) and a publication by Zabirov and Knizhnikov (1962). The entire region was covered by aerial photography in 1943 and again in 1977. [Editors' note: also in 1995]

The part of the Landsat 1 MSS image shown in figure 23A shows the deep longitudinal valley between the Severniy (northern) and Sredniy (central) mountain ranges, where the largest glacier of the massif exists — Lednik Petrova (table 9). The glaciers of the Karasay system are situated between the Sredniy (central) and Yuzhniy (southern) mountain ranges in the western part of the massif. Glaciers of the Kaindy river basin lie in the eastern part of the massif. The western slopes of Khrebet Ak-Shyyrak are nearly completely buried under ice and firn; 32.5 percent of the glacierized area is confined to the slopes that have northern and western exposures (fig. 23C).

Khrebet Ak-Shyyrak contains many types of glaciers (fig. 23B), from hanging to compound-valley. Flat summit glaciers occur on plateau-like crests of ranges. One of them is evident on the satellite image (fig. 23A). Seven compound-valley glaciers occupy more than 45 percent of the entire glacierized area. Glaciers with an area of more than 5 km^2 cover 67.5 percent of the glacierized area, although their number makes up only 13.4 percent of the total number of glaciers (see table 10). The shapes of these glaciers are evident on space images if they lie on rocky slopes. However, if the upper reaches of large glaciers merge with flat firn areas, it is practically impossible to distinguish between them.

Several surge-type glaciers have been identified in Khrebet Ak-Shyyrak. Among them, *Lednik Severniy Karasay* is of the greatest interest. It advanced at the beginning of the 20th century, creating a push moraine and a dammed lake. At the end of the 1940s, it actively retreated (Avsiuk, 1952), but by 1958 it had advanced 920 to 1,000 m, forming a push moraine up to 45 m high and about 150 m wide. L.G. Bondarev (1963) dates this surge as occurring from 1955 to 1956. Based on aerial photographs (fig. 24), the glacier began to retreat again in 1957. The Landsat 1 MSS image of 18 September 1973 (fig. 23A) shows that this retreat has continued. [Recent work includes Khromova and others (2003).]

Analysis of space images of the Tien Shan region testifies to their usefulness in studies of the general nature of glacierization of large mountain massifs and glacier complexes. Landsat MSS images are most useful for mapping glaciers at scales of 1:1,000,000 and less, and are most valuable for poorly studied mountain-glacier areas. Because of the repetitive nature of the Landsat satellite coverages, Landsat images are useful for monitoring potential natural hazards, such as glacier surges and outbursts (jökulhlaups) from dammed lakes. Higher pixel resolution and more recent Landsat spacecraft

EXPLANATION

1 LEDNIK CHOMOY
2 LEDNIK BEZYMYANNIY
3 FLAT SUMMIT GLACIER
4 LEDNIK PETROVA
5 LEDNIK DAVYDOV
6 LEDNIK SEVERNIY KARASAY
7 LEDNIK YUZHNIY KARASAY

- COMPOUND-VALLEY GLACIER
- SIMPLE-VALLEY GLACIER
- CIRQUE GLACIER
- HANGING GLACIER
- FLAT SUMMIT GLACIER
- ORIENTATION OF TOTAL GLACIER AREA–In percent
- ORIENTATION OF TOTAL NUMBER OF GLACIERS–In percent

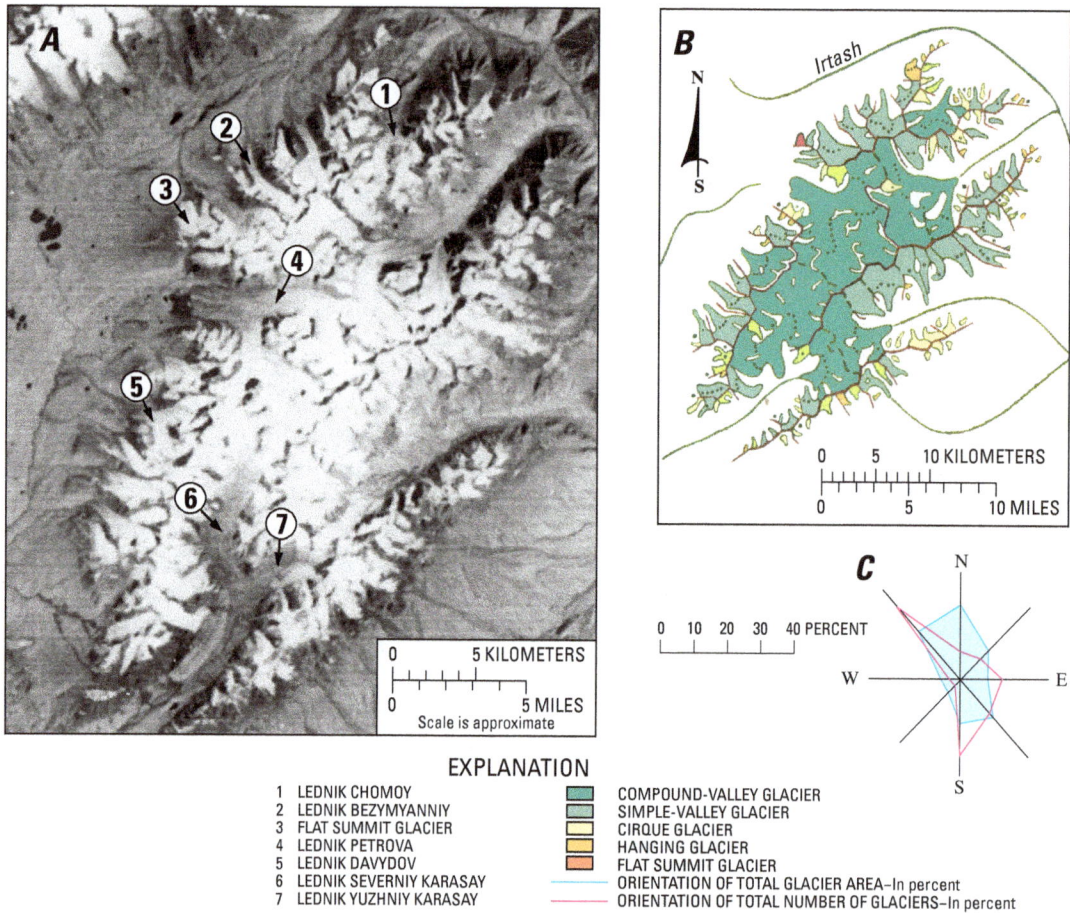

Figure 23.— (above) A, Enlargement of part of a Landsat 1 MSS image of Khrebet Ak-Shyyrak (lat 41°50'N., long 78°20'E.) taken on 18 September 1973. B, Map of the morphological types of glaciers; and C, Distribution pattern of the area (blue) and the number of glaciers (red) along the Khrebet Ak-Shyyrak.

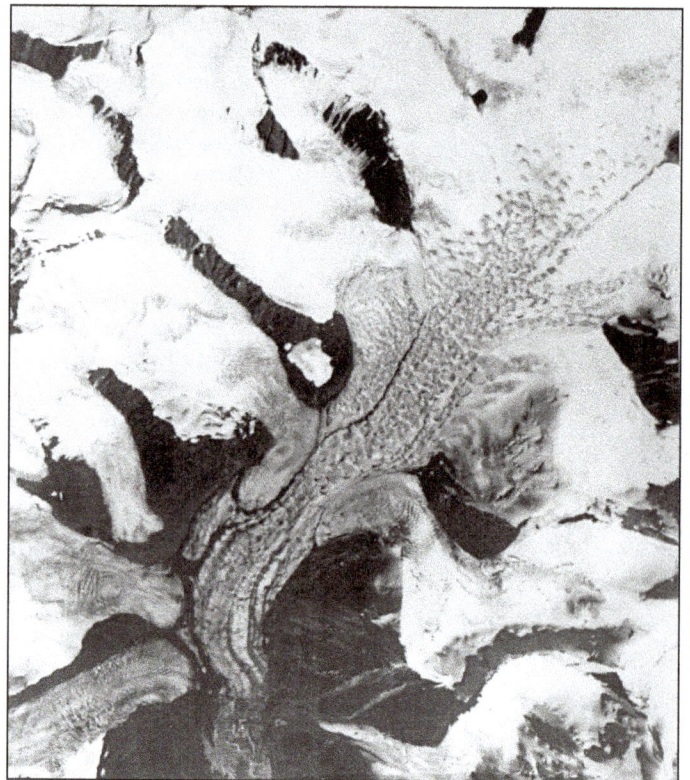

Figure 24.—Aerial photograph of Lednik Severniy Karasay in Khrebet Ak-Shyyrak.

[for example, Landsat 3 return beam vidicon (RBV, 28m), Landsats 4, 5, and 7 thematic mapper (TM, 28 m), and Landsat 7 ETM+ panchromatic (15 m), and other satellite images such as the Advanced Spaceborne Thermal Emission Radiometer (ASTER), which acquires multispectral stereoscopic images (15 m) used for the Global Land Ice Measurements from Space (GLIMS) Program] will, of course, produce data at larger scales. None of that data were available in this earlier phase of study, which was focused on the Landsat baseline period (1972–1981).

Altay (Russia and Kazakhstan)

Studies of the glaciers of Altay were initiated by V.V. Sapozhnikov (1901, 1911), who discovered about 150 glaciers during his several expeditions from 1895 to 1911. Following this, explorers turned their attention to studying the former glacierization of Altay. Aerial photographs of Altay taken in the 1940s and 1950s were important in the study of present-day glaciers, and led to the compilation of new maps that were used as the basis of the Glacier Inventory. The leading role in glaciological studies of the Altay belongs to Tomsk University. A great contribution was made by the pioneer of the Tomsk glaciological school, Professor M.V. Tronov, who wrote monographs on Altay (Tronov, 1949, 1954). A new phase of investigations began in the Altay region in 1971, when the first space surveys were performed from the first Soviet orbital space station, Salyut–1.

There are 1,499 glaciers in the Altay with a total area of more than 900 km^2 and a volume of approximately 57 km^3 of ice. More than 80 percent of the glaciers are concentrated on the highest ranges of Altay, rising 4,400 m a.s.l. — on Katunskiy Khrebet (Qatyn Zhotasy in Kazakhstan), Yuzhno-Chuyskiy Khrebet, and Severo-Chuyskiy Khrebet (fig. 25) — and also on the southern Altay. Glaciers occur mainly on slopes with northern or southern exposures because of latitudinal extension and asymmetry of the ranges, whose southern slopes are steep and northern slopes have flattened areas. In addition, more snow is concentrated on leeward slopes with northern and northeastern exposures due to the dominant transport of air masses from the south and southwest in the Altay.

In Altay, small glaciers prevail. Sixty-three percent of the glaciers are either cirque or hanging glaciers. Cirque-valley and valley glaciers are mainly concentrated in the central parts of glacier clusters, and occupy 57 percent of the total glacierized area of the Altay. There are only 15 glaciers with an area greater than 5 km^2 (table 11). The mean ice thickness of large valley glaciers in the Altay varies from 120 to 150 m.

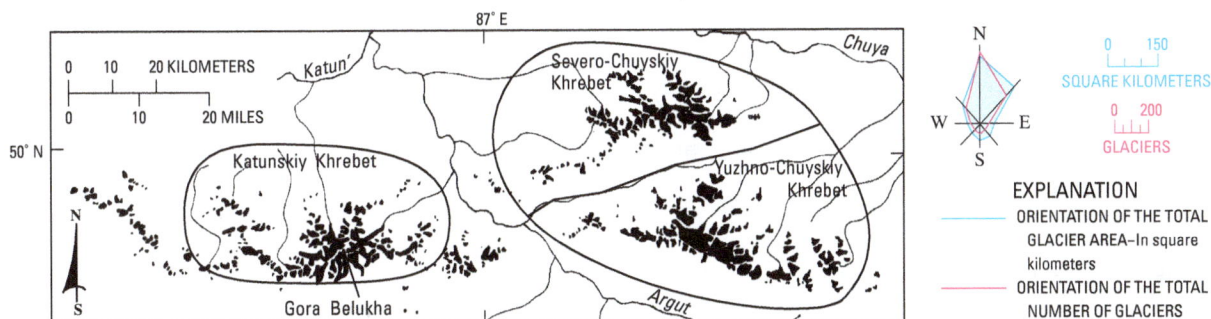

Figure 25.—*Glaciers in the main mountain-glacier areas of the Altay — on the Katunskiy, Severo-Chuyskiy, and Yuzhno-Chuyskiy Khrebet, and the distribution pattern of the area (blue) and the number (red) of glaciers.*

TABLE 11.—*Largest glaciers in the Russian Altay*

[Units: km, kilometer; m, meter]

Name of glacier	Name of river basin	Area (km²)	Length (km)	Elevation of glacier terminus (m)	Elevation of equilibrium line (m)
Bol'shoy Taldurinskiy	Taldura	28.2	7.5	2,450	2,950
Alakhinskiy	Akalakha	19.2	6.0	2,395	3,000
Sofiyskiy	Ak-Kol	17.6	7.9	2,620	2,950
Bol'shoy Maashey	Maashey	16.0	8.2	2,200	2,990
Mensu (Sapozhnikova)	Iedygem	13.2	10.5	2,120	2,800
Bol'shoy Berel'skiy	Belaya-Berel'	12.2	10.4	1,985	3,120
Maliy Berel'skiy	Belaya-Berel'	8.9	8.3	2,100	2,810
Bratjev Tronovykh	Kochurla	8.6	10.5	2,050	2,750
Katunskiy (Geblera)	Katun'	8.5	8.5	1,970	2,800
Akkemskiy (Rodzevicha)	Akkem	8.5	7.8	2,200	3,000
Yadrintsev	Karaoyuk	8.0	4.8	2,650	3,080
Ukokskiy	Akalakha	7.1	4.0	2,610	3,050
Leviy Karagem	Karagem	6.6	4.0	2,375	3,760
Leviy Aktru	Aktru	5.2	6.5	2,500	3,050
Korumdu	Eshtykol	5.0	4.2	2,235	2,950

Space observations of Altay were first undertaken onboard the Salyut–1 space station on 14 June 1971 under conditions of 20 percent cloudiness, when glaciers were still covered by seasonal snow. Despite these problems, images taken from the space station were later successfully used for delineation of glacier boundaries (fig. 26). Comparison of these Salyut–1 images with those from Landsat 2, taken less than one month later in the season, on 11 July 1977 (fig. 27A), illustrates convincingly the seasonal changes of glacier landscapes.

Characteristics of the exposure and morphology of the glaciers on Katunskiy Khrebet are evident on the Landsat 2 MSS image taken on 11 July 1977 (fig. 27A). This image illustrates the dominant cirque glaciers, curved crests of ranges, and cirques with numerous cirque lakes in the western part of the range, and the Gora Belukha with its cluster of predominant valley glaciers in the central part of the range. Differences in the morphology of Katunskiy Khrebet glaciers are well demonstrated on maps of surface morphology and morphological types of glaciers (fig. 27B, C) compiled for the World Atlas of Snow and Ice Resources from cartographic data, and also on images from the Soviet space station, Salyut–1 (fig. 26).

Three glaciers, *Ledniki Bratjev Tronovykh, Akkemskiy,* and Mensu, were easily distinguished by white tongues against the dark background of glacier valleys in the images taken from the Salyut–1 space station, but they had a dark tone on the Landsat images and blended in with the appearance of the glacier valleys. This is due to the fact that by the middle of July when the Landsat image was taken, the hot summer weather had caused the snow on the glacier tongues' surfaces to melt, and the dark, debris-covered glacier termini became visible. The ablation area of these glaciers and the snow-covered accumulation area were easily distinguished and, in some places, the firn line was clearly seen. Therefore, although a quick glance at the 11 July 1977 Landsat 2 MSS image produces the impression of a very small area of glaciers because only the white tone of their accumulation areas is visible, this is an erroneous assumption.

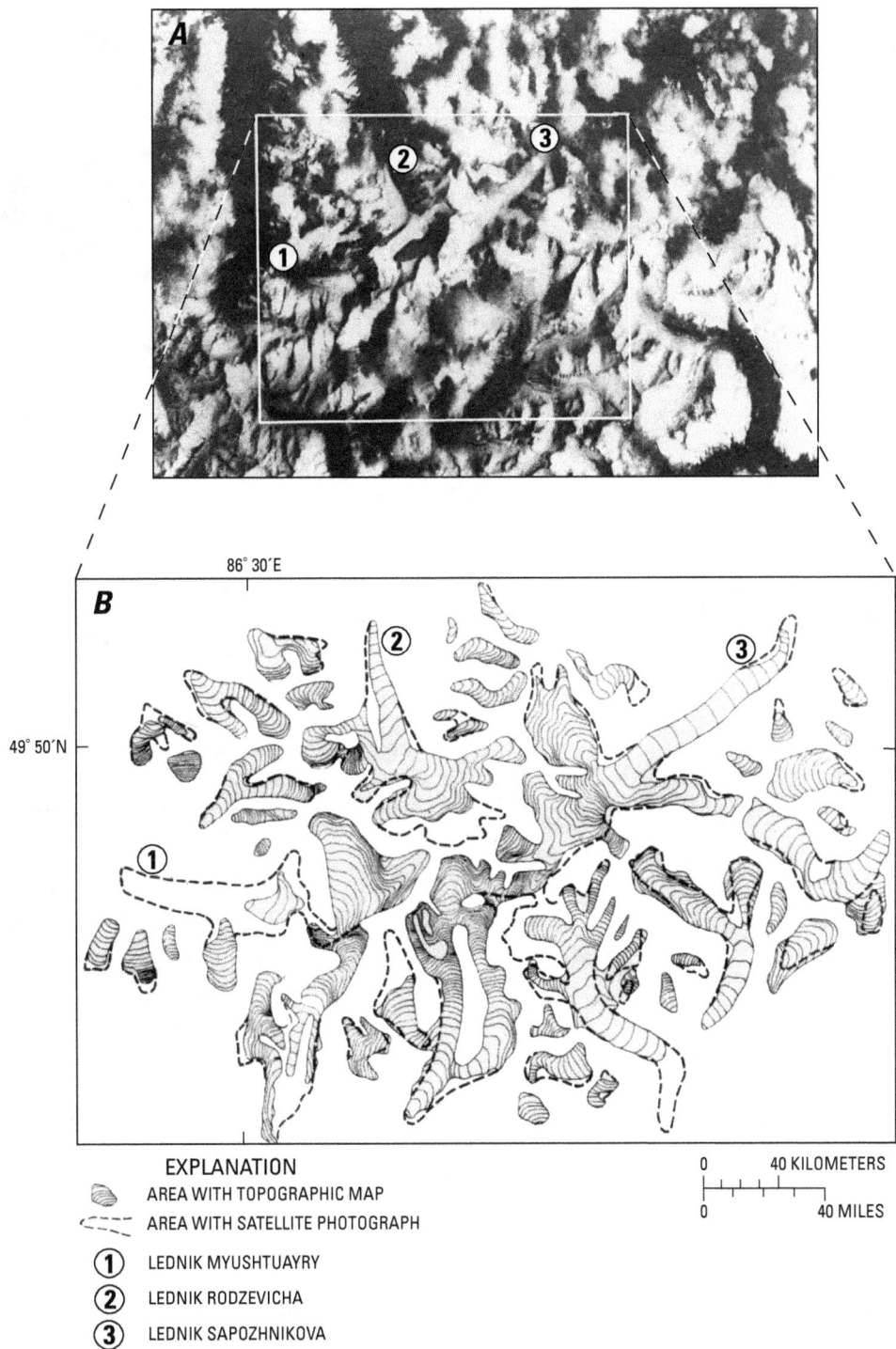

Figure 26.—A, *Image of the central Altay with the Gora Belukha glacier group taken from the Salyut–1 space station on 14 June 1971; and* **B**, *Comparison of the delineation of the glacier boundaries shown on topographic maps and derived from the space image.*

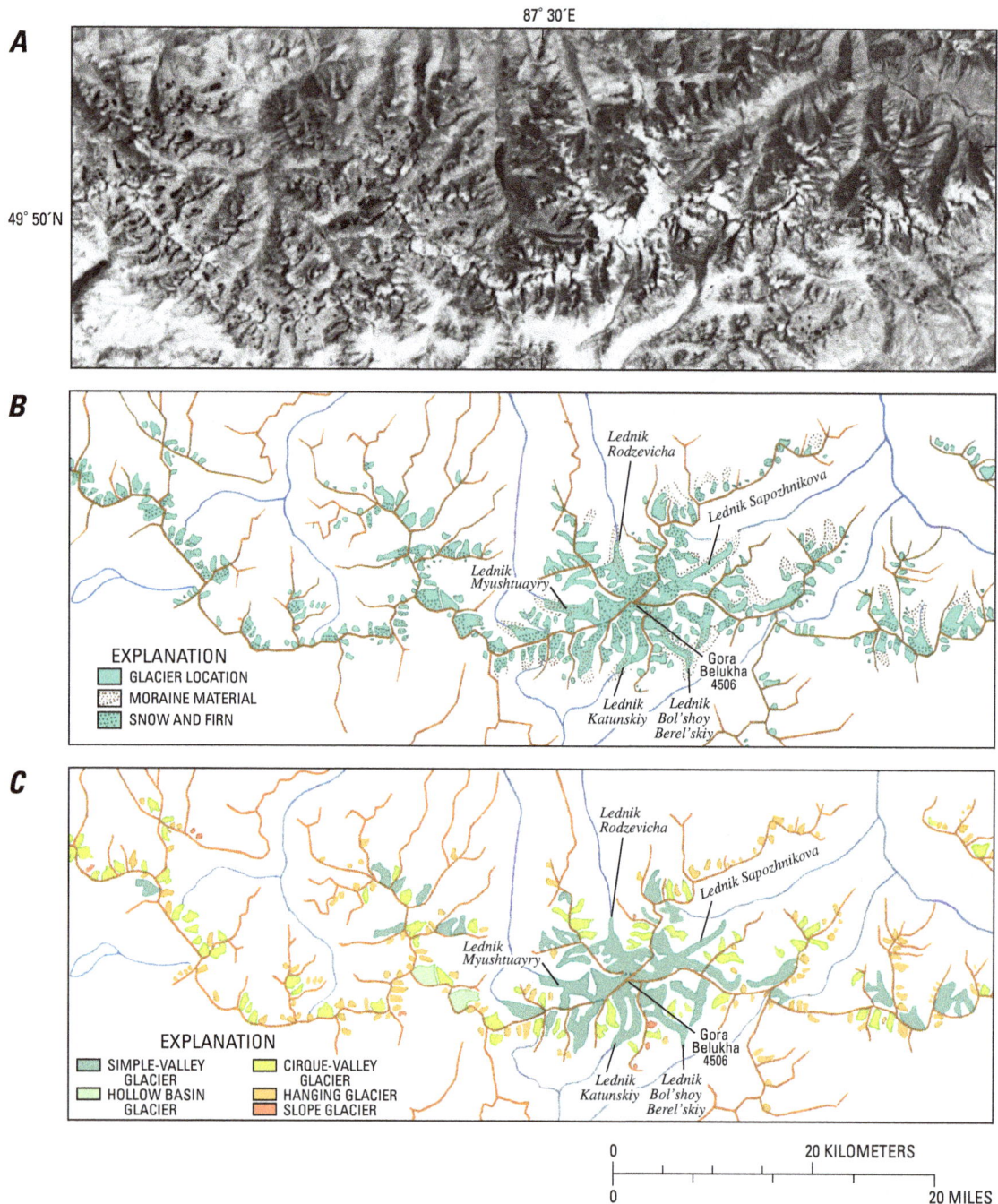

A

87° 30´E

49° 50´N

B

Lednik
Rodzevicha

Lednik Sapozhnikova

Lednik
Myushtuayry

EXPLANATION
- GLACIER LOCATION
- MORAINE MATERIAL
- SNOW AND FIRN

Gora
Belukha
4506

Lednik
Katunskiy

Lednik
Bol'shoy
Berel'skiy

C

Lednik
Rodzevicha

Lednik Sapozhnikova

Lednik
Myushtuayry

EXPLANATION
- SIMPLE-VALLEY GLACIER
- CIRQUE-VALLEY GLACIER
- HOLLOW BASIN GLACIER
- HANGING GLACIER
- SLOPE GLACIER

Gora
Belukha
4506

Lednik
Katunskiy

Lednik
Bol'shoy
Berel'skiy

0   20 KILOMETERS

0   20 MILES

Figure 27.—A, *Part of a Landsat 2 MSS image of Katunskiy Khrebet taken on 11 July 1977;* **B**, *Map of surface morphology;* and **C**, *Morphological types of glaciers in this region. Landsat image is from the U.S. Geological Survey, EROS Data Center, Sioux Falls, S. Dak. 57198.*

Comparison of the greatly enlarged part of the Landsat image of the *Bish-Irdu* glacier group with the cartographic presentation of these glaciers (fig. 28) furnishes an excellent example of the genuine relationship between the map and the Landsat image depiction of glacierization. This figure also shows the *Aktru* glaciers, well known in glaciology, that have been studied for a long time by glaciologists of Tomsk University (Tronov, 1949, 1954) [More recent work includes Galakhov and others (1987).]

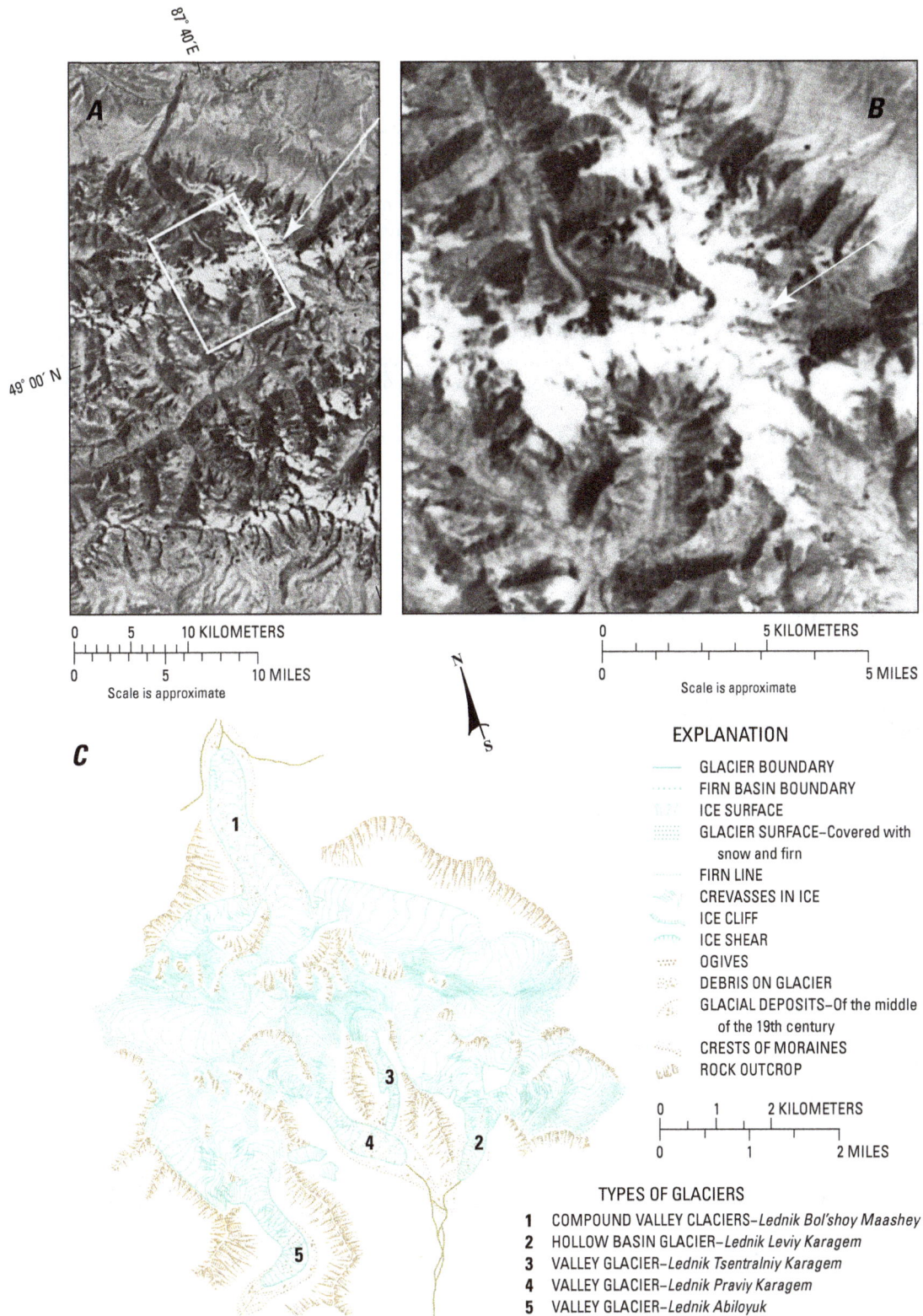

87° 40'E

49° 00' N

A

B

| 0 | 5 | 10 KILOMETERS |
| 0 | 5 | 10 MILES |

Scale is approximate

| 0 | 5 KILOMETERS |
| 0 | 5 MILES |

Scale is approximate

N
S

C

EXPLANATION

	GLACIER BOUNDARY
	FIRN BASIN BOUNDARY
	ICE SURFACE
	GLACIER SURFACE–Covered with snow and firn
	FIRN LINE
	CREVASSES IN ICE
	ICE CLIFF
	ICE SHEAR
	OGIVES
	DEBRIS ON GLACIER
	GLACIAL DEPOSITS–Of the middle of the 19th century
	CRESTS OF MORAINES
	ROCK OUTCROP

| 0 | 1 | 2 KILOMETERS |
| 0 | 1 | 2 MILES |

TYPES OF GLACIERS

1 COMPOUND VALLEY GLACIERS–*Lednik Bol'shoy Maashey*
2 HOLLOW BASIN GLACIER–*Lednik Leviy Karagem*
3 VALLEY GLACIER–*Lednik Tsentralniy Karagem*
4 VALLEY GLACIER–*Lednik Praviy Karagem*
5 VALLEY GLACIER–*Lednik Abiloyuk*

Figure 28.— *A, Part of the Landsat 2 MSS image taken on 11 July 1977 of the* Bish-Irdu *glacier complex in the Severo-Chuyskiy Khrebet;* **B**, *enlarged portion of the image; and* **C**, *depiction of this region on a map. The white arrows show* Aktru *glaciers. Landsat image is from the U.S. Geological Survey, EROS Data Center, Sioux Falls, S. Dak. 57198.*

Ural Mountains (Ural'skiye Gory)

The Ural Mountains are a typical region of small glaciers. Glaciers extend over a distance of more than 500 km — from latitude 63°53'N. to latitude 68°40'N. The glaciers are situated in the *Polar, Subpolar,* and *Northern Urals* in groups confined to the highest and most dissected parts of the mountains. The *Polar Urals* rise for 1,000 to 1,200 m, with the highest point at 1,499 m; the *Subpolar Urals* have elevations of 1,300 to 1,400 m; with the highest point at 1,894 m; the *Northern Urals* seldom exceed 1,000 m, and the highest summit, Gora Telpoziz, rises to 1,617 m a.s.l.

The first glaciers were discovered in the Ural Mountains in 1929. They were again observed during the Second IPY and the International Geophysical Year; the studies continued during the IHD. Glaciers situated in the Telpoziz mountain range and discovered quite recently (Dolgushin and Osipova, 1979) were identified on the space image of the *Northern Urals* (fig. 29). Now, regular observations are made and aerial and ground stereophoto-surveys have been conducted.

The majority of the glaciers, 91 of the 143, are situated in the *Polar Urals* and cover 20.8 km^2 of the total glacierized area of 28.7 km^2. Nearly all the glaciers lie on the western side of the Ural Mountains, facing the moisture-bearing air masses; however, they are all situated on the eastern slopes of individual ranges (fig. 30B), where snow is concentrated by wind transport. The area of present-day glaciers coincides exactly with the zone of maximum snowcover in the mountains (Troitskiy and others, 1966).

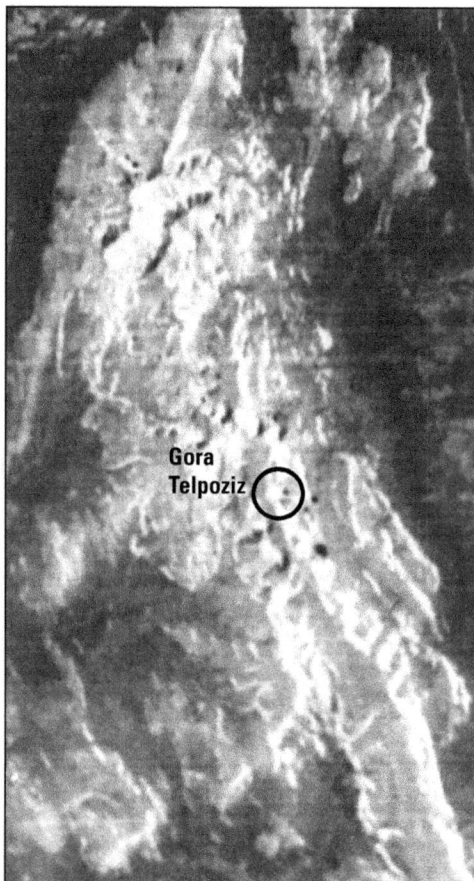

Figure 29.—Enlargement of a Landsat 1 MSS image of the Northern Urals acquired on 25 June 1973. The Gora Telpoziz glacier (lat 63°55'N., long 59°11'E.) is circled. The Landsat 1 MSS image 1182014007317590, Path 182–Row 14, is from the U.S. Geological Survey, EROS Data Center, Sioux Falls, S. Dak. 57198.

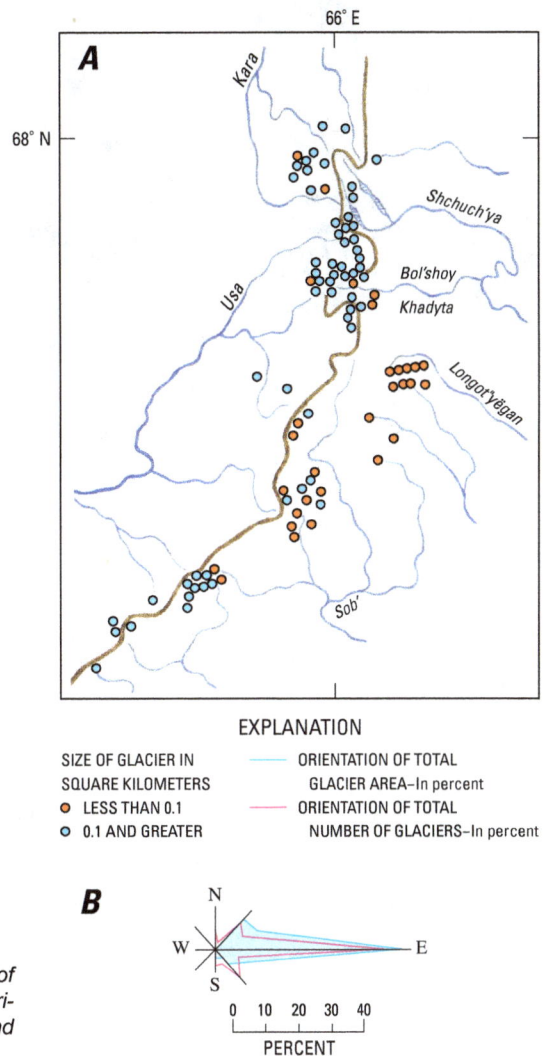

Figure 30.—A, *Distribution pattern of glaciers in the Polar Urals; and **B**, Distribution of the glaciers by area (blue) and number (red) related to orientation.*

Recent glacierization in the *Polar Urals* is represented by cirque and niche glaciers. Only two of them — *IGAN* and *MGU* glaciers — have an area more than 1 km². More than 60 percent of the glaciers have an area of 0.1 to 0.6 km²; about one third of them do not exceed 0.1 km² (fig. 30A).

Understandably, space images that have coarse resolution are not useful for studies of such small glaciers. And the available Landsat 1, 2 and 3 MSS images are not useful because of the unfavorable time of their acquisition — at the beginning of March, when all of the Ural Mountains were covered with snow. It is possible to identify the main mountain ranges and waterbodies on these images, however. The cirques of certain glaciers are visible too, but the glaciers themselves are covered with snow and long shadows from the near-by ranges and summits (fig. 31).

The end of August is the most suitable time for surveys of the glaciers of the Ural Mountains. At this time, it is possible to identify the largest cirque glaciers on images of 1:1,000,000 scale. However, for the correct interpretation of niche glaciers that greatly resemble snow patches, better resolution, larger-scale images are needed. These images would depict the details of the structure of the cirque glaciers, similar to that shown in figure 32 of Lednik Obrucheva.

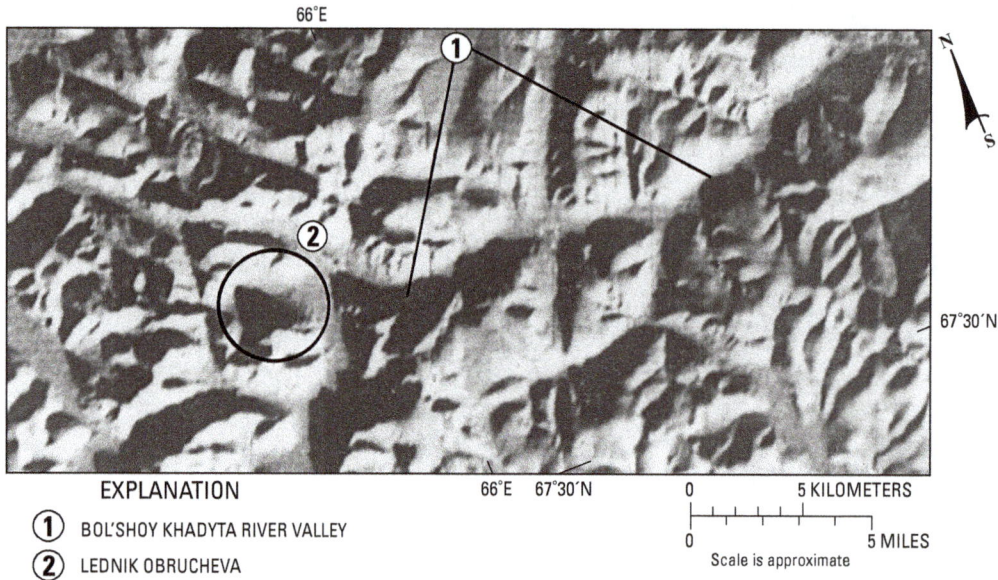

66°E

1

2

67°30′N

EXPLANATION   66°E   67°30′N   0   5 KILOMETERS

1 BOL'SHOY KHADYTA RIVER VALLEY   0   5 MILES

2 LEDNIK OBRUCHEVA   Scale is approximate

Figure 31.—*Enlargement of a Landsat 1 MSS image of part of the Polar Urals acquired on 6 March 1973. Landsat 1 MSS image 1179012007306590, Path 179–Row 12, is from the U.S. Geological Survey, EROS Data Center, Sioux Falls, S. Dak. 57198.*

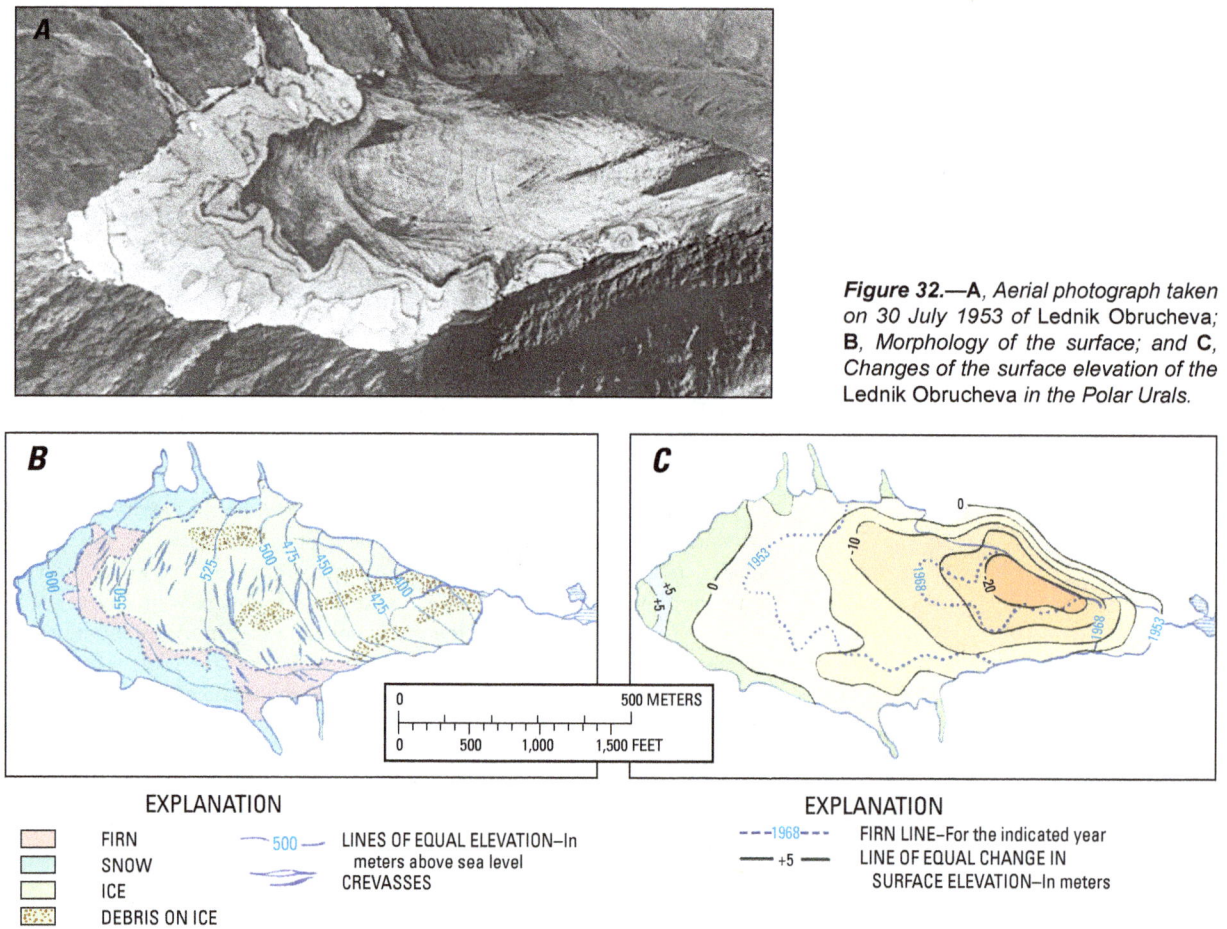

Figure 32.—**A**, *Aerial photograph taken on 30 July 1953 of Lednik Obrucheva;* **B**, *Morphology of the surface; and* **C**, *Changes of the surface elevation of the* Lednik Obrucheva *in the Polar Urals.*

EXPLANATION

FIRN
SNOW
ICE
DEBRIS ON ICE

500 — LINES OF EQUAL ELEVATION–In meters above sea level
CREVASSES

EXPLANATION

---1968--- FIRN LINE–For the indicated year
+5 — LINE OF EQUAL CHANGE IN SURFACE ELEVATION–In meters

The Byrranga Mountains (Gory Byrranga) in Northern Russia

The Gory Byrranga are situated on the Poluostrov Taymyr (peninsula) in central northern Russia. Glaciers extend over the most northern and highest part of the mountains between latitudes 75° and 76°N. and longitudes 106° and 111°E. The largest glaciers are concentrated within an area of 400 km^2. The area of the largest glacier is 4.3 km^2, and its length is approximately 3.8 km.

The first glaciers were discovered in Gory Byrranga in 1949. In the 1950s, the region was covered by aerial photography and soon the first maps of the glaciers were made. According to the glacier inventory of Gory Byrranga, the region has 96 glaciers, which occupy 30.5 km^2 (fig. 33B). Among them, there are 29 valley glaciers, 23 cirque, and 15 niche glaciers (USSR Academy of Sciences, 1965-1983). They have various orientations (fig. 33C).

There were not many Landsat 1, 2, or 3 MSS images of the Gory Byrranga area acquired. One of the few is shown in figure 33A. In this Landsat image, acquired on 26 July 1973 during the ablation period, the glacierized area lies near the edge of the image and is only partially visible. In the mountains there are traces of new snowfall, and because of this it was impossible to identify all of the glaciers in the image, although they could likely be seen under better conditions. A characteristic of the Gory Byrranga glaciers is their small accumulation areas which occupy broadened cirques.

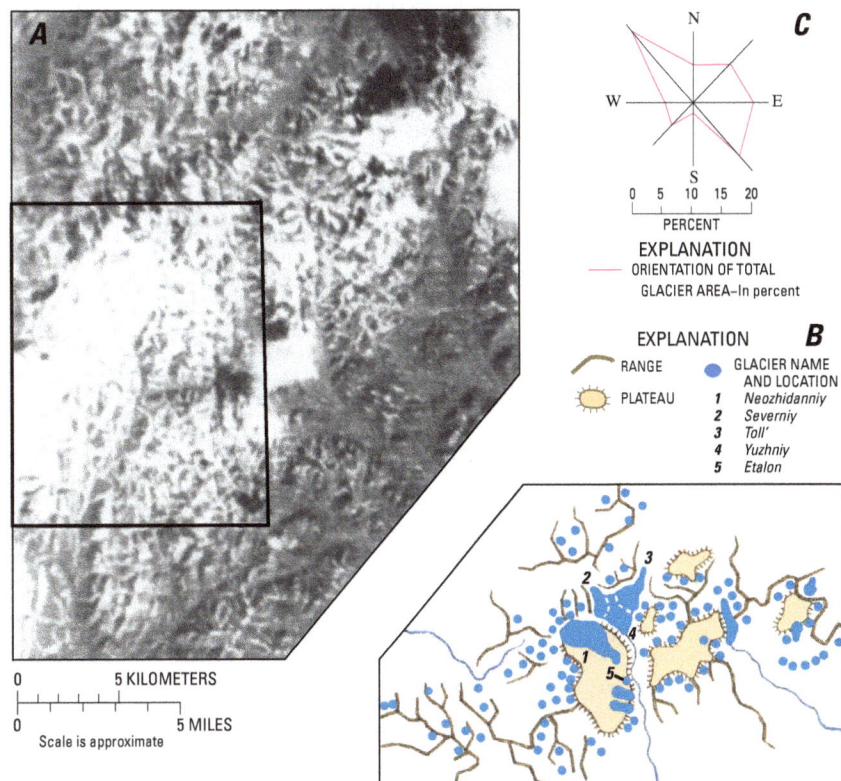

Figure 33.—**A**, *Landsat 1 MSS image acquired 26 July 1973, showing the distribution of glaciers in the Gory Byrranga in the region near lat 76°N., long 108°E.;* **B**, *Sketch of glacier location; and* **C**, *Distribution pattern of glacier orientation in the Gory Byrranga. Landsat 1 MSS image 1159006007320790, Path 159–Row 6, is from the U.S. Geological Survey, EROS Data Center, Sioux Falls, S. Dak. 57198.*

Northeastern Siberia

Over the vast territory of Northeastern Siberia, glaciers are situated in widely separated areas: Khrebet Suntar-Khayata, the Khrebet Cherskogo, and Koryakskoye Nagor'ye (upland). Khrebet Suntar-Khayata is situated near one of the coldest places in the Northern Hemisphere — Oymyakan settlement. The first mention of the permanent snow on the summits of Khrebet Suntar-Khayata was made by I.D. Chersky as early as the late 19th century. However, glaciers were not found in that area until 1939. The scale of glacierization was estimated only after aerial photographic surveys were carried out in the 1940s (USSR Academy of Sciences, 1965–1983). The first comprehensive work on the Suntar-Khayata glaciers was published by L.L. Berman (1947). About 20 glaciers were described during geological surveys at the beginning of the IGY; and during the IGY, one of the eleven Soviet glaciological stations was located here (Koreisha, 1963).

According to the data of the USSR Glacier Inventory, Khrebet Suntar-Khayata has 208 glaciers with a total area of 201.6 km^2 (fig. 34A). Similar to the Khrebet Cherskogo, nearly all of the glaciers have a northern exposure (fig. 34B, C). More than half of the total glacierized area is found in valley glaciers. Small glaciers with an area less than 1 km^2 prevail (table 12). On a Landsat 1 MSS image of 26 July 1973, it was possible to identify only one glacier that had an area of 4.4 km^2; glaciers of smaller dimensions were difficult to distinguish.

The ability to identify the small isolated glaciers that characterize northeastern Siberia depends primarily on the time of the surveys. However, very few Landsat 1, 2, and 3 MSS images were acquired of the glacierized regions of Siberia and, of these, very few were acquired during the optimum time for observing the glaciers, usually in early to mid-August. As a result, the Landsat 1, 2 and 3 MSS images have very limited usefulness in this region. For example, figure 35 shows a comparison of two enlarged Landsat images, one acquired on 2 July 1975 (fig. 35A) and the other on 26 August 1975 (fig. 35B); they both show the highest point of the Koryakskoye upland, at 2,115 m high. On the 2 July 1975 image, the glaciers are masked by the mantle of seasonal snow that is still present. And on the 26 August 1975 image, the boundaries of the glaciers are difficult to delineate. Unfortunately, there is no ideal Landsat image of this area.

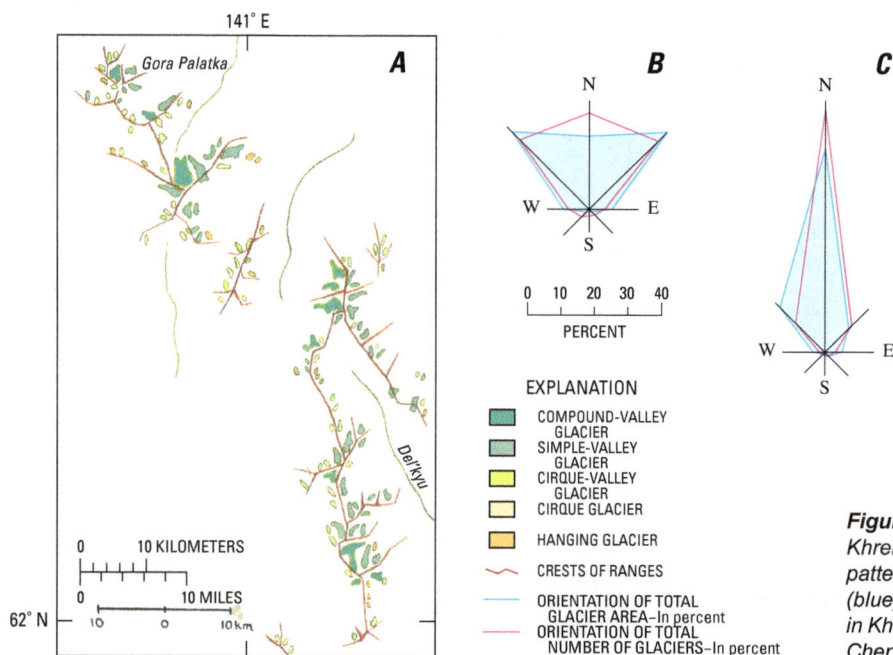

Figure 34.—A, Distribution of glaciers over Khrebet Suntar- Khayata; **B**, Distribution patterns of the number (red) and area (blue) of glaciers relating to the orientation in Khrebet Suntar-Khayata; and **C**, Khrebet Cherskogo.

TABLE 12.—*Distribution of glaciers in the Khrebet Suntar-Khayata based on their size*

[Unit: km², square kilometer]

Indices	<0.5 (km²)	0.6–1.0 (km²)	1.1–2.0 (km²)	2.1–3.0 (km²)	>3.0 (km²)
Number of glaciers	87	68	27	14	12
Percentage	41.8	32.7	13.0	6.7	5.8
Total area of glaciers (km²)	20.4	50.2	39.8	37.3	53.9
Percentage	10.2	24.9	19.7	18.5	26.7

Space images are very helpful in studies of aufeis (also called naled' or icing). Aufeis is a sheet-like mass of layered ice that forms from successive flows of ground water during freezing temperatures. In addition to showing the geographic location of aufeis, space images can reveal some of their structural elements: large faults, zones of intense crevassing, ring structures, and local depressions. The best time of image acquisition for analyzing aufeis is during the first weeks after the disappearance of snow cover. Using images larger than 1:500,000 scale, it is possible to study the morphology and geological changes in aufeis; images larger than 1:1,000,000-scale can be used to map large aufeis areas (fig. 36) and to study their variability from year to year. [See also Williams (1986, p. 536, fig. 9–24), who delineated four icing remnants on a 1:1,000,000-scale, 4 August 1973, Landsat 1 MSS image (1377-21112-6) of northeastern Alaska in the Sagavanirktok River area and environs.]

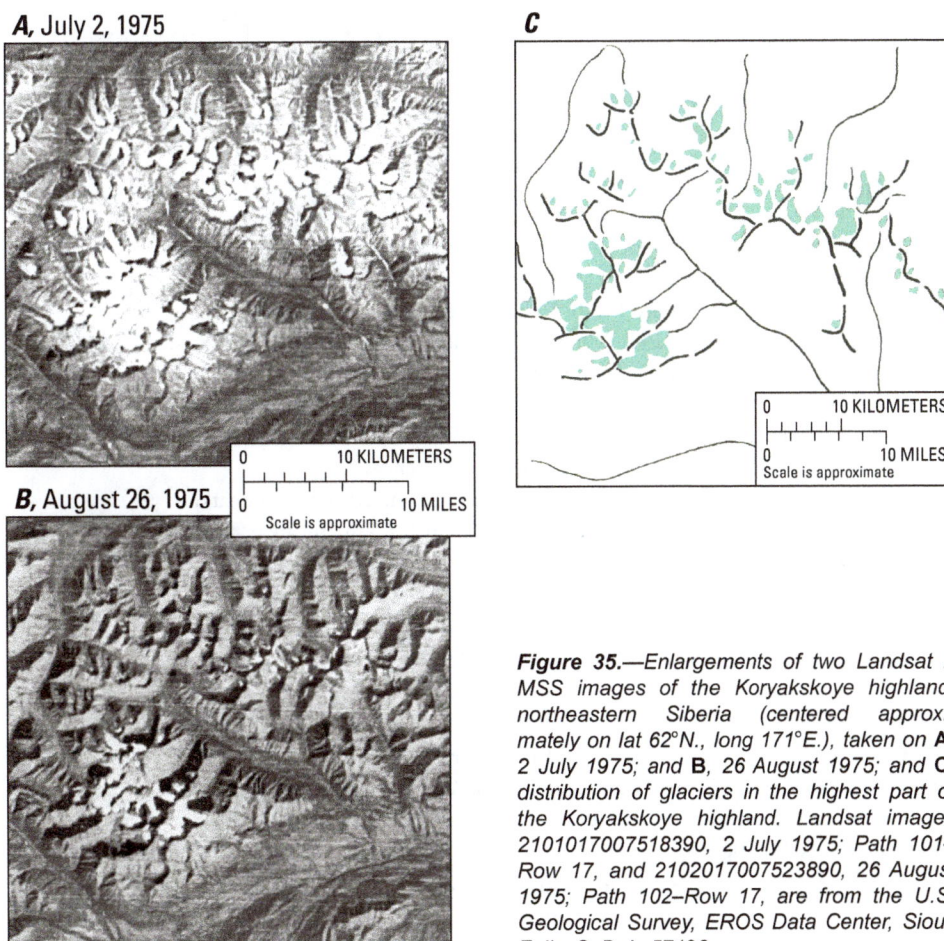

A, July 2, 1975

B, August 26, 1975

C

Figure 35.—Enlargements of two Landsat 2 MSS images of the Koryakskoye highland, northeastern Siberia (centered approximately on lat 62°N., long 171°E.), taken on **A**, 2 July 1975; and **B**, 26 August 1975; and **C**, distribution of glaciers in the highest part of the Koryakskoye highland. Landsat images 2101017007518390, 2 July 1975; Path 101–Row 17, and 2102017007523890, 26 August 1975; Path 102–Row 17, are from the U.S. Geological Survey, EROS Data Center, Sioux Falls, S. Dak. 57198.

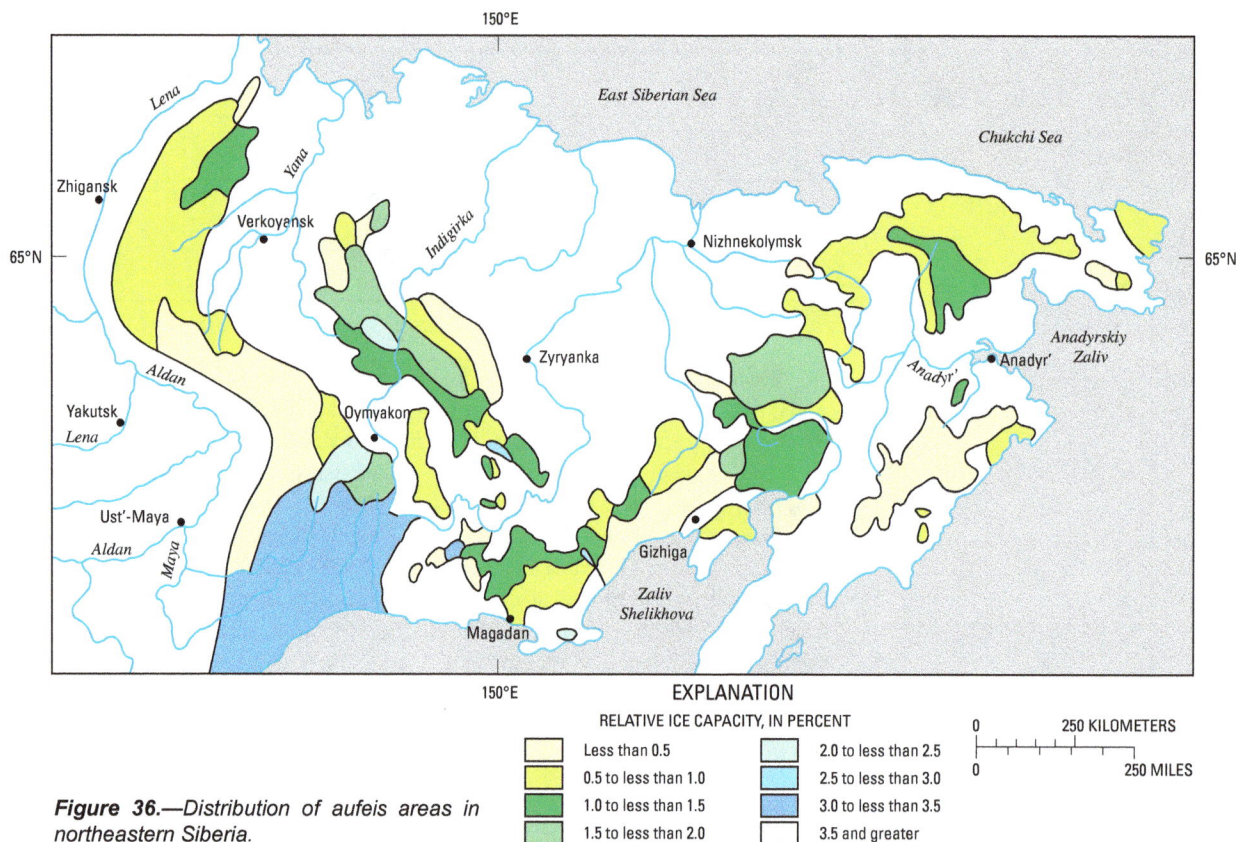

Figure 36.—*Distribution of aufeis areas in northeastern Siberia.*

Kamchatka

The Kamchatka region is located on the far eastern side of the Former Soviet Union (fig. 1). The presence of ice fields and firn fields on the volcanoes of Kamchatka has been known since the 18th century, and their characteristics had been noted by the end of the 19th century (Bogdanovich, 1899). Studies of the glaciers of Kamchatka began in the 20th century, and in the 1950s a general map of the glacierization of Kamchatka was compiled from available cartographic data (Ivan'kov, 1958). Later, glaciological investigations of Kamchatka were undertaken by the Institute of Volcanology of the USSR Academy of Sciences.

Inventories and interpretation of aerial photographs of the Kamchatka Peninsula revealed 405 glaciers that are greater than 0.1 km^2 in area; these glaciers have a total area of 874 km^2 (Vinogradov, 1968). The majority of glaciers are situated in the craters or calderas of, or on the flanks of, volcanoes, giving rise to the following specific morphological types (fig. 37):

Ice belts (fig. 37A) are confined to the highest and most active volcanoes. Snow and ice accumulate on steep (less than 30°), dissected slopes of the volcano, while glacier tongues drain the lower part of the ice belt.

Crater glaciers (fig. 37B) are located in the craters of dead, dormant, or active volcanoes (during their quiescent stage). Well-pronounced, ring-like crests of craters contribute to the identification of such glaciers on Vulkan Mutnovskaya Sopka (fig. 38).

Caldera glaciers are situated in calderas of large volcanoes. Usually they do not spill over the caldera rims containing them, and they have short outlet tongues.

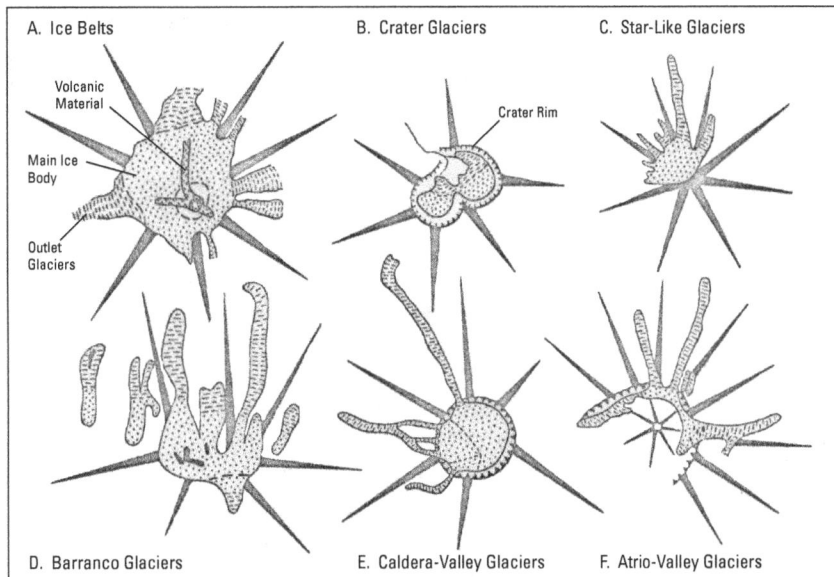

Figure 37.— *Morphological types of present-day glaciers in the volcanic regions of Kamchatka:* **A**, *Ice belt on Vulkan Klyuchevskaya Sopka;* **B**, *Crater glaciers on Vulkan Mutnovskaya Sopka;* **C**, *A star-like glacier on Vulkan Kronotskaya Sopka;* **D**, *Barranco glaciers on Vulkan Zhupanovskaya;* **E**, *Caldera-valley glaciers on* Vulkan Dal'niy Ploskiy; *and* **F**, *Atrio-valley glaciers on Vulkan Avachinskaya Sopka.*

Figure 38.—*Aerial photograph of glaciers in the crater and on the slopes of Vulkan Mutnovskaya Sopka (see fig. 37B).*

Star-like glaciers (fig. 37*C*) are situated in the near-summit area of dead or dormant volcanoes and have irregular boundaries and outlet tongues, mainly on slopes that have a northern exposure.

Barranco glaciers (fig. 37*D*) spread over recently active or dormant volcanoes. They are confined to the medial and upper parts of barrancos (steep slopes), and are usually on northern slopes. They are nourished by snow-avalanches.

Caldera-valley glaciers (fig. 37*E*) have their accumulation area in the caldera of volcanoes. They have long tongues. They are confined to the late Holocene calderas of dead, dormant, and active volcanoes; surge-type glaciers may be encountered among them.

Atrio-valley glaciers (fig. 37*F*) have their accumulation areas in large *atrio* (central parts) of calderas and ablation areas on the slopes of a volcanic somma. Glacier tongues drain the *atrio* in the form of ice-falls of several hundred meters.

Unfortunately, it is very difficult to identify Kamchatka glaciers on the available Landsat 1, 2, and 3 MSS images. Not many images have been acquired and most are cloudy or snow-covered. However, interpretation of snow-covered images can be of great interest, as they may reveal the impacts of present volcanism on glaciers. Kamchatka glaciers in the areas of active volcanism — on the Kluchevskaya and Avachinskaya group of volcanoes — differ greatly from glaciers occupying areas of Quarternary volcanism (Khrebet Sredinnyy) or non-volcanic areas (Kronotskiy Poluostrov) (Vinogradov, 1975). The morphology of these glaciers is also very different (Vinogradov and Glazyrin, 1979).

In areas of active volcanism, glaciers are mantled by the products (especially tephra) of volcanic eruptions. The thickness of the pyroclastic mantle increases from the equilibrium line to the glacier termini, and in certain areas it reaches 1 m thick. This mantle favors reduced ablation of ice and is seen more often than bare ice at the termini of glaciers. Because of this, it is difficult to identify glacier boundaries in volcanic areas from space images.

Russian Arctic[4]

The Russian Arctic consists of numerous islands and archipelagos north of mainland Russia, in the Barents and Kara Seas. This section highlights the glacierization on three of the more significant archipelagos — Franz Josef Land, Severnaya Zemlya, and Novaya Zemlya.

Franz Josef Land (Zemlya Frantsa-Iosifa)

This polar archipelago, whose existence was predicted by the Russian geographer N. Schilling (Schilling, 1865), was discovered in 1873 by the Austro-Hungarian Expedition of J. Payer and K. Weyprecht (Payer, 1876). By the time F. Nansen wintered there in 1895–1896 (Nansen, 1897), the general nature of glacierization of the archipelago had been determined. In 1931, the Swedish-Norwegian Expedition of H. Ahlmann (Ahlmann, 1933) undertook a phototheodolite survey of glacierized coasts of George Land (Ostrov Zemlya Georga). The present-day map of Franz Josef Land was compiled from aerial photography taken during 1952–1953. The map served as the basis for compiling the glacier inventory of the archipelago (USSR, Academy of Sciences, 1965 [1965–1983], Katalog lednikov SSSR, vol. 3, part 1), which,

---

[4]See also companion section in this part of the chapter, The Glaciers of the Russian High Arctic.

together with a monograph based on studies of Franz Josef Land during the IGY-IGC (Grosswald and others, 1973), is the main source of information on the glaciers of the archipelago.

Based on the nature of the glacierization, Franz Josef Land is subdivided into three regions: western, central, and eastern (fig. 39). In all parts of the archipelago, ice-cap complexes prevail (table 13). Present-day glaciers occur on 56 islands of Franz Josef Land. Glaciers cover approximately 85 percent of the archipelago. Their mean thickness is about 180 m, their area is 13,738.8 km^2, and their total ice volume is approximately 2,500 km^3. Glaciers cover 1,570 km, or approximately 60 percent of the total length of the 2,255 km coast line.

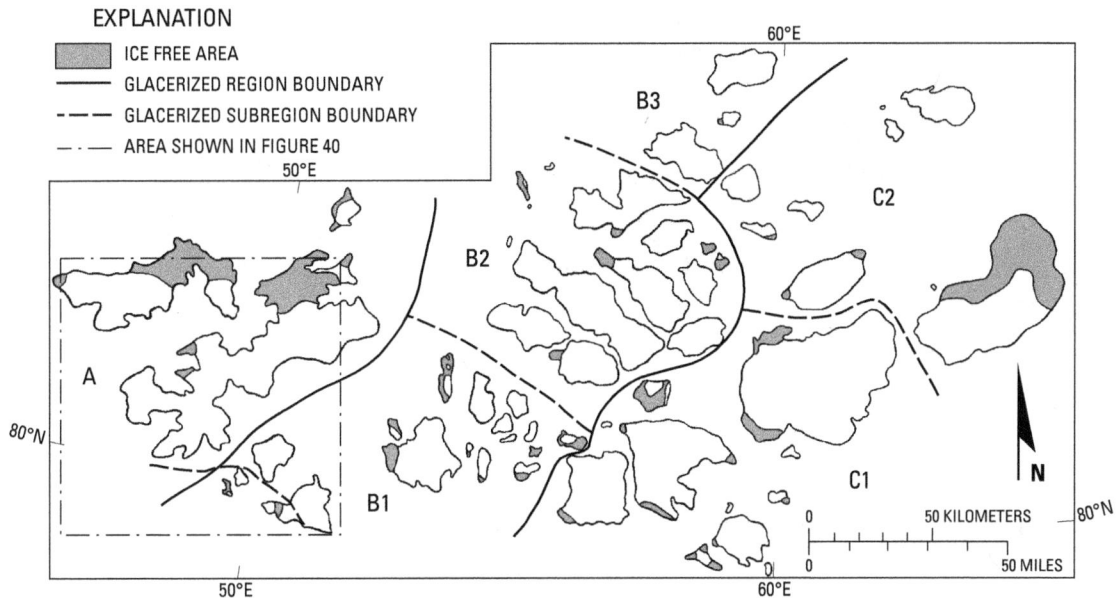

Figure 39.—*Glacierized regions of Franz Josef Land:* **A**, *western region;* **B**, *central region (1, 2, 3 — southern, central and northern areas);* **C**, *eastern region (1, 2 — southern and northern areas). The non-glacierized areas are shown in grey. The coverage of the Landsat image used for figure 40 is shown. See table 13 for glacierized areas.*

TABLE 13.—*The area of glaciers in Franz Josef Land*
[Unit: km^2, square kilometer]

Areas and subareas	Ice-cap complexes (km^2)	Ice domes (km^2)	Other glaciers (km^2)	Total glacier area (km^2)
Western region	3,085.3	3.0	17.9	3,106.2
Central region				
Southern area	1,266.1	8.4	12.4	1,286.9
Central area	2,782.3	4.6	17.3	2,804.2
Northern area	640.9	0.6	2.9	644.4
Eastern region				
Southern area	3,739.9	—	14.4	3,754.3
Northern area	2,109.5	30.1	3.2	2,142.8
Total archipelago	13,624.0	46.7	68.1	13,738.8

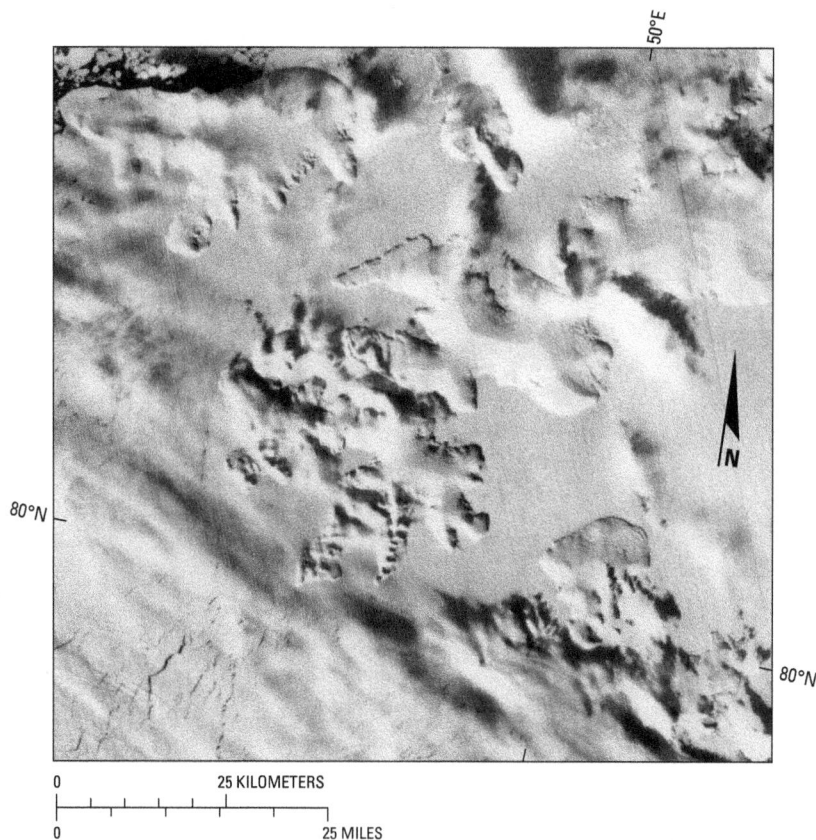

Figure 40.—*Landsat 3 MSS image of the western part of Franz Josef Land acquired on 24 March 1978. The Landsat image 30019-10411 (3222001007808390), Path 222– Row 1, is from the U.S. Geological Survey, EROS Data Center, Sioux Falls, S. Dak. 57198.*

Ice caps on islands, valley outlet glaciers, and small glaciers are the main morphological types of glaciers in Franz Josef Land. They make up four morphogenetic glacier complexes: (a) ice-cap complexes (these prevail in the eastern area), (b) ice-cap-transition complexes, (c) transition-piedmont complexes (these two complexes prevail in the central area), (d) ice plateau complexes (prevail in the western area). The amount of glacierization in the eastern and southeastern parts of archipelago and on isolated islands is, as a rule, higher than in the western and northwestern areas.

The Landsat 3 MSS image shown in figure 40 includes only the smallest, most western part of the archipelago. Because the image was acquired in winter, it is impossible to determine the equilibrium line on the glaciers, although the margins of glaciers on the land and their contact with sea ice can be traced quite distinctly. Ice divides can be seen well, and, in the majority of cases, outlet glaciers that have rough topography and zones of crevasses are easily identified.

Severnaya Zemlya

Severnaya Zemlya was discovered by a Russian Expedition in 1913 (Starokadomskiy and Barr, 1977) that mapped the eastern coast of the archipelago. The first map of the islands was compiled in 1930–1932 by the expedition of G.A. Ushakov and N.N. Urvantsev (Anonymous, 1933), who determined the extent of glaciers on the archipelago by means of over-land traverses. During the summer of 1931, an international expedition on board the Republic of Germany's LZ-127 airship *Graf Zeppelin* carried out aerial photography within the central part of the archipelago (Penck, 1931). A 1:400,000-scale map was compiled and published from these data, depicting some of the glaciers (Berson, 1933). For some areas, 1:25,000-scale maps

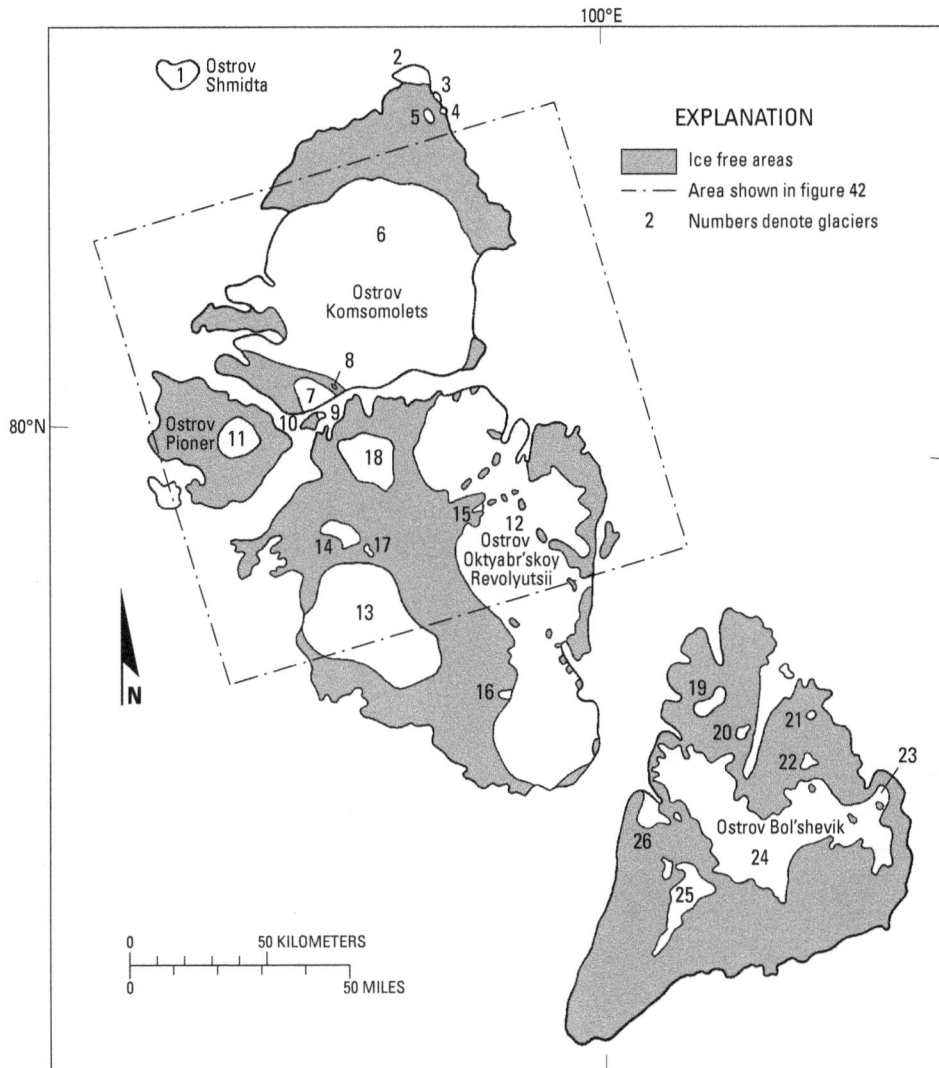

Figure 41.—*Glacierization of Severnaya Zemlya. Location of the Landsat 3 MSS image used for figure 42 is shown.*

TABLE 14.—*The area of glaciers on Severnaya Zemlya*

[Unit: km², square kilometer]

Island	Glacier numbers on fig. 41	Area (km²)
Shmidta	1	438.5
Komsomolets	2–8	6,171.6
Bol'shoy Isvestnyakoviy	9–10	6.1
Pioner	11	254.0
Oktyabr'skoy Revolyutsii	12–18	7,945.8
Bol'shevik	19–26	3,509.0
Total archipelago		18,325.0

were compiled. The map used for the compilation of glacier inventories (USSR, Academy of Sciences, 1981 [1965–1983], Katalog Lednikov SSSR, v. 10, iss. 1, pt. 1) was prepared from aerial photography acquired during the 1950s. Since 1962, glaciological expeditions from the Arctic and Antarctic Institute have been working on Severnaya Zemlya. A glaciological station was constructed on Lednik Vavilova in 1974 (Jania and Hagen, 1996).

Glaciers occupy 49.8 percent (18,325 km²) (fig. 41, table 14) of the 36,788 km² total area of the archipelago. The average thickness of glaciers is 300 m, and the ice volume is about 5,500 km³. Severnaya Zemlya contains 17 glacier complexes (occupying nearly 94 percent of the total glacierized area), 16 isolated ice domes, and 46 mountain glaciers. The extent of glacierization increases from south to north: on Ostrov Bol'shevik (island), glaciers cover 31 percent of the area; on Ostrov Oktyabr'skoy Revolutsii, glaciers occupy 58 percent of the area; and on Ostrov Komsomolets, glaciers occupy 68.5 percent of the area. The length of glacierized coasts on these three islands makes up 1.6, 13.2 and 35.3 percent, respectively, and on Ostrov Shmidta, 96.1 percent of the total length of the coast line is glacierized.

The Landsat 1 MSS image used for figure 42 was taken in July — a favorable time; however, its interpretation was hampered by the spotty snow cover in the central part of Ostrov Bol'shevik. The newly fallen snow graphically emphasizes the hydrological network and geological structure, but at the same time obscures the margins of the glaciers. The sea ice in Proliv Shokal'skogo (strait), between Ostrova Bol'shevik and Oktyabr'skoy Revolutsii, is another source of difficulty, making it impossible to delineate the termini of small tidewater glaciers. The snow also makes it impossible to determine the equilibrium line of glaciers; ablation areas can only be seen on some of the outlet glaciers draining the ice cap on Ostrov Komsomolets.

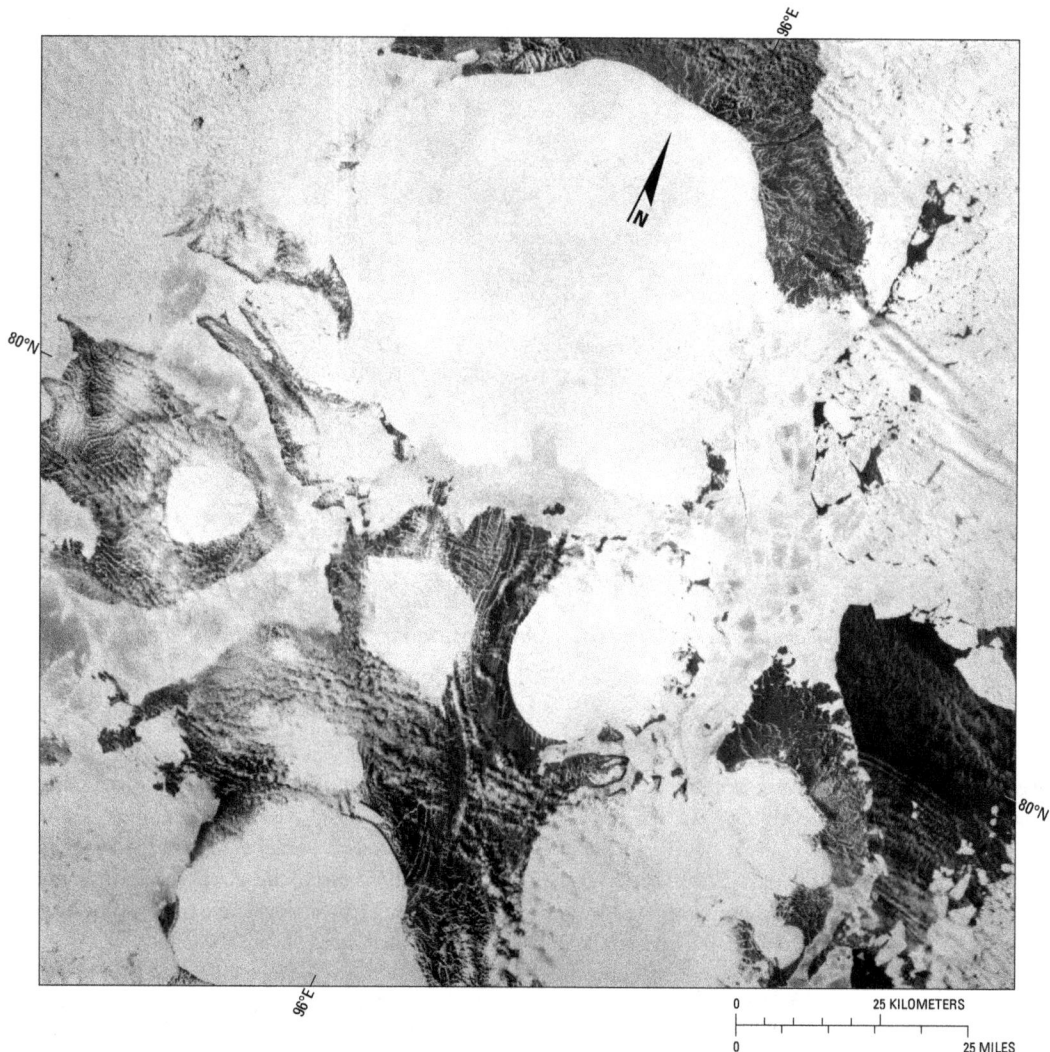

Figure 42.—Landsat 1 MSS image of the northern part of Severnaya Zemlya taken on 28 July 1973. The Landsat image 1362-07383 (1189001007320190), Path 189–Row 1, is from the U.S. Geological Survey, EROS Data Center, Sioux Falls, S. Dak. 57198.

Novaya Zemlya

Although the mapping of the coastal area of Novaya Zemlya was accomplished as early as the 19th century, the first map with a schematic presentation of glaciers on *Ostrov Severniy*, the large northern island, was compiled by V.A. Rusanov in 1910. In 1913, the glaciers were mapped by an expedition led by G.Ya. Sedov (Chizhov and others, 1968). In the 1920s and 1930s, ground surveys were done by an Arctic Research Institute expedition. Some glaciers on western coasts were mapped by the Norwegian expedition to Novaya Zemlya in 1921 (Holtedahl, 1924, 1928, 1930). The glaciers of the eastern coasts were surveyed by means of aerial photography acquired in 1931, during the flight of the Republic of Germany's LZ-127 *Graf Zeppelin* airship (Penck, 1931). A map based on these aerial photographs was published in the proceedings of the expedition (Berson, 1933). A more recent map was compiled in the 1950s from subsequent aerial photography. This map served as the basis for the Glacier Inventory of the archipelago (USSR, Academy of Sciences, 1978 [1965–1983], Katalog Lednikov SSSR, v. 3, pt. 2). The inventory and a monograph, based on studies during the IGY-IGC (Chizhov and others, 1968), are the main sources of information on the glaciers of Novaya Zemlya.

The total area of glacierization on Novaya Zemlya is 24,413 km^2, the largest glacierized area in the Russian Arctic, and the ice volume is estimated to be about 6,800 km^3. All of the known types of glaciers exist here. The submeridional position of Novaya Zemlya produces a gradual increase in extent of glacierization from south to north and makes it possible to distinguish four areas based on the type of glacier located in each area: small glaciers, mountain glaciers, transitional glaciers, and ice caps (fig. 43, table 15).

The prevailing western transport of moisture-laden Atlantic air masses produces large areas of glacierization along the coast of the Barents Sea. Here, the amount of glacierization is 1.5 times greater, and the equilibrium line lies 200 m lower than on the coast of the Kara Sea. The ice divide is also shifted eastward in the direction of the Kara Sea.

The Landsat 1 MSS image shown in figure 44 characterizes the central part of the ice cap on *Ostrov Severniy* of Novaya Zemlya. The system of outlet glaciers reaching the sea can be seen well. The margin of the ice cap is very distinct. On the western coast, it is sometimes difficult to determine the position of glacier termini in the sea because of the presence of fast ice. The Landsat 1 MSS image spectacularly demonstrates the asymmetry of the ice-cap regime that has been induced climatically. The area of ablation in the Kara Sea basin is much larger than in the Barents Sea basin.

Conclusions

The preceding discussion has shown that Landsat 1, 2, and 3 MSS images have been used successfully in some areas to study the glaciers of the Former Soviet Union. In other areas, due to the small size of the glaciers, the spatial resolution of the satellite scanner, or the time of image acquistion, the images are less useful. Table 16 and figure 45 list the optimum Landsat images of the Former Soviet Union that were acquired during the decadal Landsat baseline period (1972–1981).

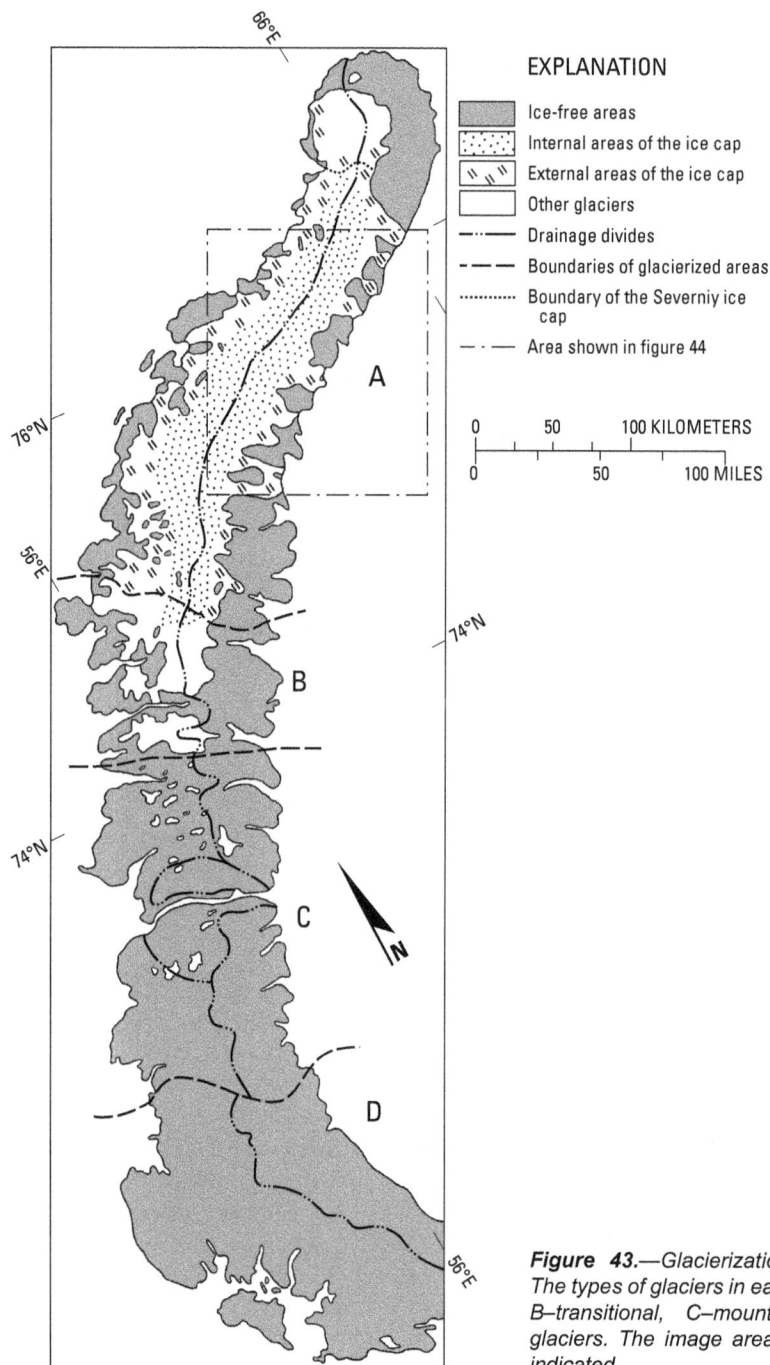

EXPLANATION

Ice-free areas
Internal areas of the ice cap
External areas of the ice cap
Other glaciers
—··— Drainage divides
——— Boundaries of glacierized areas
········ Boundary of the Severniy ice cap
—·—· Area shown in figure 44

Figure 43.—Glacierization of Novaya Zemlya. The types of glaciers in each area are: A–ice caps, B–transitional, C–mountain glaciers, D–small glaciers. The image area shown on figure 44 is indicated.

Acknowledgments

We want to gratefully acknowledge the contributions of the following scientists to this part of the chapter, who assisted in the analysis of space images and/or photographs, aerial photographs, and in the preparation of scientific illustrations and maps of the glacierized regions indicated in parentheses after their names: V.G. Loseva (Caucasus and Altay), K.P. Rototayev and N.M. Zverkova (Caucasus), V.I. Kravtsova and N.F. Chaikina (Altay), and M.M. Koreisha (Siberia).

TABLE 15.—*The glaciers on Novaya Zemlya*

[Unit: km², square kilometer]

Glacier type	Number of glaciers	Total area (km²)	Glacier type	Number of glaciers	Total area (km²)
AREA OF ICE CAPS			AREA OF MOUNTAIN GLACIERS		
Ice cap	1	18,911.3	Mountain	393	1,263.1
Mountain	73	300.1	Small	417	119.4
Small	611	398.9	**Total**	810	1,382.5
Total	685	19,610.3	AREA OF SMALL GLACIERS		
AREA OF TRANSITIONAL GLACIERIZATION			Small	771	204.4
Transitional forms	8	2,866.1	EXTENT OF GLACIERIZATION (A) AND LENGTH OF GLACIER FRONTS (B), IN PERCENT		
Mountain	94	296.1	Glacierization zone	A	B
Small	208	53.6	Ice cap	65.8	14.1
Total	310	3,215.8	Transitional	41.5	2.0
			Mountain	11.0	—
			Small glaciers	2.0	—

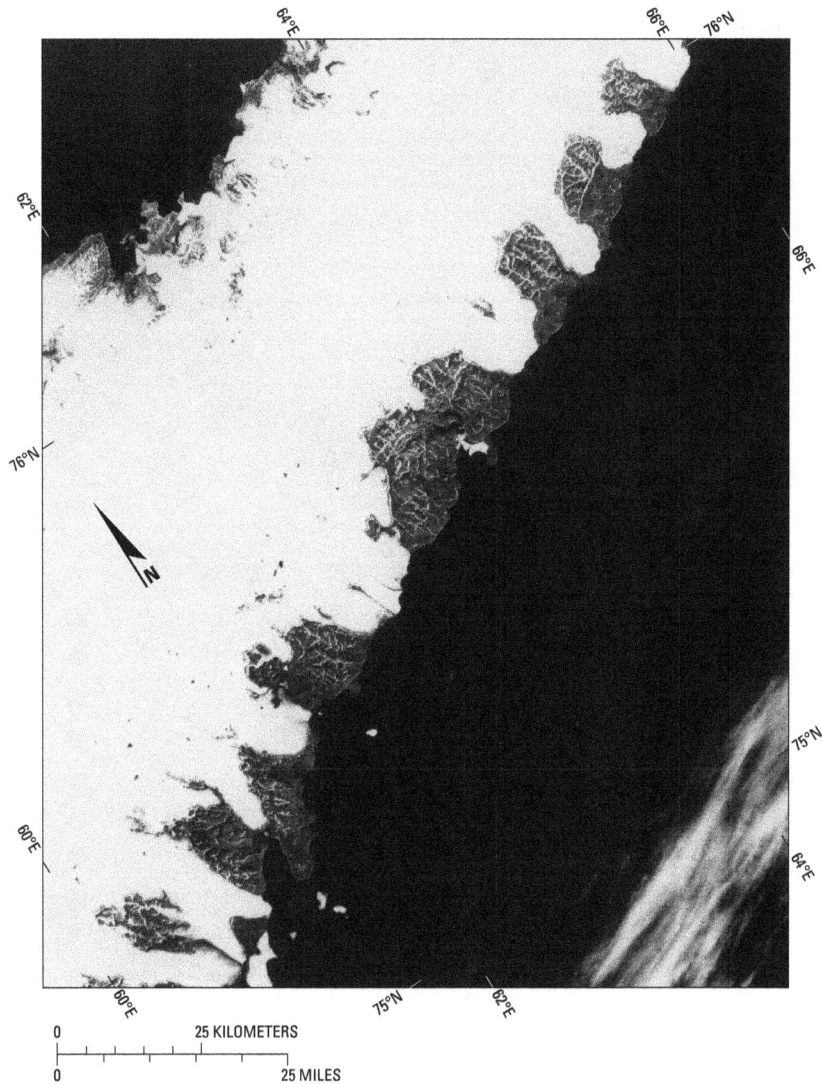

Figure 44.—*Landsat 1 MSS image of the central part of* Novozemel'skiy *ice cap taken on 23 July 1973. Landsat image 1365-07574 (1192006007320490), Path 192–Row 6, is from the U.S., Geological Survey, EROS Data Center, Sioux Falls, S. Dak. 57198.*

TABLE 16.—*Optimum Landsat 1, 2, and 3 MSS images of glaciers of the Former Soviet Union*
[Code column indicates usability for glacier studies; see figure 45 for explanation]

Path–row	Nominal scene-center latitude and longitude		Landsat identification number	Date	Code	Cloud cover (percent)	Remarks
CAUCASUS							
180–31	41°40′N.	48°56′E.	2180031007620490	22 Jul 76	◑	30	
180–32	40°14′N.	48°26′E.	2180032007620490	22 Jul 76	◑	30	
181–31	41°40′N.	47°30′E.	2181031007624190	28 Aug 76	●	0	
182–30	43°05′N.	46°36′E.	2182030007819580	14 Jul 78	◑	50	
182–31	41°40′N.	46°04′E.	2182031007819590	14 Jul 78	◑	20	
183–30	43°05′N.	45°10′E.	2183030007723790	25 Aug 77	●	0	Gora Kazbek
183–31	41°40′N.	44°38′E.	2183031007723790	25 Aug 77	●	0	
184–30	43°05′N.	43°44′E.	2184030007725690	13 Sep 77	◑	15	Gora Kazbek
185–30	43°05′N.	42°18′E.	2185030007619190	09 Jul 76	◑	15	Gora El'bruz
186–29	44°30′N.	41°24′E.	3186029008018090	28 Jun 80	●	10	
186–30	41°40′N.	40°20′E.	2186030007718690	05 Jul 77	◑	70	
187–29	44°30′N.	39°58′E.	3187029008122090	08 Aug 81	◕	10	
PAMIRS AND ALAYSKIY KHREBET							
162–32	40°14′N.	74°15′E.	1162032007227090	26 Sep 72	●	0	Fresh snow
162–33	38°49′N.	73°46′E.	1162033007225290	08 Sep 72	◐	10	
162–34	37°24′N.	73°18′E.	2162034007622290	09 Aug 76	●	0	
163–32	40°14′N.	72°49′E.	2163032007526490	21 Sep 75	◑	35	Fresh snow
163–33	38°49′N.	72°20′E.	2163033007727190	28 Sep 77	●	0	Lednik Fedshenko
163–34	37°24′N.	71°52′E.	2163034007622390	10 Aug 76	●	5	
164–32	40°14′N.	71°23′E.	2164032007720090	19 Jul 77	◑	15	
164–33	38°49′N.	70°54′E.	2164033007225490	10 Sep 72	◑	20	
165–32	40°14′N.	69°57′E.	2165032007521290	31 Jul 75	●	0	
165–33	38°49′N.	69°28′E.	2165033007526690	23 Sep 75	●	5	
166–32	40°14′N.	68°31′E.	1166032007227490	30 Sep 72	●	0	
166–33	38°49′N.	68°02′E.	2166033007720290	21 Jul 77	●	5	
TIEN SHAN							
158–30	43°05′N.	81°01′E.	2158030007723090	18 Aug 77	●	0	
158–31	41°40′N.	80°30′E.	2158031007629090	16 Oct 76	●	0	
158–31	41°40′N.	80°30′E.	2158031007723090	18 Aug 77	●	0	
159–30	43°05′N.	79°35′E.	2159030007522490	12 Aug 75	●	0	
159–31	41°40′N.	79°04′E.	2159031007522490	12 Aug 75	●	0	
159–31	41°40′N.	79°04′E.	1159031007326190	18 Sep 73	●	0	
160–30	43°05′N.	78°09′E.	2160030007522590	13 Aug 75	●	0	
160–31	41°40′N.	77°37′E.	2160031007522590	13 Aug 75	●	0	
160–32	40°14′N.	77°07′E.	2160032007522590	13 Aug 75	●	0	
161–30	43°05′N.	76°43′E.	2161030007622190	08 Aug 76	●	0	
161–31	41°40′N.	76°11′E.	1161031007225190	07 Sep 72	◐	10	
161–32	40°14′N.	75°41′E.	2161032007529890	25 Oct 75	●	5	Snow
162–30	43°05′N.	75°17′E.	2162030007727090	27 Sep 77	●	0	

TABLE 16.—*Optimum Landsat 1, 2, and 3 MSS images of glaciers of the Former Soviet Union—Continued*

[Code column indicates usability for glacier studies; see figure 45 for explanation]

Path–row	Nominal scene-center latitude and longitude		Landsat identification number	Date	Code	Cloud cover (percent)	Remarks
162–31	41°40′N.	74°45′E.	2162031007727090	27 Sep 77	◑	15	
162–32	40°14′N.	74°15′E.	1162032007227090	26 Sep 72	●	0	Snow
163–30	43°05′N.	73°51′E.	2163030007723590	23 Aug 77	●	0	
163–31	41°40′N.	73°19′E.	2163031007522890	16 Aug 75	●	0	
164–30	43°05′N.	72°25′E.	2164030007723690	24 Aug 77	●	0	
164–31	41°40′N.	71°53′E.	2164031007720090	19 Jul 77	◑	20	
165–31	41°40′N.	70°27′E.	2165031007723790	25 Aug 77	●	0	
ZHONGGHAR ALATAÜ ZHOTASY							
158–29	44°30′N.	81°34′E.	2158029007723090	18 Aug 77	●	0	
159–29	44°30′N.	80°08′E.	2159029007723190	19 Aug 77	◕	10	
160–29	44°30′N.	78°42′E.	2160029007522590	13 Aug 75	●	0	
RUSSIAN ALTAY							
155–25	50°09′N.	88°17′E.	1155025007224590	01 Sep 72	●	0	
155–26	48°44′N.	87°38′E.	1155026007224590	01 Sep 72	●	0	
156–25	50°09′N.	86°50′E.	1156025007405590	24 Feb 74	◑	NA	Snow
157–25	50°09′N.	85°24′E.	3157025007922890	16 Aug 79	●	0	
157–26	48°44′N.	84°46′E.	3157026007922890	16 Aug 79	●	0	
SAÜYR ZHOTASY							
156–27	47°20′N.	85°35′E.	2156027007518590	04 Jul 75	●	0	
156–27	47°20′N.	85°35′E.	2156027007621690	03 Aug 76	◕	10	
KUZNETSKIY ALATAU							
156–21	55°45′N.	89°47′E.	No image		○		
156–22	54°22′N.	88°59′E.	2156022007619890	16 Jul 76	◑	35	
157–21	55°45′N.	88°21′E.	3157021007921090	29 Jul 79	●	0	
157–22	54°22′N.	87°33′E.	3157022007921090	29 Jul 79	●	0	
SAYAN MOUNTAINS							
148–23	52°58′N.	99°42′E.	1148023007330490	31 Oct 73	●	0	Snow
149–22	54°22′N.	99°01′E.	3149022007820790	26 Jul 78	◕	10	
150–22	54°22′N.	97°35′E.	215022007516190	10 Jun 75	◑	35	Cloudy and Snow
150–23	52°58′N.	96°50′E.	215023007516190	10 Jun 75	◑	50	Cloudy and Snow
KHREBET KODAR							
137–20	57°08′N.	117°53′E.	No usable image		◕		
138–20	57°08′N.	116°27′E.	No usable image		◕		
GORA KHIBINY							
202–12	67°59′N.	34°24′E.	No image		○		
203–12	67°59′N.	32°58′E.	No usable image		◕		
URAL MOUNTAINS							
179–12	67°59′N.	67°23′E.	1179012007306590	06 Mar 73	◕	10	
179–13	66°40′N.	65°45′E.	1179013007306590	06 Mar 73	◑	20	
179–14	65°20′N.	64°16′E.	1179014007306590	06 Mar 73	◑	20	
180–12	67°59′N.	65°57′E.	No image		○		

TABLE 16.—*Optimum Landsat 1, 2, and 3 MSS images of glaciers of the Former Soviet Union*—Continued

[Code column indicates usability for glacier studies; see figure 45 for explanation]

Path–row	Nominal scene-center latitude and longitude	Landsat identification number	Date	Code	Cloud cover (percent)	Remarks
180–13	66°40′N. 64°19′E.	1180013007317490	23 Jun 73	◑	60	
180–14	65°20′N. 62°50′E.	1180014007322890	16 Aug 73	●	0	
180–15	63°59′N. 61°29′E.	1180015007322890	16 Aug 73	●	0	
180–16	62°38′N. 60°16′E.	1180016007322890	16 Aug 73	●	0	
181–12	67°59′N. 64°31′E.	1181012007317590	24 Jun 73	◑	70	
181–13	66°40′N. 62°52′E.	1181013007317590	24 Jun 73	◑	60	
181–14	65°20′N. 61°24′E.	1181014007317590	24 Jun 73	◑	70	
181–15	63°59′N. 60°03′E.	1181015007317590	24 Jun 73	◑	90	
182–14	65°20′N. 59°58′E.	1182014007317690	25 Jun 73	●	0	
182–15	63°59′N. 58°37′E.	1182015007317690	25 Jun 73	◕	10	
PLATO PUTORANA						
157–12	67°59′N. 98°56′E.	1157012007324190	29 Aug 73	◑	50	
157–13	66°40′N. 97°18′E.	1157013007324190	29 Aug 73	◑	40	
158–11	69°17′N. 99°19′E.	1158011007324290	30 Aug 73	◕	10	
158–12	67°59′N. 97°30′E.	1158012007324290	30 Aug 73	◑	40	
158–13	66°40′N. 95°52′E.	1158013007324290	30 Aug 73	◑	40	
159–11	69°17′N. 93°53′E.	1159011007320790	26 Jul 73	◑	20	Snow
159–12	67°59′N. 96°04′E.	No usable image		◔		
159–13	66°40′N. 94°26′E.	No usable image		◔		
160–11	69°17′N. 96°27′E.	No usable image		◔		
160–12	67°59′N. 94°38′E.	1160012007322690	14 Aug 73	◑	40	Not seen
160–13	66°40′N. 93°00′E.	1160013007322690	14 Aug 73	◑	20	Not seen
161–11	69°17′N. 95°01′E.	No image		○		
161–12	67°59′N. 93°12′E.	1161012007322790	15 Aug 73	◑	30	
161–13	66°40′N. 91°34′E.	1161013007322790	15 Aug 73	◑	30	
162–11	69°17′N. 93°35′E.	No image		○		
162–12	67°59′N. 91°46′E.	No image		○		
163–11	69°17′N. 92°09′E.	No image		○		
164–11	69°17′N. 90°43′E.	No image		○		
GORY BYRRANGA						
158–6	75°25′N. 112°39′E.	1158006007320690	25 Jul 73	◕	10	
159–6	75°25′N. 111°13′E.	1159006007320790	26 Jul 73	◑	60	
160–6	75°25′N. 109°47′E.	No image		○		
161–6	75°25′N. 108°21′E.	No image		○		
162–6	75°25′N. 106°55′E.	No image		○		
163–6	75°25′N. 105°29′E.	No usable image		◔		
164–6	75°25′N. 104°03′E.	No usable image		◔		
KHREBET ORULGAN						
135–13	66°40′N. 128°51′E.	No image		○		
135–14	65°20′N. 127°22′E.	No image		○		

TABLE 16.—*Optimum Landsat 1, 2, and 3 MSS images of glaciers of the Former Soviet Union—Continued*

[Code column indicates usability for glacier studies; see figure 45 for explanation]

Path–row	Nominal scene-center latitude and longitude		Landsat identification number	Date	Code	Cloud cover (percent)	Remarks
136–12	67°59′N.	129°03′E.	1136012007323890	26 Aug 73	◐	70	
136–13	66°40′N.	127°25′E.	1136013007323890	26 Aug 73	◐	80	
136–14	65°20′N.	125°56′E.	1136014007323890	26 Aug 73	◐	20	
137–12	67°59′N.	127°37′E.	1137012007323990	26 Aug 73	◐	70	
137–13	66°40′N.	125°59′E.	1137013007314990	29 May 73	◐	20	Snow
KHREBET SUNTAR–KHAYATA							
122–16	62°38′N.	143°27′E.	1122016007230290	28 Oct 72	●	0	Snow
122–17	61°16′N.	142°19′E.	1122017007230290	28 Oct 72	●	0	Snow
123–16	62°38′N.	142°01′E.	1123016007220790	26 Jul 72	◐	50	Not seen
123–17	61°16′N.	140°53′E.	1123017007220790	26 Jul 72	◐	20	Not seen
124–16	62°38′N.	140°35′E.	1124016007228690	12 Oct 72	◐	20	Snow
124–17	61°16′N.	139°27′E.	1124017007228690	12 Oct 72	◐	20	Not seen
125–16	62°38′N.	139°09′E.	1125016007319190	10 Jul 73	◐	40	Not seen
125–17	61°16′N.	138°01′E.	1125017007324590	02 Sep 73	◐	20	Not seen
KHREBET CHERSKOGO							
122–14	65°20′N.	146°01′E.	1122014007230290	28 Oct 72	●	0	Snow
123–13	66°40′N.	146°04′E.	1123013007308190	22 Mar 73	●	0	Snow
123–14	65°20′N.	144°35′E.	1123014007308190	22 Mar 73	●	0	Snow
124–13	66°40′N.	144°38′E.	1124013007228690	12 Oct 72	◐	20	Snow
124–14	65°20′N.	143°09′E.	1124014007228690	12 Oct 72	◐	30	Snow
125–13	66°40′N.	143°12′E.	1125013007319190	10 Jul 73	◐	40	Not seen
125–14	65°20′N.	141°43′E.	1125014007319190	10 Jul 73	◐	30	Not seen
126–13	66°40′N.	141°46′E.	1126013007308490	25 Mar 73	●	0	Snow
126–14	65°20′N.	140°17′E.	1126014007308490	25 Mar 73	●	0	Snow
127–13	66°40′N.	140°20′E.	1127013007308590	26 Mar 73	●	0	Snow
127–14	65°20′N.	138°51′E.	1127014007308590	26 Mar 73	●	0	Snow
128–12	67°59′N.	140°32′E.	1128012007225490	10 Sep 72	◐	15	Snow
128–13	66°40′N.	138°54′E.	1128013007225490	10 Sep 72	◐	20	Snow
128–14	65°20′N.	137°25′E.	1128014007321290	31 Jul 73	◐	70	
129–12	67°59′N.	139°06′E.	1129012007227390	29 Sep 72	◐	60	Snow
129–13	66°40′N.	137°27′E.	1129013007227390	29 Sep 72	◐	20	Snow
130–12	67°59′N.	137°40′E.	1130012007307090	11 Mar 73	●	0	Snow
KORYAKSKOYE NAGOR'YE (UPLAND)							
101–16	62°38′N.	173°34′E.	1101016007426990	26 Sep 74	◐	30	Snow
101–17	61°16′N.	172°26′E.	1101017007322090	08 Aug 73	◐	70	
101–17	61°16′N.	172°26′E.	2101017007518390	02 Jul 75	◐	20	
102–16	62°38N.	172°08′E.	1102016007421690	04 Aug 74	◐	30	
102–17	61°16′N.	171°00′E.	1102017007421690	04 Aug 74	◐	30	
102–17	61°16′N.	171°00′E.	2102017007523890	26 Aug 75	◐	30	
103–17	61°16′N.	169°34′E.	1103017007409190	01 Apr 74	●	0	Snow

TABLE 16.—*Optimum Landsat 1, 2, and 3 MSS images of glaciers of the Former Soviet Union—Continued*

[Code column indicates usability for glacier studies; see figure 45 for explanation]

Path–row	Nominal scene-center latitude and longitude		Landsat identification number	Date	Code	Cloud cover (percent)	Remarks
KAMCHATKA							
106–20	57°08′N.	162°21′E.	2106020007524290	30 Aug 75	◐	50	
106–21	55°45′N.	161°30′E.	2106021007609390	02 Apr 76	◐	60	Snow
107–19	58°31′N.	161°49′E.	No usable image		◔		
107–20	57°08′N.	160°55′E.	No image		○		
107–21	55°45′N.	160°04′E.	1107021007331690	12 Nov 73	●	0	Snow
107–22	54°22′N.	159°16′E.	1107022007302890	28 Jan 73	◐	50	Snow
107–23	52°58′N.	158°31′E.	1107023007225090	06 Sep 72	◐	70	
107–24	51°34′N.	157°48′E.	1107024007232290	17 Nov 72	◐	20	Snow
108–19	58°31′N.	160°23′E.	No usable image		◔		
108–20	57°08′N.	159°29′E.	1108020007304890	17 Feb 73	●	0	Snow
108–21	55°45′N.	158°38′E.	1108021007225290	08 Sep 72	◐	80	
108–21	55°45′N.	158°38′E.	1108021007304890	17 Feb 73	●	0	Snow
108–22	54°22′N.	157°50′E.	1108022007225290	08 Sep 72	◐	80	
108–22	54°22′N.	157°50′E.	1108022007304890	17 Feb 73	◐	30	Snow
108–23	52°58′N.	157°05′E.	1108023007225290	08 Sep 72	◕	10	
108–24	51°34′N.	156°22′E.	1108024007225290	08 Sep 72	◕	10	
109–19	58°31′N.	158°57′E.	No image		○		
109–20	57°08′N.	158°03′E.	3109020007923390	21 Aug 79	◐	70	
109–21	55°45′N.	157°12′E.	3109021007923390	21 Aug 79	◐	80	
109–22	54°22′N.	156°24′E.	3109022007923390	21 Aug 79	◐	50	
FRANZ JOSEF LAND							
210–1	80°01′N.	65°28′E.	No image		○		
211–1	80°01′N.	64°02′E.	No image		○		
212–1	80°01′N.	62°35′E.	No image		○		
213–1	80°01′N.	61°09′E.	No image		○		
214–1	80°01′N.	59°43′E.	No image		○		
215–1	80°01′N.	58°17′E.	No image		○		
216–1	80°01′N.	56°51′E.	No image		○		
217–1	80°01′N.	55°25′E.	No image		○		
218–1	80°01′N.	53°59′E.	No image		○		
219–1	80°01′N.	52°33′E.	No image		○		
220–1	80°01′N.	51°07′E.	3220001197813790	17 May 78	◐	90	
221–1	80°01′N.	49°41′E.	No image.		○		
222–1	80°01′N.	48°15′E.	No image		○		
223–1	80°01′N.	46°49′E.	No image		○		
224–1	80°01′N.	45°23′E.	No image		○		
225–1	80°01′N.	43°57′E.	3225001197817690	25 Jun 78	◐	60	

TABLE 16.—*Optimum Landsat 1, 2, and 3 MSS images of glaciers of the Former Soviet Union—Continued*

[Code column indicates usability for glacier studies; see figure 45 for explanation]

Path–row	Nominal scene-center latitude and longitude		Landsat identification number	Date	Code	Cloud cover (percent)	Remarks
SEVERNAYA ZEMLYA							
171–3	78°29'N.	107°59'E.	1171003197412490	04 May 74	◑	80	
172–3	78°29'N.	106°33'E.	No image		○		
173–3	78°29'N.	105°07'E.	No image		○		
174–3	78°29'N.	103°40'E.	No image		○		
175–2	79°19'N.	108°29'E.	1175002197320690	25 Jul 73	◐	10	
175–3	78°29'N.	102°14'E.	1175003197320690	25 Jul 73	◐	10	Line drops
176–2	79°19'N.	107°03'E.	1176002197320690	25 Jul 73	◐	10	
176–3	78°29'N.	100°48'E.	1176003197320690	25 Jul 73	●	0	
177–2	79°19'N.	105°37'E.	No image		○		
177–3	78°29'N.	99°22'E.	1177003197322590	13 Aug 73	◑	70	
178–2	79°19'N.	104°11'E.	No image		○		
178–3	78°29'N.	97°56'E.	No image		○		
179–2	79°19'N.	102°45'E.	No image		○		
179–3	78°29'N.	96°30'E.	No image		○		
180–2	79°19'N.	101°19'E.	1180002197519290	11 Jul 75	◑	60	Line drops
180–3	78°29'N.	95°04'E.	No image		○		
181–2	79°19'N.	99°53'E.	No image		○		
182–2	79°19'N.	98°27'E.	No image		○		
183–2	79°19'N.	97°01'E.	No image		○		
184–2	79°19'N.	95°35'E.	2184002197610190	10 Apr 76	●	10	Snowy
185–1	80°01'N.	101°19'E.	No image		○		
185–2	79°19'N.	94°09'E.	No image		○		
186–1	80°01'N.	99°53'E.	No image		○		
186–2	79°19'N.	92°43'E.	No image		○		
187–1	80°01'N.	98°27'E.	No image		○		
187–2	79°19'N.	91°17'E.	No image		○		
188–1	80°01'N.	97°01'E.	1188001197320190	20 Jul 73	◑	40	
189–1	80°01'N.	95°35'E.	1189001197320190	20 Jul 73	◑	40	
190–1	80°01'N.	94°09'E.	No image		○		
191–1	80°01'N.	92°43'E.	1191001197520390	22 Jul 75	◐	10	Line drops
192–1	80°01'N.	91°17'E.	No image		○		
193–1	80°01'N.	89°51'E.	No image		○		

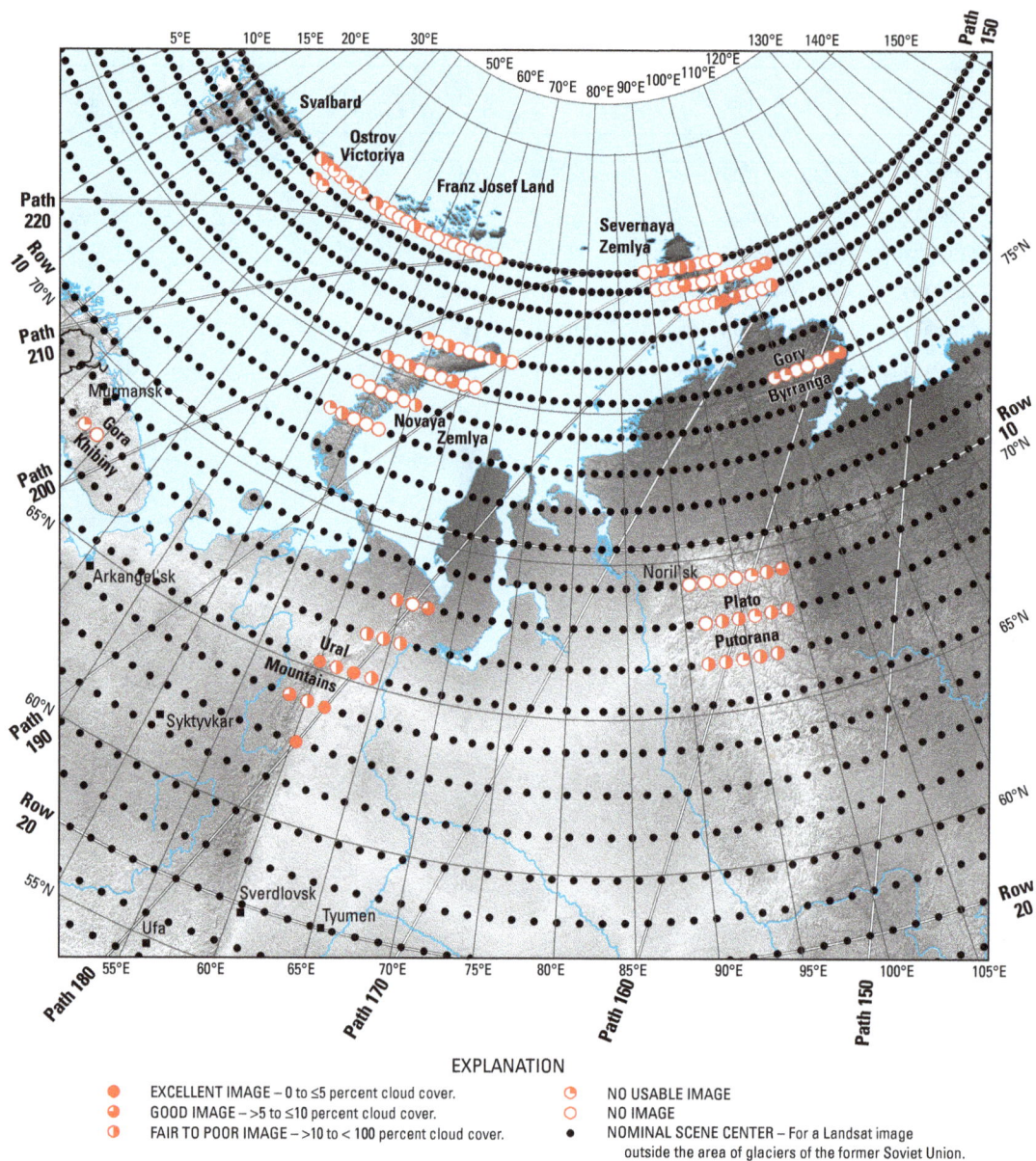

Figure 45.—*Index map of the optimum Landsat 1, 2, and 3 MSS images of the Former Soviet Union.*

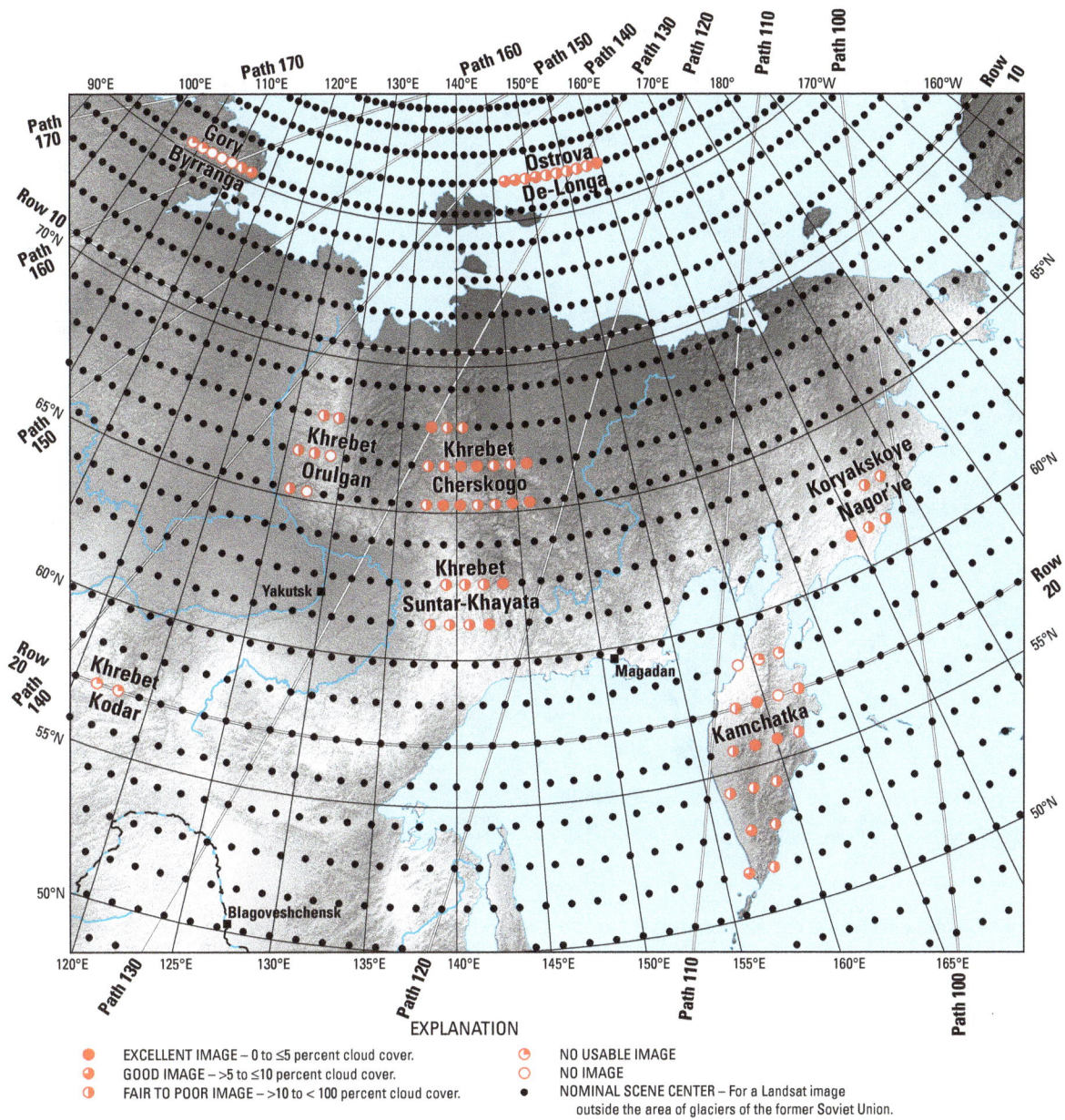

Figure 45.—*Index map of the optimum Landsat 1, 2, and 3 MSS images of the Former Soviet Union. (Continued)*

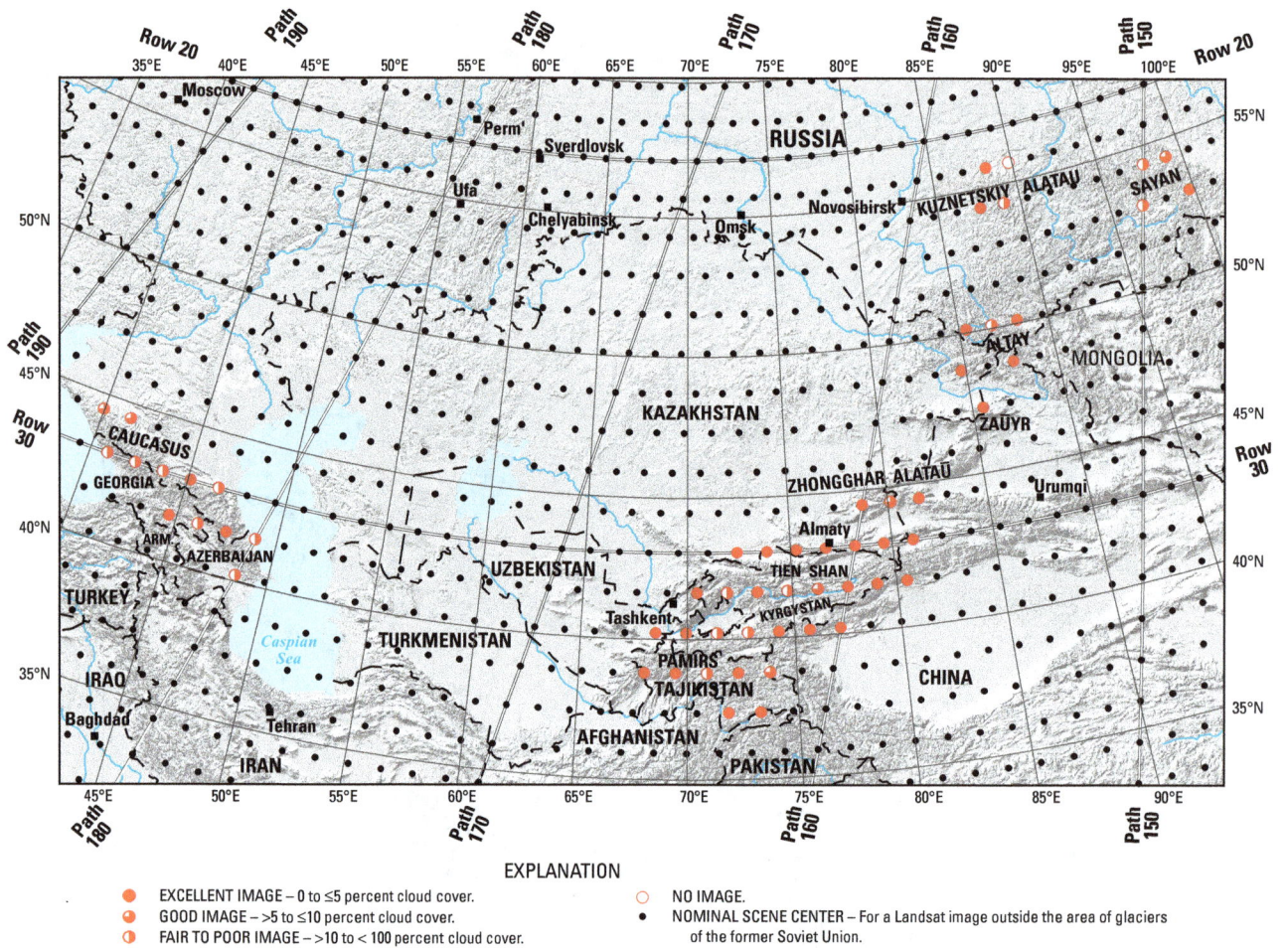

Figure 45.—*Index map of the optimum Landsat 1, 2, and 3 MSS images of the Former Soviet Union. (Continued)*

EXPLANATION

- ● EXCELLENT IMAGE – 0 to ≤5 percent cloud cover.
- ◗ GOOD IMAGE – >5 to ≤10 percent cloud cover.
- ◖ FAIR TO POOR IMAGE – >10 to < 100 percent cloud cover.
- ○ NO IMAGE.
- • NOMINAL SCENE CENTER – For a Landsat image outside the area of glaciers of the former Soviet Union.

Fluctuations of Glaciers of the Central Caucasus and Gora El'brus, with a Section on the Glaciological Disaster in North Osetiya[1]

By V.M. Kotlyakov[2], O.V. Rototaeva[2], *and* G.A. Nosenko[2]

Introduction

The volcanic massif of Gora El'brus in the central Caucasus has 16 large glaciers and 9 smaller, peripheral ones (figs. 46, 47, 48). The largest glaciers are: *Lednik Bol'shoy Azau* (10.2 km long, 19.6 km^2 in area) on the southern slope and the vast *Dzhikiugankez Floe* on the northern slope formed by two glaciers, *Ledniki Birdzhalychiran* and *Chungurchatchiran*, that cover a total area of 27.8 km^2. At present, the total area of glaciers on Gora El'brus is 125.2 km^2 (Zolotarev and Khar'kovets, 2000).

Glacier observations began here in the middle of the 19th century. The first instrumental survey of glaciers was done by military topographers from 1887 to 1890; a map at a scale of 1:42,000 was compiled. During the International Geophysical Year, an expedition of the Department of Geography of Moscow University compiled a detailed map of the entire massif at a scale of 1:10,000; it was based on phototheodolite surveys and aerial photography. Repeated surveys were done in 1981, 1987, and 1997.

Glacier retreat has been a continuous general trend for the El'brus massif for more than 150 years, although the rate of retreat has not been constant (table 17). During the first century of observations (1850–1957), the average annual retreat of all glaciers, including area and length, was slightly greater during the first 37 years (1850–1887) than during the following 70 years. During the 70-year period (1887-1957), glaciers on the northern slope of El'brus retreated more than 600 m; those on the southern slope retreated more than 1,100 m, and glaciers in the western sector decreased in length by at least 300–400 m (Kravtsova and Loseva, 1968). Small advances of glacier termini were recorded during the period 1910 to 1914 and at the end of the 1920s.

During the subsequent 30-year period, 1957–1987, there was an increase in winter precipitation, and summer temperatures remained close to the norm; this caused the average annual retreat of El'brus glaciers to slow by 1.2 times. One exception, however, was the continued dramatic retreat of *Lednik Bol'shoy Azau* (see table 17). This was because a great mass of stagnant ice that was at the terminus of the glacier in 1959 melted and had completely disappeared by 1969. If one subtracts the melting of the stagnant ice, the retreat of the glacier terminus was only about 350 m. The tongues of *Dzhikiugankez Floe* exhibited the greatest rate of retreat during this time (see the measurements of the *Ledniki Birdzhalychiran* and *Chungurchatchiran* in table 17). In contrast to the general retreat of *Lednik Bol'shoy Azau* during this period, other glaciers, including Ledniki Kukurtli, *Bityuktyube*, Ullu-kam, Ulluchiran, and *Ullumalienderku*, were temporarily advancing.

During the following decade (1987 to 1997), aerial photography of Moscow State University (MSU) documented the advance of 8 glaciers out of the 16 (Zolotarev and Khar'kovets, 2000) and the other 8 showed much slower retreat. To determine the glacial fluctuation from 1997 to 2004, the locations of glacier termini on large-scale aerial photographs taken in 1987

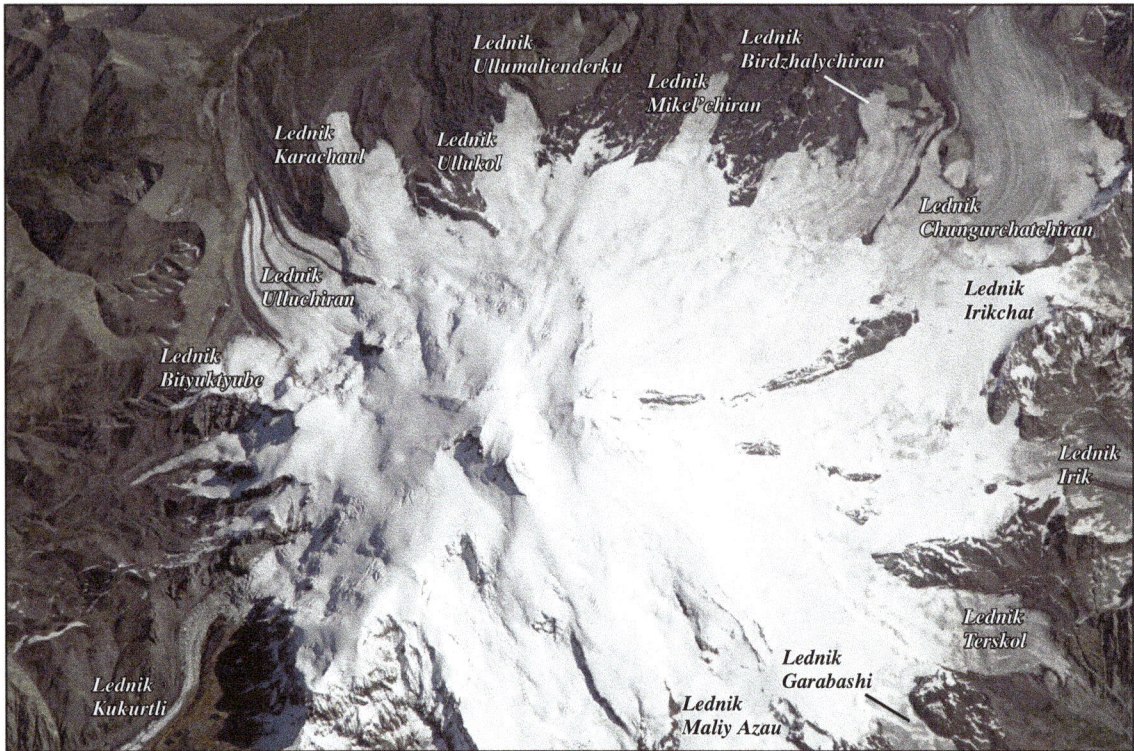

Figure 46.—Image of Gora El'brus (lat 43°21'N., long 42°26'E.) and its glaciers taken on 25 August 2002. Image from the International Space Station. Mission ISS005, Roll E., Frame 11193.

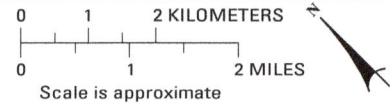

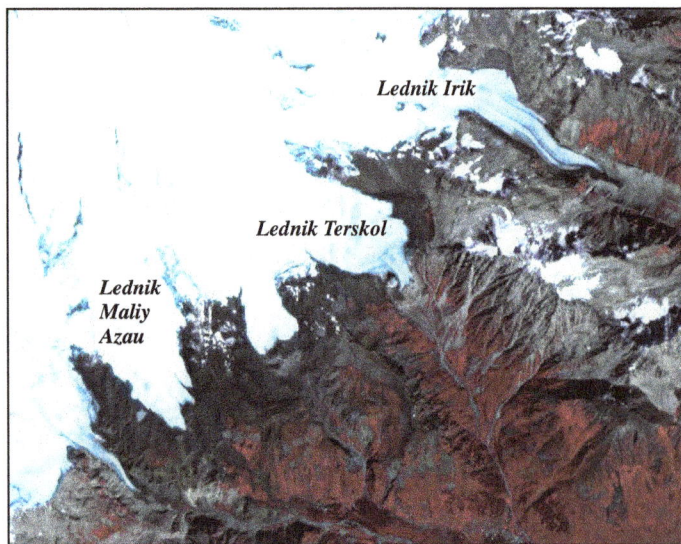

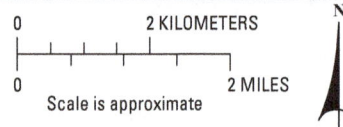

Figure 47.—ASTER Image of the glaciers on the southern slope of Gora El'brus taken on 15 September 2001. Compare with Figures 46 and 48. ASTER image AST_LIB.003:2004260219.

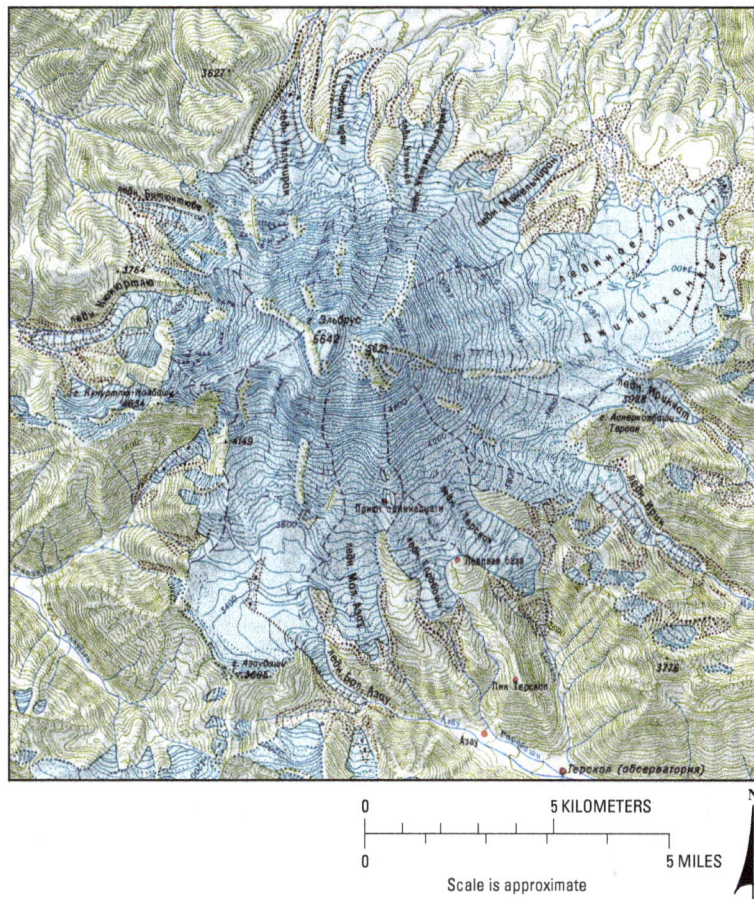

Figure 48.—Map of the glaciers of Gora El'brus (lat 43°21'N., long 42°26'E.). From the World Atlas of Ice and Snow Resources (Kotlyakov, 1997).

TABLE 17.—*Changes in the position of the termini of El'brus glaciers during different periods of observation*

[Unit: m a⁻¹, meters per year]

Name of glacier	1850–1887[1]	1887–1957[2]	1957–1987[2]	1987–1997[3]	1987–2001/04[4]
	37 years	70 years	30 years	10 years	15 years
Bol'shoy Azau	-550	-1,130	-1,150	-150	-300
Maliy Azau	-410	-637	-200	0	-70
Garabashi	-290	-1225	-70	0, -20	-50 right
Terskol	-430	-675	-60	0	-80 left
Irik	-270	-1,162	-355	-60	-75
Irikchat	-470	-750	-250	-170	-180, -200
Chungurchatchiran	-110	-925	-590	-200, +160	-165
Birdzhalychiran	-1,440	-850	-845	+90	-260
Mikel'chiran	-510	-525	-165	0	-130
Ullumalienderku	-460	-175	-120	+100	0
Ullukol	-380	-620	-205	-50	-50
Karachaul	-270	-275	-30	0, +3	-150
Ulluchiran	-240	-1,225	-220	+100	0
Bityuktyube	-80	-500	-20	+130	0
Kukurtli	-190	-275	+110	-90	-100
Ullu-Kam	-430	-320	+40	0	0
Average velocity of termini retreat (m a⁻¹)	-11	-10.1	-8.6	-1.2	-6.5

[1]Kravtsova and Loseva (1968).   [2]Zolotarev (1997).   [3]Zolotarev and Khar'kovets (2000).   [4]Nosenko and Rototaeva (this study).

were compared with those on space images from ASTER in 2001 and 2004, and with digital camera images from the International Space Station (ISS) in 2002. The ASTER images, having a picture element (pixel) resolution of 15 m in three spectral ranges (Band 1, 0.52–0.60 μm; Band 2, 0.63 –0.69 μm; and Band 3, 0.78–0.86 μm), were transformed to the WGS-84 projection. They are quite accurate for interpreting the boundaries of most large- and medium sized glaciers and for determining the conditions of the glacial surfaces based on differences in spectral reflectance in their accumulation and ablation areas. For more accurate identification of the position of the termini, very distinct perspective images from the ISS, with a pixel resolution of 5 m, as well as 2004 photographs taken from a helicopter supplied by the Moscow Institute of Electrical Engineering (MIEE), were used.

The difference between data obtained by MSU for the period 1987 to 1998 and our data for 1987 to 2001/2004 is very evident. From 1987 to 1997, conditions were very favorable for the existence of glaciers on Gora El'brus and in the whole Caucasus region. Lednik Garabashi had positive mass balances during 6 of the 10 years; and for 3 years there were slightly negative mass balances. The period from 1997 to 2001/2004 included the anomalous years of 1998 to 2001, in which rapid melting of glaciers occurred throughout the entire Caucasus and there was a great loss of ice on El'brus glaciers (Rototaeva and others, 2003). As a result, almost all of the El'brus glaciers retreated from 1987 to 2002, and the ones which had been advancing until 1997 had retreated approximately to their 1987 positions by 2002 (see table 17). Substantial melting occurred on *Dzhikiugankez Floe*. Part of the terminus of *Lednik Chungurchatchiran* lost an area of 100×60 m of ice, and the neighboring *Lednik Birdzhalychiran* changed its shape completely. The central part of the terminus preserved features of the former flow and retreated only slightly, but the right and the left parts of the tongue retreated 260–300 m. The marginal zone of the right side of the tongue melted most rapidly, and many lakes formed in the recessional moraines; some of the lakes were as long as 400 to 500 m (fig. 49).

Aerial photographs and space images have made it possible to identify surging glaciers and individual glacier surges on El'brus. Historically, many accounts exist of surge-like occurrences observed during field investigations. In 1849, G. Abikh discovered an intrusion of *Lednik Bol'shoy Azau* into a 100-years-old pine forest; broken pine trees and ice debris were piled together before a high glacier front. It was probably a glacier surge, a conclusion made more likely by later observations. In 1911, H. Burmester documented another surge of *Lednik Bol'shoy Azau*. The front of the tongue had a steep slope, there was strong fissuring of the lower icefall, and there was swelling of the glacier surface between 2,600 and 2,700 m. By 1920, the glacier had advanced by 50 m. In the 1970s, *Lednik Bol'shoy Azau* advanced again, this time by 120 m; the greatest rates of terminus advance were noted in 1975, 1977, and 1979 — up to 34 m a^{-1} (Panov, 1993). In addition, the height of the glacier surface increased by 8.5 m, and the tongue acquired the form of a drop (Martyshev, 1980). The interval between each of these three surges of *Lednik Bol'shoy Azau* was approximately 60 years.

Recently, *Lednik Bol'shoy Azau* has been continually retreating; however, one more surge has been noted. On aerial photographs taken in 1987, it can be seen that the glacier surface in the lower part of the icefall expanded, swelled, and acquired a convex form (fig. 50). A series of wide ogives shows that the ice is pushed through the narrowest gap in sections. A kinematic wave is traveling down the glacier, and a "new" tongue has started to advance on the surface of the tongue covered by a continuous mantle cover. Its frontal part is dissected by a dense network of oblique and arc-like fissures that illustrate the strong ice pressure.

In the following years, the advancing white tongue became more noticeable, and, during 1991, the "new" tongue advanced by 40 m; its surface in some places became higher by 10–12 m (Zolotarev, 1997). In addition, the new ice was separated from the old ice by a pressure ridge up to 5 m high. Similar ramparts on the tongue's surface had been noted during previous surges — in 1973, as well as 1956. Finally, on a 2002 image from the ISS, the white tongue extends even farther and has become narrower (see fig. 50).

This change in the longitudinal profile of the glacier is typical of surging ice; it is caused by the pulsation as well as the great decrease in the thickness and volume of the tongue in the first few years after advance begins — in this case, from 1987 until 1991. Later, the glacier became 4–5 times slower and a great volume of stagnant ice formed on the rapidly melting tongue (Seinova and Zolotarev, 2001).

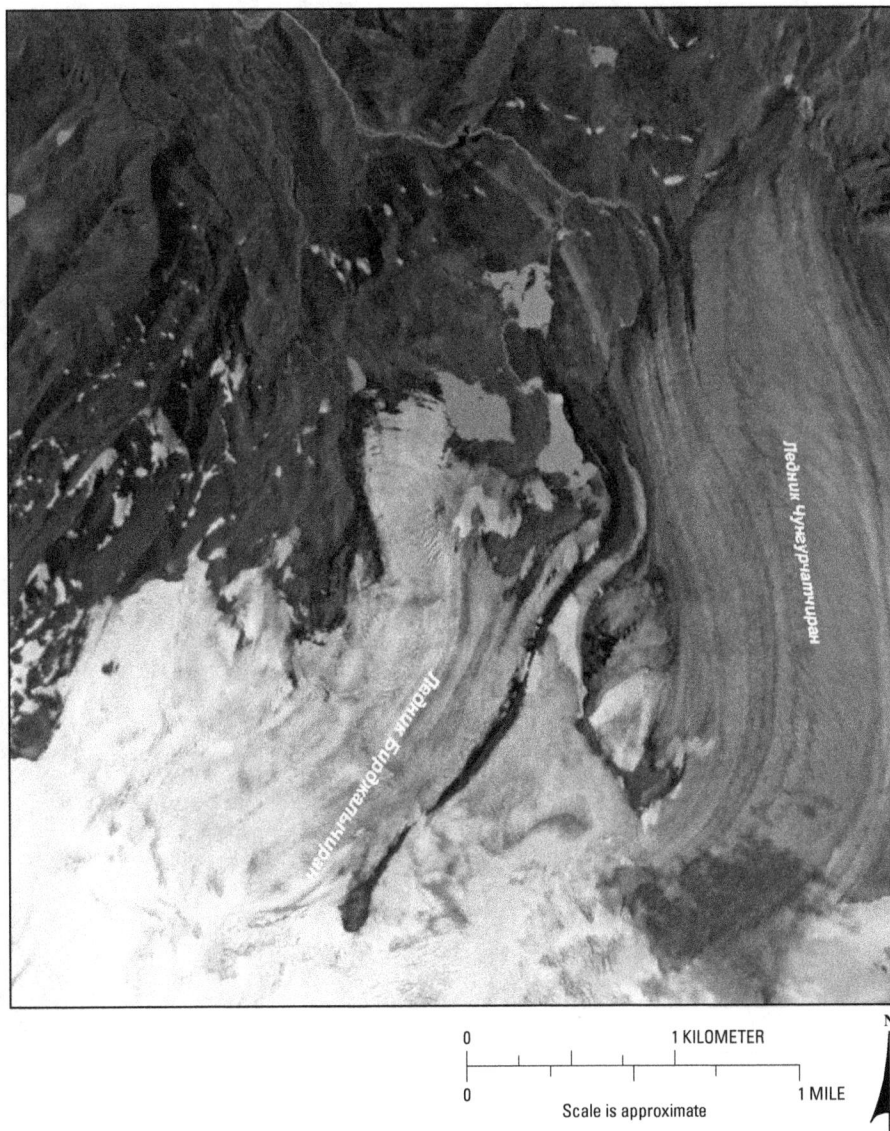

Figure 49.—*Part of the 25 August 2002 International Space Station image (fig. 46) enlarged to show the lake-moraine complex near the terminus of the retreating* Lednik Birdzhalytchiran, *Gora El'brus (in the center).* Lednick Chungurchatchiran *is to the right.*

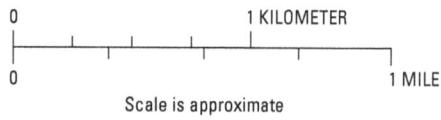

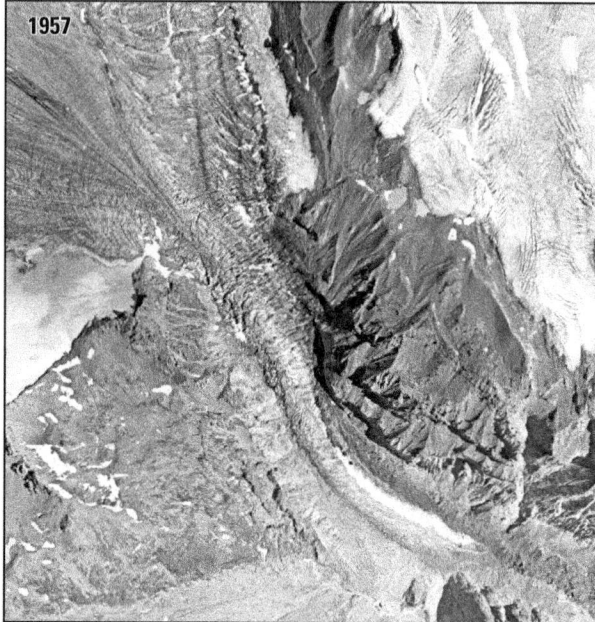

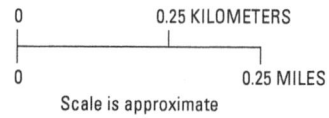

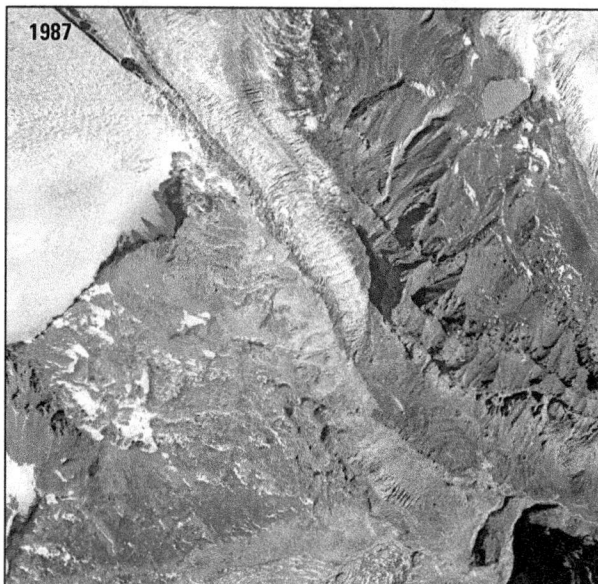

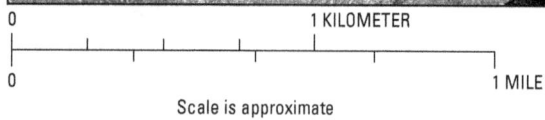

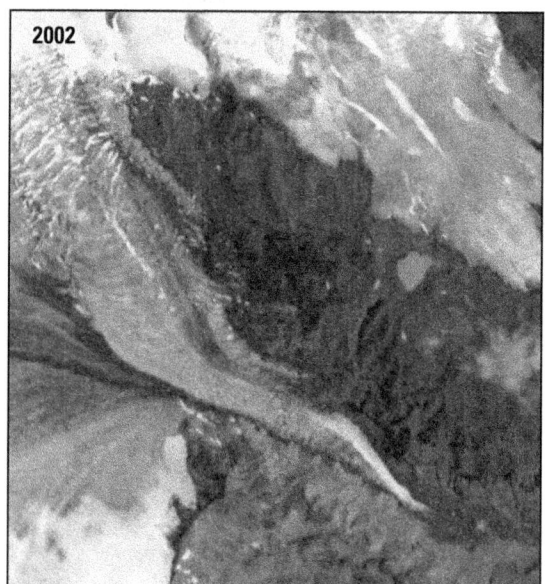

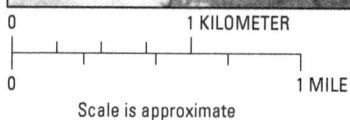

Figure 50.—Aerial and space photographs of Lednik Bol'shoy Azau, *Gora El'brus, showing changes in the terminus shape and position. 1957 and 1987 scenes are aerial photographs, the 1990 image was photographed from a helicopter on 20 August, and the 2002 image was taken from the International Space Station. See figure 48 for location.*

Obviously, such surges of ice occur on the glacier regularly and are caused by ice being constricted in the rock gap of *Bol'shoy Azau*. Huge amounts of ice from two glacier flow directions enter the narrowest part of the gap at an elevation of 3,200 m, and the release of stress results in pulsations.

Other glaciers of the El'brus complex are also notable for unstable dynamics. A 1987 survey noted a kinematic wave at the very end of *Lednik Ullumalienderku*; the terminus showed considerable cracking (fig. 51). The height of its front had increased by 40 m in comparison to its height in 1957, although during this period the glacier terminus retreated by 150 m. At the beginning of the 1970s, Lednik Terskol was advancing (Martyshev, 1983). In 1980, a surge of the right side of Lednik Ulluchiran took place. Its surface rose by 20–25 m and was dissected by fissures, and its terminus had a steep front that had advanced by 150–200 m. Its shape became like a "paw," a characteristic of other surge-type glaciers (Voloditcheva and others, 1983; Voitkovsky and others, 1989).

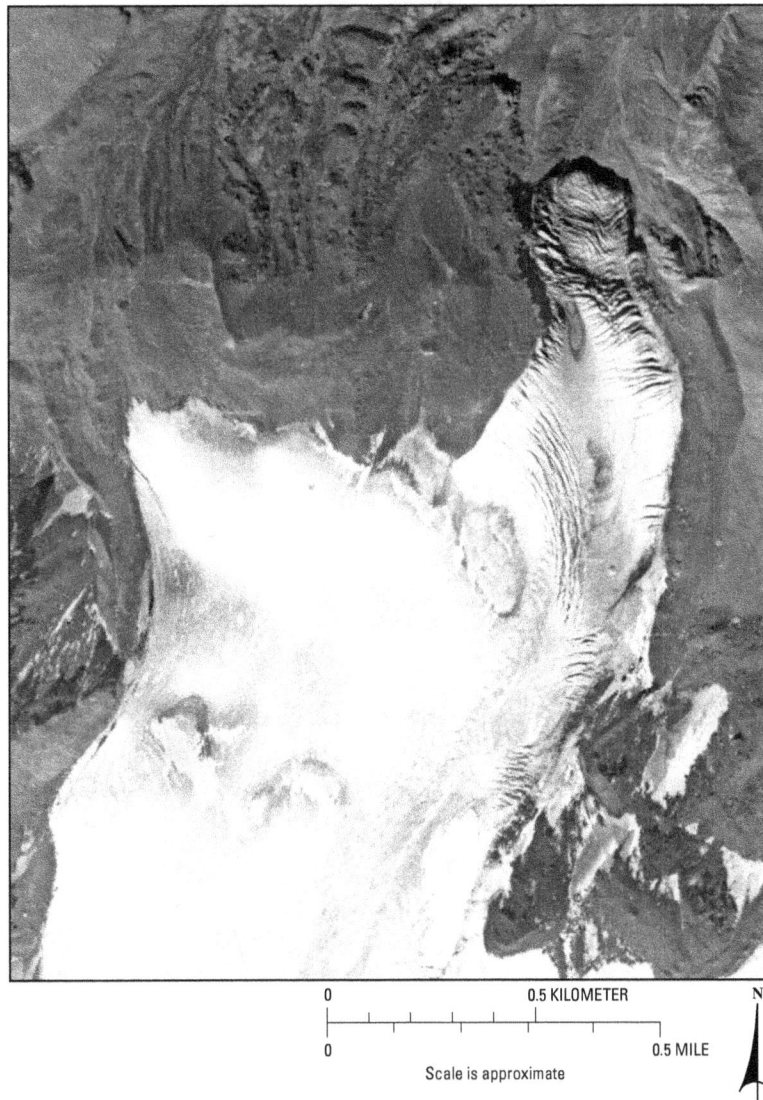

Figure 51.—*Aerial photograph (1987) of the terminus of* Lednik Ullumalienderku, *Gora El'brus, during a surge event. See figure 46 for location.*

Pulsations are also characteristic of the Kukurtli glacier complex on the western slope of El'brus (fig. 52). *Lednik Maliy Kukurtli* (part of glacier No. 316) experienced a large surge in the 1960s. In 1981, a sheer high front of a small revived glacier (No. 317), dissected by large fissures, advanced and overhung the left moraine of the Kukurtli tongue, and, in the following 1 to 2 years, it reached the edge of the main glacier. In 2002, glacier No. 317 was completely backed by moraine and showed no signs of activity at all.

A surge of the main Lednik Kukurtli (no. 318) took place at the beginning of the 1980s. By 1983, the glacier had advanced 60 to 100 m. The surface near its terminus rose by 30–40 m; along a distance of 1.5 km, the ice volume increased by 7.2×10^6 m^3. Farther up the glacier, 2 km from the end of the tongue, the surface lowered by 5 to 10 m (Vinnikov and Labutina, 1987). Space images acquired from 2002 to 2004 show that the glacier terminus has retreated approximately 100 m since 1987, but this retreat cannot be accurately measured because the entire glacier terminus was buried under the moraine (see fig. 52).

The shape of Lednik Kukurtli favors its periodic pulsations. Its vast firn fields are positioned so that orographic uplift of moisture-laden air masses bring frequent periods of precipitation; as a result, great amounts of snow accumulate. The glacier tongue is separated from the firn area by a lava escarpment and a 500-m icefall. The large amounts of firn and ice that accumulate on the steep stepped slopes are periodically pushed through the icefall towards the long tongue.

Measurements made on space images of the El'brus glaciers and of 65 other glaciers of the Central Caucasus showed considerable retreat of almost all large glaciers from 1987 to 2001/2004 (fig. 53, table 18). The tongues of the following glaciers retreated the most: Karaugom (almost 600 m) (fig. 54); Shaurtu, *Rtsyvashki*, and Midagrabin (about 300 m); *Skhelda* and Bashil' (more than 200 m); and Tseya and Skazka (160 m). In addition, *Lednik Ailama* — the former tributary of *Lednik Dykh-Su* — became detached and isolated; it was separated from the tongue of the main glacier by large amounts of stagnant ice (figs. 53, 55).

Fluctuations of even adjacent glaciers are not always synchronous, however, and sometimes can have an opposite sign. It depends on characteristics of morphology, exposure, type of nourishment of a glacier, its internal dynamics, and so forth. Thus, for example, the largest glacier of the Caucasus — Bezengi — continues to retreat, whereas the adjacent *Lednik Mizhirgichiran* continues to advance, bulldozing a push moraine before it (figs. 53, 56).

Glaciological Disaster in North Osetiya

On 20 September 2002, a disaster took place in the Genaldon river valley of North Osetiya, north of Gora Kazbek (lat 42°42'N., long 44°31'E.). Huge masses of ice, water, and debris burst from the head of the valley, rushed swiftly 16 km down the valley, sweeping everything before it, shearing off forest, and scraping off unconsolidated deposits on slopes up to a height of 100 m. The ice and water-saturated debris avalanche (jökulhlaup) was stopped by a transverse ridge of Skalistyy Khrebet just before the river entered a narrow gap. The entire bottom of Karmadon basin was under glacial debris about 4 km long, 60 m thick on the average, and in some places up to 130 m (fig. 57, 58A, 58B). Downstream in the gap, a water-saturated mudflow, including ice debris, caused further destruction along 17 more kilometers. Several recreation sites were destroyed, a tunnel on a highway was blocked, and the *Nizhni Karmadon* settlement was completely demolished. More than 100 people died.

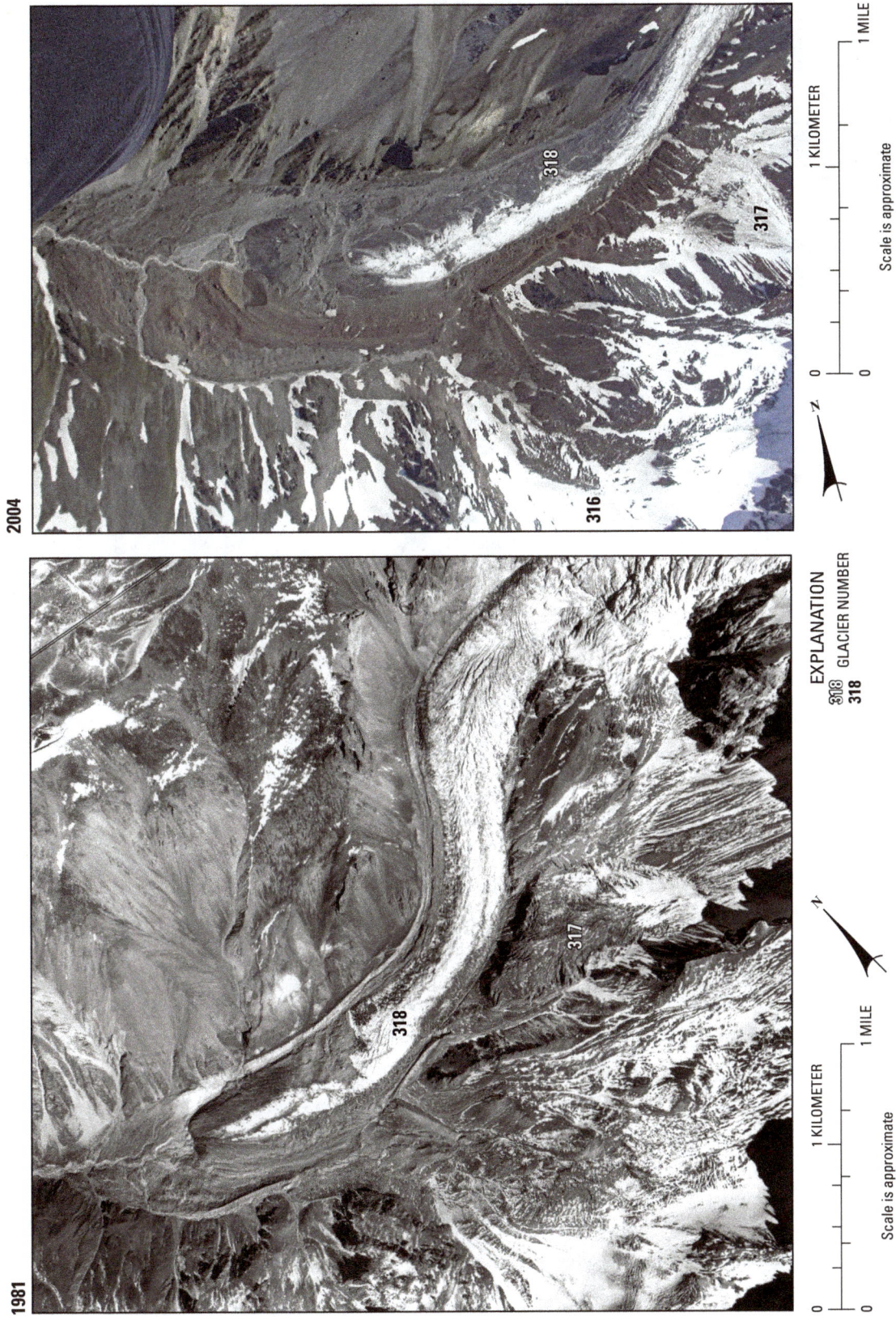

1981

2004

EXPLANATION

318   GLACIER NUMBER

Figure 52.—Photographs (1981 and 2004) of the Lednik Kukurtli, Gora El'brus. The 1981 image is an aerial photograph; the July 2004 photograph was taken from a helicopter by I.V. Galushkin, Info Terra. Numbers refer to discussion in the text on p. F66. See figure 46 for location.

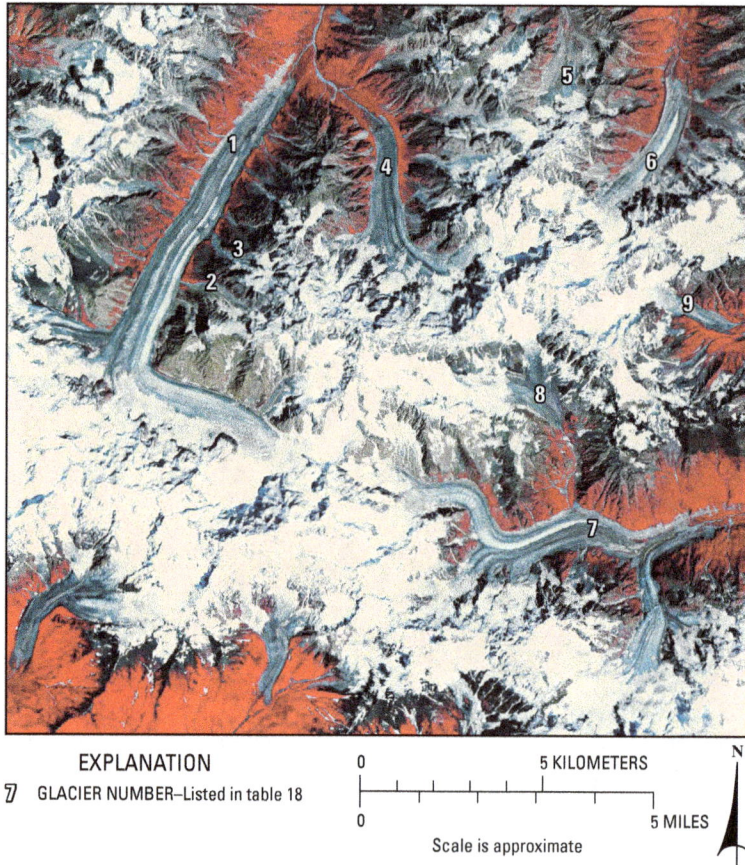

EXPLANATION

7 GLACIER NUMBER–Listed in table 18

0   5 KILOMETERS

0   5 MILES

Scale is approximate

N

Figure 53.—ASTER image taken on 15 September 2001, showing glaciers of the central Caucasus (approximate lat 43°N., long 43°E.). Numbers correlate with glaciers listed in table 18. ASTER image AST_LIB.003: 2004260219.

TABLE 18.—*Changes in the position of glacier termini from 1987 to 2001/04 in the Tcherek River basin*

[Abbreviations: km, kilometer; masl, meters above sea level; m, meter]

	Name of glacier	Morphological type (of glacier)	Exposure	Length (km)	Area (km²)	Height of terminus (masl)	The highest point (masl)	Change of length (m)
1	Bezengi	Compound-valley	NE	17.6	36.2	2,080	5,050	-190
2	Misses	Hanging corrie	NW	1.9	1.1	2,980	4,070	0
3	No. 32	Corrie-valley	NW	3.1	1.4	2,720	4,650	-90
4	Mizhirgitchiran	Compound-valley	N	8.8	9.9	2,380	4,670	150
5	Ukyu	Corrie-valley	NE	2.4	1.6	3,260	3,740	-80
6	Ulluazna	Compound-valley	NE	7.0	6.0	2,600	4,760	0
7	Dykh-su	Compound-valley	E, NE	13.3	34	2,070	5,150	-200
8	Khrumkol	Valley	SE	5.6	7.2	3,060	5,150	0
9	Tyutyun	Valley	SE	5.0	2.8	2,540	3,850	-180

Figure 54.—Image taken from the International Space Station in 2002, showing the retreating Lednik Karaugom (lat 42°48'N., long 43°44'E.). Note that cirques of tributary glaciers are becoming partially ice free.

0   3 KILOMETERS

0   3 MILES

Scale is approximate

N

Figure 55.—Aerial photograph taken in July 2004 of the terminus of Lednik Ailama (top right) and the stagnant terminus of Lednik Dykh-su (approximate lat 43°N., long 43°E.) covered with debris. This area is also shown on figure 53. Lednik Dykh-su is number 7. Photograph was taken from a helicopter by I.V. Galushkin, Info Terra.

Figure 56.—*Photograph taken in August 2003 of the advancing front of* Lednik Mizhirgitchiran *(approximate lat 43°N., long 43°E.) taken by Yu.G. Iljitchev. See glacier number 4 on figure 53.*

Figure 57.—*Oblique aerial photograph of avalanche debris in the Karmadon basin caused by the 20 September 2002 catastrophic collapse of the* Lednik Kolka. *The view is looking north toward the narrow gorge in Skalistyy Khrebet that halted the avalanche. Photograph from the Geological Survey of North Osetiya.*

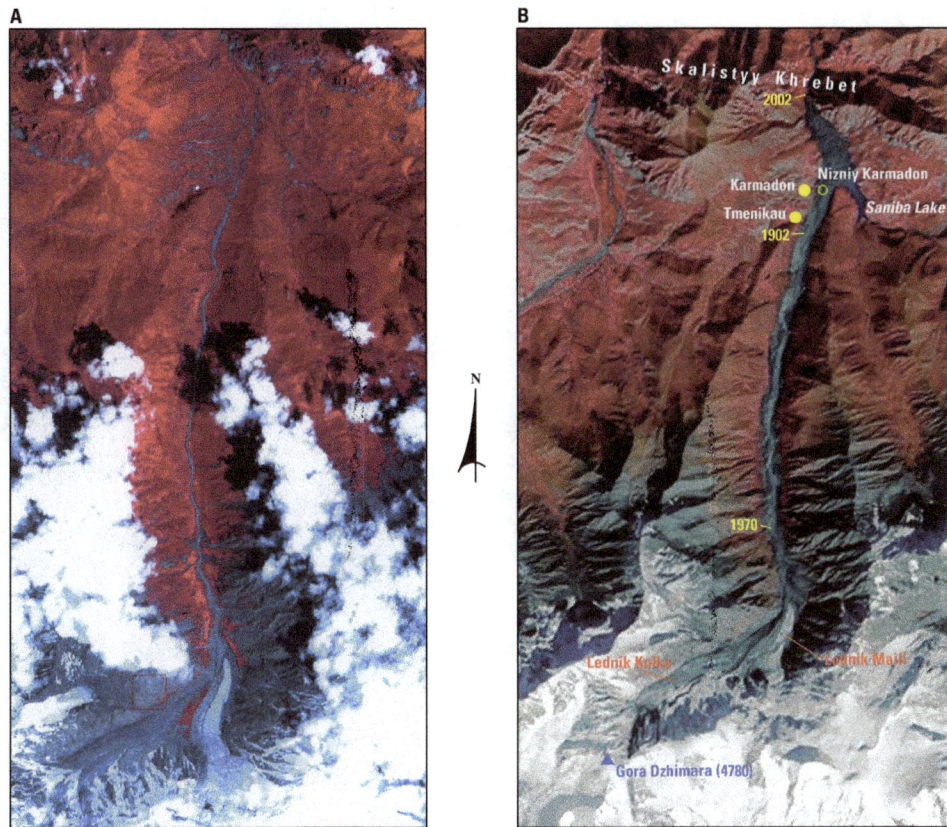

Figure 58.—*Two ASTER images of the Genaldon river valley, scene of the* Lednik Kolka *disaster. The distance from Gora Dzhimara to Skalistyy Khrebet is about 18 km.* **A**, *Section of the 22 July 2001 image taken before the disaster. ASTER image AST_LIB. 003: 2003583778.* **B**, *Annotated 27 September 2002 image taken after the 2002 disaster, showing locations of the* Lednik Kolka *terminus after earlier surges. ASTER image AST_LIA. 003:2008557907.*

In a tributary valley near the *Saniba* settlement, the glacier avalanche debris dammed the valley, forming a lake; its level continued to increase during the month after the glacial outburst flood (jökulhlaup) (fig. 59).

Lednik Kolka, a surge-type glacier noted for its historical record of repeated surges, had caused the disaster. Previous surges took place in 1835, 1902, and 1969 at intervals of about 70 years. The surge of 1902 ended in a similar glacial outburst flood that traveled 9 km, but it was not as destructive. In 1969/1970, a surge occurred that followed the "typical" scenario, the tongue advanced 4 km during a three month period without disastrous aftereffects. The glacier was studied in detail for many years by expeditions of the Institute of Geography, Russian Academy of Science (IGRAS) (Rototaev and others, 1983).

In contrast with earlier surges, only 32 years passed between the 1970 surge and the 2002 disaster. Interestingly, the glacier left its cirque completely, so that its bed became completely empty. In the empty glacial cirque (fig. 60) there was a cloud of vapor and gas with a strong smell of hydrogen sulphide for many days. An icy scarp was detected on the right margin of the glacier cirque — a steep rocky slope with hanging glaciers. In addition, a large part of the firn-ice fields under the ridge crest had disappeared. It was first thought that a great fall of ice, which struck the glacier and made it collapse, was the

Figure 59.—*Oblique aerial photograph taken on 21 September 2002, looking southwest at the avalanche debris saturated with water. The debris flowed into a tributary valley near the settlement of Saniba and a dammed lake is beginning to form. Photograph from the Geological Survey of North Osetiya.*

cause of the disaster. However, evidence and photographs of mountaineer witnesses showed that the unusually active destruction of ice and rock on the slope had already begun in summer, and by the beginning of September, this ice scarp had already been completely formed.

To investigate the causes, mechanism, and possible aftereffects of the September 2002 destructive process, many data sources were used, including multispectral Advanced Spaceborne Thermal Emission Radiometer (ASTER) images from the Terra spacecraft, images taken with a digital camera on the International Space Station (ISS), and observations from aerial and surface investigations of the glacier region, made by an expedition of IGRAS in the summer of 2003 (Kotlyakov, Kerimov, and others, 2004; Nosenko and others, 2004).

Structural features of a surge-type glacier preparing for a surge can be identified from space. These features include an increase in glacier thickness, a change in the glacier's surface structure, the appearance of new characteristic fissure fields, and changes in the position of the glacier terminus. *Lednik Kolka* is 3 km long and completely covered by moraine, so it was thought that the picture element (pixel) resolution of available ASTER images (15 m) taken a year before the disaster would not be sufficient to identify these pre-surge features (see fig. 58a). However, even on an ISS image with 5-m pixel resolution taken a month before the surge (13 August 2002), they are definitely absent (fig. 61). That means that the glacier had not yet accumulated critical mass and had not reached the threshold of a surge by the "classical" method.

Figure 60.—*Oblique aerial photograph taken on 24 September 2002, looking south at the mountain crest in the source region of* Lednik Kolka. *Rock outcrops are visible where the former snow and ice cover has slid onto the glacier below. Meltwater and fumeroles are present. Photograph from the Geological Survey of North Osetiya.*

Lednik Kolka is situated in a region where earthquakes occur often. It is in a fault zone of large fractures where movement of separate blocks is possible. Summer earthquakes which preceded the disaster were especially strong on 14 July and 22 August, and probably triggered the beginning of unusual ice, snow, and rock falls in the head wall of the glacier cirque. Both sporting groups and an ASTER image of 18 July 2002 confirmed that it was possible to see black patches of rock on snow-covered slopes where snowfall had occurred. The earthquakes caused not only the increase of falls, but probably the destruction of the internal structure of the glacier and its connection with the bed as well. Over time, these falls increased the mass of the glacier, considerably overloading its upper part, and increasing stress to the maximum throughout the whole body of ice.

From the very beginning, it was obvious that such a swift and extensive outburst of the glacier could occur only as the result of accumulation of a very great amount of water under the glacier, which "catastrophically separated" the glacier from its bed (Anonymous, 2002; Kotlyakov, Rototaeva, and others, 2004). Space images showed that there was a strong outburst of water, ice, and debris from the right margin of *Lednik Kolka* which moved over the moraine of Lednik Maili and down the right slope of the valley. This was confirmed by photography from a helicopter and surface observations. The surface structure of Lednik Maili itself did not change very much, but there are some well-defined traces of flooding on the upper edge of the flow, which had passed over ice (fig. 62). Talus on the slope under the tongue of Lednik Maili was not significantly disturbed.

Water accumulation in the glacier basin had increased in volume during preceding years not only due to anomalous melting (Lebedeva and Rototaeva, 2005), but probably by volcanic activity as well. Evidence of volcanic activity included gas emission, actively collapsing slope, and the appearance of fumaroles on the cirque wall (see fig. 60). The volcanic activity created special thermal conditions under the glacier that caused additional melting and accumulation of water. Analysis of the chemical composition of snow, ice, and water in the *Kolka* cirque proved that the hydrothermal factor played a role in the disaster. In Genaldon, the presence of a near-surface magma chamber was shown by two independent geophysical methods; its appearance in 2002 was also proven by analysis of thermal infrared space imagery from NOAA satellites (Gurbanov and others, 2004; Kornienko and others, 2004).

The large amount of water formed many lakes after the disaster. They even appeared in the floor of the cirque after the glacier surge (figs. 63, 64). The size of the main lake, identified on an ASTER image of 6 October 2002, was 300×80 m (Popovnin and others, 2003). Our surface observations in

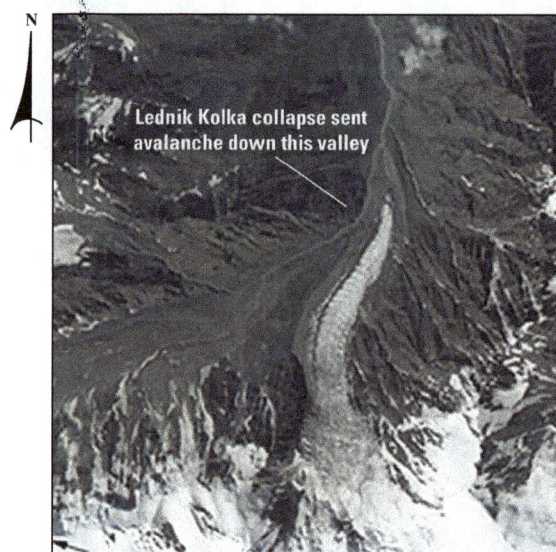

Figure 61.—*International Space Station image taken on 13 August 2002 of Lednik Maili (white tongue in center) and* Lednik Kolka *(dark, debris-covered tongue on left). Although it is a month before the disaster, there is no visible evidence of the coming event. Mission ISS005, Roll E, Frame 9691.*

Figure 62.—*Oblique aerial photograph taken September 2002, looking south at the empty cirque of* Lednik Kolka *(on right) after the disaster. Lednik Maili is on the left. Gora Dzhimara (4,780 m) is in the background on the right. Photograph from the Geological Survey of North Osetiya.*

2003 showed that the margins of the lake were composed of unconsolidated material; in June 2003, its size decreased to 150×50 m (fig. 65), and by the end of September 2003, the lake had disappeared completely.

The debris in Karmadon basin, deposited by the debris avalanche, was completely saturated with water, and chains of lakes appeared immediately along its margins (see fig. 59). The largest one — the dammed *Sanibanskoe Lake* — can be seen well on all space images. It began to fill immediately after the disaster, and in a month, more than 3 million cubic meters of water had accumulated in the lake, with a maximum depth of 40 m (Chernomorets, 2005). The avalanche debris gradually melted and was eroded by the river (fig. 66). The lake slowly drained, but at the beginning of September 2003, the appearance of the lake had changed very little. By late September 2003, however, only a small part of the lake remained.

Figure 63.—*International Space Station image taken on 19 October 2002, looking south at Gora Kasbek volcanic massif and the cirques of Ledniki Maili and Kolka. Lakes are forming in the empty bed of* Lednik Kolka *and rock outcrops are visible on the valley wall. Mission ISS005, Roll E., Frame 17830.*

Figure 64.—*Enlarged and rotated section of 27 September 2002 ASTER image (fig. 58B). Arrows show the lakes in the empty bed of* Lednik Kolka. *It is also possible to see the traces of the mudflow on the surface of Lednik Maili. Compare with figure 63.*

In conclusion, the combined analysis of space, aerial, and surface photography made it possible to determine the causes and probable scenario of the September 2002 disaster (Rototaeva and others, 2005). The surge of *Lednik Kolka* was brought about prematurely and was much larger than previous surges because of a combination of factors. These were the unstable dynamic state of the surging glacier, significant accumulation of englacial and subglacial water, collapses of ice and rock which created an overload in the upper part of the glacier, and the growing activity of endogenous processes in the region of Gora Kazbek.

Figure 65.— (at left) Terrestrial photograph taken on 28 June 2003 of one of the lakes in the Kolka bed dammed by a combination of ice and debris. Photograph by O.V. Rototaeva.

Figure 66.— (below) Oblique aerial photograph taken on September 2003, looking southeast at the collapsing avalanche debris in Karmadon basin and the dammed lake near Saniba settlement one year after the disaster. Compare with figure 59. Photograph by G.A. Nosenko.

Investigations of the Fluctuations of Surge-Type Glaciers in the Pamirs based on Observations from Space

By V.M. Kotlyakov,[2] G.B. Osipova,[2] and D.G. Tsvetkov[2]

Introduction

Recently, investigations on glacier fluctuations in the Pamirs were carried out at the Institute of Geography of the Russian Academy of Sciences (IGRAS). A variety of space imagery was analyzed, including: (1) photographs from 1972 to 1991 (up to 13 repeated surveys) at a scale of 1:200,000–1:270,000 taken with the KFA-1000 camera from Russian Resurs-F1 satellites at a resolution of about 5 m, (2) Advanced Spaceborne Thermal Emission Radiometer (ASTER) images acquired from 2001 to 2004 from the Terra satellite, with a pixel resolution of 15 m, and (3) images taken from 2001 to 2004 by astronauts with a hand-operated digital camera from the International Space Station (ISS). Using this imagery, as well as historical data (aerial photographic surveys, topographic maps, surface observations, etc.), the following work was accomplished: (1) compilation of an inventory of surging glaciers of the Pamirs; (2) monitoring of the dynamics of large compound glaciers; (3) estimation of the present-day status of two surge-type glaciers, Ledniki Medvezhiy and Geograficheskogo Obshchestva; and (4) investigation of fluctuations of "normal" non-surging glaciers in the Eastern Pamirs. This paper summarizes the main results of these investigations.

Compilation of an Inventory of Surging Glaciers of the Pamirs

In 1997, compilation of the Inventory of Surging Glaciers of the Pamirs was finished (Osipova and others, 1998). It was the first complete catalogue of unstable glaciers of a large mountain region. The Inventory was based on stereoscopic interpretation of space photographs of 1972 to 1991. To identify surge-type glaciers and assign them to a group, a set of criteria was developed, including both direct (rapid changes of the form and structure of the glacier surface) and indirect criteria. Regular movement of surge-type glaciers and their interactions with the surrounding environment were considered. All glaciers that showed dynamic instability were subdivided into three groups. Glaciers with one or more observed surges were assigned to the first group. The second group consisted of glaciers with indications of periodic activity which could be rather confidently classified as "surging," Glaciers with some indication of recent unstable dynamics were assigned to the third group.

The analysis identified 630 glaciers in the Pamirs that showed indications of instability at the beginning of the 1990s (table 19). Results indicated that between 1960 and the middle of the 1990s, 17 large-scale surges occurred in the Pamirs. By comparing 1991 space photographs with a 2003 ASTER image, the surges of three more glaciers were detected after the Inventory of Surging Glaciers had been compiled (table 20).

TABLE 19.—*The number of surge-type glaciers in the large river basins of the Pamirs*

River basin	Group 1	Group 2	Group 3
Surkhob	1	2	—
Kyzylsu	5	23	21
Muksu	30	60	47
Obikhingob	5	26	22
Vanj	4	17	19
Yazgulem	2	20	22
Bartang	1	38	60
Murghob	—	4	54
Gunt	—	7	57
Panj (headwaters and eastern [right] confluents)	—	2	44
Qarokŭl	1	6	14
Markansu	1	7	4
Eastern Kyzylsu	1	3	—
Total	51	215	364

TABLE 20.—*The largest observed surges of the glaciers of the Pamirs from 1960 to 2003*

[Unit: km, kilometer]

No.	Glacier name	River basin	Latitude and longitude of center of the tongue		Area (km²)	Length (km)	Years of the surges or intervals when recorded	Movement of tongue (km)
1	Gando-*Dorofeeva*	Obikhingob	38°54′N.	71°45′E.	22.0	44.6	1985–1991	[1](8.5)
2	*Vanchdara*	Obikhingob	38°45′N.	71°57′E.	11.5	9.5	1977–1980	2.5
3	No. 88 (*Vaizirek*, west)	Obikhingob	38°58′N.	71°18′E.	4.5	7.6	1975–1980	2.2
4	*Petra Pervogo*	Obikhingob	39°00′N.	71°25′E.	12.2	10.5	1976–1980	[1](?)
5	*Burs*	Obikhingob	38°42′N.	71°48′E.	6.9	5.3	1991–2003	3.6
6	*Sytargi*	Muksu	38°41′N.	71°46′E.	5.8	5.1	1991–2003	1.0
7	*Byrs*	Muksu	39°03′N.	71°45′E.	9.9	11.8	1981–1983	1.3
8	*Shini-Bini*	Muksu	38°59′N.	71°50′E.	16.4	10.3	1959–1961 about 1991	1.5
9	*Sugran*	Muksu	38°56′N.	71°44′E.	47.1	22.0	1976–1980	[1](5.0)
10	*Oshanina* (*Muzgazy*)	Muksu	39°03′N.	71°30′E.	15.6	10.7	1961 (1962) 1983–1984	7.5
11	*Valy*	Muksu	39°18′N.	72°45′E.	11.9	7.6	1975–1977	1.3
12	*Dzerzhinskogo*	Muksu	39°17′N.	72°49′E.	19.0	14.9	1972–1976	1.2
13	*Maliy Saukdara*	Muksu	39°16′N.	72°51′E.	23.5	14.3	1972–1975	2.8
14	*Shapak*	Muksu	39°02′N.	71°50′E.	13.7	8.0	1991–2003	1.0
15	*Bivachniy*	Muksu	38°56′N.	72°10′E.	61.2	29.6	1975–1978	[1](6.0)
16	Medvezhiy	Vanj	38°40′N.	72°12′E.	25.3	15.8	1963 1973 1988–1989	1.8 1.9 1.1
17	*Ravak*	Vanj	38°38′N.	72°04′E.	1.9	4.1	1967	0.2+ collapse
18	*Didal'*	Surkhob	38°59′N.	70°43′E.	1.5	4.9	1974	2.2+ collapse
19	Lenina (Eastern part)	Kyzylsu	38°25′N.	72°56′E.	12.4	9.6	1969–1970	1.1
20	*Oktyabr'skiy*	Qarokŭl (lake)	39°15′N.	73°00′E.	88.2	12.7	1985–1990	3.3

[1]Internal surges.

Surge-type glaciers are situated mainly in high mountain groups of the Central Pamirs near the junction of Akademii Nauk Khrebet (range) with Khrebet Petra Pervogo, *Darvozskiy Khrebet* (Qatorkŭhi Darvoz) and *Vanjskiy Khrebet* (Qatorkŭhi Vanj) (fig. 67), and in the central part of *Zaalayskiy Khrebet* (Qatorkŭhi Pasi Oloy in Tajikistan, Chong Alay Kyrka Toosu in Kyrgyzstan) (fig. 68). Surges on some of the glaciers are internal and occur only within the glacier itself; they do not cause the stagnant morphological tongue to advance. These so-called "internal" surges occurred on Ledniki Gando, Garmo, Bivachniy, *Petra Pervogo*, and other glaciers. However, in the Pamirs there are many surge-type glaciers whose tongues advance along ice-free valleys with velocities of up to 100 m d^{-1} during surge events; examples are the Ledniki Medvezhiy, *Byrs, Muzgazy, Vanchdara,* and other glaciers. Surges of the Lednik Medvezhiy in 1963 and 1973 caused disastrous aftereffects connected with outburst floods (jökulhlaups) from the lake dammed by the glacier. Also, surges of glaciers whose tongues break off and form a water-ice-debris surge can travel down valley with a high velocity and cause catastrophic damage. Such surges have occurred on the *Ledniki Ravak* and *Didal'*.

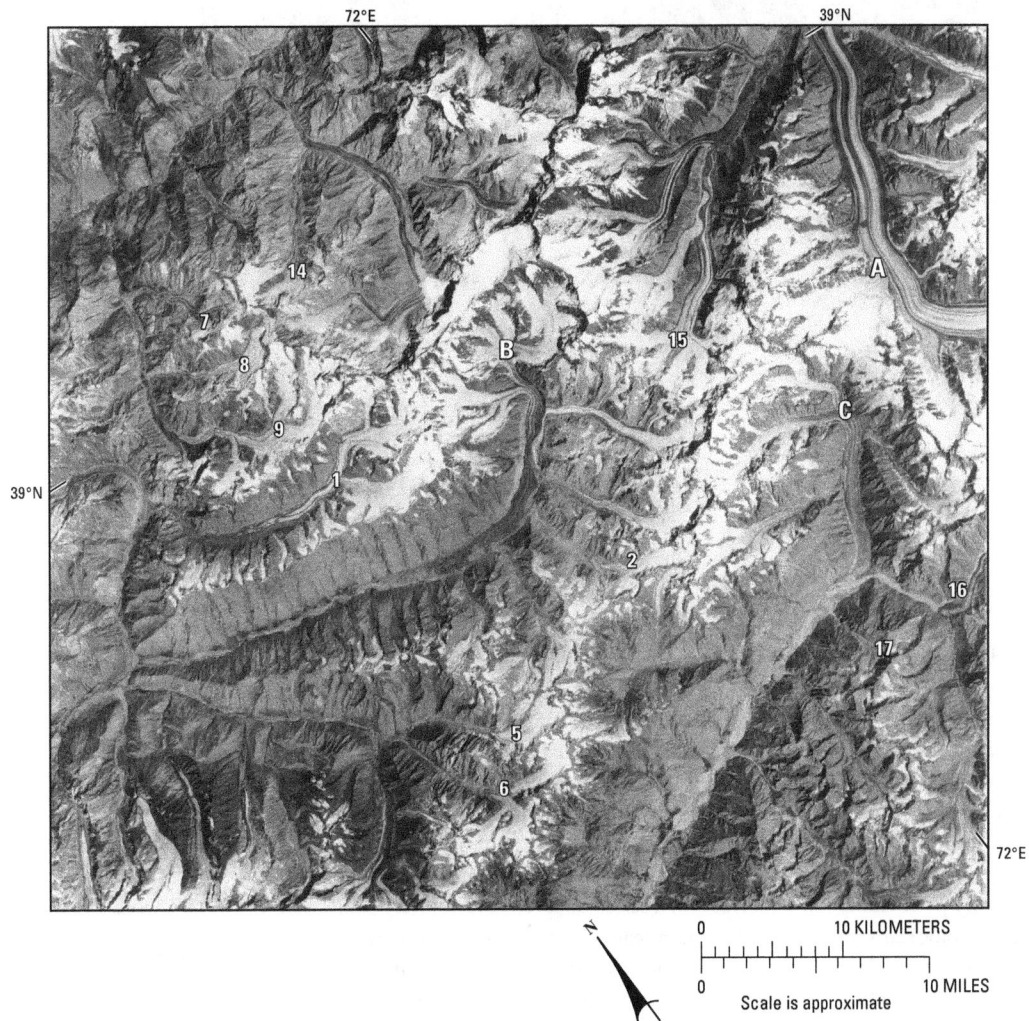

Figure 67.—*KFA-1000 photograph taken in 1973 from a Resurs-F1 satellite showing surge-type glaciers in the Central Pamirs. The numbers of the glaciers correspond to the numbers in table 20. The largest glaciers of the Pamirs are:* **A**, *Lednik Fedchenko;* **B**, *Lednik Garmo; and* **C**, *Lednik Geograficheskogo Obshchestva.*

Figure 68.—*KFA-1000 photograph taken in 1985 from a Resurs-F1 satellite showing surge-type glaciers in the central part of Zaalayskiy Khrebet (Qatorkŭhi Pasi Oloy in Tajikistan, Chong Alay Kyrka Toosu in Kyrgyzstan). The numbers of the glaciers correspond to the numbers in table 20.*

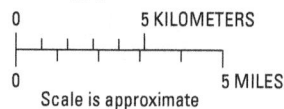

Investigation of the Fluctuation Regime of Compound Glaciers

The Institute of Geography has developed methods of monitoring fluctuations of compound surge-type mountain glaciers from space (Osipova and Tsvetkov, 2000, 2002b). For many years, analysis of the dynamic regime of glaciers was done using data on the velocity of ice movement and changes in glacier morphology. The velocity of ice movement was measured by pseudo-parallax using space photography. The method has been improved so that it is now possible to determine the shift of identical contour points on photographs of different dates with an accuracy of 5–10 m. As a result of stereoscopic interpretation of space images, sketch maps reflecting the status of the glaciers and the morphology of their surfaces were compiled for the dates of each survey.

By performing joint analyses on a series of successive velocity profiles and morphological sketch maps, the main features of the glacial systems' dynamic regimes were identified. This, in turn, allowed for an assessment of the character of interaction of their main components. Remote investigations of the dynamics of several compound, surge-type glaciers were conducted over a long time interval. Results indicate that the role of mutual damming of tributary glaciers and the main trunk glacier at different stages of their evolution is a determining factor in dynamic instability of glaciers (Osipova and Tsvetko, 2002b). Several examples follow.

The most detailed investigations of change were carried out on Lednik Oktyabr'skiye for an extended period (Osipova and Tsvetkov, 2002a). The last strong surge of this glacier occurred in the mid-1930s, and its aftereffects are easily seen on 1946 aerial photography (fig. 69A). After 1946, activity gradually increased on the upper part of the tongue, the lower part of the tongue wasted away, and tributaries in the lower course of the glacier advanced into the valley of the main trunk (fig. 69B–69C). Toward the end of the recovery stage (1978), the cross section of the glacier tongue was approximately one third narrower (figs. 69D and 70). As a result, a region with considerable longitudinal compression formed a higher ice "dam." In the following years, the release of the accumulated stresses took place, and during 1988–1990, active ice advanced over this "dam," cut or stretched tongues of tributary glaciers on the lower course of the glacier, advanced 3 km down the valley, and reached within 0.3 km of the maximum tongue extent in the prior advance (figs. 69F, G, and 70) (1990). Average ice velocity from 1988 to 1990 was 1.6 km a^{-1} (fig. 71), but at the surge peak it was considerably higher.

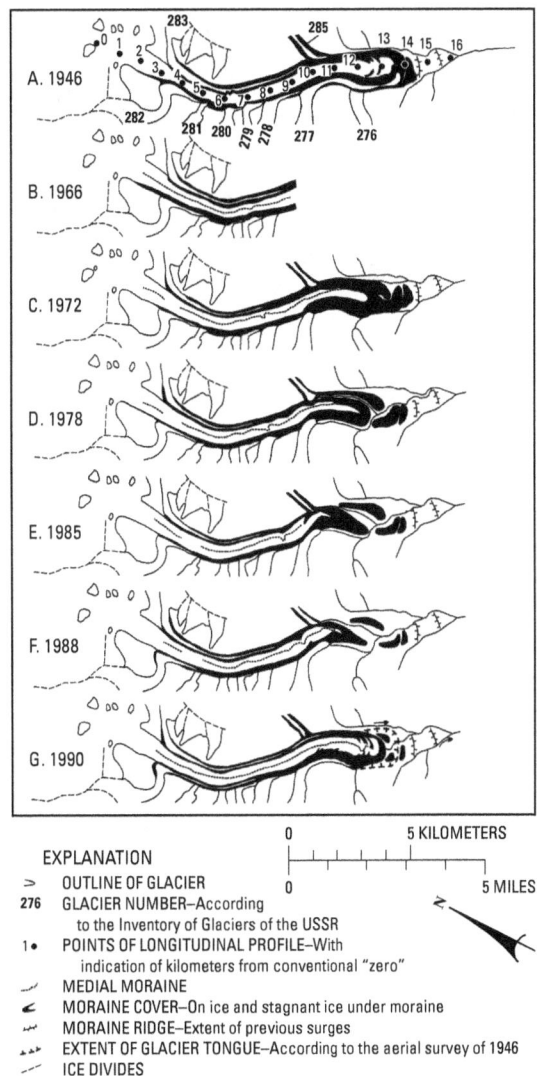

EXPLANATION

$\Rightarrow$	OUTLINE OF GLACIER
276	GLACIER NUMBER–According to the Inventory of Glaciers of the USSR
1•	POINTS OF LONGITUDINAL PROFILE–With indication of kilometers from conventional "zero"
	MEDIAL MORAINE
	MORAINE COVER–On ice and stagnant ice under moraine
	MORAINE RIDGE–Extent of previous surges
	EXTENT OF GLACIER TONGUE–According to the aerial survey of 1946
	ICE DIVIDES

Figure 69.—Morphology of the lower part of Lednik Oktyabr'skiye from 1946 to 1990.

1978, Pre-surge

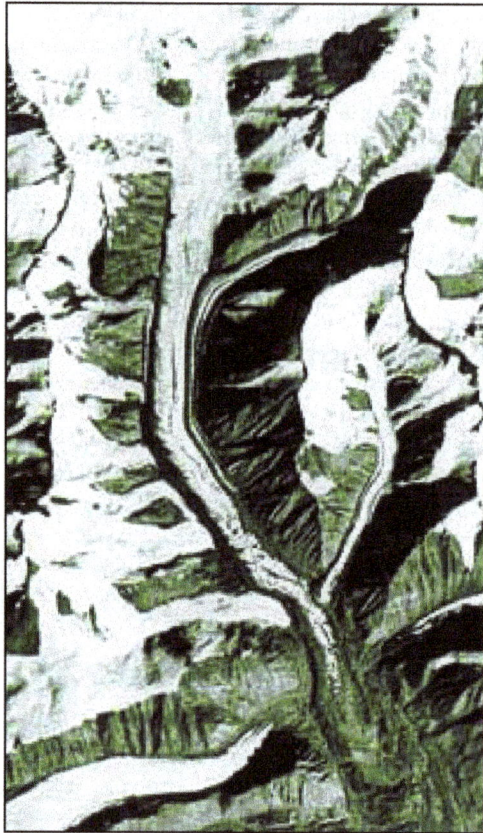

1990, Post-surge 73°E

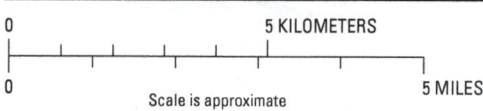

39°20′N

N

73°E

0   5 KILOMETERS

0   5 MILES
Scale is approximate

Figure 70.—*Photographs from space of Lednik Oktyabr'skiye. The photographs of 1978 (pre- surge) and 1990 (post-surge) were taken by the KFA-1000 camera (1000-mm lens) from the Resurs-F1 satellites.*

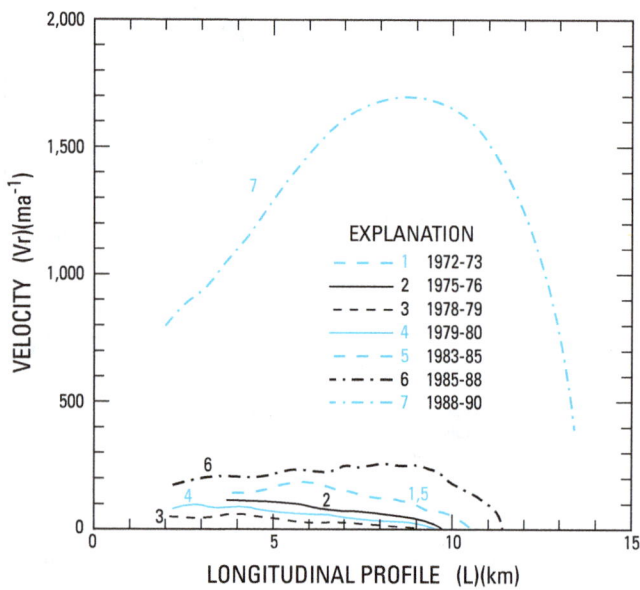

EXPLANATION

	1	1972-73
	2	1975-76
	3	1978-79
	4	1979-80
	5	1983-85
	6	1985-88
	7	1988-90

VELOCITY $(V_r)(ma^{-1})$

LONGITUDINAL PROFILE (L)(km)

Figure 71.—*Graph of the velocity (V_r) of the ice movement along longitudinal profile L of Lednik Oktyabr'skiye, showing the stage of recovery (1972–1985) and the surge (1985–1990). The location of the longitudinal profile is shown on figure 69A.*

The impact of glacier tributaries and the main glacier trunk on each other during surges was investigated on *Lednik Bivachniy*, the largest, western tributary of Lednik Fedchenko. Figure 72 shows the change of the surface structure of *Lednik Bivachniy* from 1973 to 2001 in the region of the confluence of *Lednik Moskovskogo Gosudarstvennogo Universiteta* (abbreviated *MGU*) (Moscow State University), the large western (left) tributary. During this time period, two surges of *Lednik Bivachniy* occurred — one between 1975 and 1977 and the other probably in the late 1990s (Desinov and others, 2001).

Lednik MGU continuously advanced into the valley of *Lednik Bivachniy* with rather constant velocity. By 1973, the ice of *Lednik MGU* had formed a "bulge," displacing the ice of the main trunk of *Lednik Bivachniy* to the opposite side of the valley and acting as a dam to *Lednik Bivachniy* (fig. 72A). As a result, a compression zone formed behind the dam in *Lednik Bivachniy* and, in 1976–1977, the release of accumulated stresses caused a sharp increase in ice movement, increasing in velocity by almost an order of magnitude — a surge event. As a result of the rapid movement of ice in the main trunk of *Lednik Bivachniy*, the noticeable ice bulge of the *Lednik MGU* tongue was transported 3.5 km downstream from the place of its confluence (fig. 72B). After this surge, the tongue of *Lednik Bivachniy* became dormant and *Lednik MGU* started to advance into the main valley again. By 1990, a new "bulge" had formed in the region of the confluence of the two glaciers, and the situation was analogous to the one in 1973.

On figure 72C, it can be seen that *Lednik Bivachniy* experienced one more surge not long before 2001. It developed according to the same scenario, and the bulge formed by *Lednik MGU* in 1990 moved downstream, taking the place of the bulge of the previous surge. The bulge from the previous surge in turn moved considerably farther down-valley and became heavily covered with debris.

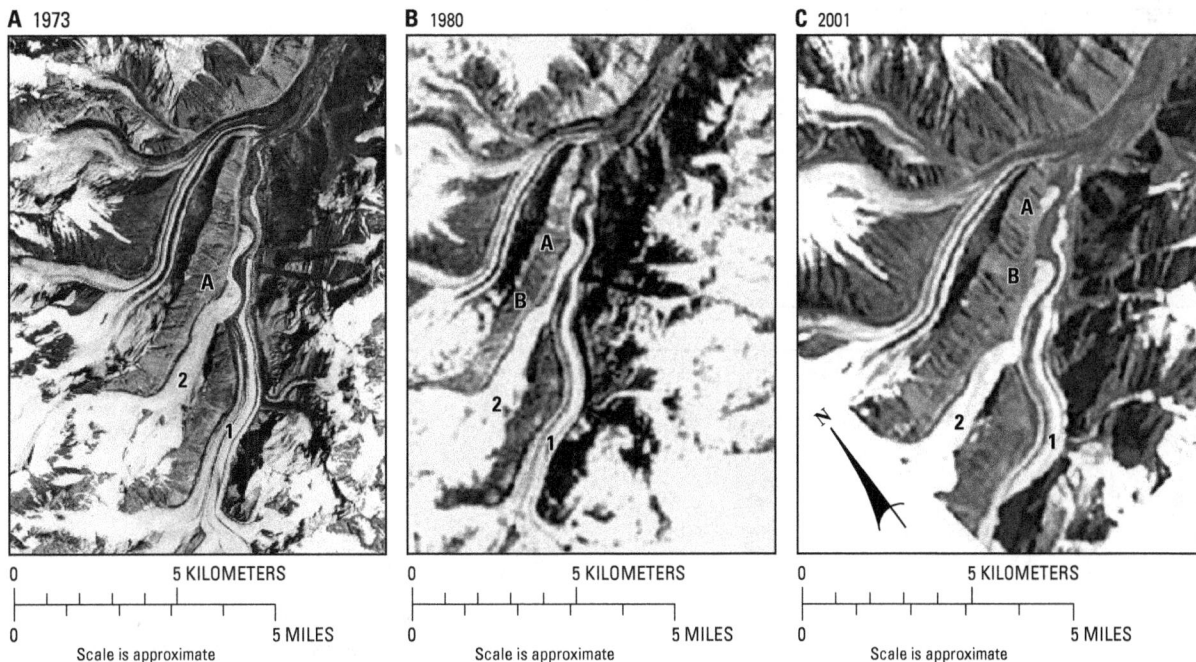

Figure 72.—Photographs of Lednik Bivachniy (1) (approximate lat 39°N., long 72°E.) and Lednik Moscow State University (MGU) (2). **A**, 1973, pre-surge; **B**, 1980, post-surge; and **C**, 2001 soon after the successive surge. On the photographs, **A** is the bulge formed before the 1976–1977 surge; **B** is the bulge formed before the late 1990s surge. **A** and **B** were taken by the KFA-1000 camera of a Resurs F-1 satellite. **C** was imaged from the International Space Station with an 80-cm lens.

As a result of such periodicity, several bulges of *Lednik MGU* can be seen on the main trunk of *Lednik Bivachniy*, transported during successive surges. On figure 72, the bulges of *Lednik MGU* are identified by letters: A is the bulge formed before the surge of 1976–1977; B is the bulge formed before the surge of the late 1990s.

At present, the surges of *Lednik Bivachniy* are internal — despite high internal ice velocity (over 1,200 m a^{-1}), the surge events have taken place within the limits of the glacier. However, it has been reported that, at the beginning of the past century, the surges of *Lednik Bivachniy* activated the lower part of Lednik Fedchenko (Osipova and Tsvetkov, 1999).

On figure 73, three images from space show the condition of the complex system of Ledniki Gando-*Dorofeeva*: (A) before the 1989–1990 surge, (B) during the surge, and (C) after the surge. After a previous surge of this glacier at the end of the 1950s, there was degradation — the main trunk of the glacier was covered with debris, and the main tributary, *Lednik Dorofeeva,*

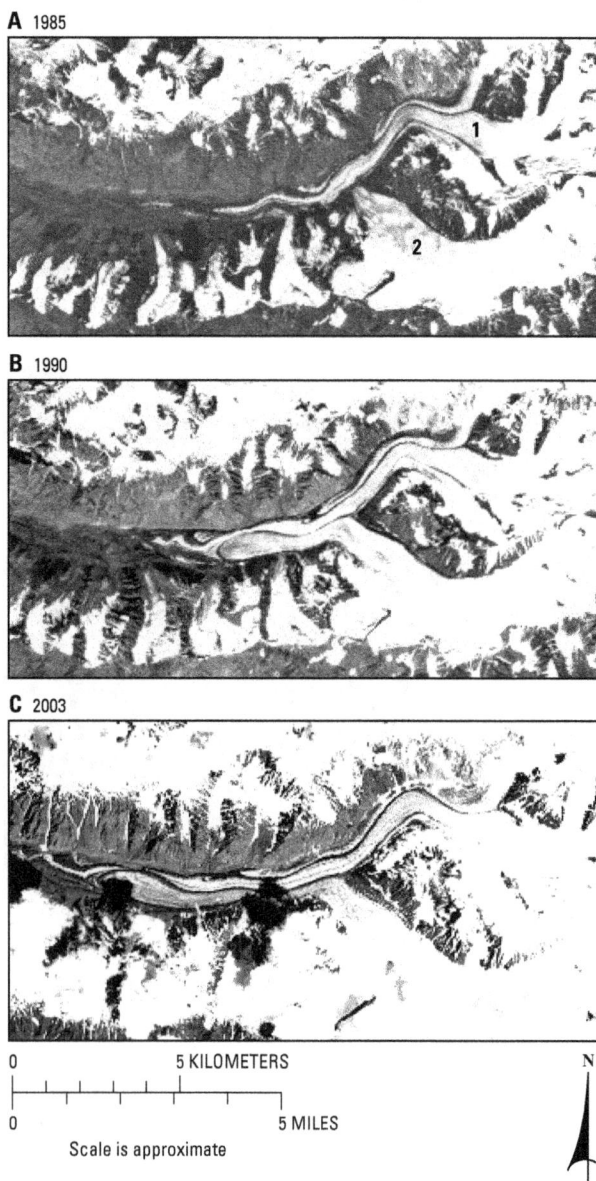

A 1985

B 1990

C 2003

0   5 KILOMETERS

0   5 MILES

Scale is approximate

N

Figure 73.—Images from space of the Gando-Dorofeeva glacier system (approximate lat 38°55'N., long 71°45'E.) before, during, and after the 1989–1990 surge. **A**, August 1985 on the eve of the surge. **B**, September 1990 in the midst of the surge. **C**, 23 August 2003 after the surge. 1, Lednik Gando, 2, Lednik Dorofeeva. **A** and **B** are photographs from the KFA-1000 camera on the Resurs-F1 satellite series. **C** is an Advanced Spaceborne Thermal Emission Radiometer (ASTER) image from the Terra satellite.

separated. At the beginning of the 1980s, *Lednik Dorofeeva* became active again and its terminus advanced and became dammed by ice from the main trunk. As a result of strong longitudinal compression, the width and thickness of the lower part of the *Lednik Dorofeeva* tongue increased considerably, testifying to its critical state (fig. 73A).

On the space photograph of 1990 (fig. 73B), the Ledniki Gando-*Dorofeeva* system is in the midst of a large surge event. In 1989–1990, ice velocity was 1.4 km a^{-1} in the middle part of the tongue of Lednik Gando, and by 1991, the active ice front had advanced by 8 km. And, judging by velocity curves and the surface condition of the glacier, the surge was not over yet (Osipova and Tsvetkov, 2001).

A 2003 space survey (fig. 73C) documented the Ledniki Gando system after the surge. The total advance of the dynamic terminus was 9.5 km; however, the lower 1.5 km of the tongue was not affected. Therefore, the surge of this glacier, and that of some of the other compound glaciers observed by us (Ledniki Bivachnyy, *Sugran*, and others), is considered an internal surge.

This analysis shows the effectiveness of repeated space surveys (especially those acquired over a period of many years) for monitoring fluctuations of large compound glaciers. No other method makes it possible to observe the entire area, to repeat investigations with short time intervals, and to provide the opportunity for any investigator to make simple, accurate stereoscopic measurements, especially of ice velocity.

The State of Surging Glaciers — Ledniki Medvezhiy and Geograficheskogo Obshchestva

Data acquired by Earth-orbiting satellites make it possible to investigate potentially catastrophic natural hazards and associated phenomena that could not be studied otherwise. At present, several satellites are orbiting the Earth, carrying a variety of remote sensors which can provide solutions to various scientific problems and practical needs in conjunction with well-planned programs. One such program, "Uragan", was planned jointly by the Institute of Geography of the Russian Academy of Sciences and the Russian "Energiya" Space Rocket Corporation, and is being carried out on the International Space Station (ISS). Visual-instrumental observations are being made, and digital images are being acquired by astronauts to study changes in the geographical environment, including natural disasters. In particular, Ledniki Medvezhiy and Geograficheskogo Obshchestva, situated near each other in the upper course of the Vanj river, are objects of constant observation and photography from the ISS (see fig. 67).

Lednik Medvezhiy (approximate lat 38°40′N., long 72°12′E.) is a typical surge-type glacier. Scientists of the Institute of Geography carried out annual surface investigations on this glacier from 1963 to 1987. From 1988 to 1991, observations were made based on multiple repeat aerial photography surveys. These investigations made it possible to comprehensively study the evolution of the glacier during its three surges (1963, 1973, and 1989) and its two stages of recovery (Dolgoushin and Osipova, 1982; Osipova and Tsvetkov, 1991; Kotlyakov and others, 1997). As a result of these multi-year investigations, the method of forecasting glacier surges based on monitoring the changes in the surface altitude and ice movement was developed (Dolgoushin and Osipova, 1975).

Figure 74A, B, and C are parts of three of the eight aerial photographs taken of Lednik Medvezhiy during its surge in 1988–1989. At that time, investigations were based on photogrammetric analysis of aerial photographs and comparison of successive topographic maps of the glacier. At the beginning of

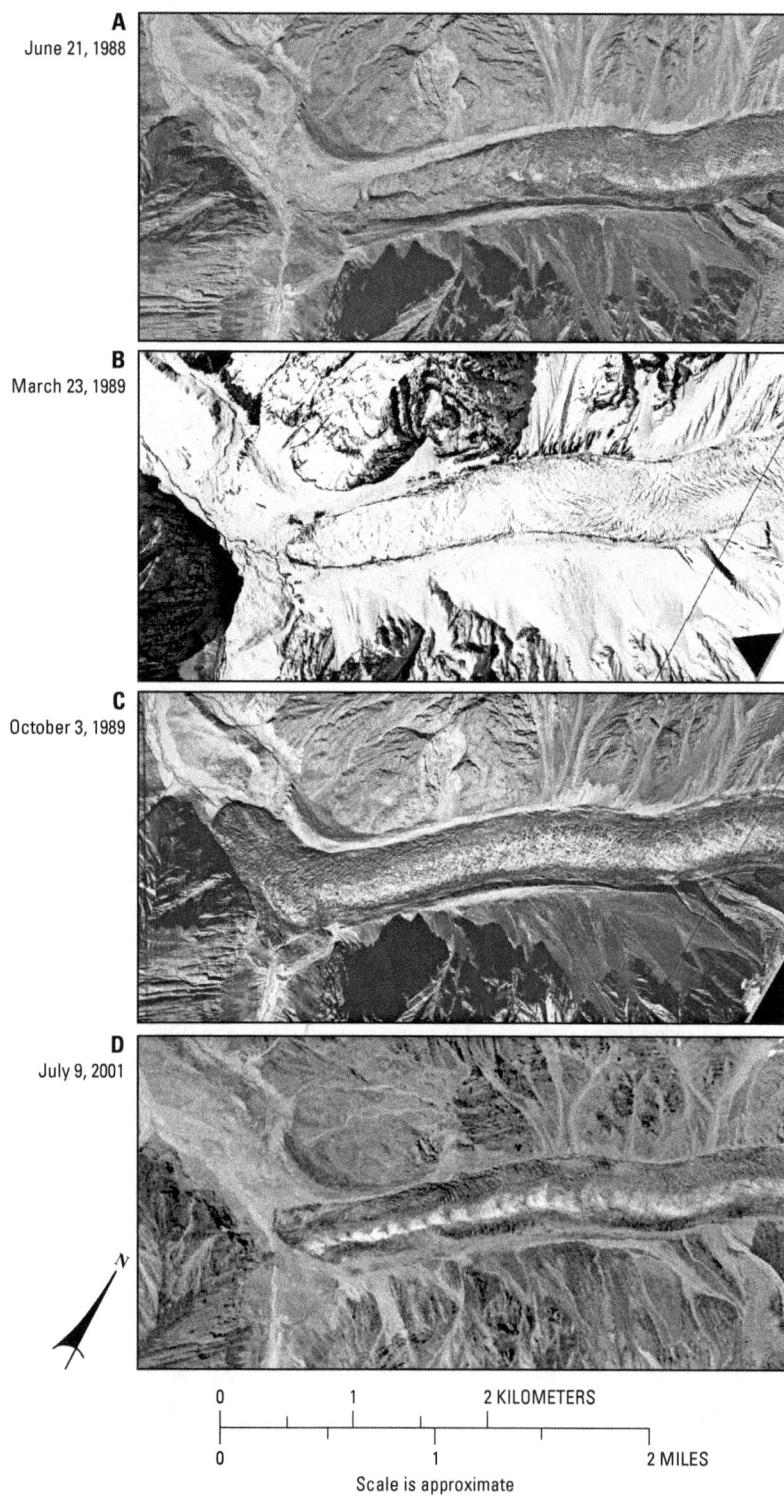

A
June 21, 1988

B
March 23, 1989

C
October 3, 1989

D
July 9, 2001

N

0   1   2 KILOMETERS

0   1   2 MILES

Scale is approximate

Figure 74.—*The tongue of Lednik Medvezhiy (approximate lat 38°40'N., long 72°12'E.) during the surge of 1988 to 1989, and in 2001. Aerial photographs:* **A**, *21 June 1988;* **B**, *23 March 1989;* **C**, *3 October 1989;* **D**, *an image from the International Space Station on 9 July 2001 with a digital camera with a 80-cm lens.*

the movement, the glacier terminus was situated at a distance of 1.2 km from the *Abdukagor* river (fig. 74A). Between the surveys of 21 June and 13 September 1988, it advanced 20–40 m. Figure 74B shows that the terminus of the tongue advanced 200 m more during winter, and the tongue assumed the shape typical of advance; it was fringed by a marginal fracture (a black line on the background of snow-covered slopes). The maximum rapid movement — up to 70 m d^{-1} — was measured by aerial pseudoparallax on surveys of 20 and 22 June 1989. The end of the surge occurred abruptly in July, and on an image of 3 October 1989 (fig. 74C), the surface structure of the tongue already had an "after-surge" appearance (Kotlyakov and others, 1997).

From 1996 to 2000, astronauts of the Russian Mir Orbital Space Station watched the dynamics of the glacier with the aid of binoculars. In February 2001, astronauts of the ISS transmitted information of renewed activity of the glacier — a distinct dark boundary fringing its snow-covered tongue. Such a boundary, as was seen on figure 74B, is a marginal fracture, and testifies to the fact that the glacier is in a state of rapid movement. Since that time, constant observations and images using hand-operated digital cameras were made from the ISS to track the condition of the glacier and to forecast a possible surge event.

Conclusions of the visual observations of the astronauts were passed to government officials of Tajikistan. Scientists from Glavtadjikgidromet, the Department of Hydrometeorology in Tajikistan, made surface observations on the glacier from 22 to 29 June 2001 and determined that, in comparison with 2000, the tongue had advanced by 450 to 500 m, but was still 200 to 250 m from the channel of the *Abdukagor* river. More than a week of observations determined velocity of the glacier — it was equal to 10–15 cm d^{-1}. Based on this, it was concluded that the glacier's advance was nearly complete, and the *Abdukagor* river would not be dammed.

By comparing these findings with previous investigations, we estimated that the peak of the glacier's activity occurred in May and early June. The first successful images (without cloud cover obscuring the tongue) were received from the ISS from a digital camera with 80-cm lens (pixel resolution of ~10 m) at the end of May and at the beginning of July (fig. 74D). By comparing ISS images taken from then (July 2001) until 2003, it was determined that during the two years, the glacier terminus advanced less than 100 m. However, during all this time, the structure of its surface, smooth alternating white and dark ramparts (ogives?), remained unchanged. But the most significant fact is that the tongue's movement from 2001 to 2003 was not as large or as destructive as during previous surges. For example, compare the structure of the surface and the overall morphology of the tongue on figs. 74C and D.

We concluded that a strong microsurge most likely took place on Lednik Medvezhiy in 2000–2001. As a result, the glacier front advanced down valley. It did not meet resistance from deposits of stagnant ice that were remnants of the former surge, because they had melted by this time. However, interpretations of ISS images indicated that the surface in the upper course of the surging part of the glacier (in the area below the icefall) did not lower. Therefore, it is likely that the glacier's state is critical, and in the near future, the next, and possibly major, surge of Lednik Medvezhiy will take place.

Lednik (Russkogo) Geograficheskogo Obshchestva, abbreviated RGO (lat 38°45′N., long 72°07′E.), is one of the six largest glaciers of the Pamirs. At the beginning of the 20th century, traces of its recent surge were detected — the tongue of the glacier had moved to the *Abdukagor* river valley and crossed it, and on the valley sides, the outline of a dammed lake was preserved (fig. 75). When the next surge of the glacier takes place, a dammed lake might reform, and its outburst flood may cause large-scale damage downstream in the Vanj river valley.

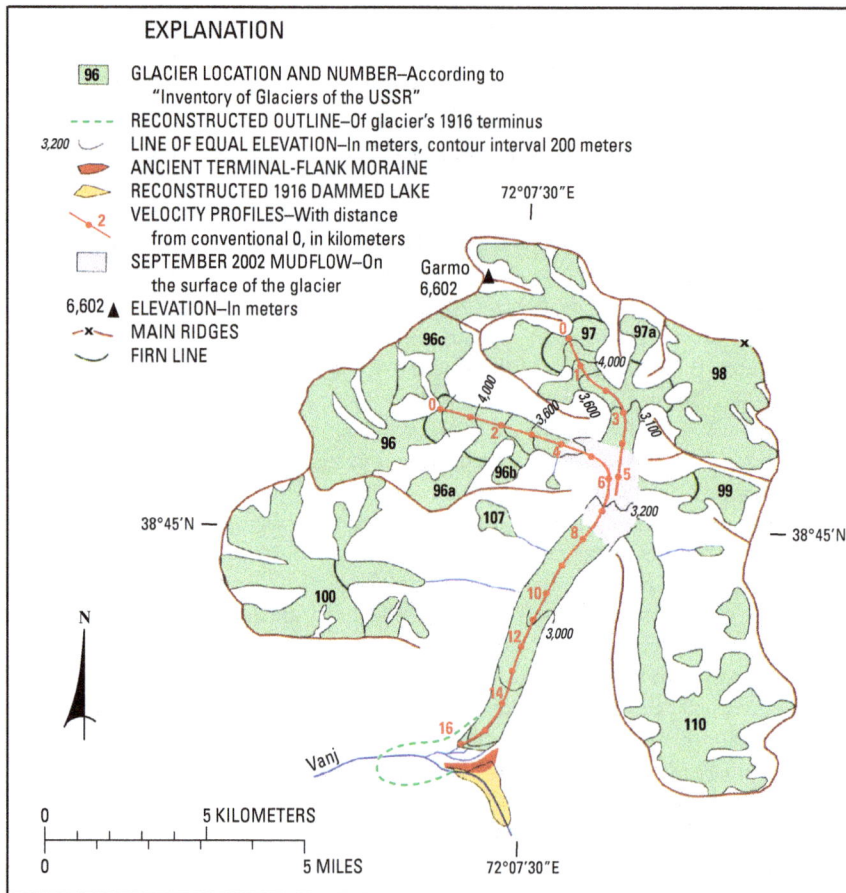

Figure 75.—*Sketch map of the Lednik Geograficheskogo Obschestva (RGO Glacier) basin showing glaciers, the reconstructed outline of the glacier's terminus in 1916, a reconstructed dammed lake in 1916, locations of velocity profiles, and a mudflow on the surface of the glacier in September 2002.*

Therefore, the increasing activity of Lednik RGO tongue, detected in August–October 2002 with surface and helicopter surveys (fig. 76), drew the attention of scientists of the Institute of Geography (Kotlyakov and others, 2003). Analysis of ISS images and helicopter photographs detected a recent mudflow on Lednik RGO in the region of the confluence of the main flows of Numbers 96 and 97 glaciers (figs. 75, 77). Previously, such large-scale mudflows had not been observed on the glaciers of the Pamirs, although large-scale mudflows are common in other glacierized mountain regions.

Later, 2003 ASTER images and photographs taken from a helicopter with a hand-held camera were analyzed. It was determined that the glacier's terminus had started to degrade, and that the activity of the glacier had decreased. It was also found that the source of the mudflow was the slopes of the upper course of the left tributary — Number 96C glacier, 2.5-3 km southwest of Khrebet Garmo. The mudflow had traveled about 7-8 km, and it covered an area of about 10 km^2.

Whatever the cause of the mudflow, it could have had two major impacts on the glacier. On the one hand, it could have caused a dynamic impact that traveled down the glacier and increased activity at the Lednik RGO terminus. However, this probably did not happen, because no changes were found in the surface structure of the glacier either at the location of the mudflow or below it.

Figure 76.—*Aerial photograph taken on 13 October 2002 of the terminus of Lednik RGO, looking northeast. Longitudinal fissures and outcrops of ice are observed along the edges of the tongue and in other places. The transverse profile of the glacier is convex. Collapsed pieces of ice can be seen at the foot of the steep terminus of the tongue. Photo taken by Prof. Zh. F. Shneider.*

Most probably, the mudflow caused a change in ablation conditions. By covering the surface, the resulting decrease of ablation could have locally increased the thickness of the glacier, and eventually advanced its terminus.

In both of the examples cited here, data acquired from space images and photography provided invaluable help in scientifically observing and monitoring the evolution of surge-type glaciers and in forecasting catastrophic surge events. The space imagery periodicity, type, availability, and processing capabilities all affect the ability of scientists to perform regional- and global-scale glaciological monitoring.

A 2001

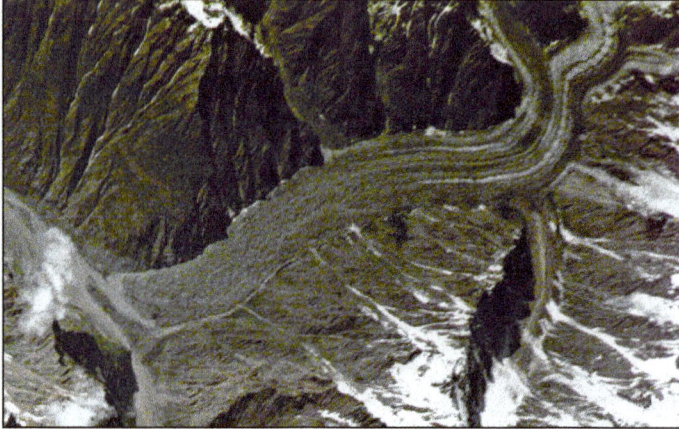

B 2002

*Figure 77.—International Space Station images of Lednik RGO (lat 38°45'N., long 72°07'E.) taken on **A**, 25 June 2001 and **B**, 30 September 2002. On the second image, a mudflow is visible that post-dated the 2001 image. Compare with figure 75 for scale.*

Investigations of the Fluctuations of "Normal" Non-Surging Glaciers in the Eastern Pamirs

The Eastern Pamirs is the most continental part of the Pamirs. The area is characterized by high elevation, low dissection of relief, low values of accumulation and ablation, and few changes in the glaciers from year to year.

To estimate the contemporary evolution of the glacierization of the Eastern Pamirs, extensive data from remote monitoring of fluctuations of glaciers' termini were used (Osipova and Tsvetkov, 2000). The shifts of termini of more than 160 glaciers in different parts of the Eastern Pamirs were measured by stereocomparator and the pseudoparallax method on images and photographs from space acquired between 1972 and 2001 (fig. 78). In optimum circumstances, data for three time intervals were obtained and analyzed: 1972–1980, 1980–1990, and 1990–2001.

Figure 78.—*Landsat image of the Eastern Pamirs where glacier-retreat measurements listed in table 21 were made.*

All investigated glaciers either retreated or were stationary. The greatest retreat velocities, up to or more than 20 m a^{-1}, were registered on tongues of large compound glaciers having termini that ended at low elevations. With a decrease in glacier size and an increase in the elevation of the glacier termini, ice retreat velocity decreased. Valley glaciers with areas from 2 to 5 km^2 correspond to the retreat value averaged by all investigated glaciers. An increase in the area of moraine cover on glacier tongues was documented everywhere.

The total results of the measurements are given in table 21. From this table we can see that the average retreat velocity (V_1) during the second time interval increased on an average of about 1.5 times. The minimum increase of 1.2 times was measured at *Zaalayskiy Khrebet* (Qatorkühi Pasi Oloy/Chong Alay Kyrka Toosu), the maximum — a two-fold increase — was measured on the western slope of Khrebet Sarykol'skiy. Average retreat velocity of 30 glaciers of Khrebet Zulumart from 1990 to 2001 was 14 percent less than on the same glaciers during the previous interval.

TABLE 21.—*Retreat velocity of the termini of glaciers of the Eastern Pamirs from 1972 to 2001*

[Retreat velocity (V_1) shown in meters per year (m a^{-1})]

	Region	1972–1980		1980–1990		1990–2001	
		V_1 (m a^{-1})	Number of glaciers	V_1 (m a^{-1})	Number of glaciers	V_1 (m a^{-1})	Number of glaciers
I	Khrebet Zulumart	-4.4	49	-6.5	51	-6.0	30
	Western slope	—	—	—	—	—	—
	Eastern slope	-3.0	11	-5.8	12	—	—
II	Zaalayskiy Khrebet	-5.5	8	-6.8	8	—	—
III	Khrebet Sarykol'skiy, western slope	-1.3	5	-2.6	19	—	—
IV	Khrebet Muzkol	—	—	-5.0	9	—	—
V	Severo-Alichursky Khrebet, northern slope	-1.3	7	-2.5	7	—	—
	Eastern Pamirs as a whole	—	80	—	106	-6.0	30

On *Lednik Severniy Zulumart* (lat 39°06′N., long 72°50′E., length 11.6 km, area 29.1 km², according to data for 1966) investigations of fluctuations of its area and terminus location were carried out for several years on the basis of comparison of an aerial photographic survey of 1946, imagery and photography of 1972, 1978, 1990 (KFA-1000 camera on the Resurs-F1 satellite) and ASTER imagery of 2001. Results are presented on figure 79 and in table 22.

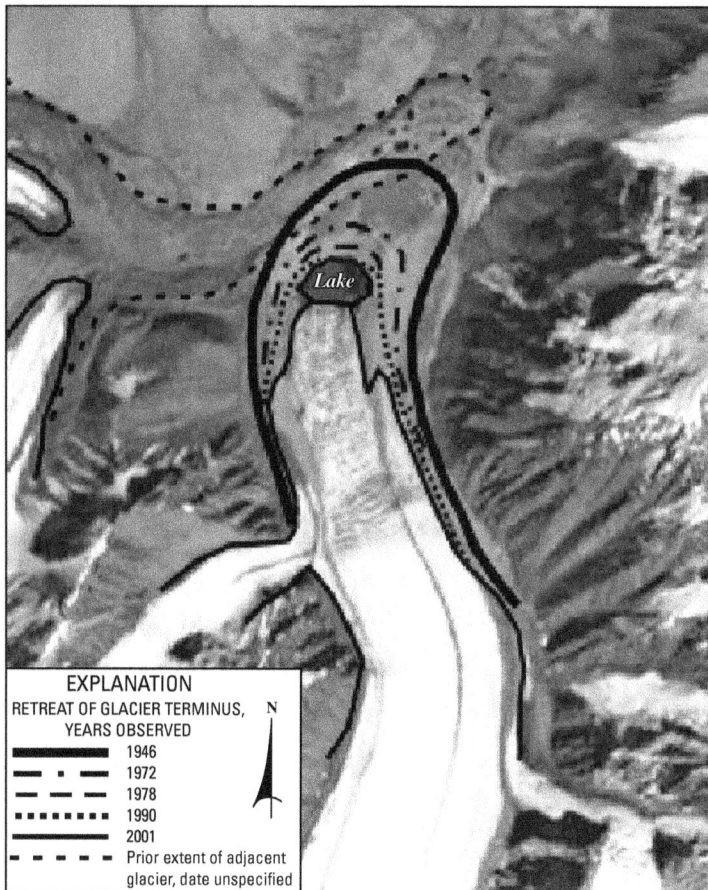

EXPLANATION
RETREAT OF GLACIER TERMINUS, YEARS OBSERVED
— 1946
—·— 1972
— — 1978
········ 1990
— 2001
– – – Prior extent of adjacent glacier, date unspecified

Figure 79.—Sketch map showing the retreat of the terminus of Lednik Severniy Zulumart (Eastern Pamirs) (lat 39°06′N., long 72°50′E.) based on interpretation of a 1946 aerial photograph and 1972, 1978, 1990, and 2001 space images and photographs.

TABLE 22.—*Retreat of the terminus and decrease in the area of the tongue of* Lednik
Severniy Zulumart *from 1946 to 2001*

[Abbreviations: m, meter; m a^{-1}, meters per year; km^2, square kilometer]

Intervals between surveys	1946–1972	1972–1978	1978–1990	1990–2001	1946–2001
Retreat of terminus (m)	-500	-100	-200	-200	-1,000
Retreat velocity (m a^{-1})	-19	-17	-18	-18	-18
Change of area (km^2)	-0.5	-0.1	-0.2	-0.2	-1.0

From 1946 until 2001, the glacier retreated with approximately the same average velocity — 18 m a^{-1} — and shortened by 1.0 km (10 percent). During the same time period, the area of the tongue decreased by 1.0 km^2, 3 percent of the initial area of the entire glacier. By 2001, a proglacial lake had formed in front of its terminus. Detailed stereoscopic interpretation of the structure and morphology of the surface of the firn area of *Lednik Severniy Zulumart* did not reveal any substantial change of boundaries in its upper area during the entire 55 years of remote observations. Later studies compared the changes in glacier extent using historical data and ASTER imagery (Khromova and others, 2006).

Investigations of glacier fluctuations carried out in several areas have shown that there is a real possibility of replacing labor-intensive field measurements with remote observations based on simple stereomeasurements of large-scale images obtained at intervals of 5–10 years, as well as using other remote sensing tools and methods.

The Glaciology of the Russian High Arctic from Landsat Imagery

By J.A. Dowdeswell,[5] E.K. Dowdeswell,[5] M. Williams,[6]
and A.F. Glazovskii[2]

Abstract

The Russian High Arctic archipelagos of Franz Josef Land (Zemlya Frantsa-Iosifa), Severnaya Zemlya, and Novaya Zemlya, fringing the Barents and Kara (Karaskoye) Seas, contain more than 56,000 km² of glaciers and ice caps, or about 10 percent of the Earth's ice outside Antarctica and Greenland. In the Russian Far East, ice covering an area of less than 100 km² occurs in the DeLong[7] archipelago (Ostrova De-Longa), east of the ice-free New Siberian Islands (Novosibirskiye Ostrova). Relatively little is known about the glaciology of some of these islands. However, Landsat satellite imagery has provided a valuable descriptive and interpretative tool for acquiring baseline glaciological data, despite problems of cloud cover which restrict the number of high-quality images available. Drainage basins are a basic unit for glaciological measurements, and for modeling ice dynamics and responses to climate change. Ice divides throughout the Russian High Arctic have been mapped from Landsat imagery to define catchment basin geometry. Major ice-cap outlet glaciers have been identified from Landsat data, and sources and magnitudes of iceberg production have been investigated. The largest icebergs observed were more than 2 km long at the Znamenity Glacier (Lednik Znamenityy), located on Vilchek Land (Zemlya Vil'cheka) in the Franz Josef Land archipelago, and more than 4 km long off the Matusevich Ice Shelf (Lednik Matusevicha) in Severnaya Zemlya. Fringing ice shelves were also identified from Landsat imagery of Franz Josef Land. Relatively few surge-type glaciers are present in the Russian High Arctic, but looped medial moraines indicate that a number of ice-cap outlets in Novaya Zemlya, and possibly Severnaya Zemlya, are of surge-type. Summer Landsat imagery displays marked zones of different brightness which are interpreted to represent, with increasing altitude, bare ice, slush, and snow. Some low-elevation ice caps in Franz Josef Land in particular may be undergoing rapid decay, with little or no net accumulation even at their crests. This decay is probably related to the termination of the cold "Little Ice Age" in the Eurasian Arctic about 100 years ago, combined with recent warming.

---

[5]Scott Polar Research Institute, University of Cambridge, Cambridge CB2 1ER, U.K.

[6]School of Civil Engineering and Geosciences, University of Newcastle, Newcastle NE1 7RU, U.K.

[7]U.S. Government publications require that official geographic place-names for foreign countries be used to the greatest extent possible. In the Glaciology of the Russian High Arctic section, the use of geographic place-names is based on the U.S. Board on Geographic Names (BGN) website: *http://earth-info.nga.mil/gns/html/index.html*. The website lists some conventional names for the areas such as Franz Josef Land, Barents and Kara Seas, and Wrangel Island. All other names will be introduced in the text by the common English equivalent with the Russian name in parenthesis. The Russian placename will then be used throughout the text. Names not listed in the BGN website will be shown in italics.

Introduction

Glaciers and ice caps are present on the islands of the Russian High Arctic between latitudes 73° and 82°N., and longitudes 36° and 158°E., providing a high latitude west-east transect of about 2,000 km across Eurasia (fig. 80) (Sharp, 1956). These ice masses have a total glacierized area of more than 56,000 km^2 (Kotlyakov and others, this volume), which represents about 10 percent of the Earth's ice cover other than Greenland and Antarctica (Meier, 1984; Dyurgerov and Meier, 1997; Dowdeswell and Hagen, 2004). The three large archipelagos — Franz Josef Land, Novaya Zemlya and Severnaya Zemlya — account for more than 99 percent of this ice. Small isolated islands, for example Victoria Island (Ostrov Viktoriya) in the west and Ostrova De-Longa in the far east of the Eurasian High Arctic, contain ice caps up to only a few tens of square kilometers in areal extent (Govorukha, 1964; Verkulich and others, 1992).

In a companion paper, Kotlyakov and others (this chapter) describe the distribution and dimensions of the ice masses within the Russian High Arctic. This paper describes briefly what is known about the glaciology of the archipelagos and islands in the Russian Arctic sector, and demonstrates the utility of Landsat satellite imagery for studying the glaciers and ice caps in this remote and little-known region of the Arctic. The topics discussed here have been investigated in other glacierized regions using Landsat imagery (for example, Bindschadler and others, 2001); they include: (1) ice-surface topography, and in particular the definition of ice divides and drainage basins; (2) ice dynamics; and (3) snow-and-ice facies, with their implications for glacier mass balance.

The Russian High Arctic is cloud-covered for many days each year, which substantially impacts the acquisition of usable visible-band satellite imagery. Marshall and others (1994) analyzed cloud-cover statistics for a number of Landsat path-rows over each of the main ice-covered areas of the Eurasian Arctic sector, including Franz Josef Land, Novaya Zemlya and Severnaya Zemlya (fig. 81). They showed that, in general, very few high-quality images with low cloud cover are available for the Russian High Arctic. The probability of obtaining relatively cloud-free scenes is greatest for Novaya Zemlya (fig. 81A). Cloud-free images of Franz Josef Land are most likely to be obtained in May (fig. 81B), whereas July is the optimum month for Novaya Zemlya (fig. 81C), and September imagery is of uniformly poor quality, except for some opportunities in Severnaya Zemlya (fig. 81D).

Glaciological Background

The following section describes general glacier characterisitics, including meteorological conditions, for each of the three primary archipelagos in the Russian High Arctic — Franz Joseph Land, Severnaya Zemlya, and Novaya Zemlya. Only one percent of the glacierized Russian High Arctic lies outside these archipelagos; however, these small glacierized islands are also discussed here.

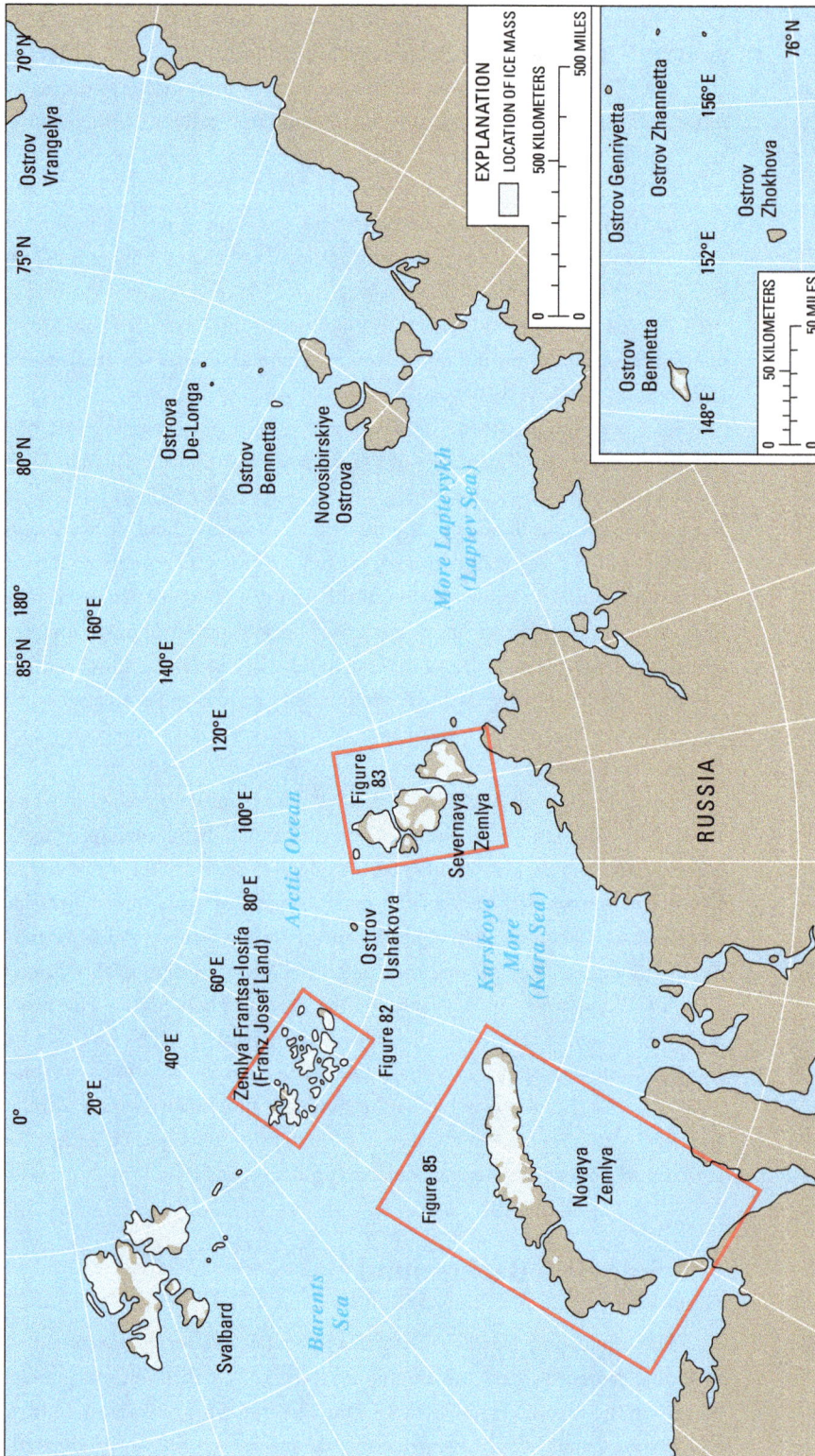

Figure 80.—Map of the Russian High Arctic showing the major ice-covered archipelagos of Franz Josef Land, Novaya Zemlya, and Severnaya Zemlya, together with the smaller glacierized islands further east (inset).

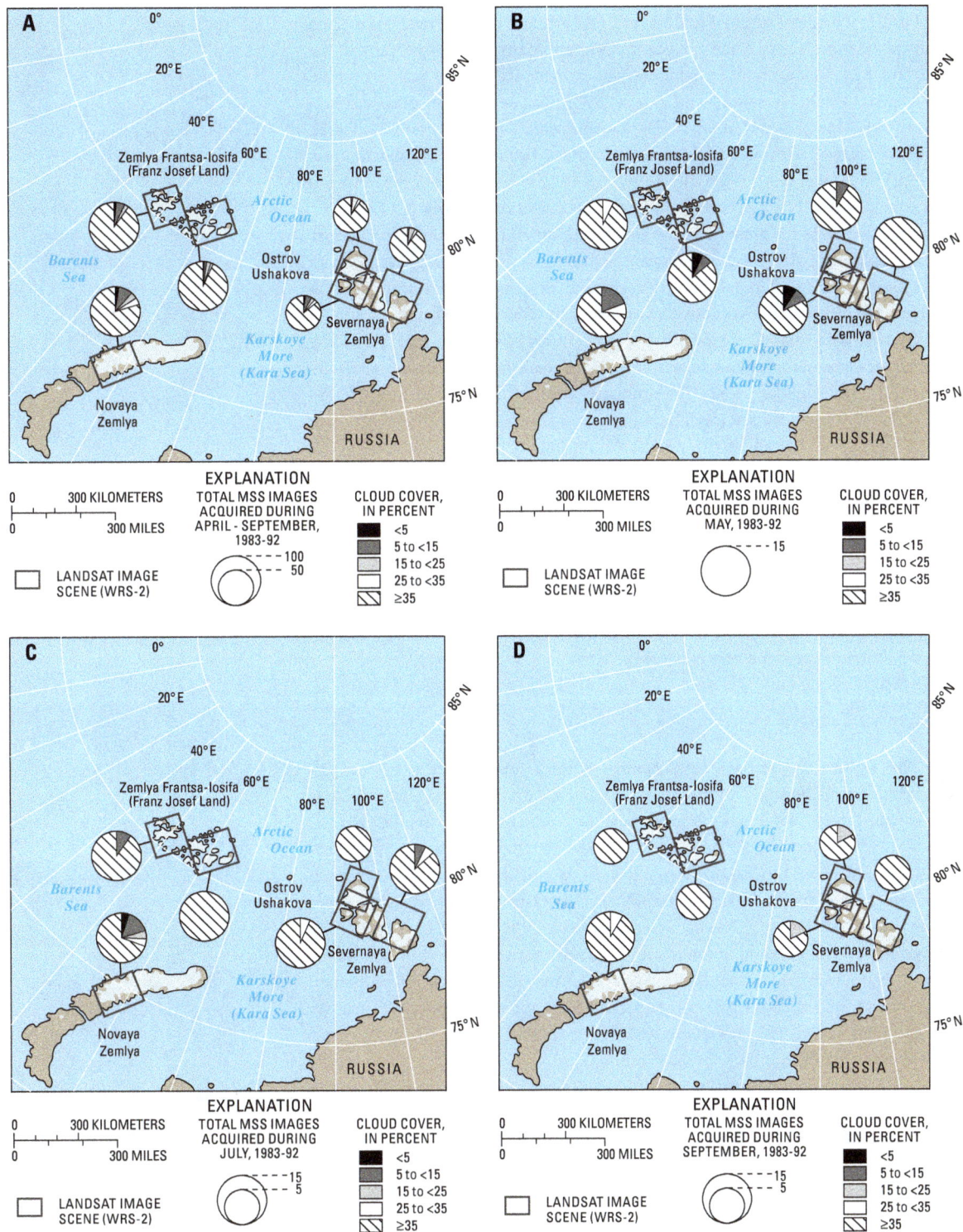

Figure 81.—Diagrams showing the effects of cloud cover on Landsat image quality for selected Path–Rows in the Russian High Arctic, including Franz Josef Land (Path 203–Row 001, Path 197–Row 001), Novaya Zemlya (Path 178–Row 007) and Severnaya Zemlya (Path 174–Row 001, Path 170–Row 002, Path 161–Row 003): **A**, April to September; **B**, May; **C**, July; and **D**, September. Modified from Marshall and others (1994).

Franz Josef Land

Glacier ice covers about 13,700 km^2 or 85 percent of the Franz Josef Land archipelago, and glaciers reach tidewater along approximately 60 percent of the coast (fig. 82). Relatively little is known about these ice masses (cf. Grosswald and Krenke, 1962; Grosswald and others, 1973; Sin'kevich and others, 1991; Dowdeswell, Gorman, and others, 1994; Dowdeswell, Glazovskii, and Macheret, 1995; Dowdeswell, Gorman, Glazovskii, and others, 1996). Mean annual temperature in Franz Josef Land is about −12 to −13 °C and mean July temperature is approximately 0 to 1 °C (Dowdeswell, 1995). Precipitation is about 0.3–0.5 m a^{-1} (ice equivalent), and is greatest in the southeast part of the archipelago, because of the direction of the prevailing moisture-bearing winds. The highest point in Franz Josef Land is 670 m, but many glaciers and ice caps have extensive areas at relatively low elevations (Dowdeswell, Gorman, and others, 1994).

Glacier mass balance is generally negative (mean of −0.26 m a^{-1}), according to field measurements linked to meteorological data for the period 1930–59 (Grosswald and Krenke, 1962). The modern equilibrium line altitude (ELA) is at about 300 m above sea level, based on an analysis of aerial photographs of the archipelago acquired in the 1950s. According to these studies, the ELA varies relatively little across Franz Josef Land (Grosswald and others, 1973). The summits of a number of smaller ice caps lie below 300 m, and these ice masses appear to be downwasting rapidly (Grosswald and Krenke, 1962; Dowdeswell, 1995). Many glaciers also show evidence of retreat during the present century from a "Little Ice Age" maximum position (Miller and others, 1992). This retreat applies particularly to tidewater glaciers, where mass is lost by significant iceberg calving as well as by melting (Grosswald and Krenke, 1962; Dowdeswell, Gorman, and others, 1994).

Severnaya Zemlya

The archipelago of Severnaya Zemlya is made up of four main islands: Komsomolets, Pioneer (Pioner), October Revolution (Oktyabr'skoy Revolyutsii), and Bol'shevik Islands, together with several smaller islands (fig. 83). Approximately 50 percent of the 36,800 km^2 archipelago is glacierized (Kotlyakov and others, this chapter). The largest ice mass is the Academy of Sciences Ice Cap (Lednik Akademii Nauk) on Ostrov Komsomolets, which has an area of 5,575 km^2 (Dowdeswell and others, 2002) (fig. 84). Floating ice shelves are fed from several of the ice caps on Severnaya Zemlya; the best-documented example is the *Matusevich Ice Shelf* on Ostrov Oktyabr'skoy Revolyutsii, which has an area of about 200 km^2 and a surface slope of less than 0.2° near its margin (Govorukha, 1988a; Dowdeswell, Gorman, and others, 1994; Williams and Dowdeswell, 2001).

Mean annual temperature recorded at Fëdorova at the northern tip of the Taymyr Peninsula (Poluostrov Taymyr), about 40 km south of Severnaya Zemlya, is approximately −15 °C, and mean July temperature is about 1.5 °C (Dowdeswell, 1995). Precipitation is about 0.25–0.3 m a^{-1} (ice equivalent). Mass-balance data for the Vavilov Ice Cap (Lednik Vavilova) (Barkov and others, 1992), a 1,820 km^2 ice cap on Ostrov Oktyabr'skoy Revolyutsii, show that the mean net balance was slightly negative between 1974 and 1988 (−0.03 m a^{-1} water equivalent), but that there was significant interannual variability about this value (standard deviation ±0.36 m a^{-1}). The equilibrium line altitude (ELA) varies from more than 600 m in southeastern Severnaya Zemlya, to about 300 m on Ostrov Komsomolets, to 200 m or less on Schmidt Island (Ostrov Shmidta) in the northwest (Govorukha, 1970), suggesting a moisture source and precipitation gradient from the southeast.

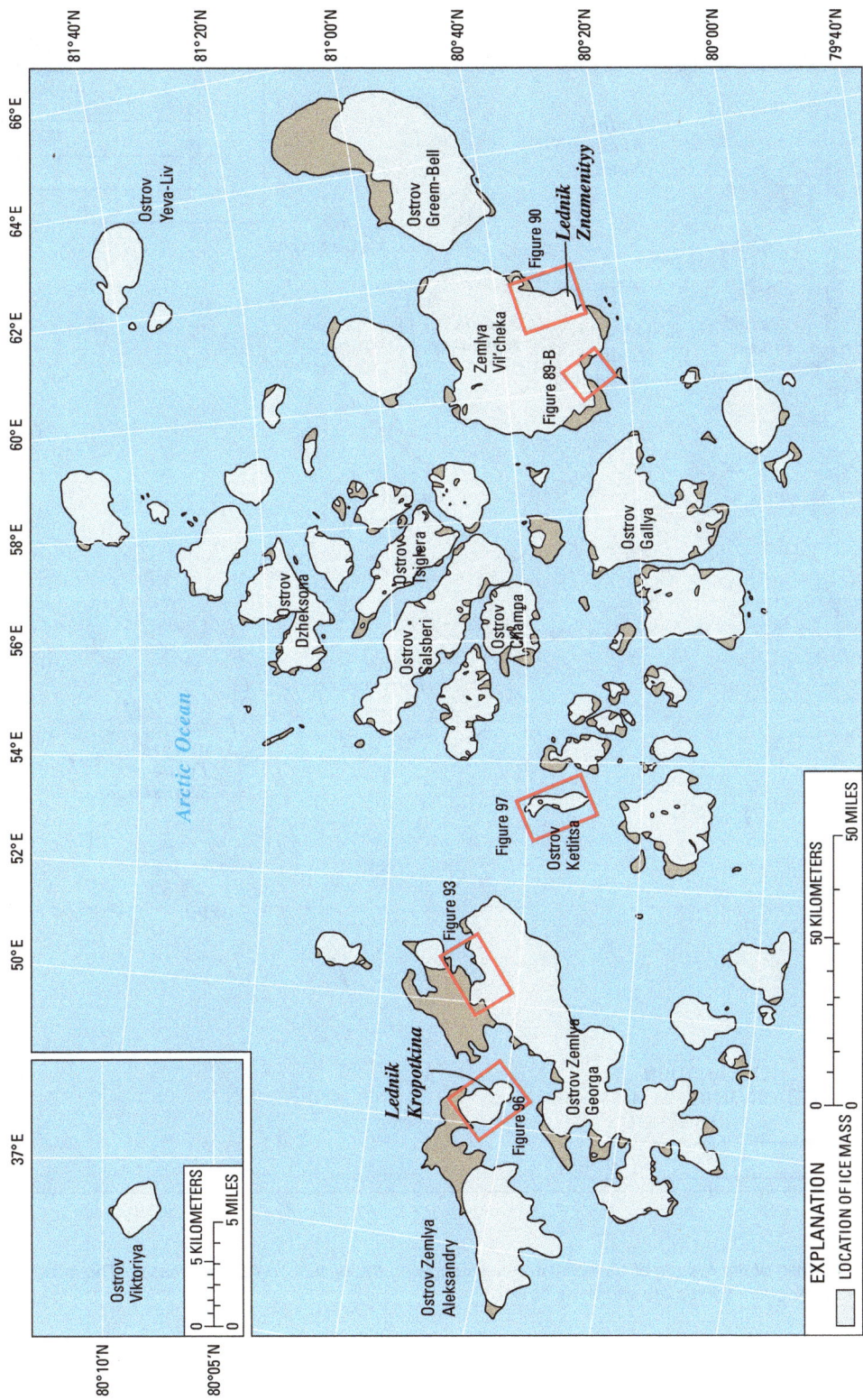

Figure 82.—*Map of the Franz Josef Land archipelago showing the distribution of ice masses. The locations of Landsat images in subsequent figures are shown as boxes.*

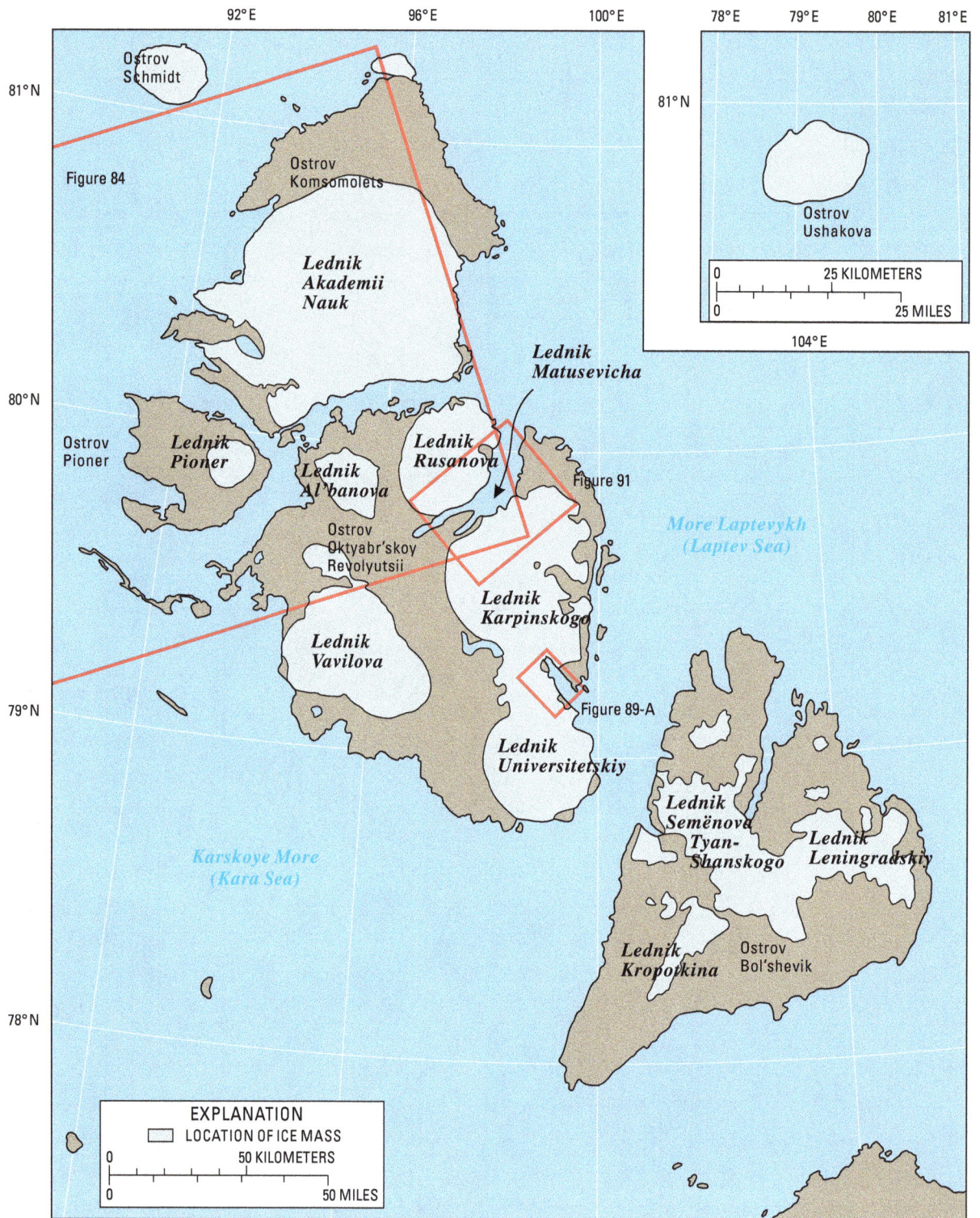

Figure 83.—Map of the islands of Severnaya Zemlya showing the distribution of ice masses. The locations of Landsat images in subsequent figures are shown as boxes.

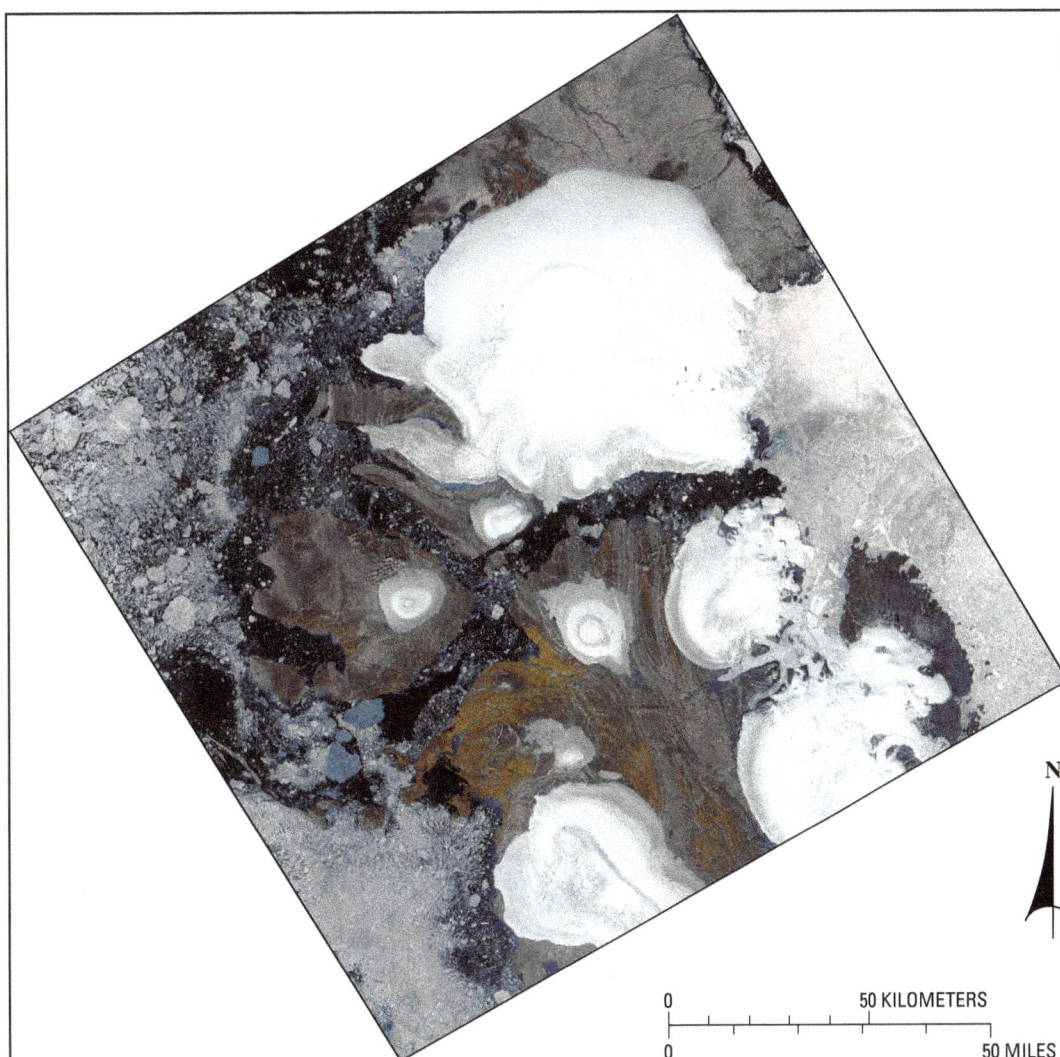

Figure 84.—Landsat false-color Multispectral Scanner (MSS) image of the ice caps in northern Severnaya Zemlya (located in fig. 83). The 185 x 185 km full scene was acquired on 22 July 1975. It shows the zones of melting on several ice caps. The 5,575 km² Lednik Akademii Nauk is the largest in the archipelago. The grayish zones closest to the ice-cap margins are interpreted to represent bare glacier ice. The upper, white regions are snow-covered ice, and the thin blue/gray zone separating them is interpreted as a slush zone. A summer cover of shorefast sea ice is present around the islands.

The oxygen-isotope and melt-layer records in deep ice cores (550–760 m deep) from the Ledniki Vavilova and Akademii Nauk indicate that temperatures have risen markedly in the last 120–140 years (Vaikmyae and Punning, 1984; Kotlyakov and others, 1989, 1990), but interpretation of the older parts of each core is hampered by poor chronological control associated with strong surface melting (Tarussov, 1992). Analysis of aerial photographs of the archipelago, acquired since the 1930s, shows a general retreat of glacier margins, interpreted to be linked to the termination of the "Little Ice Age," as indicated in the ice-core record. A loss of about 500 km² of ice-covered area in Severnaya Zemlya between 1931- 84 was reported by Govorukha and others (1987), although Koryakin (1986) suggests that this may be an over-estimate resulting from errors in interpretation of aerial photographs.

Novaya Zemlya

About 24,400 km^2 or 30 percent of the two major islands making up Novaya Zemlya is ice covered (fig. 85), with the larger proportion on the northern island (Kotlyakov and others, this volume). There appears to be little recent glaciological information available on Novaya Zemlya, perhaps reflecting its role as a major site for nuclear weapons testing (Lamb, 1991). Mean annual temperature on the northern tip of Novaya Zemlya is -9 °C, with a July mean of 2 °C; on the milder west coast, the July mean is 6.5 °C. Precipitation is also higher on the west coast, reflecting the influence of relatively moist westerly air masses derived from the Norwegian Sea. The late summer snowline elevation decreases toward the north; the elevation is about 600 m near Matochkin Shar, the strait separating the main islands (fig. 85), it falls to 450 m at about lat 75°N., and is at 325 m in the northernmost part of Novaya Zemlya (McCauley, 1958). The equilibrium line is about 200 m lower on the western Barents Sea margin than on the eastern Kara Sea side of the islands. Calculations of net mass balance for the ice cover on Novaya Zemlya, based on empirical relationships between temperature, accumulation, and ablation, suggest that the net balance has been negative, by a mean of about 0.16 m a^{-1}, for most years since the 1920s (Chizhov and Koryakin, 1962). Measurements on a number of glaciers, some of which drain the central ice field on the north island, indicate that ice masses shrank by about 2–3 percent between the 1930s and the 1950s, and some tidewater glaciers retreated by up to 7 km (Chizhov and Koryakin, 1962; Chizhov and others, 1968). Zeeberg (2001) and Zeeberg and Forman (2001) also report that most of the 20th century glacier retreat in northern Novaya Zemlya was complete by the 1950s, with slower retreat since that time.

Other Glacierized Islands in the Russian High Arctic

Less than 1 percent of the glacierized area in the Russian High Arctic occurs outside Franz Josef Land, Severnaya Zemlya, and Novaya Zemlya (fig. 80). Ostrov Viktoriya, between Svalbard and Franz Josef Land at 80°N., 37°E., is the western-most island in the Russian Arctic and lies only about 60 km east of the 99 percent ice covered Norwegian island of Kvitøya (Bamber and Dowdeswell, 1990). The 11 km^2 ice cap on Ostrov Viktoriya has a maximum elevation of 110 m. It has a consistently negative mass balance; its entire surface exhibits summer melting, and old glacier ice is exposed at the surface (Govorukha, 1964, 1988b).

Ushakov Island (Ostrov Ushakova), with a completely glacierized surface of 325 km^2, is located between Franz Josef Land and Severnaya Zemlya at lat 81°N., long 79.5°E. (fig. 80). The ice cap has a negative mass balance, but retains a zone of net accumulation above an equilibrium line at 200–250 m in altitude (Govorukha, 1988b). It has a maximum ice thickness of almost 300 m, measured from radio-echo sounding (Bogorodskii and Federov, 1971).

In the Asian Far East sector of the Russian High Arctic, ice masses totaling about 60 km^2 occur on Ostrova De-Longa, located northeast of the ice-free Novosibirskiye Ostrova between lat 76–77°N. and long 149–158°E. (fig. 80). On Ostrova De-Longa, the largest ice mass is the *Tollya Ice Cap* on Bennett Island (Ostrov Bennetta), with an area of about 55 km^2, a maximum elevation of 380 m, and a firn line at about 180 m. Smaller ice masses occur on Henrietta Island (Ostrov Genriyetta) and Jeanette Island (Ostrov Zhannetta), with areas of 6 and 0.4 km^2, respectively (Bazheva, 1981; Verkulich and others, 1992). Mass balance investigations on Ostrov Bennetta in 1986–87 showed net losses on each of the four ice masses and, combined with meteorological data for the previous 30 years, suggested negative balances for about 70 percent of this period (Verkulich and others, 1992). Wrangel Island (Ostrov Vrangelya), at 71°N., 180°E., contains a few semi-permanent snow patches but no glaciers.

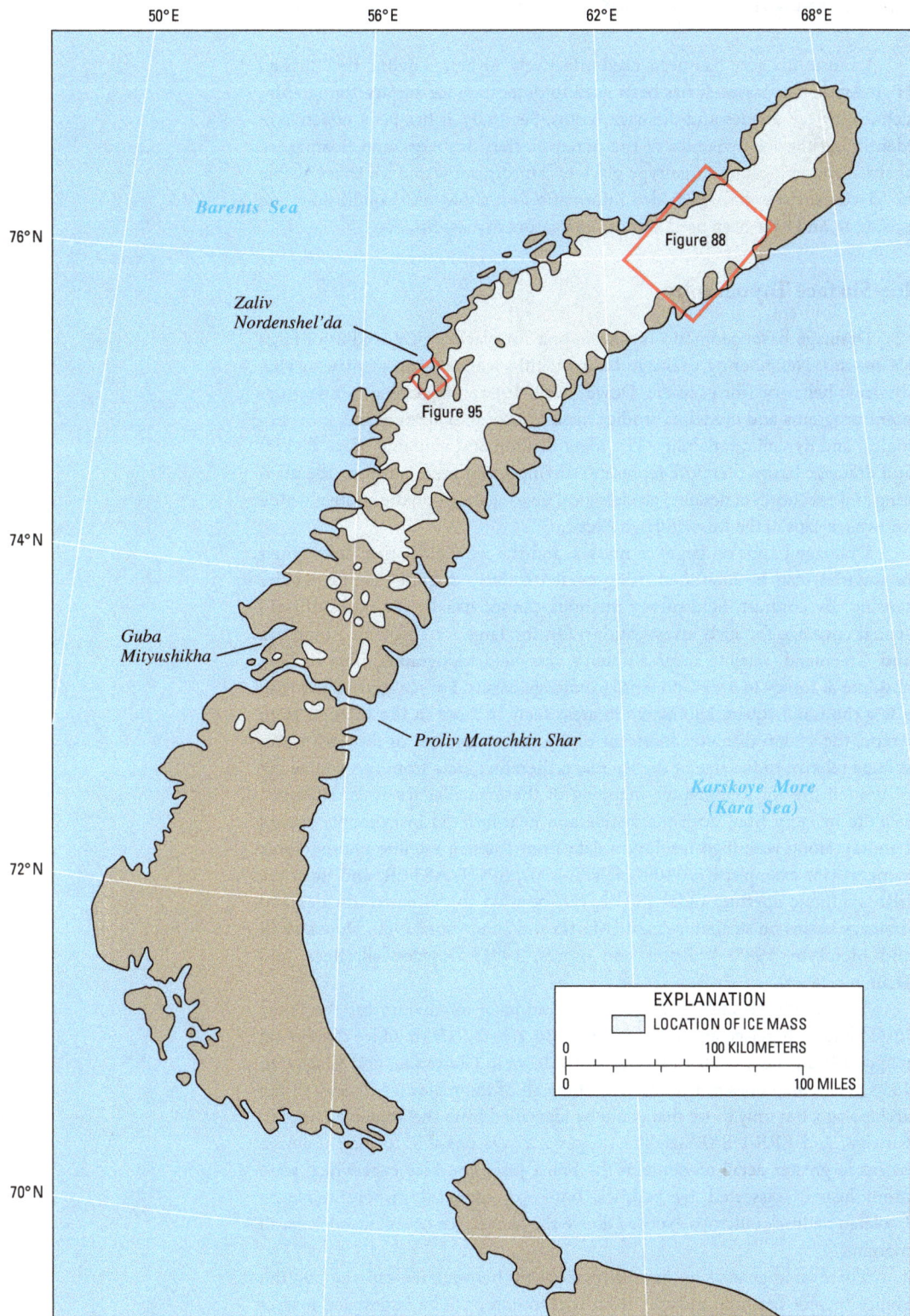

Figure 85.—Map of Novaya Zemlya showing the distribution of ice masses. The locations of Landsat images in subsequent figures are shown as boxes.

Glaciological Investigation of Russian High Arctic Ice Masses from Landsat Imagery

Landsat imagery has been used effectively to better define the Russian High Arctic ice masses. It has been used to determine ice-surface topography, delineating ice divides and drainage basins. Secondly, it has been effective in identifying the ice dynamics of the outlet glaciers, icebergs, and floating ice shelves, and in locating surge-type glaciers. Thirdly, snow and ice facies visible on glacier surfaces have provided information on snow- and equilibrium-line positions, and have been used as proxy indicators of mass balance.

Ice-Surface Topography

Drainage basin geometry is a basic unit for glaciological studies of large ice masses. Neighboring drainage basins within ice caps can display varying dynamic behavior (for example, Dowdeswell, 1986a). Glaciological measurement programs and modeling studies need to take into account this morphological and dynamic variability. The identification and mapping of ice divides and drainage basins therefore represents an important prerequisite to the planning and execution of detailed glaciological investigations on the relatively little known ice caps of the Russian High Arctic.

Drainage basins on large ice masses, and the ice divides that define their boundaries, can be identified using evidence derived from satellite remote sensing. By contrast, field-survey methods cannot usually provide sufficient spatial coverage for such investigations. On the large ice sheets of Antarctica and Greenland, satellite radar altimetry can yield topographic data of high absolute accuracy in order to identify drainage basins. For ice caps of less than a few thousand square kilometers in area, such as those in the Russian High Arctic, the 18-km diameter footprint of the radar altimeter of the ERS series is large relative to the size of the ice mass; therefore, data from this source are of limited use for topographic mapping of these smaller ice caps. However, satellite imagery from both the visible and near-infrared instruments aboard Landsat, along with high-resolution data from Russian satellite photographic cameras (for example, KFA-1000, KATE-200), SPOT, ASTER, and from the ERS synthetic aperture radar (SAR), can be used to define ice divides and drainage basins on smaller ice caps (Martin and Sanderson, 1980; Dowdeswell and McIntyre, 1987; Williams and others, 1991; Dowdeswell, Rees, and Diament, 1994).

The results of satellite image identification of ice divides for the Franz Josef Land archipelago are shown in figure 86. About 915 km of ice divides are mapped from the available imagery (Dowdeswell, Glazovskii, and Macheret, 1995). The map does not necessarily show all of the major ice divides in the archipelago, but only those that could be identified from the available Landsat, Russian, and ERS-1 SAR satellite imagery. It was possible to map drainage basins in greater detail on some of the Franz Josef Land ice caps which were more heavily dissected by bedrock outcrops, especially where nunataks (isolated bedrock outcrops exposed above the general ice cover) acted as basin margins.

There is, in general, an inverse relationship between ice-cap size and the ability to plot basin boundaries from rock outcrops. The largest ice masses in the Franz Josef Land archipelago, on Vilchek (Vil'cheka) and Graham Bell (Greem-Bell) islands (figs. 82, 86), have very few nunataks; therefore, basin extent can, for the most part, be mapped only from the identification of ice divides on satellite imagery, as illustrated on the satellite image of the Hall

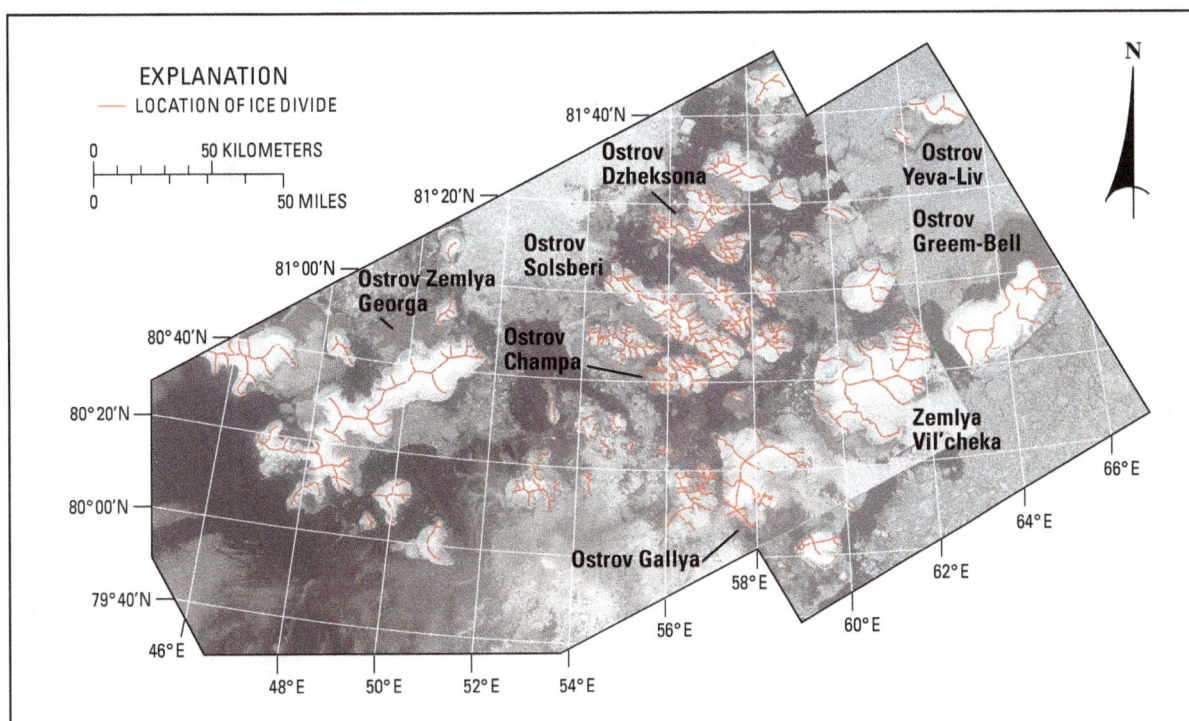

Figure 86.—Map of the ice divides on the ice caps of Franz Josef Land, shown on a georeferenced mosaic of Landsat TM images from Path 199–Rows 001 and 002, acquired on 25 July 1986. From Williams (2003).

Island (Ostrov Gallya) ice cap in figure 87. By contrast, the ice caps on islands such as Salisbury (Solsberi), Champ (Champa), and Jackson (Dzheksona) can be subdivided extensively on the basis of large numbers of nunataks. On a few ice caps, particularly those on Ostrova Greem-Bell and Eva-Liv (Yeva-Liv) (fig. 86), few ice divides were identified, probably because these ice masses have a particularly well-defined shape with a single summit, rather than having elongate crests with similar summit elevations extending for a number of kilometers, such as George Land (Ostrov Zemlya Georga), where the crest extends for tens of kilometers (fig. 86).

Similar analyses of the locations of ice-cap divides and drainage basins have been conducted for the archipelagos of Severnaya Zemlya and Novaya Zemlya (for example, Williams and Dowdeswell, 2001). On the northern island of Novaya Zemlya (fig. 85), the position of the major ice divide separating ice flowing west to the Barents Sea and east to the Kara Sea is shown in figure 88. The ice ridges between the drainage basins of individual outlet glaciers ending in tidewater can also be seen. The outlet glaciers generally have less smooth surfaces than the intervening ridges.

Ice Dynamics

Landsat imagery has been used to determine the ice dynamics of Russian High Arctic outlet glaciers, icebergs, and floating ice shelves, or ice tongues. It has also been valuable in identifying surge-type glaciers in the region.

Outlet Glaciers and Icebergs

Examination of Landsat imagery of many of the larger ice caps in the Russian High Arctic archipelagos shows that they are composed of a number of clearly identifiable drainage basins which are likely to behave as dynamically distinct entities (for example, Dowdeswell and others, 2002). Marked

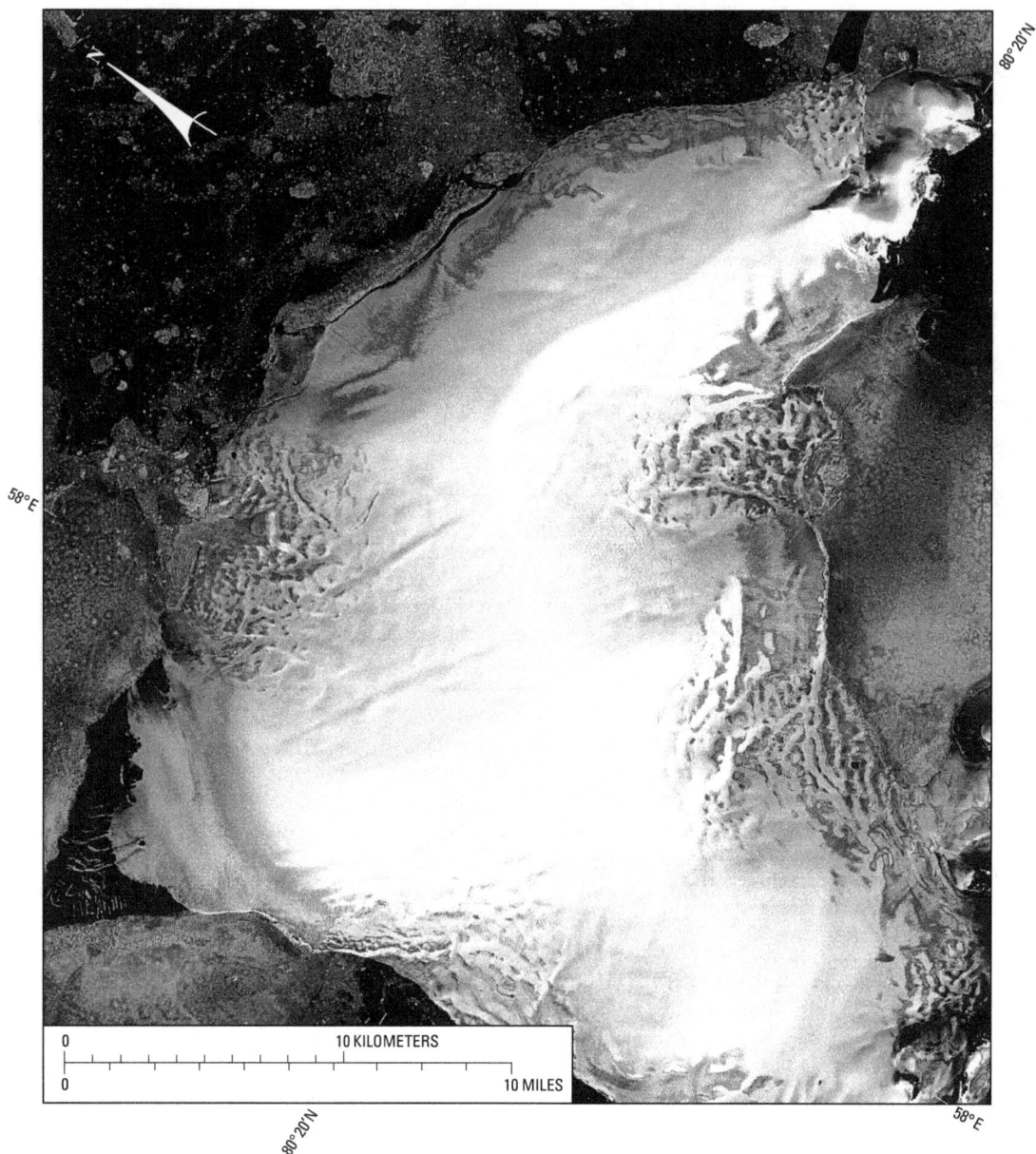

Figure 87.—*An example of the major ice divides on the Ostrov Gallya ice cap in Franz Josef Land. The digitally enhanced Landsat TM sub-scene is from Path 197–Row 001, acquired on 1 August 1988. Ostrov Gallya is located in figure 82.*

differences in the behavior of individual drainage basins within Arctic ice caps have been demonstrated previously for the two largest ice caps on Nordaustlandet in the Norwegian archipelago of Svalbard, west of Franz Josef Land (Dowdeswell, 1986a; Dowdeswell and Collin, 1990; Dowdeswell and others, 1999, 2008), and for the Devon Island Ice Cap in Arctic Canada (Dowdeswell and others, 2004; Burgess and others, 2005).

A number of the ice caps and drainage basins on Franz Josef Land, Severnaya Zemlya, and Novaya Zemlya have outlet glaciers which appear to have relatively rapid flow speeds and significant production of large tabular icebergs. Landsat images of the seaward margins of several outlet glaciers are shown

Figure 88.—*Digitally-enhanced Landsat TM sub-scene of the major ice divide running along the northeast-southwest trending spine of Novaya Zemlya. Outlet glaciers descending from the central divide to end in tidewater are shown on both sides of the island. The TM image is from Path 178–Row 006, acquired on 6 July 1986. It is located in figure 85.*

in figure 89. Each image shows linear boundaries, delineating ice-flow units from different parts of the upper drainage basin. In each case, the outlet glacier or ice stream appears on Landsat imagery to be the site of iceberg production (Dowdeswell and others, 2002). Close to several of the tidewater ice margins, transverse and sometimes chaotic crevassing is observed on the satellite imagery. The transverse crevasses indicate longitudinal tension at the unconfined ice cliffs at the glacier terminus (Hodgkins and Dowdeswell, 1994), and provide fractures along which iceberg calving takes place. The fast-flowing ice streams within the Lednik Akademii Nauk on Severnaya Zemlya can be clearly identified using both Landsat imagery and synthetic aperture radar interferometry (Dowdeswell and

EXPLANATION

→ EXAMPLE OF LARGE ICEBERG – Embedded within sea ice.

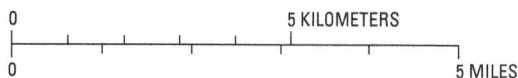

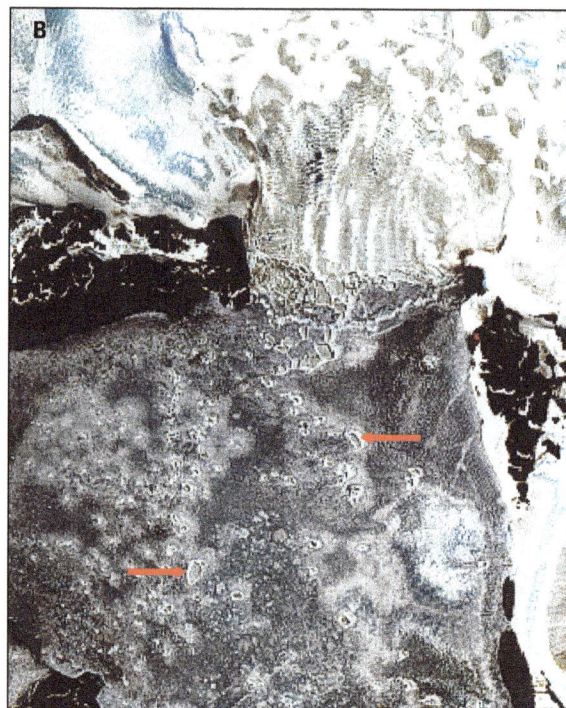

0   5 KILOMETERS

0   5 MILES

Figure 89.—*Landsat TM images of iceberg-producing outlet glaciers draining ice caps in the Russian High Arctic. The sea is usually covered by sea ice a meter or two thick in these images, and many icebergs (some larger bergs arrowed) can be seen embedded within this ice, which is often shorefast.* **A**, *Outlet glaciers and tabular icebergs from Severnaya Zemlya (located in figure 83). The scene is from Path 164–Row 003, acquired on 26 August 1988.* **B**, *Outlet glacier of the Zemlya Vil'cheka ice cap in Franz Josef Land (located in figure 82). The scene is from Path 197–Row 001, acquired on 1 August 1988.*

others, 1999); these ice streams produce large numbers of tabular icebergs (up to 1.7 km in length and about 2 km^2 in area) and a total flux of about 0.5 km^3 a^{-1} (Dowdeswell and others, 2002).

Several authors, including Sandford (1955), Voevodin (1972), and Abramov (1992), have reported the occurrence of tabular icebergs in Franz Josef Land waters, and icebergs in the Barents Sea which are inferred to be derived from the ice caps of Franz Josef Land. The calving of tabular icebergs from several low-gradient margins of ice caps on Franz

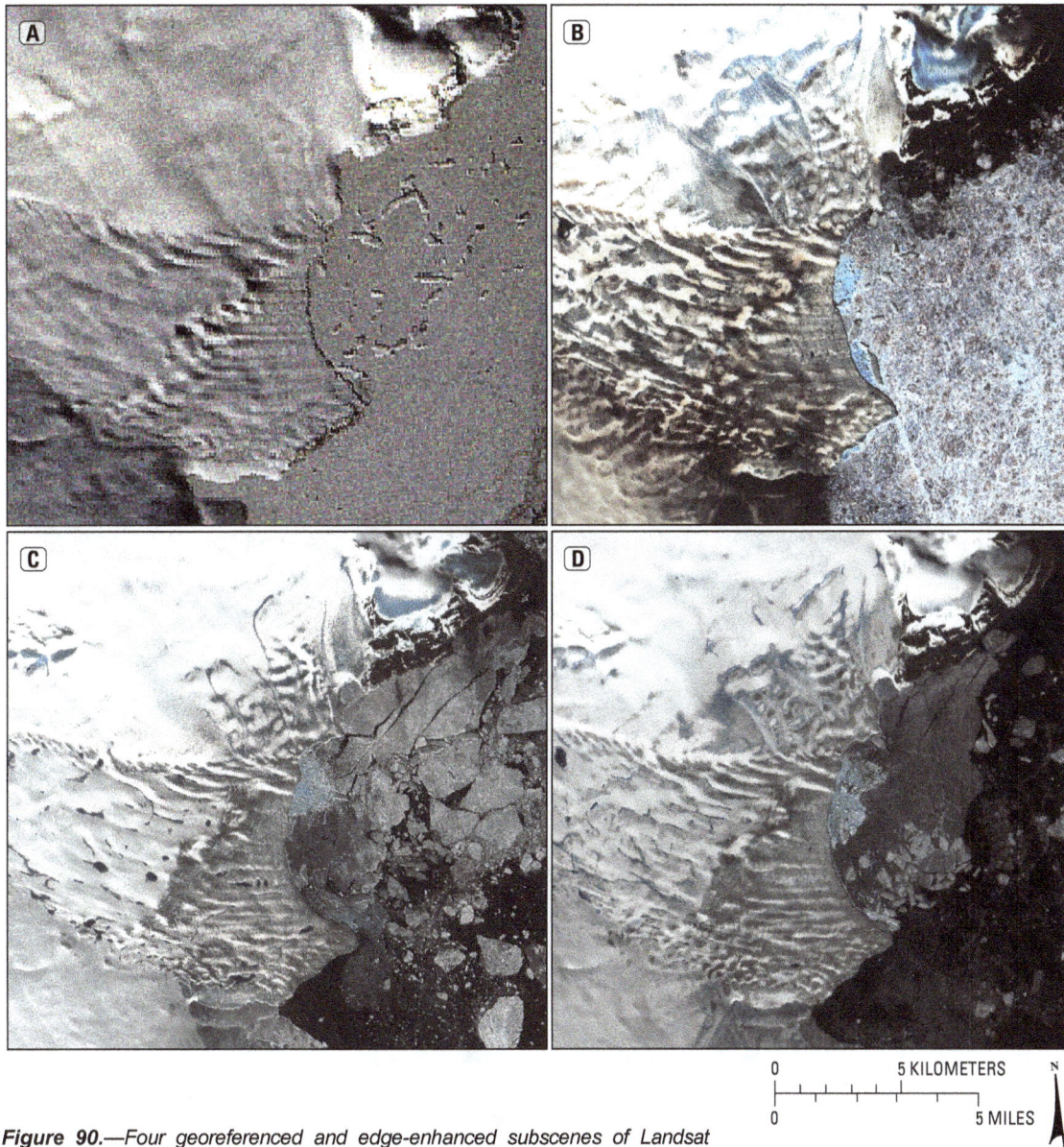

Figure 90.—*Four georeferenced and edge-enhanced subscenes of Landsat images of the Lednik Znamenityy on Zemlya Vil'cheka, Franz Josef Land (located in figure 82) showing large numbers of tabular icebergs. The tabular icebergs calved from this outlet glacier are among the largest produced in Franz Josef Land. The large iceberg in image **B** is >1.9 km in length. **A**, MSS image, of 8 May 1983 (Path 197–Row 001). **B**, TM scene of 25 July 1986 (Path 199–Row 001). **C**, TM scene of 8 August 1987 (Path 196–Row 001) and **D**, TM scene of 1 August 1988 (Path 197–Row 001) (From Williams, 2003).*

Josef Land has also been observed on Landsat imagery. The largest tabular icebergs observed on available satellite scenes are found offshore of Lednik Znamenityy, on Zemlya Vil'cheka (fig. 90). The icebergs appear to have smooth upper surfaces and are up to about 2 km in length and about 3 km^2 in area. Tabular icebergs longer than 200 m have also been observed on Landsat TM imagery, offshore of other drainage basins of Zemlya Vil'cheka, and offshore of Ostrova Zemlya Georga, Ziegler (Tsiglera), Solsberi, and Gallya in Franz Josef Land (fig. 82). Upon examination of photographic products from the Russian KATE-200 satellite camera system, the largest tabular iceberg observed in Franz Josef Land was 2.3 km in length (Dowdeswell, Gorman, Macheret and others, 1996).

Floating Ice Shelves or Ice Tongues

Ice shelves have been noted only infrequently in the little-known Eurasian Arctic sector, and the largest northern hemisphere ice shelf is the Ward Hunt off Ellesmere Island, Canada (Jeffries, 1987, 2002). Three ice shelves have, nonetheless, been reported from the Severnaya Zemlya archipelago (Dibner, 1955; Zinger and Koryakin, 1965; Govorukha, 1988a). These occur on Ostrova Komsomolets and Oktyabr'skoy Revolyutsii (fig. 83). The largest, the *Lednik Matusevicha*, is illustrated on the Landsat image in figure 91.

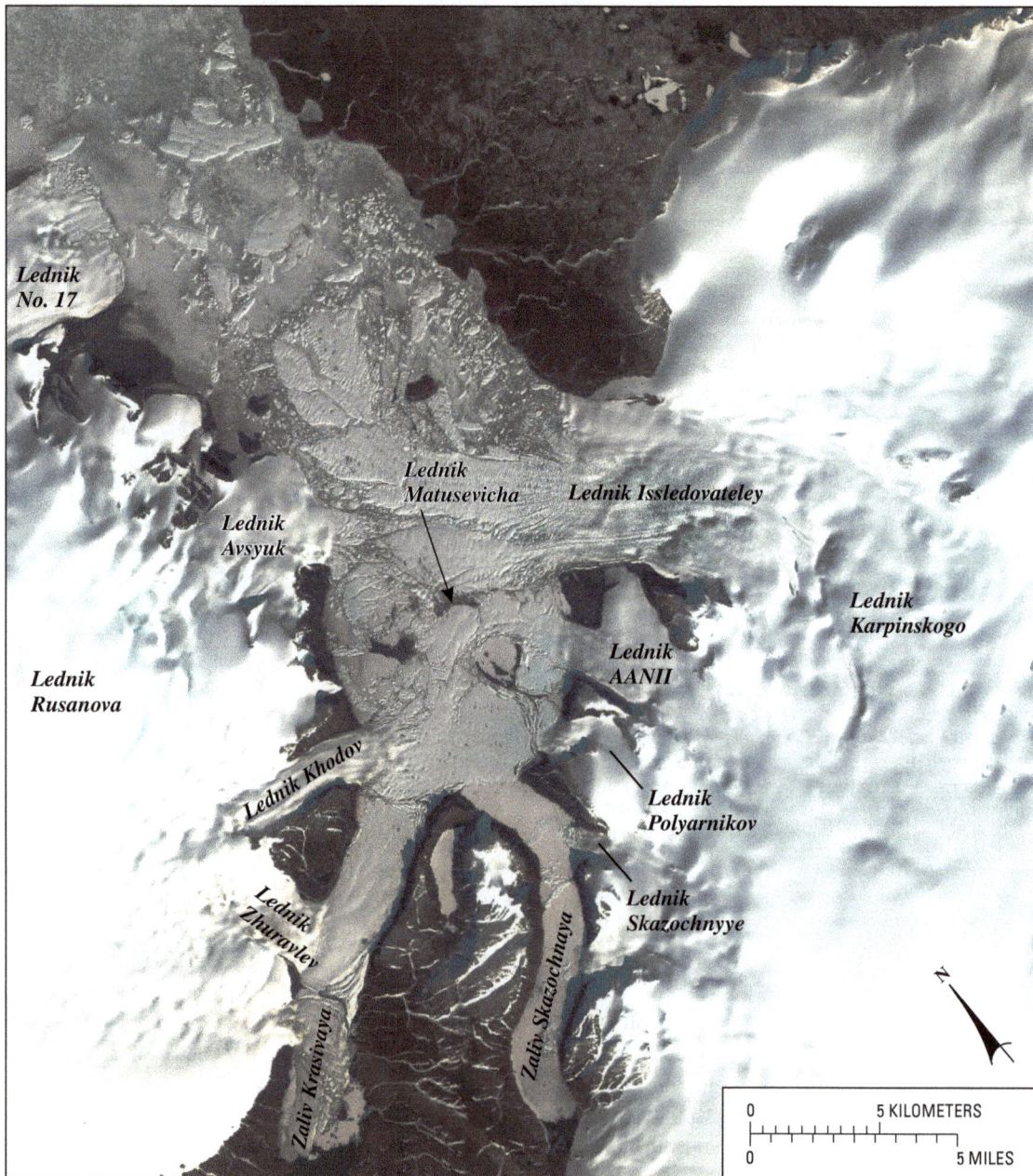

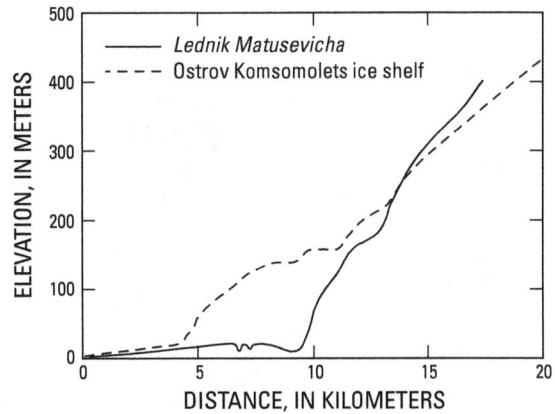

Figure 92.—*Ice-surface profiles from two ice shelves and their parent drainage basins on Severnaya Zemlya, derived from analysis of Russian aerial photographs from the 1950s. The* Lednik Matusevicha *is imaged in figure 91, and the smaller ice shelf on Ostrov Komsomolets is fed by an eastern outlet of the Lednik Akademii Nauk. Modified from Dowdeswell, Gorman, and others (1994)*

This ice shelf is about 200 km² in area, with a surface slope of less than 0.2° near its margin (Govorukha, 1988a). Although this feature is referred to as an ice shelf, according to the definition of Armstrong and others (1966), it is similar to a floating ice tongue. Ice-surface profiles of (1) the *Lednik Matusevicha* and its drainage basin, and (2) an outlet glacier of the Lednik Akademii Nauk, on Ostrov Komsomolets, are given in figure 92. The marginal 9- and 4-km elevations of these profiles, respectively, represent floating ice shelves. Williams and Dowdeswell (2001) measured terminus fluctuations and the drift of icebergs calved from *Lednik Matusevicha* using Landsat and earlier imagery. Between 1931 and 1994, the floating margin of the ice shelf underwent at least 2 cycles of retreat followed by advance. Most icebergs calved from the ice shelf remain trapped within the adjacent fjord for 10 to 20 years.

There is some controversy in the early Russian literature as to the occurrence of ice shelves in Franz Josef Land. Spizharskiy (1936), Shumskiy (1949), and Govorukha (1968) each suggest that some small ice shelves may be present, whereas Grosswald and others (1973) argue that none exist. Evidence in support of either case is ambiguous, although Govorukha (1968) states specifically that an ice shelf is present in the northeast of Ostrov Zemlya Georga (fig. 82).

A number of ice caps with smooth and apparently very low surface gradients at their seaward margins were observed on Landsat Thematic Mapper (TM) and Multispectral Scanner (MSS) satellite imagery of Franz Josef Land (fig. 93) (Dowdeswell, Gorman, and others, 1994). These areas are often associated with the production of relatively large tabular icebergs. The areas are dynamically part of the parent ice mass and have a marked break of slope at their inner margins. Most, although not all, of the low gradient margins are located in relatively protected embayments, rather than along sections of open coastline, and there is often relatively deep water offshore. These flat features account for 315 km² or 2.3 percent of the total area of the ice caps in the archipelago and the largest of these low gradient areas is 45 km² (fig. 94).

Figure 91.— *(at left) Annotated Land-sat TM band 3 subscene of the* Lednik Matusevicha *on Ostrov Oktyabr'skoy Revolyutsii, Severnaya Zemlya (located in figure 83). The ice shelf is composed of units derived from outlet glaciers of the Rusanova (west) and Karpinskogo (east) ice caps. Tabular icebergs up to about 4 km long are being calved from the ice-shelf margin. The Landsat TM sub-scene was acquired on 26 August 1988 from Path 164–Row 003. From Williams (2003).*

EXPLANATION

IS at A, C, and E   INTERPRETED AS POSSIBLE ICE SHELVES
B, D, and F   ICE RIDGES OR CAPS – Separating flat marginal areas
⟶   INDICATES EXAMPLE OF TABULAR ICEBERG

Figure 93.—*Landsat TM image of three flat ice-marginal glacier termini on the ice cap on Ostrov Zemlya Georga, western Franz Josef Land (located in figure 82). IS denotes ice shelves. Ice ridges or caps separate the flat marginal areas. Tabular icebergs (examples arrowed) can be seen in the sea beyond the ice shelf margins among sea-ice floes. The scene is from Path 199–Row 002, acquired on 25 July 1986.*

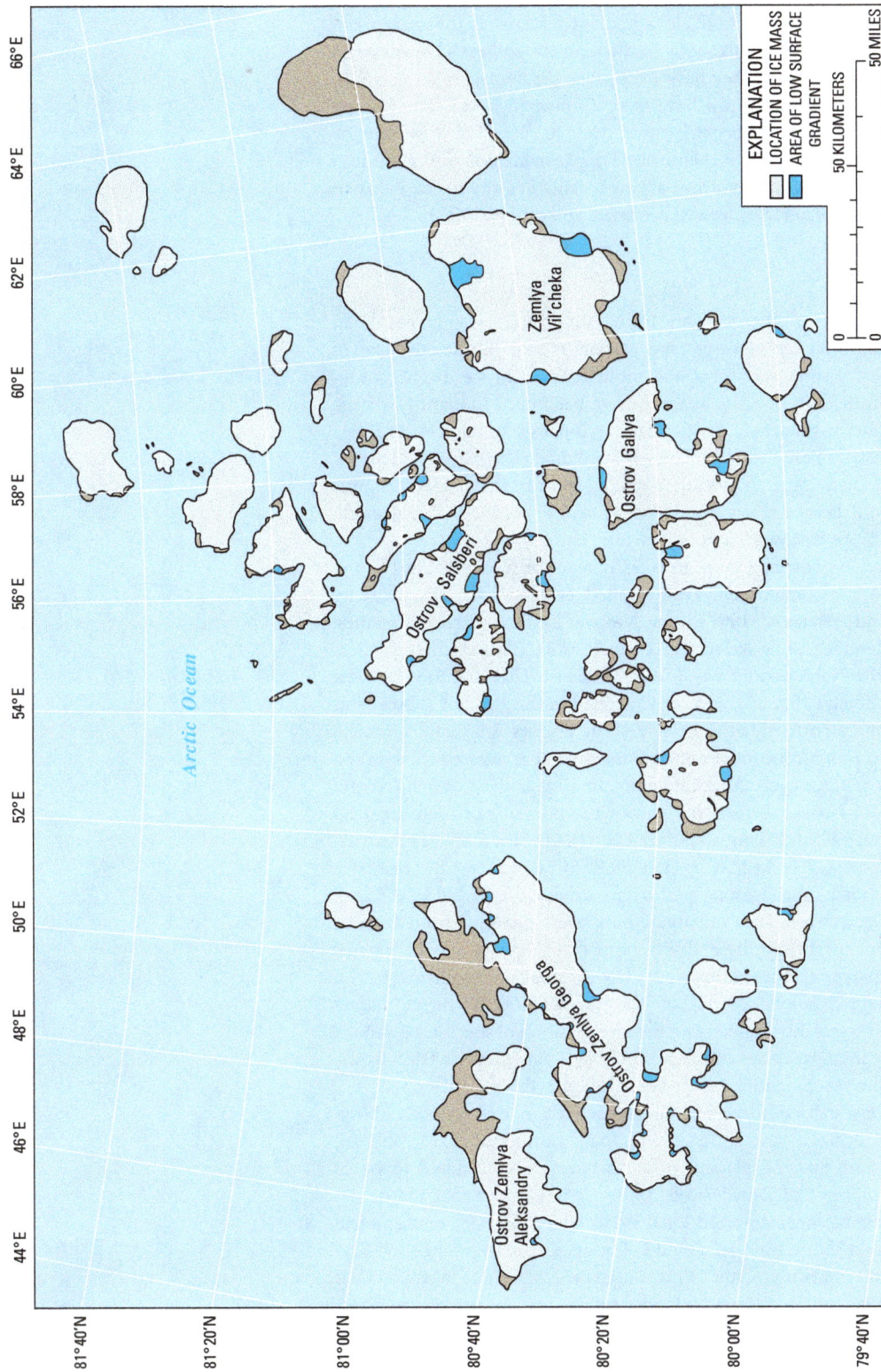

Figure 94.—Map of the ice caps on Franz Josef Land, with marginal areas of low surface gradient shaded dark blue. Some of these ice-cap margins may be floating or near to full buoyancy. (Modified from Dowdeswell, Gorman, and others, 1994.)

EXPLANATION

LOCATION OF ICE MASS

AREA OF LOW SURFACE GRADIENT

50 KILOMETERS

50 MILES

Arctic Ocean

Zemlya Vil'cheka

Ostrov Gallya

Ostrov Salsberi

Ostrov Zemlya Georga

Ostrov Zemlya Aleksandry

They make up a total of 175 km or 6 percent of the ice-ocean interface from which iceberg calving takes place. Ice-surface profiles, derived from analysis of vertical aerial photographs, show slopes of 0.5° on these features, as compared with 3.5–5° slopes on other ice caps in Franz Josef Land.

At least some of these smooth, low-gradient features at the margins of Franz Josef Land ice caps are likely to be floating ice shelves (Dowdeswell, Gorman, and others, 1994). They have ice surface gradients similar to gradients on the known ice shelves on Severnaya Zemlya (fig. 92). There is no requirement for deep water to occur beneath these features, but they must be buoyant over a significant part of their base. They provide one of the modern sources of tabular icebergs to both the eastern Barents Sea and to the Eurasian Basin of the Arctic Ocean (Dowdeswell, Gorman, and others, 1994).

Surge-Type Glaciers

Landsat imagery of the Russian High Arctic archipelagos has been examined for the presence of surge-type glaciers within the 56,000 km^2 of glacierized land. Ice-surface features and rapid changes in ice extent, sometimes easily identified on Landsat imagery, have been used to identify a large number of surge-type glaciers in the Svalbard archipelago to the west of Franz Josef Land (fig. 80). These features include looped medial moraines and highly crevassed ice surfaces, together with the rapid advance of the glacier terminus in a time series of images (for example, Meier and Post, 1969; Dowdeswell, 1986b; Liestøl, 1993; Kotlyakov and others, this chapter).

Landsat imagery covering more than 95 percent of the ice masses in the Russian High Arctic contains only very limited evidence of glacier surging (Dowdeswell and Williams, 1997). On Novaya Zemlya, three surge-type glaciers were identified using ice-surface characteristics. Two of these glaciers are located in the Nordenskiöld Bay (Zaliv Nordenshel'da) area on the west side of Novaya Zemlya (fig. 85), and are illustrated in the Landsat subscene in figure 95. The more northerly of these surge-type glaciers is heavily crevassed, as well as having a diagnostic looped medial moraine; it is inferred to be in the active phase of the surge cycle. A recent comprehensive analysis of Landsat and ASTER images of Novaya Zemlya has shown the presence of more than 30 glaciers of likely surge-type (Grant and others, 2009). On Severnaya Zemlya, two outlet glaciers of the Karpinsky Ice Cap (Lednik Karpinskogo) (fig. 83) also have looped medial moraines, suggesting past surge activity. On Franz Josef Land, no ice-surface features indicative of surging have been identified from Landsat imagery. It is concluded from these observations that there are only very small numbers of surge-type glaciers in this region (Dowdeswell and Williams, 1997). However, having incomplete coverage of the Russian High Arctic ice masses during times of the year most suitable for the identification of specific ice-surface features, and having time series of images for only a few Landsat Path–Rows, there may be more yet unidentified surge-type glaciers in this region.

The relative scarcity of identified surge-type glaciers in the Russian High Arctic provides a significant contrast with neighboring Svalbard (Liestøl, 1993), where more than 35 percent of ice masses are considered to be of surge-type (Hamilton and Dowdeswell, 1996). It is possible that the generally negative mass balances reported for Russian High Arctic ice masses since the end of the cold "Little Ice Age," about 100 years ago, have resulted in some glaciers switching from surge-type to non-surge type behavior. In Svalbard, it has been shown that some glaciers no longer surge; their consistently negative mass balance means they are thinning, their vertical temperature gradients are increasing, and basal melting is being replaced by a cold bed without the active hydrological system necessary for fast flow (Dowdeswell, Hodgkins, and others, 1995).

Figure 95.—*Digitally enhanced Landsat TM image of the outlet glaciers entering Zaliv Nordenshel'da, Novaya Zemlya (located in figure 85). Note the looped moraines (arrowed) indicating surge activity. The image is from Path 178–Row 007, acquired on 6 July 1986.*

Snow-and-Ice Facies and Glacier Mass Balance

The patterns of snow-and-ice facies on the surfaces of glaciers (Benson, 1962) have been used in a number of satellite studies (1) to provide information on snow- and equilibrium-line positions, and (2) as proxy indicators of mass balance (for example, Williams and others, 1991). Late summer satellite images are required, and time series of observations can be constructed either within a given year, to monitor the up-glacier retreat of the transient snow-line, or for successive years, at the end of each melt season (for example, Dowdeswell and Drewry, 1989).

Three examples of the complexity of the spectral response associated with summer Landsat TM images of the Russian High Arctic are illustrated in figures 84, 96, and 97. A full 185-by-185 km Landsat TM image of the

Figure 96.—*Digitally enhanced Landsat TM sub-scene showing the snow-and-ice facies on a summer image of the Lednik Kropotkina on Ostrov Zemlya Aleksandry, Franz Josef Land (Path 199–Row 001, acquired on 25 July 1986). The image is located in figure 82.*

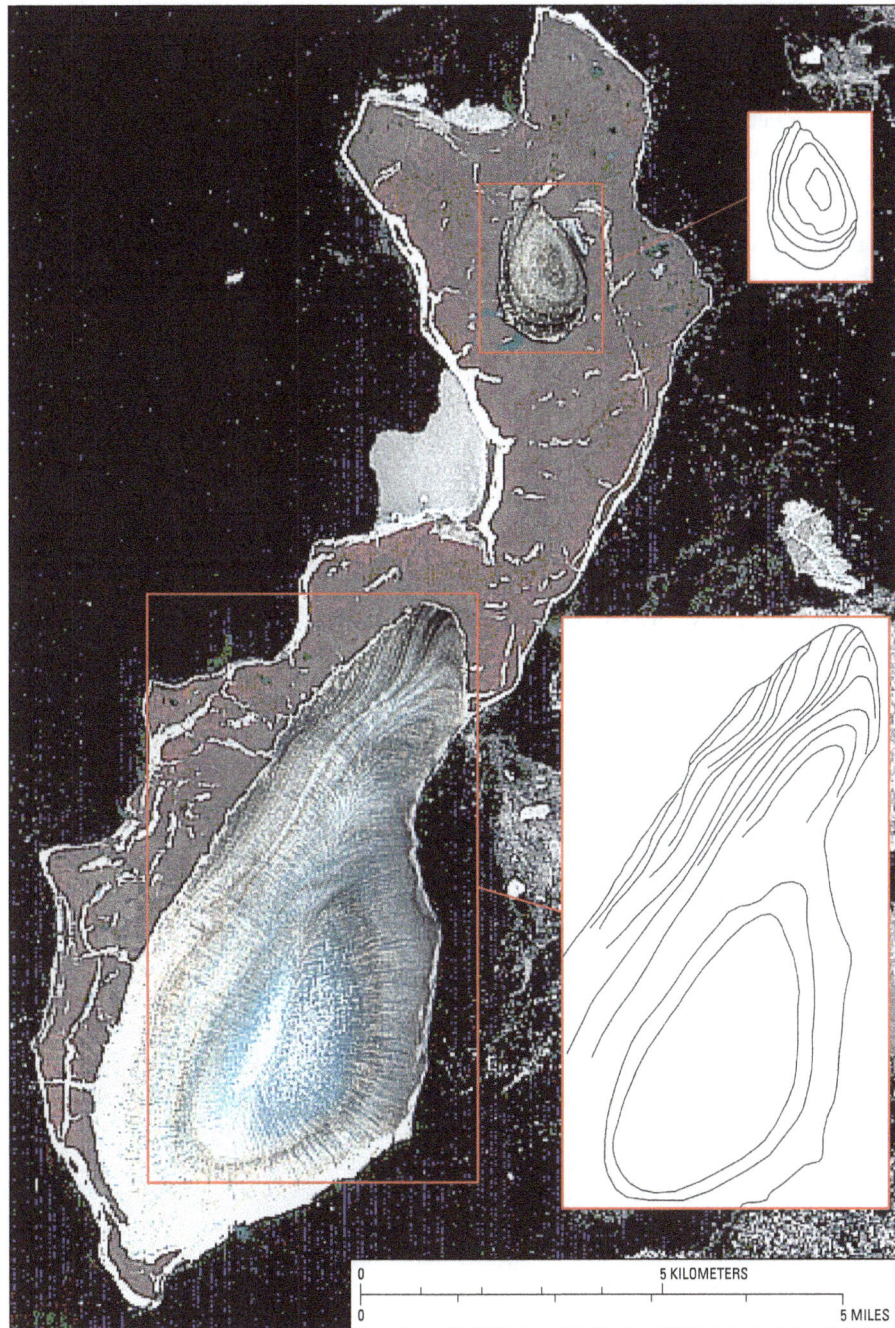

Figure 97.—*Landsat TM sub-scene of Ostrov Ketlitsa, Franz Josef Land (Path 199–Row 001, 25 July 1986) (located in figure 82). The two ice caps have bare ice exposed over most of their surface. The two insets are interpretations of the primary stratification exposed at the surface of each ice cap.*

northern part of Severnaya Zemlya, acquired during summer, shows a series of several spectral zones ranging from bare ice at the margin, through a thin slush zone, to snow on the crest of the main ice caps (fig. 84).

A second image (fig. 96) shows a TM sub-scene of the 120 km^2 Kropotkin Ice Cap (Lednik Kropotkina) on Alexandra Land (Ostrov Zemlya Aleksandry) in the Franz Josef Land archipelago. The detail in this image shows a large spectral variability within zones and the generally transitional nature of the boundaries between them. The surface of the ice cap is clearly not smooth, and snow remains in hollows close to sea level even in summer. The blue-colored slush zone, above the bare ice zone, appears to have a series of lineaments within it which extend in some cases down into the bare ice facies. These are interpreted as supraglacial drainage channels, which have been observed on glaciers in Franz Josef Land and eastern Svalbard (Rees and others, 1995). Two small supraglacial ponds are also observed. The ice cap appears to have two domes. The larger one has a continuous covering of snow at this stage of the summer melt season, but the lower one, to the south, appears to be composed largely of bare ice (fig. 96). This pattern may indicate that either accumulation is by superimposed ice only, or that net mass loss is taking place even at the crest of this smaller ice cap.

The image of the two small ice caps on Koettlitz Island (Ostrov Ketlitsa) (fig. 97), in western Franz Josef Land, illustrates an application of Landsat data to the problem of glacier response to climate change. The south and north ice caps cover a total of about 40 km^2 and have crests at 160 m and 50 m above sea level, respectively. The TM image was acquired on 25 July 1986, relatively early in the melt season, and yet the winter snow has almost completely disappeared from both ice caps to reveal ice. In the smaller (north) ice cap, ice structures assumed to represent primary depositional stratification are exposed right up to the summit, indicating that there is net mass loss over the entire ice cap (fig. 97). The northern margin of the larger (south) ice cap also contains primary structures or foliation which are clearly truncated (cf. Hambrey, 1975; Hambrey and Müller, 1978). This pattern suggests that thinning and retreat of this larger ice cap is also taking place. It is likely that most ice caps with summit elevations below about 150–200 m in the archipelago are in a state of strong negative balance, probably as a result of the warming experienced in the Eurasian High Arctic at the termination of the "Little Ice Age" and through modern warming.

Conclusions

Landsat digital and photographic products have provided important insights into several aspects of the morphology and dynamics of the relatively little-known glaciers and ice caps in the Russian High Arctic. This is so, despite the fact that this part of the Eurasian Arctic sector is most severely affected by cloud-cover problems (Marshall and others, 1994). Landsat imagery has been used at both a synoptic scale and at a more detailed level, through the selective digital enhancement of sub-scenes, to further our understanding of the surface topography, ice dynamics, and mass balance of the more than 56,000 km^2 of ice masses in the Russian Arctic archipelagos. Analysis of the imagery has demonstrated, for example, that these ice masses produce the largest icebergs in the Eurasian Arctic and that small ice caps are likely to be in negative net balance in response to recent climate change (ACIA, 2004). Information derived from these investigations has been incorporated in numerical models testing the sensitivity of Russian High Arctic ice masses to both future and past climate change (Glazovskii and others, 1996; Siegert and Dowdeswell, 1995).

Acknowledgments

Funding was provided through U.K. NERC Grants GR3/8507 and GR3/9958, INTASGrant 1010-CT93-0006, the John Ellerman Foundation, Conoco Phillips, and the Newton Trust. The Royal Society provided financial support for satellite image purchase and for collaborative meetings in Cambridge and Moscow with Russian colleagues. This is a contribution to the European Science Foundation program on the Quaternary Environment of the Eurasian North (QUEEN).

References Cited

Abramov, V.A., 1992, Russian iceberg observations in the Barents Sea, 1933–1990: Polar Research, v. 11, no. 2, p. 93–97.

ACIA, 2004, Impacts of a warming Arctic: Arctic climate impact assessment: Cambridge, U.K., Cambridge University Press, 139 p. [http://www.acia.uaf.edu]

Ahlmann, H.W., 1933, Scientific results of the Swedish-Norwegian Arctic expedition in the summer of 1931: Geografiska Annaler, v. 15, nos. 1–4, 348 p.

Aizen, V.B., Aizen, E.M., and Kuzmichenok, V.A., 2007, Glaciers and hydrological changes in the Tien Shan – Simulation and prediction: Environmental Research Letters, v.2, 045019, 10 p., doi:10.1088/1748-9326/2/4/045019.

Aizen, V.B., and Kuzmichenok, V.A., 2007, Glacier changes in the Tien Shan as determined from topographic and remotely sensed data: Global and Planetary Change, v. 56, no. 3, p. 328-340.

Aizen, V.B., Kuzmichenok, V.A., Surazakov, A.B., and Aizen, E.M., 2006, Glacier changes in central and northern Tien Shan during the last 150 years based on surface and remote sensing data: Annals of Glaciology, v. 43, p. 202–213.

Anonymous, 1933, Soviet work in the Arctic: Geographical Journal, v. 81, no. 6, p. 532–535.

Anonymous, 2002, Lednik Kolka: snova katastrofa [Kolka Glacier: disaster again]: Materialy Glyatsiologicheskikh Issledovanii, Pub. 93, p. 221–228. (In Russian)

Armstrong, T., Roberts, B., and Swithinbank, C., 1966, Illustrated glossary of snow and ice, (2d ed.): Cambridge, U.K., Scott Polar Research Institute Special Publication No. 4, 60 p.

Avsiuk, G.A., 1952, Nekotoriye dannye ob oledenenii i orografii gornogo massiva Ak-Shyyrak [Some data on the glaciers and orography of the Ak-Shyyrak Mountain Massif]: Moscow, Trudy Instituta Geografii AN SSSR, v. 49, p. 5–32. (In Russian)

Bagrationes, Vakhushti, 1997, Atlas of Georgia: Tibilsi, 95 p. (In Georgian) [Originally published in the 1740s.]

Bamber, J.L., and Dowdeswell, J.A., 1990. Remote-sensing studies of Kvitøyjøkulen, an ice cap on Kvitøya, north-east Svalbard: Journal of Glaciology, v. 36, no. 122, p. 75–81.

Barkov, N.I., Bol'shiyanov, D.Yu., Gvozdik, O.A., Klement'yev, O.L., Makeev, V.M., Moskalenko, I.G., Potapenko, V.Yu., and Yunak, R.I., 1992, Novye dannye o stroenii i razvitii lednika Vavilova na Severnoi Zemle [New data on the structure and development of the Vavilov Ice Dome, Severnaya Zemlya]: Materialy Glyatsiologicheskikh Issledovanii, Pub. 75, p. 35–41. (In Russian)

Bazheva, V.Ya., 1981, Katalog Lednikov SSSR [Catalogue of Glaciers of the USSR], Ostrova De-Longa [DeLong Islands]: Gidrometeoizdat (Leningrad), v. 17, p. 15–25. (In Russian)

Benson, C.S., 1962, Stratigraphic studies in the snow and firn of the Greenland Ice Sheet: [Hanover, New Hampshire,] Snow, Ice and Permafrost Research Establishment (SIPRE), U.S. Army Cold Regions Research and Engineering Laboratory (CRREL) Research Report No. 70, 93 p. [Published version of C.S. Benson's Ph.D. dissertation (1960), Division of Geological Sciences, California Institute of Technology, Pasadena, CA, 213 p.]

Berman, L.L., 1947, Sovremennoe oledenenie verkhov'ev reki Indigirki [Present day glaciers of the Indigirka upper reaches]: Moscow, Voprosy geografii, iss. 4. (In Russian)

Berson, Arthur, 1933, Die Arktisfahrt des Luftschiffes "Graf Zeppelin" im Juli 1931 [The Arctic expedition of the airship "Graf Zeppelin" in July 1931] Wissenschaftliche Ergebnisse: Petermanns Mitteilungen Ergänzungsheft No. 216, Gotha, 112 p. (In German)

Bindschadler, R.A., Dowdeswell, J.A., Hall, D.K., and Winther, J.-G., 2001, Glaciological applications with Landsat-7: early assessments: Remote Sensing of Environment, v. 78, nos. 1–2, p. 163–179.

Bogdanovich, K.I., 1899, Ocherk deyatel'nosti Okhotsko-Kamchatskoi gornoi ekspeditsii 1895–1898 gg [Notes on the Okhotsk-Kamchatka Mountain Expedition, 1895-1898]: Izvestiya Imperatorskogo Geograficheskogo Obshchestva, v. 35, no. 6, p. 549–600. (In Russian)

Bogorodskii, V.V., and Fedorov, B.A., 1971. Radar probing of Severnaya Zemlya glaciers, in Bogorodskii, V.V., ed., The Physics of Ice: Springfield, Virginia, National Technical Information Service, p. 1–16.

Bolch, T., 2007, Climate change and glacier retreat in northern Tien Shan (Kazakhstan/ Kyrgyzstan) using remote sensing data: Global and Planetary Change, v. 56, iss. 1–2, p. 1–12.

Bondarev, L.G., 1963, Ocherki po oledeneniyu massiva Ak-Shyyrak [Notes on the glaciers of the Ak-Shyyrak Massif]: Frunze, AN Kirgizskoy SSR, 202 p. (In Russian)

Burgess, D.O., Sharp, M., Mair, D.W.F., Dowdeswell, J.A., and Benham, T.J., 2005, Flow dynamics and iceberg calving rates of the Devon Ice Cap, Nunavut, Canada: Journal of Glaciology, v. 51, p. 219-230.

Burmester, H., 1913, Rezente glaziale Untersuchungen und photogrammetrische Aufnahmen im Baksanquellgebiete (Kaukasus) [Recent glaciological investigations and photogrammetric surveys in Baksan upper reaches [Quellgebiet] (Caucasus)]: Zeitschrift für Gletscherkunde, v. 8, no. 1, p. 1–41. (In German)

Chernomorets, S.S., 2005, Selevye ochagi do i posle katastrof [Mudflow foci before and after disasters]: Moscow, Nauchnyi Mir, 182 p. (In Russian)

Chizhov, O.P., and Koryakin, V.S., 1962, Recent changes in the regimen of the Novaya Zemlya ice sheet: International Association of Scientific Hydrology, Publication No. 58, p. 187–193.

Chizhov, O.P., Koryakin, V.S., Davidovich, N.V., Kanevskii, Z.M., Zinger, E.M., Bazheva, V.Ya., Bazhev, A.B., and Khmelevskoi, I.F., 1968, Oledenenie Novoi Zemli [Glaciers of Novaya Zemlya]: Rezul'taty Issledovanii po Programme Mezhdunarodnogo Geofizicheskogo Goda. IX Razdel Programmy MGG, v. 18,

Moscow, Nauka, 338 p. (In Russian) [Also released in translation as Glaciation of the Novaya Zemlya, 1974, Springfield, VA, NTIS TT-70-59126, 2 parts, 727 p.]

Desinov, L.V., Kotlyakov, V.M., Osipova, G.B., Tsvetkov, D.G., 2001, Snova dal o sebe znat' lednik Medvezhiy [Medvezhiy Glacier in retrospect]: Materialy Glyatsiologicheskikh Issledovanii, Pub. 91, p. 249–253. (In Russian)

Desinov, L.V., Zakharov, V.G., Kotlyakov, V.M., and Suslov, V.F., 1978, Itogy pervogo podsputnikovogo experimenta po indikatsii pul'siruyushchikh lednikov Pamira [Results of the first satellite experiment of the indication of surging glaciers of the Pamirs]: Izvestiya Vsesoyuznogo Geograficheskogo obshchestva, v. 110, no. 6, p. 505–511. (In Russian)

Dibner, V.D., 1955, O proiskhozhdenii plovuchikh ledianykh ostrovov [The origin of floating ice-islands]: Priroda, v. 3, p. 89–92. (In Russian)

Dolgoushin, L.D., and Osipova, G.B., 1975, Glacier surges and the problem of their forecasting: International Association of Hydrological Sciences, IAHS Publication No. 104, p. 292–304.

Dolgushin, L.D., and Osipova, G.B., 1979, Novye ledniki na khrebte Telpoz-Iz [New glaciers on the Telpoziz Range]: Materialy Glyatsiologicheskikh Issledovanii, Pub. 36, p. 214–218. (In Russian)

Dolgushin, L.D., and Osipova, G.B., 1982, Pul'siruiushchie ledniki [Surging glaciers]: Leningrad, Gidrometeoizdat, 192 p. (In Russian)

Dolgushin, L.D., and Osipova, G.B., 1989, Ledniki [Glaciers]: Moscow, 444 p. (In Russian)

Dowdeswell, J.A., 1986a, Drainage-basin characteristics of Nordaustlandet ice caps, Svalbard: Journal of Glaciology, v. 32, no. 110, p. 31–38.

Dowdeswell, J.A., 1986b, Remote sensing of ice cap outlet glacier fluctuations on Nordaustlandet, Svalbard: Polar Research, v. 4, no. 1, p. 25–32.

Dowdeswell, J.A., 1995, Glaciers in the High Arctic and recent environmental change: Philosophical Transactions of the Royal Society, London, Series A, v. 352, no. 1699, p. 321–334.

Dowdeswell, J.A., Bassford, R.P., Gorman, M.R., Williams, M., Glazovskii, A.F., and others, 2002, Form and flow of the Academy of Sciences Ice Cap, Severnaya Zemlya, Russian High Arctic: Journal of Geophysical Research, v. 107, No. B4, EPM5, doi: 10.1029/2000 JB000129, 17 p.

Dowdeswell, J.A., Benham, T.J., Gorman, M.R., Burgess, D., and Sharp, M.J., 2004, Form and flow of the Devon Island Ice Cap, Canadian Arctic: Journal of Geophysical Research, v. 109, no. F2, F02002, doi: 10.1029/2003 JF000095, 14 p.

Dowdeswell, J.A., Benham, T.J., Strozzi, T., and Hagen, J.O., 2008, Iceberg calving flux and mass balance of the Austfonna ice cap on Nordaustlandet, Svalbard: Journal of Geophysical Research, v. 113, F03022, doi:10.1029/2007JF000905.

Dowdeswell, J.A., and Collin, R.L., 1990, Fast-flowing outlet glaciers on Svalbard ice caps: Geology, v. 18, no. 8, p. 778–781.

Dowdeswell, J.A., and Drewry, D.J., 1989, The dynamics of Austfonna, Nordaustlandet, Svalbard: surface velocities, mass balance and subglacial meltwater: Annals of Glaciology, v. 12, p. 37–45.

Dowdeswell, J.A., Glazovskii, A.F., and Macheret, Yu.Ya., 1995, Ice divides and drainage basins on the ice caps of Franz Josef Land, Russian High Arctic, defined from Landsat, Russian KFA-1000 and ERS-1 SAR satellite imagery: Arctic and Alpine Research, v. 27, no. 3, p. 264–270.

Dowdeswell, J.A., Gorman, M.R., Glazovskii, A.F., and Macheret, Yu.Ya., 1994, Evidence for floating ice shelves on Franz Josef Land, Russian High Arctic: Arctic and Alpine Research, v. 26, no. 1, p. 86–92.

Dowdeswell, J.A., Gorman, M.R., Glazovskii, A.F., and Macheret, Yu.Ya., 1996, Airborne radio-echo sounding of the ice caps on Franz Josef Land in 1994: Materialy Glyatsiologicheskikh Issledovanii, Pub. 80, p. 248–254.

Dowdeswell, J.A., Gorman, M.R., Macheret, Yu.Ya., Glazovskii, A.F., and Moskalevski, M.Y., 1996, High resolution imagery from the Russian KATE-200 satellite camera: morphology and dynamics of ice masses in the European High Arctic: International Journal of Remote Sensing, v. 17, no. 17, p. 3343–3356.

Dowdeswell, J.A, and Hagen, J.O., 2004, Arctic ice caps and glaciers, in Bamber, J.L., and Payne, A.J., eds., Mass balance of the cryosphere: Cambridge, U.K., Cambridge University Press, p. 527–557.

Dowdeswell, J.A., Hodgkins, R., Nuttall, A-M., Hagen, J.O., and Hamilton, G.S., 1995, Mass balance change as a control on the frequency and occurrence of glacier surges in Svalbard, Norwegian High Arctic: Geophysical Research Letters, v. 22, no. 21, p. 2909–2912.

Dowdeswell, J.A., and McIntyre, N.F., 1987, The surface topography of large ice masses from Landsat imagery: Journal of Glaciology, v. 33, no. 113, p. 16–23.

Dowdeswell, J.A., Rees, W.G., and Diament, A.D., 1994, ERS-1 SAR investigations of snow and ice facies on ice caps in the European High Arctic, in Kaldeick-Schumann, B., ed., Space at the service of our environment, proceedings of the Second ERS-1 Symposium, Hamburg, Germany, 11–14 October 1993: European Space Agency Special Publication 361, p. 1171–1176.

Dowdeswell, J.A., Unwin, B., Nuttall, A.-M., and Wingham, D.J., 1999, Velocity structure, flow instability and mass flux on a large Arctic ice cap from satellite radar interferometry: Earth and Planetary Science Letters, v. 167, nos. 3–4, p. 131–140.

Dowdeswell, J.A., and Williams, M., 1997, Surge-type glaciers in the Russian High Arctic identified from digital satellite imagery: Journal of Glaciology, v. 43, no. 145, p. 489–494.

Dyurgerov, M.B., Chaohai, L., and Zichu, X., eds., 1995, Oledenenie Tian' Shanya [Glaciation of the Tien Shan]: Moscow, VINITI Publishing House, 237 p. (In Russian)

Dyurgerov, M.B., and Meier, M.F., 1997, Mass balance of mountain and subpolar glaciers: a new global assessment for 1961–1990: Arctic and Alpine Research, v. 29, no. 4, p. 379–391.

Galakhov, V.P., Narozhniy, Yu.K., Nikitin, S.A., Okishev, P.A., Sevastyanov, V.V., Sevastyanova, L.N., Shurov, V.I., 1987, Ledniki Aktru, Altai — vodnyi, ledovoi i teplovoi balansy gorno-lednikovykh basseynov [Aktru glaciers, Altai — water, ice and energy balances of glacier-mountain basins]: Leningrad, Gidrometeoizdat, 118 p. (In Russian)

Glazovskii, A.F., Ignat'eva, I.Yu., and Macheret, Yu.Ya., 1996, Modelirovanie evolyutsii lednikovogo kupola Greem-Bell, Zemlya Frantsa-Iosifa: Chustvitel'nost'modeli k zadaniyu vnutrennikh parametrov i izmeneniyam vneshnikh uslovii [Modelling the evolution of an ice cap on Graham Bell

Island, Franz Josef Land: sensitivity of the model to internal parameters and changes in external conditions]: Materialy Glyatsiologicheskikh Issledovanii, Pub. 80, p. 18–30. (In Russian)

Golubev, G.N., Dyurgerov, M.B., Markin, V.A., and others, 1978, Lednik Dzhankuat (Tsentralnyi Kavkas) [Dzhankuat Glacier (Central Caucasus)]: Leningrad, Gidrometeoizdat, 184 p. (In Russian)

Govorukha, L.S., 1964, Novye dannye o sovremennom i drevnem oledenenii ostrova Viktorii [New data on the current and former glaciations of Victoria Island]: Izvestiya Vsesoyuznogo Geograficheskogo Obshchestva, v. 96, p. 352–354. (In Russian)

Govorukha, L.S., 1968, Glyatsiologicheskie issledovaniya Arkticheskogo i Antarkticheskogo nauchno-issledovatel'skogo instituta na Zemle Frantsa-Iosifa v 1960 godu [Glaciological studies of the Arctic and Antarctic Research Institute on Franz Joseph Land in 1960]: Materialy Glyatsiologicheskikh Issledovanii, Pub. 8, p. 79–87. (In Russian)

Govorukha, L.S., 1970, Raschet srednego mnogoletnego byudzheta l'da v sisteme vneshnego massoobmena lednikovogo pokrova Severnoi Zemli [Calculation of the average multi-year budget of ice in the system of external mass exchange of the Severnaya Zemlya ice cover]: Arkticheskii i Antarkticheskii Nauchno-Issledovatel'skii Institut, Trudy, v. 294, p. 12–27. (In Russian)

Govorukha, L.S., 1988a, Sovremennoe nazemnoe oledenenie Sovetskoi Arktiki [The modern land glacierization of the Soviet Arctic]: Leningrad, Hydrometeorological Publishing House, 256 p. (In Russian)

Govorukha, L.S., 1988b, The present state of ice cap islands in the Soviet Arctic: Polar Geography and Geology, v. 12, no. 4, p. 312–316. [Translated from Materialy Glyatsiologicheskikh Issledovanii, Pub. 63, p. 114–117, 1988.]

Govorukha, L.S., Bol'shiyanov, D.Yu., Zarkhidze, V.S., Pinchuk, L.Ya., and Yunak, R.I., 1987, Changes in the glacier cover of Severnaya Zemlya in the twentieth century: Polar Geography and Geology, v. 11, no. 4, p. 300–305. [Translated from Materialy Glyatsiologicheskikh Issledovanii, Pub. 60, p. 155–158, 1987.]

Grant, K.L., Stokes, C.R., and Evans, I.S., 2009, Identification and characteristics of surge-type glaciers on Novaya Zemlya, Russian Arctic: Journal of Glaciology, v. 55, no. 124, p. 960–972.

Grosswald, M.G., and Krenke, A.N., 1962, Recent changes and the mass balance of glaciers on Franz Josef Land: International Association of Scientific Hydrology, Publication No. 58, p. 194–200.

Grosswald, M.G., Krenke, A.N., Vinogradov, O.N., Markin, V.A., Psariova, T.V., Rasumeiko, N.G., and Sukhodrovskii, V.L., 1973, Oledenenie Zemli Frantsa-Iosifa Resul'taty issledovanii po programme Mezhdunarodnogo Geofizicheskogo Goda [Glaciers of Franz Josef Land: Results of research under the programme of the International Geophysical Year]: Moscow, Nauka, 351 p. (In Russian)

Gurbanov, A.G., Kusraev, A.G., and Tchel'diev, A.Kh., 2004, First results of investigation of endogene processes in Genaldonskoe and adjoining gaps: Vestnik of Vladikavkaz Research Center, Vladikavkaz, v. 4, iss. 3, p. 2–8. (In Russian)

Hambrey, M.J., 1975, The origin of foliation in glaciers: Evidence from some Norwegian examples: Journal of Glaciology, v. 14, no. 70, p. 181–185.

Hambrey, M.J., and Müller, F., 1978, Structures and ice deformation in the White Glacier, Axel Heiberg Island, Northwest Territories, Canada: Journal of Glaciology, v. 20, no. 82, p. 41–66.

Hamilton, G.S., and Dowdeswell, J.A., 1996, Controls on glacier surging in Svalbard: Journal of Glaciology, v. 42, no. 140, p. 157–168.

Hodgkins, R., and Dowdeswell, J.A., 1994, Tectonic processes in Svalbard tidewater glacier surges: evidence from structural glaciology: Journal of Glaciology, v. 40, no. 136, p. 553–560.

Holtedahl, O., 1924, 1928, 1930, Report of the scientific results of the Norwegian expedition to Novaya Zemlya, 1921: Oslo, Det Norske Videnskaps-Akademi, 3 v., 1600 p., v.1 (1924), v. 2 (1928), v. 3 (1930).

Ivan'kov, P.A., 1958, Oledenenie Kamchatki [Glaciers of Kamchatka]: AN SSSR, Seriya geograficheskaya, no. 2, p. 42–53. (In Russian)

Jania, Jacek, and Hagen, J.O., 1996, Mass balance of Arctic glaciers: International Arctic Science Committee, IASC Report No. 5, 54 p.

Jeffries, M.O., 1987, The growth, structure and disintegration of Arctic ice shelves: Polar Record, v. 23, no. 147, p. 631–649.

Jeffries, M.O., 2002, Ellesmere Island ice shelves and ice islands, in Williams, R.S., Jr., and Ferrigno, J.G., eds., Satellite image atlas of glaciers of the world: U.S. Geological Survey Professional Paper 1386-J (Glaciers of North America), p. J147–J164. [http://pubs.usgs.gov/prof/p1386j]

Kalesnik, S.V., 1937, Gornye lednikovye rayony SSSR [Mountain glaciers of the USSR]: Leningrad-Moscow, 182 p. (In Russian)

Kaulbars, A.V., 1874, Materialy po geografii Tian'-Shanya, sobrannye vo vremya puteshestviya 1869 goda [Information on the geography of Tien Shan collected during travel in 1869]: St. Petersburg, Tipografia M. Stasyulevicha, 286 p. (In Russian)

Kharkov University, 1960–61, Materialy Kavkazskoi Ekspeditsii (po programme MGG) [Data of the Caucasus expedition (under the IGY program)]: Kharkov, Kharkov State University, v. 1, 363 p.; v. 2, 257 p.; and v. 3, 439 p. (In Russian)

Khromova, T.E., Dyurgerov, M.B., and Barry, R.G., 2003, Late twentieth century changes in glacier extent in the Ak-shirak Range, Central Asia, determined from historical data and ASTER imagery: Geophysical Research Letters, v. 30, no. 16, 1863. (doi 10.1029/2003GLO17233)

Khromova, T.E., Osipova, G.B., Tsvetkov, D.G., and Barry, R.G., 2006, Changes in glacier extent in the eastern Pamir, Central Asia, determined from historical data and ASTER imagery: Remote Sensing of Environment, v. 102, nos. 1–2, p. 24–32.

Koreisha, M.M., 1963, Sovremennoe oledenenie khrebta Suntar-Khyata [Present day glaciers of the Suntar-Khyata Range]: Moscow, AN SSSR, 167 p. (In Russian)

Kornienko, S.G., Lyashenko, O.V., Gurbanov, A.G., 2004, Identification of indications of focus magmatism within the limits of the Kazbek volcanic center from data of thermal space photography: Vestnik of Vladikavkaz Research Center, Vladikavkaz, v. 4, iss. 3, p. 25–32. (In Russian)

Koryakin, V.S., 1986, Decrease in glacier cover on the islands of the Eurasian Arctic during the twentieth century: Polar Geography and Geology, v. 10, no. 2, p. 157–165. [Translated from Materialy Glyatsiologicheskikh Issledovanii, Pub. 54, p. 103–108, 1985.]

Korzhenevskii, N. L., 1930, Katalog lednikov srednei Asii [Glacier inventory of central Asia]: Tashkent, 200 p. (In Russian)

Kotlyakov, V.M., 1997, Atlas snezhno-ledovykh resursov mira [World atlas of ice and snow resources]: Moscow, Rossiiskaya Akademiya Nauk, v. 3, 392 p. (In Russian)

Kotlyakov, V.M., ed., 2006, Oledenenie severnoi i tsentral'noi Evrazii v sovremennuyu epokhu [Glaciation in north central Eurasia at the present time]: Moscow, Nauka, 482 p. (In Russian)

Kotlyakov, V.M., Desinov, L.V., Osipova, G.B., Hauser, M., Tsvetkov, D.G., Schneider, J.F., 2003, Sobytiya 2002 goda na lednike Geograficheskogo Obshchestva (RGO), Pamir [Events in 2002 on Geographical Society Glacier (RGO), Pamirs]: Materialy Glyatsiologicheskikh Issledovanii, Pub. 95, p. 221–230. (In Russian)

Kotlyakov, V.M., Kerimov, A.M., Nosenko, G.A., Nosenko, O.A., Rototaeva, O.V., Khmelevskoi, I.F., 2004, Estimation of the state of Kolka and Maili glaciers according to surface investigation in June–September 2003. Problems of forecast of extreme situations: Third Scientific Practical Conference, October, 2003, Moscow, p. 136–143. (In Russian)

Kotlyakov, V.M., Korotkov, I.M., Nikolaev, V.I., Petrov, V.N., Barkov, N.I., and Klement'ev, O.L., 1989, Rekonstruktsiya klimata golotsena po resul'tatam issledovaniya ledyanogo kerna lednika Vavilova na Severnoi Zemle [Reconstruction of the Holocene climate from the results of ice-core studies on the Vavilov Dome, Severnaya Zemlya]: Materialy Glyatsiologicheskikh Issledovanii, Pub. 67, p. 103–108. (In Russian)

Kotlyakov, V.M., Osipova, G.B., Tsvetkov, D.G., 1997, Fluctuations of unstable glaciers: scale and character: Annals of Glaciology, v. 24, p. 338–343.

Kotlyakov, V.M., Rototaeva, O.V., Desinov, L.V., Zotikov, I.A., Osokin, N.I., 2004, Causes and effects of a catastrophic surge of Kolka Glacier in the central Caucasus: Zeitschrift für Gletscherkunde und Glazialgeologie, v. 38, no. 2, p. 117–128.

Kotlyakov, V.M., Zagorodnov, V.S., and Nikolaev, V.I., 1990, Drilling on ice caps in the Soviet Arctic and on Svalbard and prospects of ice core treatment, in Kotlyakov, V. M., and Sokolov, V.Y., eds., Arctic research: Advances and prospects, Moscow, Nauka, v. 2, p. 5–18.

Kravtsova, V.I., and Chaikina, N.F., 1980, Primenenie snimkov s kosmicheskogo korablya "Soyuz-22" dlya kartigrafirovania oledenenia Alaiskogo khrebta [Application of images taken from the Soyuz-22 for the mapping of Alai Range glaciers]: Materialy Glyatsiologicheskikh Issledovanii, Pub. 37, p. 56–62. (In Russian)

Kravtsova, V.I., and Loseva, V.G., 1968, Izmeneniya lednikov Elbrusa za sto let, in Uspekhi Sovetskoi gliatsiologii [Change of El'brus glaciation over 100 years, in Advances in Soviet glaciology]: Vsesoiuznyi gliatsiologicheskii simpozium III [Third All-Union Glaciological Symposium], Proceedings, Frunze, Kyrgyzstan, p. 262–270. (In Russian)

Krenke, A.N., Menshutin, V.M., Voloshina, A.P., and others, 1988, Lednik Marukh (Zapadniy Kavkaz) [Marukh Glacier (Western Caucasus)]: Leningrad, Gidrometeoizdat, 254 p. (In Russian)

Krimmel, R.M., 1978, Detection of surging glaciers using aerial photography and LANDSAT images: Materialy Glyatsiologicheskikh Issledovanii, Pub. 33, p. 43–46, (In Russian), p. 131–133. (In English)

Krimmel, R.M., Post, A., and Meier, M.F., 1976, Surging and nonsurging glaciers in the Pamir Mountains, U.S.S.R., in Williams, R.S., Jr., and Carter, W.D., eds., ERTS-1. A new window on our planet: U.S. Geological Survey Professional Paper 929, p. 178–179.

Kuzmichenok, V.A., and Chaohai, L., 1995, The distribution of glaciers and their morphology over Tien Shan, in Dyurgerov, M.B., Chaohai, L., and Zichu, X., eds., Oledenenie Tian' Shanya [Tien Shan glaciers]: Moscow, VINITI Publishing House. (In Russian)

Lamb, J.M., 1991, The battle of Novaya Zemlya: Arctic Circle, v. 1, p. 46–47.

Lebedeva, I.M., and Rototaeva, O.V., 2005, Klimaticheskii faktor katastroficheskoi podvizhki lednika Kolka v 2002 godu [Climatic factor of the disastrous surge of Kolka Glacier in 2002]: Materialy Glyatsiologicheskikh Issledovanii, Pub. 98, p. 155–161. (In Russian)

Liestøl, O., 1993, Glaciers of Svalbard, Norway, in Williams, R.S., Jr., and Ferrigno, J.G., eds., Satellite Image Atlas of Glaciers of the World: U.S. Geological Survey Professional Paper 1386-E (Europe), p. E127–E151. [http://pubs.usgs.gov/prof/p1386e]

Marshall, G.J., Dowdeswell, J.A., and Rees, W.G., 1994, The spatial and temporal effect of cloud cover on the acquisition of high quality Landsat imagery in the European Arctic sector: Remote Sensing of Environment, v. 50, no. 2, p. 149–160.

Martin, P.J., and Sanderson, T.J.O., 1980, Morphology and dynamics of ice rises: Journal of Glaciology, v. 25, no. 91, p. 33–45.

Martyshev, A.P., 1980, Kolebaniya yazyka lednika Bol'shoy Azau na Elbruse [Fluctuations of the tongue of Bol'shoy Azau Glacier on El'brus]: Materialy Glyatsiologicheskikh Issledovanii, Pub. 39, p. 207–209. (In Russian)

Martyshev, A.P., 1983, Sovremennye izmeneniya lednikov v verkhoviakh r. Baksan v Prielbruse [Present-day changes of glaciers in the headwaters of the Baksan River in the El'brus region]: Materialy Glyatsiologicheskikh Issledovanii, Pub. 47, p. 209–210. (In Russian)

McCauley, C., 1958, Glaciers of the Arctic islands of the Soviet Union, in Geographic Study of Mountain Glaciation in the Northern Hemisphere, pt. 6, ch. 5: New York, American Geographical Society.

Meier, M.F., 1984, Contribution of small glaciers to global sea level: Science, v. 226, no. 4681, p. 1418–1421.

Meier, M.F., and Post, A.S., 1969, What are glacier surges?: Canadian Journal of Earth Sciences, v. 6, no. 4, pt. 2, p. 807–817.

Merzbacher, Gottfried, 1905, The central Tian Shan Mountains, 1902-1903: New York, E.P. Dalton, v. 8.

Miller, G.H., Forman, S.L., Synder, J.A., Lubinski, D.J., Matishov, G.G., and Korsun, S., 1992, Neoglaciation of Franz Josef Land, Russia [abs.]: Geological Society of America, Abstracts with Programs, 1992, p. 346.

Nansen, Fridtjof, 1897, Farthest north: Westminster, Archibald Constable and Co., v. 1, 510 p.; and v. 2, 671 p.

Narama, C., Shimamura, Y., Nakayama, D., and Abdrakhmatov, K., 2006, Recent changes of glacier coverage in the western Terskey-Alatoo range (sic), Kyrgyz Republic, using Corona and Landsat: Annals of Glaciology, v. 43, p. 223–229.

Nosenko, G.A., Rototaeva, O.V., Nosenko, O.A., 2004, Possibilities of monitoring from space the dangerous glacial processes in mountain regions that are difficult to access (from the example of the Karmadon disaster of 2002): Issledovanie Zemli iz Kosmosa: no. 6, p. 24–31. (In Russian)

Osipova, G.B.,and Tsvetkov, D.G., 1991, Kinematics of the surface of a surging glacier (comparison of the Medvezhiy and Variegated Glaciers): International Association of Hydrological Sciences, IAHS Pub. 208, p. 345–357.

Osipova, G.B., and Tsvetkov, D.G., 1999, Rol' podpruzhivaniya v dinamike slozhnykh gornykh lednikov. Ledniki Bivachniy i Sugran, Tsentralinyi Pamir [Role of damming in the dynamics of compound mountain glaciers. Bivachniy and Sugran Glaciers, Central Pamirs]: Materialy Glyatsiologicheskikh Issledovanii, Pub. 86, p. 133–141. (In Russian)

Osipova, G.B., and Tsvetkov, D.G., 2000, Kosmicheskii monitoring oledeneniya Vostochnogo Pamira [Satellite monitoring of glaciation in the Eastern Pamirs]: Materialy Glyatsiologicheskikh Issledovanii, Pub. 89, p. 163–174. (In Russian)

Osipova, G.B., and Tsvetkov, D.G., 2001, Dinamika pul'siruyushchikh lednikov Gando i Dorofeeva v 1972–1991gg [Dynamics of the surging glaciers Gando and Dorofeev from 1972 to 1991]: Materialy Glyatsiologicheskikh Issledovanii, Pub. 90, p. 193–198. (In Russian)

Osipova, G.B., and Tsvetkov, D.G., 2002a, Lednik Oktyabr'skii, vostochnyi Pamir, v 1945–1990 gg; osobennasti evolyutsii v stadii vosstanovleniya i podvizhki [Oktyabr'skiy Glacier, Eastern Pamirs, from 1945 to 1990. Development trends in recovery and surge stages]: Materialy Glyatsiologicheskikh Issledovanii, Pub. 93, p. 25–34. (In Russian)

Osipova, G.B., and Tsvetkov, D.G., 2002b, Issledovanie dinamiki slozhnykh lednikov po kosmicheskim snimkam [Investigations of the dynamics of compound glaciers on the basis of space images]: Izvestiya Akademii Nauk, Seriya Geograficheskaya, No. 3, p. 29–38. (In Russian)

Osipova, G.B., Tsvetkov, D.G., Schetinnikov, A.S., Rudak, M.S., 1998, Katalog pul'siruyushchikh lednikov Pamira [Inventory of surging glaciers of the Pamirs]: Materialy Glyiatsiologicheskikh Issledovanii, Pub. 85, p. 1–136. (In Russian)

Panov, V.D., 1993, Evolutsiia sovremennogo oledeneniia na Kavkaze [Evolution of the recent glaciation of the Caucasus]: Saint Petersburg, Gidrometeoizdat, 432 p. (In Russian)

Payer, Julius, 1876, Die Osterreichisch-Ungarische Nordpol-Expedition in den Jahren 1872–1874 [The Austro-Hungarian Expedition to the North Pole in the years 1872–74]: Vienna, Alfred Hölder, 696 p. (In German)

Penck, Albrecht, 1931, Geographical and photogrammetric results from the Arctic cruise of the Graf Zeppelin: Geographical Journal, v. 78, no. 6, p. 569–570.

Podozerskii, K.I., 1911, Ledniki Kavkazskogo Khrebta [Glaciers of the Caucasus Range]: Zapiski Kavkazskogo otdela Russkogo Geograficheskogo Obshchestva, v. 29, iss. 1, Tiflis, 200 p. (In Russian)

Popovnin, V.V., Petrakov, D.A., Tutubalina, O.V., Chernomorets, S.S., 2003, The glacial disaster of 2002 in North-Ossetia: Kriosfera Zemli, v. 7, no. 1, p. 1–10. (In Russian)

Post, Austin, 1969, Distribution of surging glaciers in western North America: Journal of Glaciology, v. 8, no. 53, p. 229–240.

Post, A., and Mayo, L.R., 1971, Glacier dammed lakes and outburst floods in Alaska: U.S. Geological Survey Hydrological Investigations Atlas HA-455, 10 p., and 3 maps.

Rees, W.G., Dowdeswell, J.A., and Diament, A.D., 1995, Analysis of ERS-1 synthetic aperture radar data from Nordaustlandet, Svalbard: International Journal of Remote Sensing, v. 16, no. 5, p. 905–924.

Rototaev, K.P., Khodakov, V.G., and Krenke, A.N., 1983, Issledovanie pul'siruyushchego lednika Kolka [Investigation of the surging Kolka Glacier]: Moscow, Nauka, 169 p. (In Russian)

Rototaeva, O.V., Kotlyakov, V.M., Nosenko, G.A., Khmelevskoi, I.F., Chernov, R.A., 2005, Istoricheskie dannye o podvizhkakh lednikov na severnom Kavkaze i Karmadonskaya katastrofa 2002 g. [Historical data on glacier surges in the Northern Caucasus and Karmadon disaster of 2002]: Materialy Glyatsiologicheskikh Issledovanii, Pub. 98, p. 136–145. (In Russian)

Rototaeva, O.V., Nosenko, G.A., Khmelevskoi, I.F., Tarasova, L.N., 2003, Balansovoe sostoyanie lednika Garabashi (Elbrus) v 80-kh i 90-kh godakh stoletiya [Mass balance of Garabashi Glacier (El'brus) in the 1980s and 1990s]: Materialy Glyatsiologicheskikh Issledovanii, Pub. 95, p. 111–121. (In Russian)

Russia, Committee of the Second International Polar Year, 1936a, Trudy lednikovykh ekspeditsii [Transactions of glaciological expeditions], iss. 1: Leningrad, 485 p. (In Russian)

Russia, Committee of the Second International Polar Year, 1936b, Trudy lednikovykh ekspeditsii [Transactions of glaciological expeditions], iss. 5, Kavkas. Lednikovye rayony [Caucasus. Glacier regions] 488 p. (In Russian)

Sandford, K.S., 1955, Tabular icebergs between Spitsbergen and Franz Josef Land: Geographical Journal, v. 121, pt. 2, p. 164–170.

Sapozhnikov, V.V., 1901, Katun' i ee istoki [Katun and its headwaters]: Tomsk, 271 p. (In Russian)

Sapozhnikov, V.V., 1911, Mongol'skii Altai v istokakh Irtysha i Kobdo [Mongolian Altay in the headwaters of the Irtysh and Kobdo rivers]: Tomsk, 408 p. (In Russian)

Schilling, Nikolai, 1865, Soobrazheniya o novom puti dlya otkryti v Severnom Polyarnom okeane [The conceptualization of a new route for opening the northern polar ocean]: Moskow Sbornik, v. 78, no. 5, p. 213–224. (In Russian)

Sedov, R.V., 1997a, Glaciers of the Chukotka: Materialy Glyatsiologicheskikh Issledovanii, Pub. 82, p. 213–217. (In Russian)

Sedov, R.V., 1997b, Glaciers of the Taigonos peninsula: Materialy Glyatsiologicheskikh Issledovanii, Pub. 82, p. 218–221. (In Russian)

Seinova, I.B., and Zolotarev, E.A., 2001, Ledniki i seli Priel'brusya (Evolutsiya oledeneniya i selevoi aktivnosti) [Glaciers and mudflows of the El'brus region (Evolution of glaciation and mudflow activity)]: Moscow, Nauchnyi mir, 204 p. (In Russian)

Semenov, P.P., 1858, Pervaya poezdka na Tien'Shan' ili Nebesny khrebet [The first trip to Tien Shan or Sky Range]: Vestnik RGO [Proceedings of the Russian Geographical Society], v. 1, no. 24. (In Russian)

Sharp, R.P., 1956, Glaciers in the Arctic: Arctic, v. 9, nos. 1–2, p. 78–117.

Shumskii, P.A., 1949, Sovremennoe oledenenie Sovetskoi Arktiki [Modern glaciation of the Soviet Arctic]: Trudy Arkticheskogo Instituta: v. 111, p. 11–39. (In Russian)

Siegert, M.J., and Dowdeswell, J.A., 1995, Late Weichselian ice-sheet sensitivity over Franz Josef Land, Russian High Arctic, from numerical modelling experiments: Boreas, v. 24, no. 3, p. 207–224.

Sin'kevich, S.A., Korolev, P.A., and Smirnov, K.E., 1991, Glaciological reconnaissance on the Loonney ice cap, Alexandra Land, Franz Josef Land: Journal of Glaciology, v. 37, no. 125, p. 183–185.

Solomina, O.N., Barry, R.G., Bodnya, V., 2004, The retreat of Tien Shan glaciers (Kyrgyzstan) since the Little Ice Age estimated from aerial photographs, lichenometric, and historical data: Geografiska Annaler, v. 46A, no. 2, p. 205–216.

Spizharskiy, T.N., 1936, Oledenenie Zemli Frantsa-Iosifa [Glaciation of Franz Josef Land]: Trudy Arkticheskogo Instituta, v. 41, p. 5–37. (In Russian)

Starokadomskiy, L.M., and Barr, William (translator), 1977, Charting the Russian northern sea route — The Arctic Ocean hydrographic expedition, 1910–1915: Montreal, McGill-Queens University Press, 332 p.

Tarussov, A., 1992, The Arctic from Svalbard to Severnaya Zemlya: climatic reconstructions from ice cores, in Bradley, R.S., and Jones, P.D., eds., Climate since A.D. 1500: London, Routledge, p. 505–516.

Troitskii, L.S., Khodakov, V.G., Mikhalev, V.I., Gus'kov, A.S., Lebedeva, I.M., Adamenko, V.N., and Zhivkovich, L.A., 1966, Oledenenie Urala [Glaciers of the Urals]: Moscow, Nauka, 307 p. (In Russian)

Tronov, M.V., 1949, Ocherki oledenenia Altaya [Notes on the glaciers of Altay]: Moscow, Geografizdat, 375 p. (In Russian)

Tronov, M.V., 1954, Voprosy gornoi glyatsiologii [Problems of mountain glaciation]: Moscow, Geografizdat, 276 p. (In Russian)

Tushinskii, G.K., 1968, Oledenenie Elbrusa [Glaciers of the Elbrus]: Izdatel'stvo Moskovskogo Universiteta, 344 p. (In Russian)

USSR Academy of Sciences, 1965–1983, Katalog lednikov SSSR [Glacier inventory of the USSR]: Leningrad, Gidrometeoizdat, 20 v. [Also available in digital form as part of the World Glacier Inventory at http://nsidc.org/data/glacier_inventory

USSR Academy of Sciences, 1966, Rukovodstvo po sostavleniyu Kataloga lednikov SSSR [Instructions for the compilation of the glacier inventory of the USSR]: Leningrad, Gidrometeoizdat, 154 p. (In Russian)

Vaikmyae, R.A., and Punning, Ya.M.K., 1984, Isotope and geochemical investigations on the Vavilov Glacier Dome, Severnaya Zemlya: Polar Geography and Geology, v. 8, no. 1, p. 73–79. [Translated from Materialy Glyatsiologicheskikh Issledovanii, Pub. 44, p. 145–149, 1982.]

Various authors, 1964, Nekotorye itogi issledovanii Instituta Geografii AN SSSR na Elbruse [Some results of the investigations of the Institute of Geography, Academy of Sciences, USSR, on Elbrus]: Materialy Glyatsiologicheskikh Issledovanii, Pub. 10, p. 55–103. (In Russian)

Various authors, 1967, Atlas lednikov Elbrusa [Atlas of the Elbrus glaciers]: Materialy Glyatsiologicheskikh Issledovanii, Pub. 13, p. 125–186. (in Russian)

Verkulich, S.R., Krasanov, A.G., and Anisimov, M.A., 1992, The present state of, and trends displayed by, the glaciers of Bennett Island in the past 40 years: Polar Geography and Geology, v. 16, no.1, p. 51–57. [Translated from Materialy Glyatsiologicheskikh Issledovanii, Pub. 70, p. 111–115, 1990.]

Vilesov, E.N., and Uvarov, V.N., 2001, Evolutsiya sovremennogo oledeneniya Zailiiskogo Alatau v XX veke [The evolution of modern glaciation of the Zailiyskiy Alatau in the 20th century]: Almaty, Kazakh State University, 252 p. (In Russian)

Vinnikov, L.P., and Labutina, I.A., 1987, Izmeneniya lednika Kiukiurtliu na Elbruse za chetvert' veka [Changes of Kukurtli Glacier on El'brus for more than a quarter of a century]: Materialy Glyatsiologicheskikh Issledovanii, Pub. 60, p. 147–152. (In Russian)

Vinogradov, V.N., 1968, Katalog lednikov SSSR, Kamchatka [Inventory of glaciers of the USSR, Kamchatka], v. 20: Leningrad, Gidrometeoizdat, 2–4. (In Russian)

Vinogradov, V.N., 1975, Sovremennoe oledenenie rayonov aktivnogo vulkanizma [Present day glaciers in the areas of the active volcanism]: Moscow, Nauka, 103 p. (In Russian)

Vinogradov, V.N., and Glazyrin, G.E., 1979, Statisticheskii podkhod k izucheniyu morfologii lednikov Kamchatki [Statistical approach to the studies of the morphology of Kamchatka glaciers]: Isvestiya Vsesoyuznogo Geograficheskogo Obshchestva, v. 111, iss. 4, p. 325–329. (In Russian)

Vinogradov, O.N., Konovalova, G.I., and Psareva, T.V., 1976, Novye dannye o glaziogeomorfologii sovremennykh lednikov Kavkasa i ikh evolyutsii v. xx veke [New data on glaciogeomorphology of modern glaciers of the Caucasus and their evolution in the twentieth century]: Materialy Glyatsiologicheskikh Issledovanii, Pub. 27, p. 44–50. (In Russian)

Vinogradov, O.N., Konovalova, G.I., and Psareva, T.V., 1980, Morfologicheskie kharakteristiki lednikovoi sistemy Tien'-shanya na kartakh masshtaba 1:1500000 v Atlase snezhno-ledovykh resursov mira [Morphological characteristics of the Tien Shan glacier system on the 1:1,500,000-scale maps in the World Atlas of Snow and Ice Resources: Materialy Glyatsiologicheskikh Issledovanii, Pub. 37, p. 56–62. (In Russian)

Voevodin, V.A., 1972, O razmerakh aisbergov v rayone Zemli Frantsa-Iosifa i Shpitsbergena [Dimensions of icebergs in the region of Franz-Josef Land and Spitsbergen]: Problemy Arktiki i Antarktiki, v. 39, p. 138–140. (In Russian)

Voitkovsky, K.F., Benkevich, V.V., Volodicheva, N.A., 1989, Sovremennye tendentsii razvitiya lednikovoi sistemy Elbrusa [Present-day trends of development of the El'brus glacial system]: Materialy Glyatsiologicheskikh Issledovanii, Pub. 67, p. 73–80. (In Russian)

Volodicheva, N.A., Evteev, O.A., Kirpichenkov, S.Ya., Knizhnikov, Yu.F., Kravtsova, V.I., Martyshev, A.P., and Myagkov, S.M., 1983, Izuchenie izmenenii oledeneniya Elbrusa za poslednie 25 let [Study of changes of El'brus glaciation over the past 25 years]: Materialy Glyatsiologicheskikh Issledovanii, Pub. 47, p. 130–137. (In Russian)

Williams, M., 2003, Remote sensing of glacierized area ice dynamics in the Russian High Arctic: Aberystwyth, University of Wales Aberystwyth, Ph.D. thesis, 428 p.

Williams, M., and Dowdeswell, J.A., 2001, Historical fluctuations of the Matusevich Ice Shelf, Severnaya Zemlya, Russian High Arctic: Arctic, Antarctic, and Alpine Research, v. 33, no. 2, p. 211–222.

Williams, R.S., Jr., 1986, Glaciers and glacial landforms; Chapter 9, *in* Short, N.M., and Blair, R.W., Jr., editors, Geomorphology from space. A global overview of regional landforms: NASA Special Publication, SP-486, p. 521–596. [*http://disc.gsfc.nasa.gov/geomorphology/index.shtml*]

Williams, R.S., Jr., Hall, D.K., and Benson, C.S., 1991, Analysis of glacier facies using satellite techniques: Journal of Glaciology, v. 37, no. 125, p. 120–128.

Zabirov, R.D., 1955, Oledenenie Pamira [Glaciers of the Pamirs]: Moscow, Geografizdat, 372 p. (In Russian)

Zabirov, R.D., and Bakov, E.K., 1981, Ozero Mertsbakhera. Glyatsiologicheskie issledovaniya v Kirgizii [Lake Mertsbakhera. Glaciological studies in Kyrgyzstan]: Frunze, p. 82–92. (In Russian)

Zabirov, R.D., and Knizhnikov, Yu. F., 1962, Fototeodolitnaya s'emka lednikov Tian Shania v period MGG [Photogrammetric survey of the Tien Shan glaciers during the International Geophysical Year]: Frunze, Izd-vo Akademii nauk Kirgizskoi SSR, 99 p. (In Russian)

Zeeberg, J.J., 2001, Climate and glacial history of the Novaya Zemlya Archipelago, Russian Arctic: Amsterdam, Rozenberg, 174 p.

Zeeberg, J.J., and Forman, S.L., 2001, Changes in glacier extent on northern Novaya Zemlya in the twentieth century: The Holocene, v. 11, p. 161–175.

Zinger, E.M., and Koryakin, V.S., 1965, Yest' li shel'fovye ledniki na Severnoi Zemle? [Are there ice shelves on Severnaya Zemlya?]: Materialy Glyatsiologicheskikh Issledovanii, Pub. 11, p. 250–253. (In Russian)

Zolotarev, E.A., 1997, Izmeneniya lednikov Elbrusa v poslednem stoletii [Changes of El'brus glaciers in the past century]: Materialy Glyatsiologicheskikh Issledovanii, Pub. 83, p. 146–153. (In Russian)

Zolotarev, E.A., and Khar'kovets, E.G., 2000, Oledenenie Elbrusa v kontse XX v. (tsifrovaya ortofotokarta Elbrusa na 1997 g.) [Glaciation of El'brus at the end of the 20th century (digital orthophotomap of El'brus for 1997)]: Materialy Glyatsiologicheskikh Issledovanii, Pub. 89, p. 175–181. (In Russian)

Glaciers of Asia—

GLACIERS OF CHINA

By Shi Yafeng, Mi Desheng, Yao Tandong, Zeng Qunzhu, *and* Liu Chaohai

SATELLITE IMAGE ATLAS OF GLACIERS OF THE WORLD

Edited by RICHARD S. WILLIAMS, JR., *and* JANE G. FERRIGNO

U.S. GEOLOGICAL SURVEY PROFESSIONAL PAPER 1386–F–2

CONTENTS

Introduction -- 127

TABLE 1. Commonly used transliterated Chinese terms for geographic features -- 128

TABLE 2. Glacierized areas in China -- 128

FIGURE 1. Map of glaciers and snowline altitude (elevation) in the mountain ranges of western China and adjacent countries -------------- 129

TABLE 3. The parameters for three types of glaciers in West China ------------------ 130

TABLE 4. Locations of glaciers, glacierized regions, and other geographic features in China ------------------------------ 131–132

Geographic Distribution of Glaciers in China -------------------------------------- 132

Distribution of Glaciers in the Altay Shan ------------------------------ 132

Distribution of Glaciers in Tian Shan ----------------------------------- 133

Distribution of Glaciers in the Qilian Shan ---------------------------- 133

Distribution of Glaciers in the Chinese Pamir ------------------------ 134

Glaciers in the Karakorum -- 134

Distribution of Glaciers in the Kunlun Shan ------------------------- 134

Distribution of Glaciers on the *Qiantan Gaoyuan* ----------------- 135

Distribution of Glaciers in the Tanggula Shan ----------------------- 135

Distribution of Glaciers in the Nyainqêntanglha Shan ----------- 135

Distribution of Glaciers in the Gangdisê Shan ---------------------- 136

Glaciers of the Himalaya --- 136

Distribution of Glaciers in Hengduan Shan -------------------------- 137

Ablation of Glaciers in China and Discharge of Meltwater into the Glacierized Drainage Basins -- 137

Landsat MSS Images Used to Illustrate Selected Glacierized Areas ------------------ 138

Central Chinese Tian Shan Region --------------------------------------- 139

FIGURE 2. Annotated Landsat 2 MSS image of the central Chinese Tian Shan region -- 139

TABLE 5. Selected place-names of glaciers and other geographic features of the central Chinese Tian Shan region ------------------------- 140

Hantengri Feng and *Tomur Feng* Region of the Tian Shan --------------------- 141

FIGURE 3. Annotated Landsat 2 MSS false-color composite image of the western part of the Chinese Tian Shan region, including the *Hantengri Feng -Tomur Feng* area ------------------------------- 142

TABLE 6. Selected place-names of glaciers and other geographic features of the western part of the Chinese Tian Shan region, including the *Hantengri Feng* and *Tomur Feng* regions ------------------------- 143

Western Qilian Shan Region --- 144

FIGURE 4. Annotated Landsat 2 MSS false-color composite image of the western Qilian Shan region --- 144

TABLE 7. Selected place-names of glaciers and other geographic features of the western Qilian Shan region -- 145

Qogir Feng (K2) of the Karakorum Shan Region ----------------------------- 147

FIGURE 5. Annotated Landsat 3 MSS false-color composite image of the Qogir Feng (K2) area of the Karakorum Shan region -------------------- 148

TABLE 8. Selected place-names of glaciers and other geographic features of the Qogir Feng (K2) of the Karakorum Shan region ---------------- 149

Western Kunlun Shan Region -- 150

FIGURE 6. Annotated Landsat 2 MSS image of the western Kunlun Shan region --- 150

TABLE 9. Selected place-names of glaciers and other geographic features of the western Kunlun Shan region ------------------------------------- 151

Tanggula Shan Region--- **152**

    FIGURE   7. Annotated Landsat 1 MSS image of the Tanggula Shan region------------ **152**

    TABLE 10. Selected place-names of glaciers and other geographic features
          of the Tanggula Shan region -------------------------------------- **153**

Western Nyainqêntanglha Shan Region--------------------------------------- **154**

    FIGURE   8. Annotated Landsat 2 MSS image of the western Nyainqêntanglha
          Shan region --- **154**

    TABLE 11. Selected place-names of glaciers and other geographic features
          of the western Nyainqêntanglha Shan region ------------------------ **155**

Eastern Nyainqêntanglha Shan Region --------------------------------------- **156**

    FIGURE   9. Annotated Landsat 2 MSS false-color composite image of the
          eastern Nyainqêntanglha Shan region ----------------------------- **156**

    TABLE 12. Selected place-names of glaciers and other geographic features
          of the eastern Nyainqêntanglha Shan region ----------------------- **157**

Central Himalaya Region-- **158**

    FIGURE 10. Annotated Landsat 2 MSS false-color composite image of the
          central Himalaya--- **158**

    TABLE 13. Selected place-names of glaciers and other geographic features
          of the central Himalaya region------------------------------------- **159**

Optimum Landsat 1, 2, and 3 Images of the Glacierized Regions
   of China -- **160**

    TABLE 14. Optimum Landsat 1, 2, and 3 images of the glaciers
          of China --- **160–163**

    FIGURE 11. Index map of optimum Landsat 1, 2, and 3 images of the
          glacierized areas of China-- **164**

References Cited --- **165**

GLACIERS OF ASIA—

GLACIERS OF CHINA[1]

By Shi Yafeng, Mi Desheng, Yao Tandong, Zeng Qunzhu, *and*
Liu Chaohai[2]

Introduction

China is home to many of the major mountain-glacier systems of the world. Figure 1 shows the distribution of glaciers in China. Western China includes a series of major mountain ranges and plateaus, namely the Himalaya, Karakorum Shan, Kunlun Shan, Qilian Shan, Tian Shan, and Altay Shan, and the *Qinghai-Xizang Gaoyuan*[3] (Qing Zang Gaoyuan) (Tibetan Plateau) where glaciers are located (table 1). Glaciers are widely scattered and cover an area of about 59,425 km[2] (fig. 1). Table 2 and figure 1 show the generalized snowline elevations[4] in China. The elevation of the snowline ranges from 2,800–3,000 m in the Altay Shan to 6,000–6,200 m in southern and western Xizang Zizhiqu (Tibet Autonomous Region). Isolines of snowline elevation make irregular, concentric circles centered near western Xizang Zizhiqu, where the snowline may reach the highest elevation in the world. This peculiar form of variation of snowline elevations is caused by the decrease in precipitation from the peripheral mountains to the interior of the Qing Zang Gaoyuan. The annual precipitation is 800 to 3,000 mm in some of the outer mountains, but is only 200 to 300 mm in the interior of the plateau. High elevations result in greater solar radiation on the Qing Zang Gaoyuan than on the surrounding lowlands and causes an increase in elevation of the interior snowlines.

---

[1]This manuscript was originally written in 1981 to describe the glaciers of China during the benchmark period of the late 1970s and early 1980s. Because there was a delay in printing, the authors substantially updated the manuscript in 2005.

[2]Cold and Arid Regions Environmental and Engineering Research Institute, Chinese Academy of Sciences, 260 West Donggang Road, Lanzhou 730000, China.

[3]Chinese place-names are used throughout the Glaciers of China subchapter and transliterated English words have been added to assist those unfamiliar with Chinese terms for geographic features: Table 1 provides a list of frequently used words. U.S. Government publications require that official geographic place-names for foreign countries be used to the greatest extent possible. In the Glaciers of China subchapter, the spelling of geographic place-names is derived from the "Gazetteer of China — An Index to the Atlas of the People's Republic of China" published by the Gazetteer Laboratory of Geodesy and Mapping Institute, National Bureau of Geology and Mapping, Beijing, China. Gazetteers of place-names in the Xizang Zizhiqu (Tibet Autonomous Region) and the Xinjiang Uygur Zizhiqu (Uygur Autonomous Region) were consulted. In addition, note that the Government of China employs a different spelling system for geographic place-names in regions of China with large Han, Mongol, Tibetan, and Uygur nationalities; the spelling used combined the Chinese phonetic alphabet and the Wade System. The Chinese place-names were compared with the Gazetteer of the People's Republic of China (U.S. Defense Mapping Agency, 1990) and the U.S. Board on Geographic Names website *http://earth-info.nga.mil/gns/html/index.html*. Names not listed on the website are shown in italics and the spelling of the authors is used.

[4]Snowline elevation is defined in this chapter as equivalent to the equilibrium line altitude (ELA) at which the annual glacier balance is zero, averaged over many consecutive recent years.

TABLE 1.—*Commonly used transliterated Chinese terms for geographic features*

Chinese word	Transliterated[*] English word
bingchuan	glacier
co (Tibetan term)	lake
feng	summit or peak of mountain, mountains
gaoyuan	plateau
gou	stream
he	river, stream
hu	lake
jiang	large river
kangri	summit or peak of mountain, mountains
kou	pass
ling	mountain range
pendi	basin, depression
qu	stream
shan	mountain range
shankou	mountain pass

[*]Chinese Pinyin characters transliterated into English.

Figure 1.— (at right) *Map of glaciers and snowline altitude (elevation) in the mountain ranges of western China and adjacent countries. The location of the Landsat images used for figures 2–10 is shown. The national boundaries and the geographic place names are taken from the 1:4M "Map of the People's Republic of China," published by China Cartographic Publishing House in 1989.*

TABLE 2.—*Glacierized areas in China*

[Units: m, meter; km², square kilometer]

Mountains, region, or glacier	Elevation of the snowline (m)	Area of glaciers (km²)	Source[1]	Investigator
Altay Shan (including *Muztau Ling*)	2,800–3,200	297	GIC 2, 1982	Liu Chaohai and others
Tian Shan	3,600–4,400	9,225	GIC 3, 1986–1987	Liu Chaohai and others (1987); Shi Yafeng and others (2005)
Qilian Shan (including Altun Shan)	4,300–5,200	2,206	GIC 1, 1981	Wang Zongtai and others
Pamir	5,500–5,700	2,696	GIC 4, 1988, (revised) 1999	Mi Desheng and others
Karakorum Shan	4,700–5,600	6,262	GIC 5, 1989	Yang Huian and others (1989); Shi Yafeng and others (2005)
Kunlun Shan	5,800–6,100	12,267	GIC 6, 1992–1994	Yang Huian and others
Qiantan Gaoyuan (Plateau)	5,600–6,100	1,802	GIC 7, 1988	Jiao Keqin and others
Tanggula Shan	5,300–5,800	2,213	GIC 8, 1994	Pu Jianchen and others
Nyainqêntanglha Shan	4,200–5,700	10,700	GIC 9, 2001	Pu Jianchen and others (2001); Mi Desheng and others (2002)
Gangdisê Shan	5,800–6,000	1,760	GIC 7, 1988	Jiao Keqin and others (1988); Shi Yafeng and others (2005)
Himalaya	4,300–6,200	8,418	GIC 11, 2002	Mi Desheng and others
Hengduan Shan	4,600–5,600	1,579	GIC 8, 1994	Pu Jianchen and others
Total		59,425		Shi Yafeng and others (2005)

[1]Data for glacier area are taken from Glacier Inventory of China (GIC) which was compiled by the staff of Lanzhou Institute of Glaciology and Geocryology and published by Science Press, Beijing, China. The inventory is also available digitally at World Data Center D for Glaciology and Geocryology, Lanzhou, China. The glacier areas were later revised and published in Concise Glacier Inventory of China (Shi Yafeng and others, 2005; Shi Yafeng, 2008).

RUSSIA

KAZAKHSTAN

MONGOLIA

ULAAN BAATAR

Altay Shan

3,200

3,400

Junggar Pendi

3,600

3,800

3,800

154/30

Urumqi

Shan

4,000

CHINA

4,000

158/31

BISKEK

KIRGHIZSTAN

ALMA-ATA

Tian

4,200

4,000

4,400

Pamir

Xinjiang

Tarim Pendi

4,600

4,400

4,800

Qilian   Shan

5,000

146/33

Qaidam Pendi

Xining

Lanzhou

Gansu

5,400

Kunlun   Shan

5,000

159/35

Karakorum Shan

5,200

156/35

5,600

5,200

Kunlun   Shan

5,400

149/37

Qinghai

4,600

5,800

Tanggula   Shan

6,000

Qiantan   Gaoyuan

Xizang

(Tibet)

Sichuan

Chengdu

6,000

6,000

Gangdise   Shan

Nyainqentangiha

149/39

144/40

Shan

Hengduan

4,800

5,000

4,800

5,800

149/39

Lasa

5,000

NEW DELHI

Himalaya   Shan

5,000

6,000

151/40

4,400

4,600

Shan

Chang

Jiang

NEPAL

KATHMANDU

THIMBU 4,800

Himalaya Shan

4,800

Kunming

BHUTAN

INDIA

Yunnan

INDIA

BANGLADESH

DHAKA

VIET NAM

EXPLANATION

Bay of Bengal

MYANMAR

GLACIER

LAOS

APPROXIMATE PROVINCE BOUNDARY

—6,000— SNOW-LINE – Altitude, in meters

LANDSAT IMAGE – Path/Row number

VIENTIAN

151/40

0   250 KILOMETERS

THAILAND

Chief Editor, Shi Yafeng;   Editor, Mi Desheng;
Cartographic fairdrawing and Platemaking, He Baoshan

Mountain glaciers in western China and central Asia may be classified into three types: maritime, subcontinental, and extreme continental (table 3). Glaciers of southeastern Xizang Zizhiqu (Tibet Autonomous Region), including the eastern Himalaya and Hengduan Shan, belong to the maritime type. The majority of glaciers in Altay Shan, Tian Shan, eastern Qilian Shan, eastern Kunlun Shan, eastern Tanggula Shan, north slope of the middle and western Himalaya, Karakorum Shan, etc., belong to the subcontinental type. Glaciers in the western part of Qing Zang Gaoyuan including western Kunlun Shan, eastern Pamir, western Tanggula Shan, western Qilian Shan, *Qiantan Gaoyuan*, etc., belong to the extreme continental type. Their differences are mainly due to regional variation in climate (fig. 1, table 3).

TABLE 3.—*The parameters for three types of glaciers in West China (Shi and Liu, 2000)*

[Abbreviations: E., eastern; N., northern; W., western; S.E., southeastern; km², square kilometer; m, meter; mm, millimeter; ~, approximate]

Parameter	Glacier type		
	Maritime	Subcontinental	Extreme continental
Annual precipitation	1,000–3,000 mm	500–1,000 mm	200–500 mm
Average air temperature at snowline			
a. Summer	1 ~ 5 °C	0 ~ 3 °C	<-1 °C
b. Annual	-2 ~ -6 °C	-6 ~ -12 °C	<-10 °C
Ice temperature within 20-m depth	-1 ~ 0 °C	-1 ~ -10 °C	<-10 °C
Distribution region	S.E. part of *Qinghai-Xizang Gaoyuan*, Hengduan Shan, E. Himalaya Mid E. Part of Nyainqêntanglha Shan	Tian Shan, Altay Shan, E. Qilian Shan, E. Kunlun Shan, E. Tanggula Shan, N. slope of midwest Himalaya, N. slope of Karakorum Shan	W. part of *Qinghai-Xizang Gaoyuan*, W. Kunlun Shan, E. Pamir, Tanggula Shan, W. Gangdisê Shan, W. Qilian Shan
Approximate area	13,200 km²	27,200 km²	19,000 km²

The earliest description of glaciers in Chinese literature was made by the Buddhist monk Xuazang during the Tang Dynasty (618–907 A.D.). Xuazang crossed the *Muzart Bingchuan* in 630 A.D, on the way to India in search of Buddhist Scriptures. In the late 19th century and early 20th century, some western and Chinese explorers observed glaciers in western China. H. von Wissman (1959) made a comprehensive study of the glaciers and snowline elevation in *High Asia (Gaoya)*. In 1958, an expedition called the "Alpine Ice and Snow Utilization Team" (1959), which was the forerunner of what was to become the Institute of Glaciology and Geocryology, was organized under the leadership of Shi Yafeng. In 2000, this institute was renamed the Cold and Arid Regions Environmental and Engineering Research Institute, Chinese Academy of Sciences. This team, and later the Institute, investigated glaciers in the glacierized regions of China: the Qilian Shan, the Tian Shan, the western Kunlun Shan, the Karakorum Shan, the Himalaya, the Hengduan Shan and other areas. Monitoring stations were set up on *Ürümqi He No. 1 Bingchuan* in the Tian Shan, on the *Hailuogou Bingchuan* in the Gongga Shan, on the *Tanggula Bingchuan* in the Tanggula Shan and in the Yulongxue Shan. In the 1980s, glaciologists from the United States, Japan, Europe, Australia, Russia, and many other countries came to China to cooperate and exchange information with Chinese glaciologists. From the middle 1980s, ice-core studies were performed in the *Qinghai-Xizang Gaoyuan* (Tibetan Plateau) to reconstruct past environments and climates (Thompson and others, 1989; Nakawo and others, 1990; Yao and others, 1995, 1996). A comprehensive glacier inventory, begun in 1979 and completed in 2002, provides a detailed account of the distribution of glaciers in China. Most of the data contained in this report are derived from that inventory, the Glacier Inventory of China (table 2) (also available digitally from World Data Center D for Glaciology and Geocryology, Lanzhou, China). A summary of the glacier inventory was published by Shi (2008).

This report first discusses the distribution of glaciers within each of 12 glacierized regions. It then describes the ablation of the glaciers, and discharge of meltwater into the glacierized drainage basins. In the last part of the report, Landsat 2 MSS images and aerial photographs were used as a visual base to describe numerous glaciers in several of the glacierized regions and discuss their retreat or advance. The names of numerous geographic features are used in this report, and the locations (latitude and longitude) of many of the glaciers and glacierized regions, as well as a few other geographic features, are presented on table 4. Additionally, Landsat images (figs. 2-10) and their associated tables (tables 5-13) provide the locations of many of the glaciers, mountains, mountain ranges, lakes, rivers, and other geographic features.

TABLE 4.—*Locations of glaciers, glacierized regions, and other geographic features in China*

Name	Latitude	Longitude
Altay Shan (Chinese part)	45°30′N.–50°10′N.	85°00′E.–91°10′E.
Youyi Feng	49°27′N.	87°56′E.
Kuitun Feng	47°00′N.	85°30′E.
Karaxi Bingchuan	49°25′N.	87°53′E.
Tian Shan (Chinese part)	41°00′N.–45°30′N.	76°00′E.–95°00′E.
Hantengri Shan	42°15′N.	80°15′E.
Harlertau Shan	42°00′N.	82°00′E.
Eren Habirga Shan	43°30′N.	85°00′E.
Bogda Feng	43°50′N.	88°40′E.
Muzart Bingchuan	42°18′N.	80°45′E.
Qilian Shan	36°00′N.–40°00′N.	94°00′E.–103°00′E.
Dangjin Shankou (pass)	39°05′N.	94°08′E.
Wushao Ling	37°30′N.	102°42′E.
Tuanjie Feng	38°30′N.	98°20′E.
Shule Nanshan	38°53′N.	97°16′E.
Laohugou Bingchuan	39°03′N.	96°40′E.
Daxue Shan	39°24′N.	96°36′E.
Lenglong Ling Bingchuan	37°38′N.	101°38′E.
Tergun Daba Shan	38°08′N.	96°21′E.
Chinese Pamir	38°00′N–39°30′N.	74°00′E.–76°00′E.
Muztagata (peak)	38°15′N.	75°10′E.
Kongur Shan	38°40′N.	75°20′E.
Karakorum Shan	35°30′N–36°30′N.	76°00′E.–77°30′E.
Qogir Feng (K2)	35°55′N.	76°30′E.
Yengisogat Bingchuan	36°03′N.	76°17′E.

TABLE 4.—*Locations of glaciers, glacierized regions, and other geographic features in China*—Continued

Name	Latitude	Longitude
Kunlun Shan	33°15′N.–38°20′N.	75°00′E.–105°00′E.
A'nyêmaqên Shan	34°30′N.–33°40′N.	98°00′E.–102°10′E.
Bayan Har Shan	33°15′N.–35°40′N.	96°00′E.–99°40′E.
Yulong Bingchuan	35°27′N.	81°08′E.
Muztag Shan	36°30′N.	87°30′E.
Yaherung Bingchuan	34°54′N.	99°10′E.
***Qiantan Gaoyuan* (Plateau)**	32°20′N.–34°30′N.	80°00′E.–90°00′N
Purog Kangri	34°20′N.	85°30′E.
Caixong Kangri	34°15′N.	80°05′E.
Tanggula Shan	31°30′N.–34°00′N.	89°00′E.–98°00′E.
Mitijangzhanm Co	33°24′N.	90°19′E.
Yushu	32°45′N.	97°00′E.
Zhag'yab	31°40′N.	97°55′E.
Geladaindong (peak)	33°29′N.	91°13′E.
Bija Shan	31°48′N.	94°40′E.
Bija Glacier Group	31°40′N.–31°55′N.	94°35′E.–94°50′E.
Puogou Bingchuan	31°43′N.	94°35′E.
Nyainqêntanglha Shan	28°00′N.–31°30′N.	90°00′E.–100°00′E.
Qungmo Kangri	29°52′N.	90°03′E.
Margyang He	29°48′N.	89°58′E.
Anjiula	29°40′N.	97°45′E.
Rawu	29°30′N.	97°45′E.
Yarlung Zangbo Jiang	29°20′N.	89°30′E.
Kyaggen Bingchuan	30°31′N.	94°42′E.
Yagnung Bingchuan	29°19′N.	96°42′E.

Geographic Distribution of Glaciers in China

There are 12 primary glacierized regions in China (fig. 1, table 2): Altay Shan, Tian Shan, Qilian Shan, Pamir, Karakorum Shan, Kunlun Shan, *Qiantan Gaoyuan* (Plateau), Tanggula Shan, Nyainqêntanglha Shan, Gangdisê Shan, Himalaya, and Hengduan Shan. Altay Shan, located in the northernmost part of China, has the smallest glacier coverage (297 km²), and Kunlun Shan has the largest (12,267 km²).

Distribution of Glaciers in the Altay Shan

The Chinese Altay Shan (fig. 1, table 4) extends from the border between China and Russia on the west to the border between China and Mongolia on the east, between latitude 45°30′N. and 50°10′N. The highest peaks are *Youyi Feng* (4,374 m) and *Kuitun Feng* (4,104 m). It is the northernmost glacierized area in China. The glaciers are concentrated in the upper course of Burqin He, a tributary of the *Irtish He* (Irtysh). There are 424 glaciers with a total area of 297 km² on the Chinese side of this range, also including *Muztau Ling*, to the south of Altay Shan (table 2, Liu and others, 1982). The *Karaxi Bingchuan*, at the head of Burqin He, is the longest, with a length of 10.8 km. Its terminus descends to 2,416 m, the lowest elevation of any glacier in China.

Distribution of Glaciers in Tian Shan

The Tian Shan (fig. 1, table 4) lies across the middle part of Asia; the part in China is called eastern Tian Shan, extending about 1,700 km in an east-west direction; it is higher in the west and lower in the east. Under the influence of the westerly circulation, precipitation in this region decreases gradually, and the snowline elevation rises gradually, from the west to the east.

A glacier inventory using aerial and ground photographs shows that there are 9,035 glaciers in eastern Tian Shan that cover an area of 9,225 km² (table 2, Liu and others, 1987; Shi and others, 2005). This is 15.5 percent of the total glacierized area of China. The *Hantengri-Tomur Feng* area (fig. 3) is the largest glacierized center in the Tian Shan; 4,093 km² of the glacierized area is concentrated here, including glaciers of Kyrgyzstan. Eren Habirga Shan (fig. 2) is the second glacierized center in the Tian Shan, and the glaciers cover an area of about 1,400 km². Bogda Feng rises abruptly in the eastern part of the Tian Shan and is also a major center of glacierization.

From the beginning of the 20th century to the 1950s, the glaciers of the Tian Shan were in a state of recession. For example, the *Muzart Bingchuan* was first described by the travelers of the Tang Dynasty in the 7th century and, at that time, its spectacular ablation scenery and dangerous passage were vividly written about in the famous book *Records of the Western Regions of the Great Tang Dynasty*. From 1909 to 1950, the glacier retreated 750 m (Shi and Wang, 1979). Analysis of aerial photographs and Landsat images since the beginning of the 1970s indicates that the general glacier recession has been reduced; some glaciers are stagnant but some are advancing.

Distribution of Glaciers in the Qilian Shan

The Qilian Shan (fig. 1, table 4) is situated to the northeast of the *Qinghai-Xizang Gaoyuan* (Tibetan Plateau), extending from Dangjin Shankou (pass) in the west to the Wushao Ling (mountains) in the east. It is about 800 km long. The elevation of the mountains increases from east to west. The highest peak is *Tuanjie Feng* (5,808 m) in the Shule Nanshan (fig. 4). Under the influence of the monsoon, the annual precipitation decreases from the eastern section (800 mm) to the western section (300 mm). The elevation of the snowline is 4,300 m in the eastern section and 5,200 m in the west.

A glacier inventory shows 2,815 glaciers in the Qilian Shan with an area of 1,931 km² (Wang and others, 1981). Glaciers in Altun Shan are sometimes also included in Qilian Shan; when included, the number and area of glaciers increases to 3,015 and 2,206 km², respectively (table 2). The largest valley glacier is *Laohugou Bingchuan* of Daxue Shan, 10.1 km long, with an area of 21.9 km² (fig. 4). *Laohugou Bingchuan* has been studied in more detail than other glaciers in the Qilian Shan. A deep ice core from the *Dunde Binggai* (ice cap) (lat 38°06'N., long 96°24'E.), which has an area of 57 km² and is located at an elevation of 5,325 m, was analyzed to determine the glacial and climate history from the last glaciation to the present (Thompson and others, 1989; Yao and Thompson, 1992).

Field investigations integrated with analyses of Landsat images during a period of 20 years shows that the *Lenglong Ling Bingchuan,* in the eastern section of Qilian Shan is receding at a rate of 12.5 to 22.5 m a⁻¹. However, in the Tergun Daba Shan (fig. 4) in the southwestern section of the Qilian Shan, several valley glaciers >5 km long advanced 200–400 m from 1966 to 1977. During the past 5 years the recession of glaciers in the Qilian Shan has been reduced. On many glaciers the mass balance was positive during the 1970s (Xie Zichu, 1980) but became negative again during the 1980s and 1990s.

Distribution of Glaciers in the Chinese Pamir

The Chinese Pamir (fig. 1) are called the eastern Pamir; they are located to the south of the Tian Shan, to the east of the western Pamir in Kyrgyzstan and Tajikistan, to the west of Yarkant He, and to the north of the border between Afghanistan and Pakistan with China. The glacierized area in this region is 2,696 km^2 [table 2; Mi and others, 1988, 1999 (revised)]; 80 percent is located in the region of Muztagata peak and Kongur Shan.

In this part of the Pamir, glaciers are mostly situated on the leeward slopes. The snowline elevation reaches 5,500–5,700 m and the termini end at an elevation of 4,000–5,000 m. Ice ablation is relatively minor and the glacier motion is relatively slow. The glaciers are typically of the extreme continental type.

Glaciers in the Karakorum Shan

The Karakorum Shan (fig. 1, table 4) combines extreme ruggedness, high elevation, and extensive glacier-ice cover to a greater degree than any other mountain system in the world. There are four peaks more than 8,000 m in elevation, including Qogir Feng (K2) (8,611 m), the second highest peak in the world (fig. 5). The glaciers are larger than those in the Himalaya and rank among the longest valley glaciers in the world outside the polar regions. In 1959, the total area of Karakorum Shan glaciers was about 15,000 km^2 (von Wissman, 1959), which is 37 percent of the mountain area. The snowline elevation lies at about 4,700–5,600 m. There are six glaciers longer than 50 km in the Karakorum Shan.

The Chinese Karakorum Shan occupies a wide region in the northern part of the Karakorum Shan (fig. 1). Landsat images and aerial photographic maps show the area of glaciers in the Chinese Karakorum Shan to be about 6,262 km^2 (table 2; Yang and others, 1989; Shi and others, 2005). Several large valley glaciers stretch into the upper Yarkant He and its tributary, the *Klejin He*. The *Yengisogat Bingchuan* is 42.0 km long, with an area of 380 km^2 and a terminus at 4,000 m; it is the longest glacier in China. Using data from a map by Shipton (1938), a 1968 aerial photomap, and 1978 Landsat images, it is seen that the debris-covered terminus of the *Yengisogat Bingchuan* was stable from 1937–1978. However, the *Skyang Bingchuan* (northeast of Qogir Feng, 18.8 km long) receded 5.2 km from 1937 to 1978. Some glaciers in the Chinese Karakorum Shan are advancing; Landsat images of 1978 clearly show that a glacier has dammed the *Klejin He* and impounded a lake; serious flooding along the *Klejin He* occurred in 1961 and 1978.

Distribution of Glaciers in the Kunlun Shan

The Kunlun Shan (fig. 1, table 4), extending from the Pamir eastward to the A'nyêmaqên Shan and the Bayan Har Shan, has a total length of about 2,500 km, and its elevation is higher in the west and lower in the east.

The area of the glaciers in this region is about 12,267 km^2 as determined from the aerial photomaps and Landsat images (table 2; Yang and others, 1992–94). About 70 percent of the glaciers are located between the upper reaches of the Kaxgar He and the *Keliyar He* in the western Kunlun Shan. There are 10 large glaciers more than 100 km^2 in area, of which *Dufeng Bingchuan*, in the upper reach of *Yulongkay He*, 31 km long and 252 km^2 in area, is the largest. The snowline elevations in western Kunlun Shan are 5,800–6,100 m above sea level. The *Guliya Binggai* (ice cap) (fig. 6) is the coldest glacier in central Asia. The lowest measured ice temperature was –19 °C at a depth of 10 m. The longest ice core, 309 m long, was analyzed to determine the glacial and climate history for about 0.7 million years (Thompson and others, 1997).

Aerial photomaps and Landsat images show that 20 of the 37 glaciers measured have advanced in the 1960s and 1970s. Some glaciers in the Muztag Shan and A'nyêmaqên Shan have experienced major advances; for instance, the *Yaherung Bingchuan* in the A'nyêmaqên Shan advanced 1,800–2,000 m from November 1966 to July 1977.

Distribution of Glaciers on the *Qiantan Gaoyuan*

The *Qiantan Gaoyuan* region (fig. 1, table 4), a wide inland plateau, lies to the south of the Kunlun Shan, to the west of the Tanggula Shan, and to the north of the Gangdisê Shan. The relative elevation of the mountains in this region is small. The snowline elevation is as high as 5,600–6,100 m. The area of the glaciers amounts to 1,802 km^2 (table 2; Jiao and others, 1988). The glaciers are small and are scattered and distributed mainly near the highest peaks (>6,000 m); for example, *No. 41 Bingchuan* of Purog Kangri (6,482 m) is 13 km long and 5.3 km^2 in area.

Distribution of Glaciers in the Tanggula Shan

The Tanggula Shan (fig. 1, table 4) extends for 500 km from *Chilbuzhang Co* in the west, eastward to the Yushu and Zhag'yab regions in Qinghai Sheng (Province) and Xizang Zizhiqu (Tibet Autonomous Region). The general orientation of this mountain range is from west-northwest to east-southeast. The crest of the range is the divide between the drainage areas of the Chang Jiang (Yangtze River) to the north and the Nu Jiang (Salween River) and some small inland water systems to the south.

The glaciers of Tanggula Shan have a total area of 2,213 km^2 (table 2), and are concentrated around Geladaindong peak (6,621 m) (fig. 7) in the western section and around the *Bija Shan* in the eastern section (Pu and others, 1994). The snowline is higher in the west (5,820 m) and lower in the east (5,300 m).

The *Bija* glacier group exists on the transitional belt between the subcontinental- and maritime-type glaciers; mean temperature near the snowline is –6.0 to –7.0 °C. The *Puogou Bingchuan* on the southern slope of *Bija Shan* advanced from the beginning of the 19th century to the 1920s; since the 1930s, the glacier has retreated.

Distribution of Glaciers in the Nyainqêntanglha Shan

The Nyainqêntanglha Shan (fig. 1, table 4) extends westward from *Qungmo Kangri* (7,048 m) to the north to the *Margyang He* and ends in the east at *Anjiula*, north of Rawu, with a total length of 740 km in both the western (fig. 8) and eastern (fig. 9) sections. The eastern section of the Nyainqêntanglha Shan is adjacent to the great bend of Yarlung Zangbo Jiang (fig. 9) (the upper reach of the Brahmaputra River), which is the passage for the warm, damp southwest monsoon of the Bay of Bengal into *Qinghai-Xizang Gaoyuan* (Tibetan Plateau) in summer (April to September). Precipitation here is abundant and hence the glaciers, all of the maritime type, are well developed.

The total area of all the glaciers in the Nyainqêntanglha Shan amounts to 10,700 km^2 (Pu and others, 2001; Mi and others, 2002). The elevation of the snowline ranges from 4,200 to 5,700 m. The *Qiaqin Bingchuan* is one of the largest glaciers in this region, 35 km long, with the snowline at an elevation of 4,850 m. *Qiaqin Bingchuan* terminates at an elevation of 2,900 m, in an evergreen broadleaf forest amidst pawpaw and vegetable gardens. At present, most of the glaciers located in Nyainqêntanglha Shan are retreating. For example, during the late Holocene glacial advance, prior to 1000 B.P., the *Zepu Bingchuan* extended more than 5 km downvalley from the present-day position of its terminus; it extended 2 km downvalley of its present terminus during most of the past 600 years (Iwata and Jiao, 1993). The *Roguo Bingchuan* has retreated about 20 km from its furthest glacial advance. Between 1940 and 1975, it retreated 2 km, 62 percent of which occurred in 1959–1975 (Li and others, 1986). Since 1975, it has been stable.

Distribution of Glaciers in the Gangdisê Shan

The Gangdisê Shan (fig. 1, table 4), extending from Bangong Hu on the west to the west of Nyingchi Xian (county) in the east, has a glacierized area of 1,760 km² (table 2; Jiao and others, 1988; Shi and others, 2005). The glaciers are small and mostly of the cirque-hanging type; they are concentrated in the district of the *Saint Shan* around Kangrinboqê Feng (6,714 m). Precipitation is low and the snowline is high, around 5,800–6,000 m.

Glaciers of the Himalaya

The Himalaya (fig. 1, table 4) consists of high mountains lying along the southern border of the *Qinghai-Xizang Gaoyuan* (Tibetan Plateau) from Namjagbarwa Feng (7,782 m in China) to Nanga Parbat (8,114 m in Pakistan) and extends for a distance of about 2,400 km. Ten of the world's 13 highest mountains (greater than 8,000 m in elevation) are in the Himalaya; Qomolangma Feng (Mount Everest) (8,848 m) (fig. 10), the highest, is on the border between China and Nepal. According to von Wissman (1959), glaciers cover 30,000 km² or 17 percent of the total mountain area. More recent work by Qin (1999) increased the glacierized area of the Himalaya to about 20 percent.

Qin (1999), from analysis of 1975–1978 Landsat MSS 7 and false-color images and some aerial photographs, determined that the total glacierized area of the Himalaya was 35,110 km² with a total volume of 3,734 km³ of glacier ice. A total of 18,065 glaciers were inventoried. The glacierized area of the Himalaya within China is 8,418 km², or 24 percent of the total (table 2; Mi and others, 2002). The snowline elevation on the glaciers is at 4,300 m in the eastern Himalaya and 6,200 m on *East Rongpu Bingchuan* on the northern slope of Qomolangma Feng. The climate of the Himalaya consists essentially of two seasons: a dry, cold winter and a wet, warm summer. From late May to September, the Indian Ocean monsoon affects the range, especially the eastern part and southern slope. Recent observations in the Qomolangma Feng area indicate that the glacierized area at an elevation of more than 5,000 m receives less than 1,000 mm a⁻¹ of precipitation. Glaciers of the eastern Himalaya (for example, on Namjagbarwa Feng and in the Assam Himalaya) may be of the maritime type with high monsoonal precipitation. Glaciers of the central and western Himalaya are of the subcontinental type, and glaciers are not more than 30 km long. The *Rongpu Bingchuan*, which is 22.4 km long and has an area of 85.4 km², is the largest glacier on the north slope of the Himalaya.

Distribution of Glaciers in Hengduan Shan

The Hengduan Shan region (fig. 1, table 4) extends from Chola Shan in the north to Yulongxue Shan in the south, and from the Gongga Shan in the east to the Baxoila Ling in the west. The glacierized area of this region is about 1,579 km², of which 40 percent is distributed in the Baxoila Ling (table 2; Pu and others, 1994).

Gongga Shan is the highest peak (7,556 m above sea level) in the Hengduan Shan system. Around the main peak are 74 glaciers with a total area of 256 km², among which the *Hailuogou Bingchuan* is the longest valley glacier at 13.1 km; its terminus is at 2,980 m, and the snowline elevation is at 4,900 m above sea level (Xie and Kotlyakov, 1994).

With its main peak, *Shanzidou* (5,596 m), the Yulongxue Shan is the southernmost limit (lat 27°N.) where glaciers are found in China. There are 18 glaciers in the Yulongxue Shan; most of these are distributed on eastern slopes and the snowline is about 5,000 m above sea level.

Because the Hengduan Shan is mainly affected by the Indian Ocean monsoon, precipitation is abundant and often reaches 2,500–3,000 mm a⁻¹ near the snowline. All of the glaciers are of the maritime type, with snowline elevations between 4,600 m and 5,600 m.

Ablation of Glaciers in China and Discharge of Meltwater into the Glacierized Drainage Basins

Glaciers in China are concentrated in the high mountain ranges and the *Qinghai-Xizang Gaoyuan* (Tibetan Plateau); all of the major river systems of China have their headwaters in this high plateau. In the arid regions of northwestern China, the upper courses of the inland rivers, such as the Shule He, the Manas He, and the Tarim He, are fed by meltwater from the glaciers to varying degrees. Therefore, the presence of glaciers in a drainage basin is important to agricultural production in these arid regions, a fact well known to the ancient cultures of China. According to records of more than a hundred years ago, in the spring of dry years the peasants living at the foot of the northern Qilian Shan blackened the surface of ice and snow to promote ablation to increase the discharge of the rivers (Alpine Ice and Snow Utilization Team, 1959). During 1958–1961, Chinese scientists, using coal or other black powder, by hand or aircraft, artificially blackened the surface of snow, glaciers, and river ice to promote ablation and increase the amount of glacier meltwater from Qilian Shan and Tian Shan (Shi, 1961).

Research data from Chinese scientists suggest that the energy for ablation of continental glaciers in middle latitudes, such as in the Qilian Shan, the Tian Shan, the Kunlun Shan, the Karakorum Shan, and the central and western Himalaya (Kou and others, 1974), comes mainly from two sources: solar radiation and turbulent heat exchange between the surface of ice and snow and the air layer near the ground. It follows that the volume of meltwater from the glaciers can be estimated from radiation and temperature data of glacierized areas (Shi and Xie, 1964).

Many of the modern glaciers in China are of the continental type, which may be subdivided into subcontinental and extreme-continental types, according to the annual precipitation and air temperature (table 3). The temperature at the surface of the extreme-continental type glacier is very low, and the amount of annual ablation is 200–500 mm a⁻¹ (water equivalent) at the ice tongue of *Gozhe Bingchuan* region of western Kunlun Shan. The general trend of the ablation of glacier ice decreases from southeastern

Xizang in two directions, west-northwest and northwest (Zeng and others, 1982). For example, the amount of ablation at the terminus of maritime glaciers, such as *Abzha Bingchuan*, in southeastern Xizang is 8,000 mm a^{-1}, but at the termini of the subcontinental glaciers, such as the Bogda Shan glaciers in eastern Tian Shan and the western Qilian Shan glaciers, it is reduced to 1,500 mm a^{-1} and 1,000 mm a^{-1}, respectively, the lowest extreme of the annual ablation of glaciers in China. Ablation occurs mainly in midsummer; July and August ablation is more than 65 percent of the total (as much as 85 percent during certain times) (Yang, 1980).

Meltwater from glaciers indisputably contributes significantly to the seasonal discharge of rivers originating in glacierized regions. This is especially important in the arid region of northwestern China, where precipitation is very low and extremely variable. In dry years, precipitation is low, cloud cover is minimal, and solar radiation is extreme in the mountainous regions. In wet years, the situation is reversed. For example, during 1975–76, the temperature was low, and precipitation was abundant in the Qilian Shan region; at *Laohugou Bingchuan* in the western Qilian Shan, the annual total volume of glacier melt-water was 61,420,000 m^3—only 25 percent of the mean annual runoff in the dry years (1959–1961). Based on analyses of the hydrologic data on the *Ürümqi He No. 1 Bingchuan* in the Tian Shan and the *Laohugou Bingchuan* in the Qilian Shan, the Cv (annual discharge variation) values of glacier-melt-water-fed rivers are relatively large compared to the Cv values of the rivers fed by both glacier meltwater and runoff from rainfall, but are smaller than those of rivers fed only by runoff from rainfall.

Glaciers in China cover less than 0.6 percent of the country, but the total volume of glacier meltwater is about 2 percent of the annual discharge of rivers throughout China. The percentage of glacier meltwater in the total volume of river discharge is different for each drainage basin. The general tendency, however, is for the percentage contribution from glacier-meltwater to increase from the periphery to the interior of mountain ranges. For example, it is only 3.3 percent in the Shiyang He system of the eastern Qilian Shan, but increases to 35.7 percent in the Shule He region of the western Qilian Shan.

Glacier meltwater can be hazardous. In midsummer, the volume of glacier meltwater increases the base flow of rivers, increasing the threat of a flood. Moreover, in summer the ablation of glacier ice is high, and the rivers carry large quantities of sediment, often resulting in high-volume debris flows, particularly in the maritime-glacier region of southeastern Xizang Zizhiqu (Du, 1981). For example, a high-volume glacier-debris flow burst suddenly from the *Guxiang Gully* of Xizang Zizhiqu on the night of 29 September 1953, causing serious damage. Since then, the *Guxiang Bingchuan* glacier-debris flow occurs almost every year. In 1964, 85 of these sudden glacier debris flows occurred. Glacier-debris flows are also a major hazard in the Kunlun Shan, and along the China-Pakistan Highway in the Karakorum Shan.

Landsat MSS Images Used to Illustrate Selected Glacierized Areas

Landsat 1, 2, and 3 MSS Images acquired between 16 July 1973 and 18 July 1978 were analyzed to identify glacial features in nine glacierized regions of China. These images were compared to earlier aerial photography and previous study results, to determine which glaciers are retreating, which are advancing, and the rate of change.

Central Chinese Tian Shan Region

A Landsat 2 MSS image of the central Chinese Tian Shan (fig. 2, table 5) includes Ürümqi (1), the capital of the Xinjiang Uygur Zizhiqu (Autonomous Region), and the region to its southwest. This mountainous area belongs to the central section of the Tian Shan. Elevations of features on the Landsat image (fig. 2) range from about 700 m near Ürümqi to nearly 6,000 m. The broad valley floors visible in the southwest corner of the Landsat image are about 2,700 m in elevation. Precipitation in the region occurs mainly during the summer and decreases from west to east. Air temperatures remain below 0 °C.

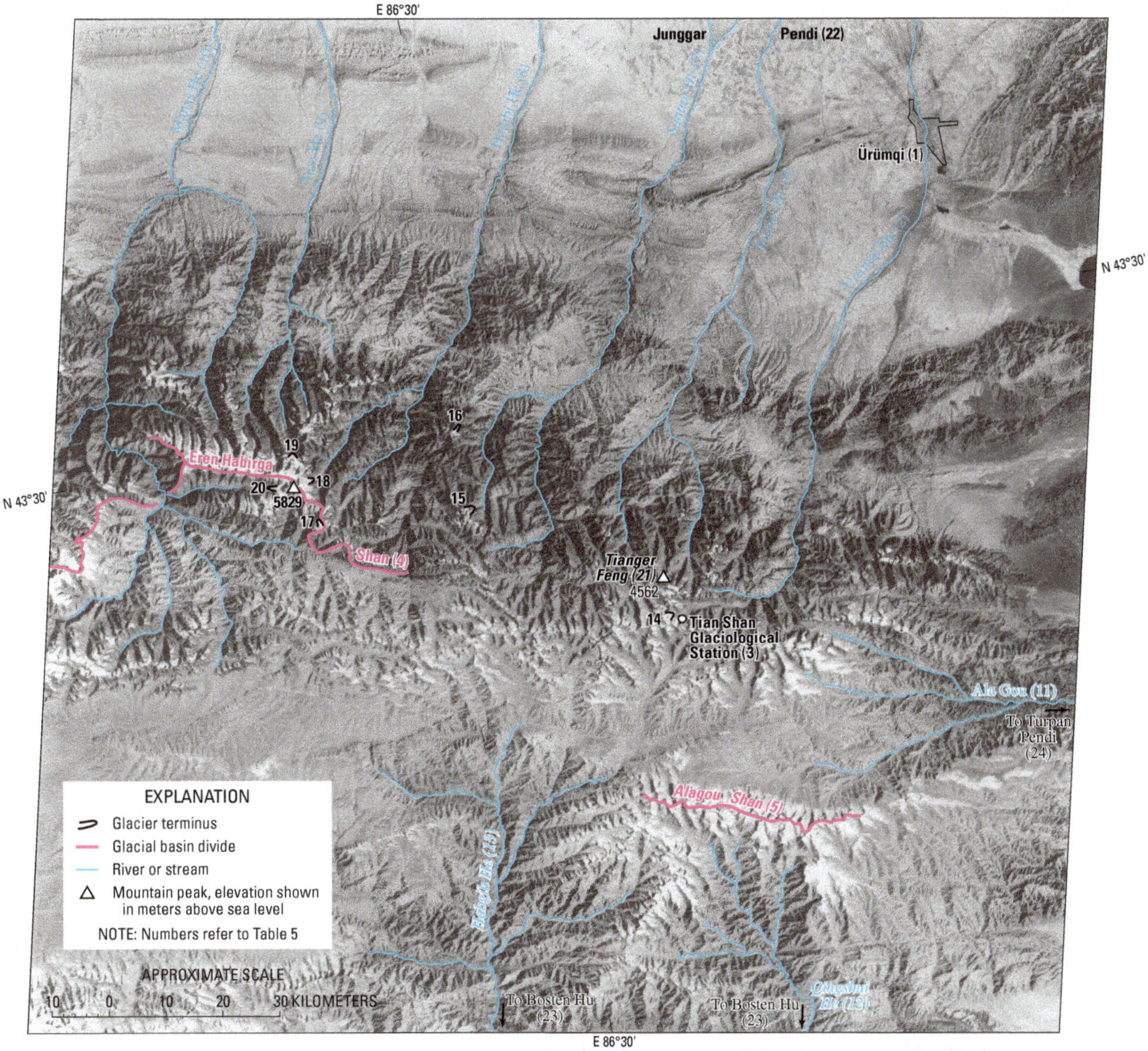

Figure 2.—Annotated Landsat 2 MSS image of the central Chinese Tian Shan region. Drainage divides between glacierized basins shown in red; selected rivers shown in blue; elevations of selected peaks shown in meters; selected positions of glacier termini shown in black; numbers indicate place-names of geographic features that are keyed to table 5. Landsat 2 MSS image (2971–03523, band 7; 19 September 1977; Path 154–Row 30) from the U.S. Geological Survey, EROS Data Center, Sioux Falls, S. Dak.

TABLE 5.—*Selected place-names of glaciers and other geographic features of the central Chinese Tian Shan region*

[See figure 2 for numbered locations of the place-names on a Landsat image of the region]

Place-name of glacier or other geographic feature	Latitude	Longitude
Ürümqi (1)	43°46'N.	87°35'E.
Ürümqi He (2)	43°30'N.	87°19'E.
Tian Shan Glaciological Station (3)	43°00'N.	86°30'E.
Eren Habirga Shan (4)	43°30'N.	85°00'E.
Alagou Shan (5)	42°47'N.	86°53'E.
Toutun He (6)	43°35'N.	87°10'E.
Santun He (7)	43°41'N.	86°55'E.
Hutubi He (8)	43°48'N.	86°36'E.
Taxi He (9)	43°51'N.	86°14'E.
Manas He (10)	43°56'N.	85°56'E.
Ala Gou (11)	42°52'N.	87°30'E.
Qingshui He (12)	42°33'N.	86°53'E.
Habgih He (13)	42°44'N.	86°21'E.
Upper Ürümqi He No. 1 Bingchuan (14)	43°05'N.	86°49'E.
Santun He No. 99 Bingchuan (15)	43°21'N.	86°26'E.
Santun He No. 102 Bingchuan (16)	43°27'N.	86°27'E.
Langtekazheng Bingchuan (17)	43°23'N.	86°06'E.
Hutubi He Urcheng No. 21 Bingchuan (18)	43°26'N.	86°06'E.
Taxi He No. 44 Bingchuan (19)	43°29'N.	86°04'E.
Manas He Cailagol No. 17 Bingchuan (20)	43°26'N.	86°01'E.
Tianger Feng (21)	43°10'N.	86°49'E.
Junggar Pendi (22)	45°30'N.	86°00'E.
Bosten Hu (23)	42°00'N.	87°00'E.
Turpan Pendi (24)	42°45'N.	89°00'E.

and glacier accumulation lasts for up to 10 months of the year. The glaciers shown on the Landsat image (fig. 2) are mainly cirque and valley glaciers of the continental type. At the present time, most of them are retreating.

Chinese glaciologists established a glacier research station known as the Tian Shan Glaciological Station (3) situated on the upper Ürümqi He (2) at an elevation of 3,539 m, near *Tianger Feng* (21). At this Station, research has been conducted on glacier flow, mass balance, physics, chemistry, hydrology; radiation and heat balances, and the extent of Quaternary glaciations. The average annual temperature on the surface of upper *Ürümqi He No. 1 Bingchuan* (14) is –8.4 °C. at an elevation of 3,840 m. The annual precipitation from 1959–1962 was 530 mm (Xie and Ge, 1965). Radiation accounts for 84.4 percent of the total summer surface heat input (Bai and Xie, 1965). Data from the

Eren Habirga Shan (4) indicates that the average annual air temperature of the glacierized areas ranges from –10 to –12 °C, and the annual precipitation is about 1,000 mm. Both conditions are favorable for glacier development, and the Eren Habirga Shan has the largest glacierized area of the Tian Shan, including 1,443 glaciers with a total of 1,566 km². Glaciers are considerably smaller in the *Alagou Shan* (5), the highest elevation of which is 4,415 m.

The Junggar Pendi (Basin) (22) lies north of the Tian Shan. There are several interior drainages running from south to north: the Ürümqi He (2), *Toutun He* (6), *Santun He* (7), Hutubi He (8), Taxi He (9), and Manas He (10). The Ala Gou (11) flows eastward into the Turpan Pendi (24); the *Qingshui He* (12) and the *Habgih He* (13) flow southward into the Bosten Hu (23).

Investigation of upper *Ürümqi He No. 1 Bingchuan* shows that it retreated 65.5 m during the period from September 1962 to August 1973, with an average recession of –6 m a⁻¹. Comparison of the aerial photograph acquired in August 1964 with the Landsat image acquired on 19 September 1977 (fig. 2) indicates that the glacier retreated about –65 m, or –5 m a⁻¹; for the same period, *No. 99 Bingchuan* (15) on the *Santun He* advanced +84 m, or +7.5 m a⁻¹; and *No. 102 Bingchuan* (16) on the *Santun He* retreated –467 m, or –36 m a⁻¹. A study of the variation of four glaciers in the Eren Habirga Shan [*Langtekazheng Bingchuan* (17), *Hutubi He Urcheng No. 21 Bingchuan* (18), *Taxi He No. 44 Bingchuan* (19), and *Manas He Cailagol No. 17 Bingchuan* (20)] using the same method shows that three glaciers (17, 18, 20) retreated with a maximum recession of –154 m, averaging –11.6 m a⁻¹; one glacier (19) advanced +100 m, averaging +6.1 a⁻¹.

Hantengri Feng and *Tomur Feng* Region of the Tian Shan

This area comprises Kyrgyzstan, the northwest corner of the Landsat 2 MSS false-color composite image (fig. 3, table 6) acquired on 16 October 1976, and the western Chinese part of the Tian Shan. The highest peaks in the area are *Tomur Feng* (2) (7,435 m) and *Hantengri Feng* (1) (6,995 m). The ridges are often 6,500 m high, with 2,500–3,500 m of relief. There are 1,357 glaciers in this region, covering a total area of 4,093 km².

Precipitation in this area is abundant; most of the water vapor comes from the west and northwest. About 70 percent of the precipitation occurs May through September, and 30 percent from October through April. Precipitation in the mountains increases with elevation and is 750–1,000 mm at the snowline (3,600–4,000 m elevation). Mean annual temperature at the snowline elevation is –7 to –11 °C, resulting in conditions favorable for glaciers (Liu and others, 1987).

The major mountain ranges are the north-south *Ziwu Ling* (3), Halik Shan (4), *Hantengri Shan* (5), and *Tomur Shan* (6). All of the south-flowing rivers drain inland into the Tarim Pendi: the *Muzart He* (7), *Karayulgun He* (8), Terang He (9), *Kokyar He* (10), and Aksu He (11).

Chinese glaciologists have investigated the *Tomur Feng* region along the *Qongterang He* watershed (Su and Wang, 1985; Su and others, 1985). There are 57 glaciers, with a total area of 266 km². The glaciers are an important water resource for agricultural areas to the south. Many of the glaciers in the area are retreating. The *Karagul Bingchuan* (12) retreated 15 m a⁻¹ between 1964 and 1978; the *Tugbeliqi Bingchuan* (13) retreated 8 m a⁻¹ between 1964 and 1976; the *Kiqikkuzbay Bingchuan* (14) retreated 15 m a⁻¹ between 1942 and 1976; the *East Qongterang Bingchuan* (17) retreated 5 m a⁻¹ between 1942 and 1978, and the *West Qongterang Bingchuan* (18) retreated 18 m a⁻¹ between 1942 and 1976.

Other glaciers in the area are stable; including the *Kiqikterang Bingchuan* (16) and North and South Engil'chek glaciers (24 and 23, respectively) on the west side of *Ziwu Ling*. The North Engil'chek glacier terminates in a large ice-dammed lake that periodically produces jökulhlaups down the Engil'chek valley. The following glaciers are advancing: the *Saysepil Bingchuan* (19),

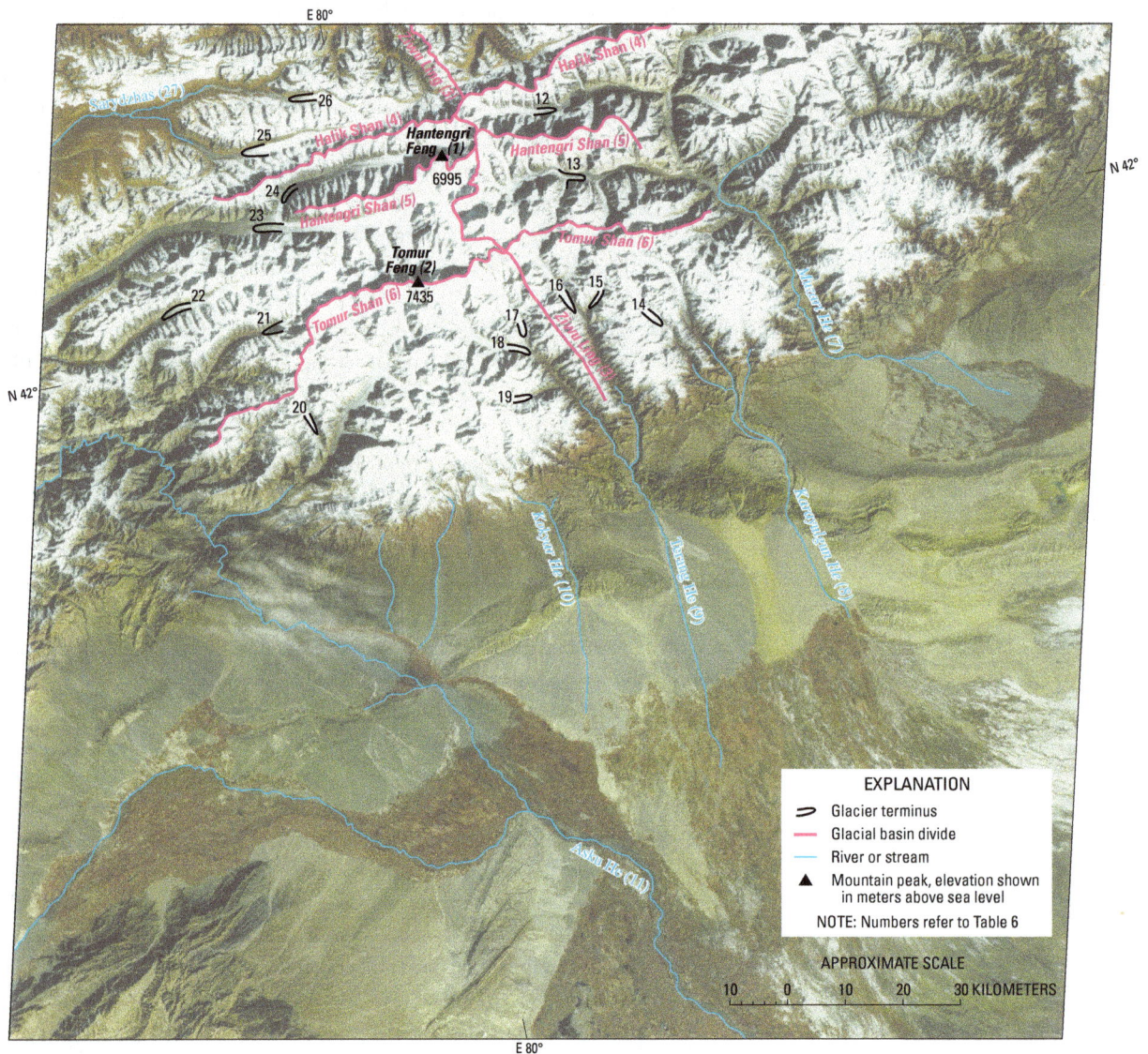

Figure 3.—*Annotated Landsat 2 MSS false-color composite image of the western part of the Chinese Tian Shan region, including the* Hantengri Feng -Tomur Feng *area. Drainage divides between glacierized basins shown in red; selected rivers shown in blue; elevations of selected peaks shown in meters; selected positions of glacier termini shown in black; numbers indicate place-names of geographic features that are keyed to table 6. Landsat 2 MSS image (2633–04315, bands 4, 5, 7; 16 October 1976; Path 158–Row 31) from the U.S. Geological Survey, EROS Data Center, Sioux Falls, S. Dak.*

TABLE 6.—*Selected place-names of glaciers and other geographic features of the western part of the Chinese Tian Shan region, including the* Hantengri Feng *and* Tomur Feng *regions*

[See figure 3 for number locations of the place-names on a Landsat image of the region]

Place-name of glacier or other geographic feature	Latitude	Longitude
Hantengri Feng (1)	42°14′N.	80°10′E.
Tomur Feng (2)	42°02′N.	80°05′E.
Ziwu Ling (3) (Meridianal ridge)	41°48′N.–42°23′N.	80°15′E.–80°20′E.
Halik Shan (4)	42°15′N.–42°26′N.	79°45′E.–80°42′E.
Hantengri Shan (5)	42°11′N.–42°17′N.	79°40′E.–80°42′E.
Tomur Shan (6)	41°51′N.–42°03′N.	79°36′E.–80°46′E.
Muzart He (7)	41°49′N.	80°54′E.
Karayulgun He (8)	41°40′N.	80°43′E.
Terang He (9)	41°41′N.	80°28′E.
Kokyar He (10)	41°38′N.	80°15′E.
Aksu He (11)	41°09′N.	80°11′E.
Karagul Bingchuan (12)	42°16′N.	80°27′E.
Tugbeliqi Bingchuan (13)	42°09′N.	80°29′E.
Kiqikkuzbay Bingchuan (14)	41°54′N.	80°36′E.
Ajen Bingchuan (15)	41°58′N.	80°27′E.
Kiqikterang Bingchuan (16)	41°57′N.	80°25′E.
East Qongterang Bingchuan (17)	41°46′N.	80°17′E.
West Qongterang Bingchuan (18)	41°44′N.	80°17′E.
Saysepil Bingchuan (19)	41°49′N.	80°15′E.
Aylonsu Bingchuan (20)	41°49′N.	79°49′E.
Telekiqi Bingchuan (21)	42°01′N.	79°34′E.
Kayind Bingchuan (22)	42°04′N.	79°33′E.
South Lednik Engil'chek (23)	42°21′N.	79°48′E.
North Lednik Engil'chek (24)	42°13′N.	79°52′E.
Lednik Mushketova (25)	42°18′N.	79°49′E.
Lednik Semënova (26)	42°21′N.	79°54′E.
Sarydzhas (river) (27)	42°23′N.	79°30′E.

which advanced 62 m a⁻¹ between 1942 and 1976; the *Aylonsu Bingchuan* (20), which advanced 53 m a⁻¹ between 1964 and 1976; and the *Kayind Bingchuan* (22), which advanced 38 m a⁻¹ between 1942 and 1976 (Su and Wang, 1985). Lednik Mushketova (25) advanced about +4,500 m in 1956 and 1957; this was likely a surge (Bakov, 1975).

Generally, glaciers in the western part of the region are stable or advancing, while those in the south are retreating. In the last few years, the snowline elevation has been increasing and as a consequence, retreat has quickened.

Western Qilian Shan Region

A Landsat 2 MSS false-color composite image (fig. 4, table 7) includes an area of the western Qilian Shan located in the boundary district between Gansu and Qinghai Provinces. Elevations range from 2,000 m in the broad valley in the northwest corner of the image to peaks of 5,600 m.

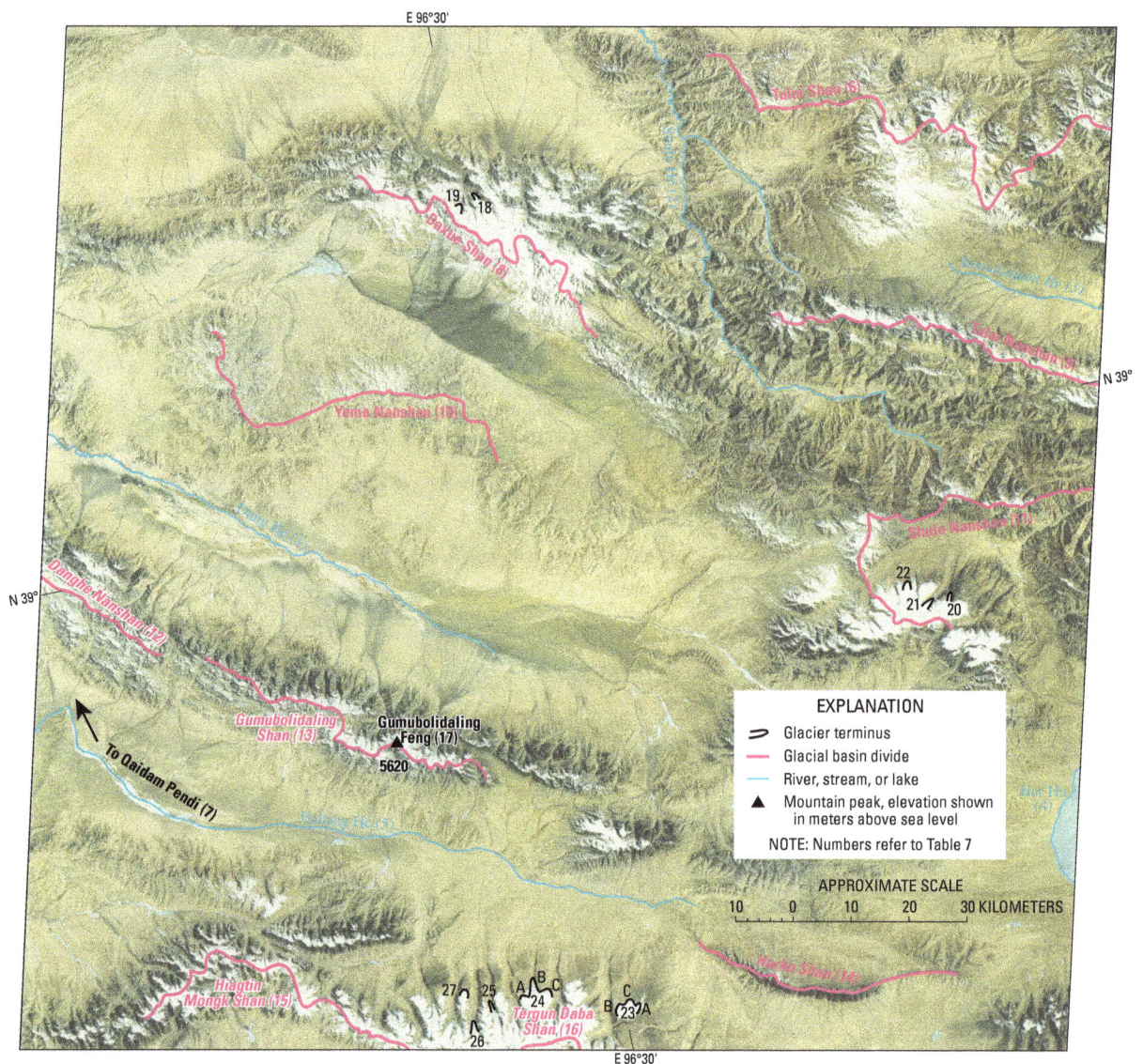

Figure 4.—Annotated Landsat 2 MSS false-color composite image of the western Qilian Shan region. Drainage divides between glacierized basins shown in red; selected rivers shown in blue; elevations of selected peaks shown in meters; selected positions of glacier termini shown in black; numbers indicate place-names of geographic features that are keyed to table 7. Landsat 2 MSS image (2819–03160, bands 4, 5, 7; 20 April 1977; Path 146–Row 33) from U.S. Geological Survey, EROS Data Center, Sioux Falls, S. Dak.

TABLE 7.—*Selected place-names of glaciers and other geographic features of the western Qilian Shan region*

[See figure 4 for number locations of the place-names on a Landsat image of the region]

Place-name of glacier or other geographic feature	Latitude	Longitude
Dang He (1)	39°00′N.	96°00′E.
Shule He (2)	39°30′N.	97°00′E.
Yemadaquon He (3)	39°09′N.	97°36′E.
Har Hu (4)	38°18′N.	97°32′E.
Haltang He (5)	38°33′N.	96°00′E.
Tulai Shan (6)	39°30′N.	97°29′E.
Qaidam Pendi (7)	37°00′N.	93°00′E.
Daxue Shan (8)	39°24′N.	96°36′E.
Tulai Nanshan (9)	39°08′N.	97°23′E.
Yema Nanshan (10)	39°15′N.	96°13′E.
Shule Nanshan (11)	38°53′N.	97°16′E.
Danghe Nanshan (12)	39°00′N.	95°36′E.
Gumubolidaling Shan (13)	38°42′N.	96°00′E.
Harko Shan (14)	38°11′N.	96°52′E.
Hiagtin Mongk Shan (15)	38°15′N.	95°53′E.
Tergun Daba Shan (16)	38°08′N.	96°21′E.
Gumubolidaling Feng (17)	38°38′N.	96°08′E.
Laohugou No. 12 Bingchuan (18)	39°29′N.	96°31′E.
Laohugou No. 18 Bingchuan (19)	39°28′N.	96°29′E.
Zahelung Bingchuan (20)	38°45′N.	97°13′E.
Lungno'odang Bingchuan (21)	38°45′N.	97°15′E.
Yematan No. 2 Bingchuan (22)	38°48′N.	97°18′E.
Ih Xahalgou Gol No. 1 Bingchuan (23)	38°09′N.	96°30′E.
Mohalxalge Bingchuan (24)	38°12′N.	96°19′E.
Gurban Gol No. 1 Bingchuan (25)	38°12′N.	96°13′E.
Horbaxahige No. 10 Bingchuan (26)	38°10′N.	96°11′E.
Horbaxahige No. 13 Bingchuan (27)	38°13′N.	96°10′E.

Under conditions of continental climate, many large continental glaciers exist and are an important water resource for the *Hexi Zoulang* (Corridor) of Gansu Province and the Qaidam Pendi (7) of Qinghai Province. The drainage systems shown on the Landsat 2 MSS image (fig. 4, table 7) include the Dang He (1), the Shule He (2), and the *Yemadaquon He* (3), a tributary to the Beida He; all drain inland and eventually northward into the *Hexi Zoulang* (Corridor). The Har Hu (4) is an independent inland drainage. The Haltang He (5) flows into the Qaidam Pendi (7).

, The mountain ranges from north to south are the Tulai Shan (6), the Daxue Shan (8), Tulai Nanshan (9), Yema Nanshan (10), Shule Nanshan (11), Danghe Nanshan (12), the *Gumubolidaling Shan* (13) the *Harko Shan* (14), the *Hiagtin Mongk Shan* (15), and the Tergun Daba Shan (16); the highest peak is *Gumubolidaling Feng* (17) at an elevation of 5,620 m.

The glaciers are concentrated in the Daxue Shan (8), the Shule Nanshan (11), the *Gumubolidaling Shan* (13), the *Hiagtin Mongk Shan* (15), and the Tergun Daba Shan (16). Most of the glaciers of the Qilian Shan retreated between 1966 and 1977. Glaciers in the *Hiagtin Mongk Shan*, however, generally advanced during that period.

Most Chinese glaciologic research has been done on the glaciers of the Daxue Shan (8). The *Laohugou No. 12 Bingchuan* (18), in the Daxue Shan, is 10.1 km long, the longest in the Qilian Shan. It is a typical continental type glacier with sub-freezing ice temperatures and a clean surface. Field surveys indicate that this glacier retreated -72 m (-5m a^{-1}) from May 1962 to August 1976. The *Laohugou No. 18 Bingchuan* (19) has a snowline at an elevation of 4,710 m and a terminus at 4,440 m; field surveys show a retreat of -30 m (-2 m a^{-1}) from May 1962 to August 1976. For the period 1957 through 1973, the *Laohugou No. 12 and 18 Bingchuan* retreated -150 m and -76 m, respectively, and exhibited a much greater rate of retreat during the late 1950s and early 1960s than later.

Three major glaciers in the Shule Nanshan all retreated between 1966 and 1977, as determined from 1966 aerial photographs and a Landsat image acquired on 20 April 1977 (fig. 4). The flat-topped *Zahelung Bingchuan* (20), with a snowline at 4,840 m and terminus at 4,540 m, retreated -250 m; the *Lungo'odang Bingchuan* (21), which heads in a 5,472 m peak, and has its snowline at 5,075 m and its terminus at 4,500 m, retreated -150 m; and the sloping *Yematan No. 2 Bingchuan* (22), with a terminus at 4,550 m, retreated -70 m.

Several termini position changes were also seen in the *Hiagtin Mongk Shan (15)* between 1966 and 1977. The *Ih Xahalgou Gol No. 1 Bingchuan* (23) has a flat accumulation zone with three advancing ice tongues (A, B, and C): A advanced $+1,000$ m, B advanced $+100$ m, and C advanced $+100$ m. The lowest terminus is at an elevation of 4,700 m. The *Mohalxalge Bingchuan* (24) also has three ice tongues which are at 4,470 m or above; the glacier's snowline elevation is 4,810 m. A advanced $+100$ m, B advanced $+200$ m, and C experienced a minor retreat. The *Gurban Gol No. 1 Bingchuan* (25) advanced $+1,300$ m, the *Horbaxahige No.10 Bingchuan* (26) was stable, and the *Horbaxahige No. 13 Bingchuan* (27), with its head at an elevation of 5,361 m and its terminus at 4,845 m, advanced $+100$ m.

Qogir Feng (K2) of the Karakorum Shan Region

The northern two-thirds of a Landsat 3 MSS false-color composite image (fig. 5, table 8) shows the Chinese Xinjiang Region; to the south is the Kashmir of Pakistan and India. Physiographically, the area includes the western Kunlun Shan (19) and eastern Karakorum Shan (20). Precipitation south of the Karakorum Shan is much greater than north of the range, and, accordingly, glaciers are larger on the southern slopes. Because of the extremely high elevation, including Qogir Feng (K2) (18) at 8,611 m, many large glaciers have developed: the *Yengisogat Bingchuan* (1), *Yulin Bingchuan* (2), *Qogir Bingchuan* (3), *Skyang Bingchuan* (4), *Gasherbrum Bingchuan* (5), *Urdok Bingchuan* (6), *Sitager Bingchuan* (7), *Teram Kangri Bingchuan* (8), *Kyagar Bingchuan* (9), *North,* Central, and *South* Rimo Glaciers (10, 11, 12), Siachen Glacier (13), Baltoro Glacier (14), *Tatrugou Bingchuan* (15), *Sudikuwadigou Bingchuan* (16), and *Qiadijileg Bingchuan* (17).

Recent changes in many of these glaciers were determined using maps, aerial photographs, and Landsat images. The *Yengisogat Bingchuan* (1) is the longest glacier in China. It is 42 km long, has a snowline elevation of about 5,450 m, and terminus at 4,000 m. Based on a map by Shipton (1938), a 1968 map, and Landsat images of 1973, the glacier was stable and the terminus did not move. From 1973 to 1978, the terminus position still did not change, but the clean ice zone moved 500 m downglacier (Shipton, 1938; Zhang, 1980).

From 1937 to 1968, the *Qogir Bingchuan* (3) retreated –1,700 m, and the *Skyang Bingchuan* (4) retreated –4,200 m. From 1973 to 1978, the *Skyang Bingchuan* retreated an additional –1,000 m. From 1973 to 1978, the western part of the *Teram Kangri Bingchuan* (8) retreated –300 m, while the eastern part advanced +300 m.

In 1926, the *Kyagar Bingchuan* (9) dammed the *Shaksgam He* (25) (Mason, 1930). An ice-dammed lake again formed between 1976 and 1978 (Hewitt, 1979), suggesting that a recession occurred between 1926 and 1976; but since 1976 the *Kyagar Bingchuan* has advanced. Between 3 September 1961 and 6 September 1978, repeated failures of the ice dam generated multiple jökulhlaups; the maximum flood discharges were 6,270 m^3 s^{-1} and 4,700 m^3 s^{-1} respectively (Zhang and others, 1990). Other glaciers have also recently advanced; from 1968 to 1978, the *Tatrugou Bingchuan* (15), *Sudikuwadigou Bingchuan* (16), and *Qiadijileg Bingchuan* (17) advanced +705 m, +500 m, and +60 m, respectively. In general, the glaciers of the Karakorum Shan region of China have been advancing. During the recent warming, however, recession of the glaciers has occurred.

Figure 5.—Annotated Landsat 3 MSS false-color composite image of the Qogir Feng (K2) area of the Karakorum Shan region. Drainage divides between glacierized basins shown in red; selected rivers shown in blue; elevations of selected peaks shown in meters; selected positions of glacier termini shown in black; numbers indicate place-names of geographic features that are keyed to table 8. Landsat 3 MSS image (30135–04492; bands 4, 5, 7; 18 July 1978; Path 159–Row 35) from U.S. Geological Survey, EROS Data Center, Sioux Falls, S. Dak.

TABLE 8.—*Selected place-names of glaciers and other geographic features of the Qogir Feng (K2) of the Karakorum Shan region*

[See figure 5 for number locations of the place-names on a Landsat image of the region]

Place-name of glacier or other geographic feature	Latitude	Longitude
Yengisogat Bingchuan (1)	36°03′N.	76°17′E.
Yulin Bingchuan (2)	35°58′N.	76°21′E.
Qogir Bingchuan (3)	36°04′N.	76°29′E.
Skyang Bingchuan (4)	36°02′N.	76°32′E.
Gasherbrum Bingchuan (5)	35°55′N.	76°46′E.
Urdok Bingchuan (6)	35°51′N.	76°45′E.
Sitager Bingchuan (7)	35°47′N.	76°49′E.
Teram Kangri Bingchuan (8)	35°46′N.	76°59′E.
Kyagar Bingchuan (9)	35°43′N.	77°11′E.
North Rimo Glacier (10)	35°32′N.	77°32′E.
Central Rimo Glacier (11)	36°22′N.	77°38′E.
South Rimo Glacier (12)	35°21′N.	77°37′E.
Siachen Glacier (13)	35°13′N.	77°10′E.
Baltoro Glacier (14)	35°42′N.	76°11′E.
Tatrugou Bingchuan (15)	36°20′N.	76°24′E.
Sudikuwadigou Bingchuan (16)	36°10′N.	76°40′E.
Qiadijileg Bingchuan (17)	36°11′N.	76°59′E.
Qogir Feng (K2) (18)	35°55′N.	76°30′E.
Kunlun Shan (19)	36°33′N.	77°13′E.
Karakorum Shan (20)	35°39′N.	76°46′E.
Karakax He (21)	36°22′N.	78°03′E.
Yarkant He (22)	36°25′N.	77°15′E.
Shyok River (23)	35°09′N.–35°15′N.	77°16′E.–76°24′E.
Tarim Pendi (24)	39°30′N.	83°30′E.
Kulqin (Shaksgam) He (25)	36°09′N.	76°24′E.

Western Kunlun Shan Region

A Landsat 2 MSS image (fig. 6, table 9) shows glaciers concentrated in the lower third of the image, in the western Kunlun Shan region (1) on the border of Xinjiang Uygur Zizhiqu and Xizang Zizhiqu (Tibet Autonomous Region). These glaciers comprise about 8,000 km² of snow and ice — the largest glacier concentration in China. *Kunlun Feng* (2), at 7,167 m, is the highest peak in the area; the valley floors in the south are at an elevation of 5,000 m, and the northern desert is at 1,500 m. The *Guliya Binggai* (ice cap) (9) is the largest (200 km²), highest (6,700 m), and coldest (–19 °C at 10 m depth) glacier in Central Asia.

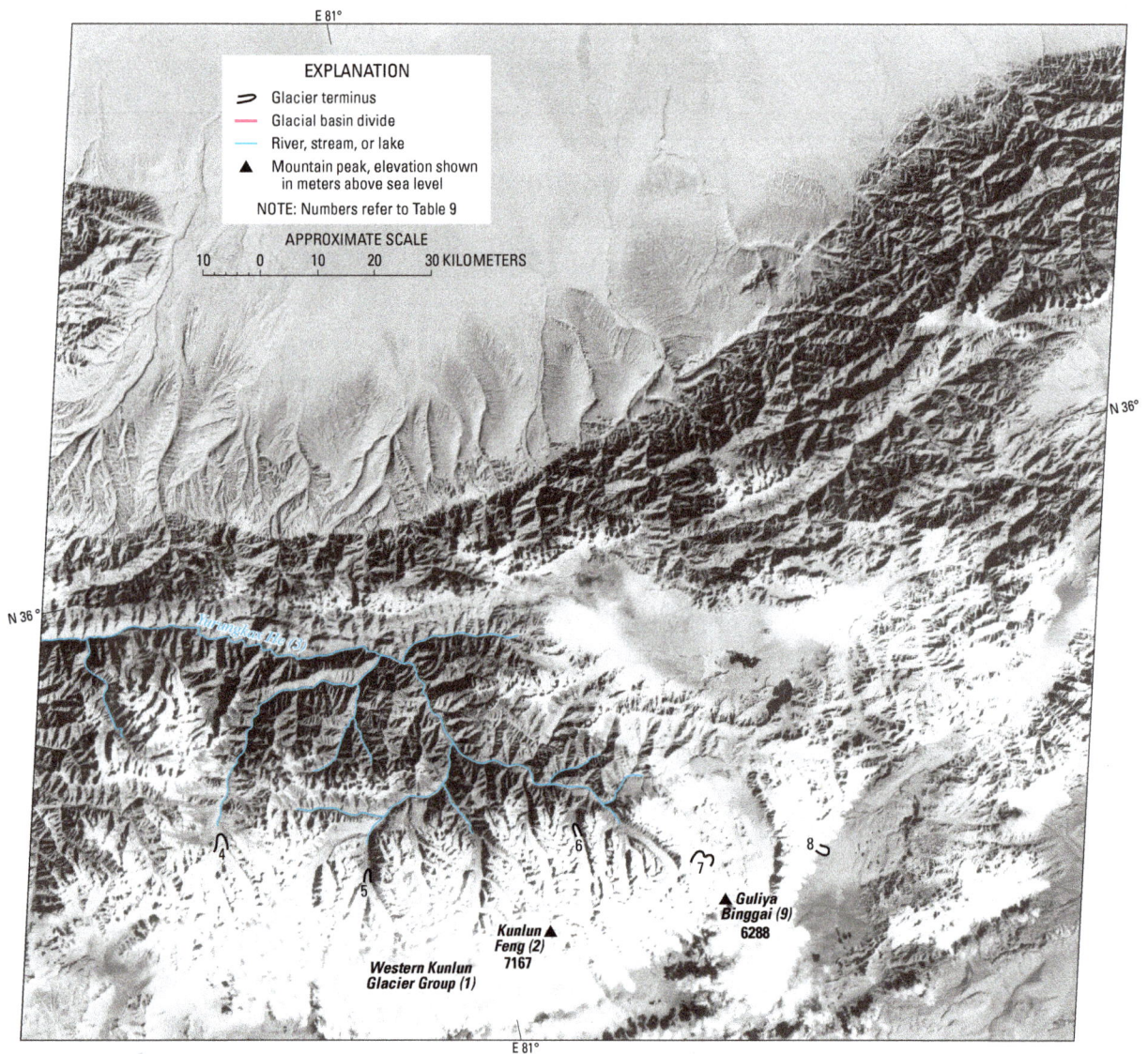

Figure 6.—Annotated Landsat 2 MSS image of the western Kunlun Shan region. Selected rivers shown in blue; elevations of selected peaks shown in meters; selected positions of glacier termini shown in black; numbers indicate place-names of geographic features that are keyed to table 9. Landsat 2 MSS image (2667–04204, band 7; 19 November 1976; Path 156–Row 35) from U.S. Geological Survey, EROS Data Center, Sioux Falls, S. Dak.

TABLE 9.—*Selected place-names of glaciers and other geographic features of the western Kunlun Shan region*

[See figure 6 for number locations of the place-names on a Landsat image of the region]

Place-name of glacier or other geographic feature	Latitude	Longitude
West Kunlun Glacier Group (1)	35°26′N.–35°13′N.	80°07′E.–81°40′E.
Kunlun Feng (2)	35°21′N.	81°05′E.
Yurungkax He (3)	35°57′N.	80°28′E.
Yurungkax He Source South No. 51 Bingchuan (4)	35°35′N.	80°29′E.
Kunlun Bingchuan (5)	35°29′N.	80°51′E.
Upper Yurungkax He No. 20 Bingchuan (6)	35°30′N.	81°11′E.
Yakabaktak He No. 10 Bingchuan (7)	35°26′N.	81°21′E.
Atedimu No. 13 Bingchuan (8)	35°24′N.	81°38′E.
Guliya Binggai (9)	35°17′N.	81°29′E.

The climate of this region is dry; annual precipitation is about 300 mm, of which the total on the southern slope is higher than that on the northern. The glacier meltwater from the southern slopes flows into the *Guozha Co* and the Tianshuihai basin, while that on the northern slopes empties mainly into the Yurungkax He (3).

The change over time in the positions of termini of several glaciers was determined using maps compiled from 1968–71 data and the 19 November 1976 Landsat 2 MSS image (fig. 6). The *Yurungkax He Source South No. 51 Bingchuan* (4) retreated –160 m between 1968–76. Its snowline elevation is at 5,800 m, and its terminus is at 5,125 m. The *Kunlun Bingchuan* (5) is a dendritic valley glacier with its snowline elevation at 5,920 m and its terminus at 4,880 m. The terminus is covered with debris. From 1970–76, the *Kunlun Bingchuan* advanced +400 m. The *Yakabaktak He No. 10 Bingchuan* (7) has its snowline at an elevation of 6,000 m; its head is at 6,288 m, and its terminus is at 5,510 m. It advanced +100 m between 1970–76. The *Atedimu No. 13 Bingchuan* (8) has a piedmont terminus at 5,610 m and a snowline elevation at 6,070 m. It retreated –90 m between 1971and 1976.

Typically, the northern slopes in this area are steep, making field investigations quite difficult; however, the southern slopes are gentle. This glacier group is an important water source for the *Hetun* region of the southern Xinjiang Region (fig. 1) (Li, 1975).

Tanggula Shan Region

A Landsat 1 MSS image (fig. 7, table 10) includes the area lying on the boundary between Xizang Zizhiqu (Tibet Autonomous Region) and Qinghai Sheng (Province) in China. This region includes the western section of the Tanggula Shan (15). The major glacier group of the region, the Geladaindong, is the source of the largest river in China, the Chang Jiang (Yangtze River). Subranges of the Tanggula Shan are seen at the northern and southern fringes of the Landsat image; the Jurhen Ul Shan (13) and *Gyotangmaryo Shan* (14), respectively.

Elevations in this region range from 3,000 m to over 6,000 m. The highest peak is Geladaindong (6) at 6,621 m. According to the data obtained from weather stations at an elevation of 4,900 m in an adjacent area, the average

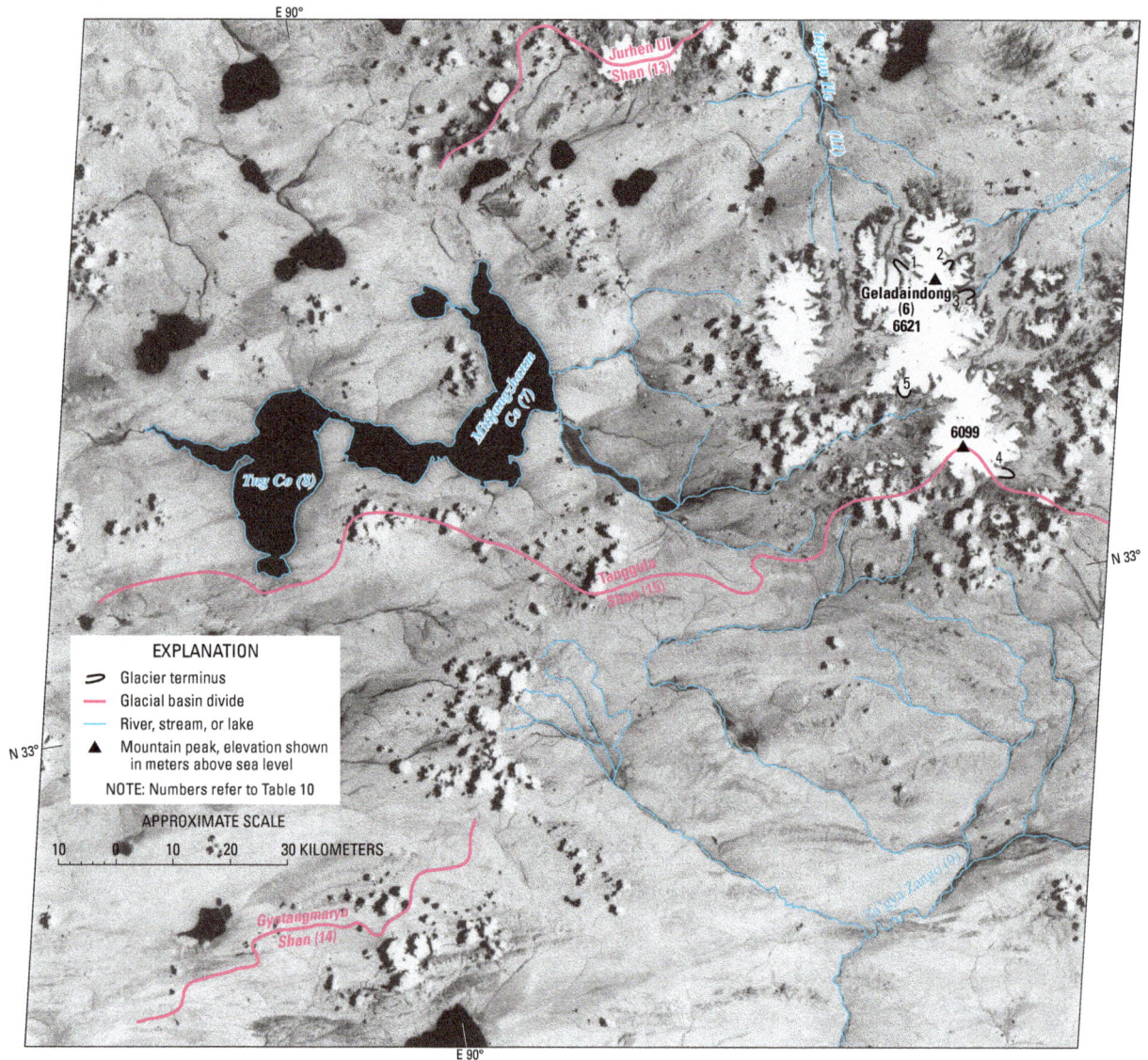

Figure 7.—Annotated Landsat 1 MSS image of the Tanggula Shan region. Drainage divides between glacierized basins shown in red; selected rivers shown in blue; elevations of selected peaks shown in meters; selected positions of glacier termini shown in black; numbers indicate place-names of geographic features that are keyed to table 10. Landsat 1 MSS image (1358–04040, band 7; 16 July 1973; Path 149–Row 37) from U.S. Geological Survey, EROS Data Center, Sioux Falls, S. Dak.

TABLE 10.—*Selected place-names of glaciers and other geographic features of the Tanggula Shan region*

[See figure 7 for number locations of the place-names on a Landsat image of the region]

Place-name of glacier or other geographic feature	Latitude	Longitude
Sumeiqu Bingchuan (1)	33°32'N.	91°03'E.
Guoriqu Bingchuan (2)	33°31'N.	91°09'E.
Gangjiaquba Bingchuan (3)	33°28'N.	91°10'E.
Goluqu Bingchuan (4)	33°09'N.	91°10'E.
Ganaqinmaqu Bingchuan (5)	33°20'N.	91°02'E.
Geladaindong (6)	33°29'N.	91°13'E.
Mitijangzhanm Co (7)	33°24'N.	90°19'E.
Tug Co (8)	33°23'N.	89°50'E.
Za'gya Zangbo (9)	32°33'N.	90°58'E.
Siling Co (south of the image)	31°50'N.	89°00'E.
Togton He (11)	33°49'N.	90°57'E.
Guor Qu (12)	33°34'N.	91°20'E.
Jurhen Ul Shan (13)	33°55'N.	90°37'E.
Gyotangmaryo Shan (14)	32°37'N.	89°47'E.
Tanggula Shan (15)	33°10'N.	90°20'E.

annual air temperature is –5.5 °C; the monthly average air temperature in January is –16 °C and the maximum monthly temperature (5.5 °C.) is in July. The duration of the sub 0 °C period is as long as 8 months. As a result, a deep layer of permafrost has developed in this region. Meteorological satellite data show that warm and moist air from the Bay of Bengal frequently moves northward and affects this area during summer. Therefore, 80 to 90 percent of the annual precipitation (about 550–650 mm) is concentrated in June, July, August, and September. The Landsat image was acquired on 16 July 1973 (fig. 7), early in the rainy season, so the ground surface was rather wet.

The lakes shown on the Landsat image are the *Mitijangzhanm Co* (7) and *Tug Co* (8). Glaciers drain southward into Za'gya Zangbo (9) and Siling Co (south of the image), westward into *Mitijangzhanm Co* (7), and northward into the *Togton He* (11) and eastward to the *Guor Qu* (12), two sources of the Chang Jiang.

The glaciers on Geladaindong (6) have a total area of 74.94 km² (Glacier Inventory Basin No. 5Z213B16). These glaciers are typical continental type, and are developed on a planation surface in the shape of small ice caps with flow radiating from a central accumulation area. The glacier surfaces are very clean, because there are no higher sources of rock debris. Most of the glaciers in the Geladaindong are currently advancing. In comparing November 1969 aerial photographs with the 16 July 1973 Landsat image (fig. 7), several changes in glacier position can be observed. The *Sumeiqu Bingchuan* (1) terminates at 5,360 m, and has a snowline elevation of 5,770 m — its terminus retreated about 260 m (–69 m a^{-1}) for this period. The *Guoriqu Bingchuan* (2) advanced 190 m (about +50 m a^{-1}). The *Goluqu Bingchuan* (4) advanced 160 m (about +43 m a^{-1}). The *Ganaqinmaqu Bingchuan* (5) advanced 140 m (about +37 m a^{-1}).

Western Nyainqêntanglha Shan Region

A Landsat 2 MSS image (fig. 8, table 11) features the western Nyainqêntanglha Shan (2) and Nam Co (1) in the Xizang Zizhiqu (Tibet Autonomous Region). The *Bopuri Shan* (3) and the Nyainqêntanglha Shan form the divide between the inland-lake system to the north and the Indian Ocean to the south. The larger rivers of the inland-lake system are the *Bo Qu* (4), which drains into Nam Co, the *Tarma Zangbo* (5), and *Yougzhu Zangbo* (19), which drain into *Muggen Co* (13), and the Chumbu Zangbo (6) which drains into *Yoqia Co*. Other lakes are *Sen Co* (9), *Quru Co* (10), *Ringco Ogma* (11), *Ringco Kungma* (12), *Muggen Co* (13), and Gomang Co (14). The rivers of the inland

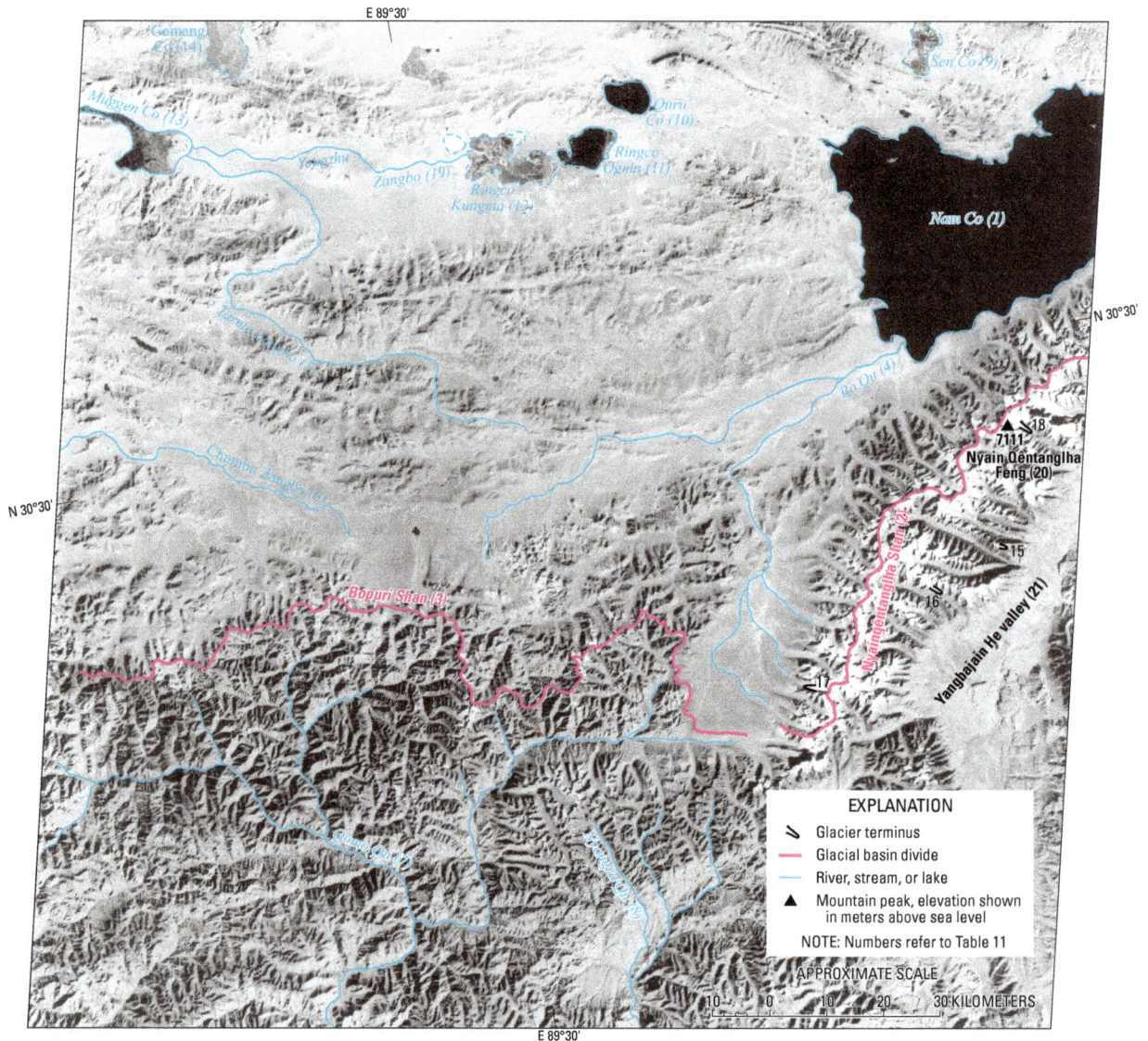

Figure 8.—*Annotated Landsat 2 MSS image of the western Nyainqêntanglha Shan region. Drainage divides between glacierized basins shown in red; selected rivers shown in blue; elevations of selected peaks shown in meters; selected positions of glacier termini shown in black; numbers indicate place-names of geographic features that are keyed to table 11. Landsat 2 MSS image (2696–03410, band 7; 18 December 1976; Path 149–Row 39) from U.S. Geological Survey, EROS Data Center, Sioux Falls, S. Dak.*

TABLE 11.—*Selected place-names of glaciers and other geographic features of the western Nyainqêntanglha Shan region*

[See figure 8 for number locations of the place-names on a Landsat image of the region]

Place-name of glacier or other geographic feature	Latitude	Longitude
Nam Co (1)	30°45'N.	90°30'E.
Nyainqêntanglha Shan (2)	30°18'N.	90°19'E.
Bopuri Shan (3)	30°13'N.	89°15'E.
Bo Qu (4)	30°29'N.	90°15'E.
Tarma Zangbo (5)	30°46'N.	89°26'E.
Chumbu Zangbo (6)	30°30'N.	89°02'E.
Qomo Qu (7)	29°45'N.	89°15'E.
Nyangra Qu (8)	29°37'N.	89°38'E.
Sen Co (9)	31°00'N.	90°28'E.
Quru Co (10)	30°59'N.	89°53'E.
Ringco Ogma (11)	30°55'N.	89°50'E.
Ringco Kungma (12)	30°56'N.	89°39'E.
Muggen Co (13)	31°02'N.	89°00'E.
Gomang Co (14)	31°11'N.	89°10'E.
Botang Bingchuan (15)	30°10'N.	90°29'E.
Linggong Bingchuan (16)	30°06'N.	90°18'E.
Tarqongqu Bingchuan (17)	30°00'N.	90°03'E.
Pagnuqu Bingchuan (18)	30°20'N.	90°33'E.
Yougzhu Zangbo (19)	30°58'N.	89°24'E.
Nyainqêntanglha Feng (20)	30°20'N.	90°31'E.
Yangbajain He valley (21)	30°05'N.	90°30'E.

regions have low erosive power. As a result, the surrounding areas have gentle relief and excellent highland pastures; the mean elevation is 5,200 m. The major rivers flowing to the Indian Ocean are the *Qomo Qu* (7) and the *Nyangra Qu* (8). Both are tributaries of the Yarlung Zangbo Jiang.

Under the influence of the Indian Ocean monsoon, the annual precipitation is 380 mm in a valley east of the Nyainqêntanglha Shan at 4,300 m; 79 percent of the precipitation occurs in June, July, and August. The mean annual temperature is 2.5 °C. The mean annual temperature at the snowline (5,700 m) is estimated to be –7.5 °C.

The major glaciers are concentrated around Nyainqêntanglha Feng (20) (7,111); most are small cirque or valley glaciers, 4 to 5 km long. A comparison of 1970 maps with the 18 December 1976 Landsat 2 MSS image (fig. 8) indicates that the *Botang Bingchuan* (15) advanced +80 m, the *Linggong Bingchuan* (16) retreated –100 m, and the *Tarqongqu* (17) and *Pagnuqu* (18) *Bingchuan* were stable.

Moraines of past Quaternary glaciation can be easily seen on the Landsat image (fig. 8). Evidence of a major prehistoric ice tongue can be seen north of the *Bopuri Shan* (3); a small lake now occurs in a depression near the terminus of the former glacier. Piedmont glaciers extended to the west and into the present Nam Co from the Nyainqêntanglha Shan, and, to the southeast, glaciers from straight valleys terminated at the edge of the Yangbajain He valley (21).

Eastern Nyainqêntanglha Shan Region

A Landsat 2 MSS false-color composite image (fig. 9, table 12) includes the area at the junction of the *Kangri Garbo Shan* (1) in the eastern Nyainqêntanglha Shan region of the Xizang Zizhiqu (Tibet Autonomous Region) and *Subola Ling* (Mountains) (2) in the Hengduan Shan. Peaks in this region rise to 6,600 m and valleys generally are at about 2,000 m, thus relief is extreme and vertical zonality of environment is very distinct.

This area is strongly affected by the southwestern monsoon (Indian Ocean monsoon); monsoonal precipitation is extremely high, and clouds and fog are common. The major rivers are the Zayü Qu (13) and its tributary, the *Gomrigarbu Qu* (3); the *Dalung Qu* (4) which joins the Yarlung Zangbo Jiang (8); the Parlung Zangbo (5); and the Sang Qu (6), a tributary of the Zayü Qu (13).

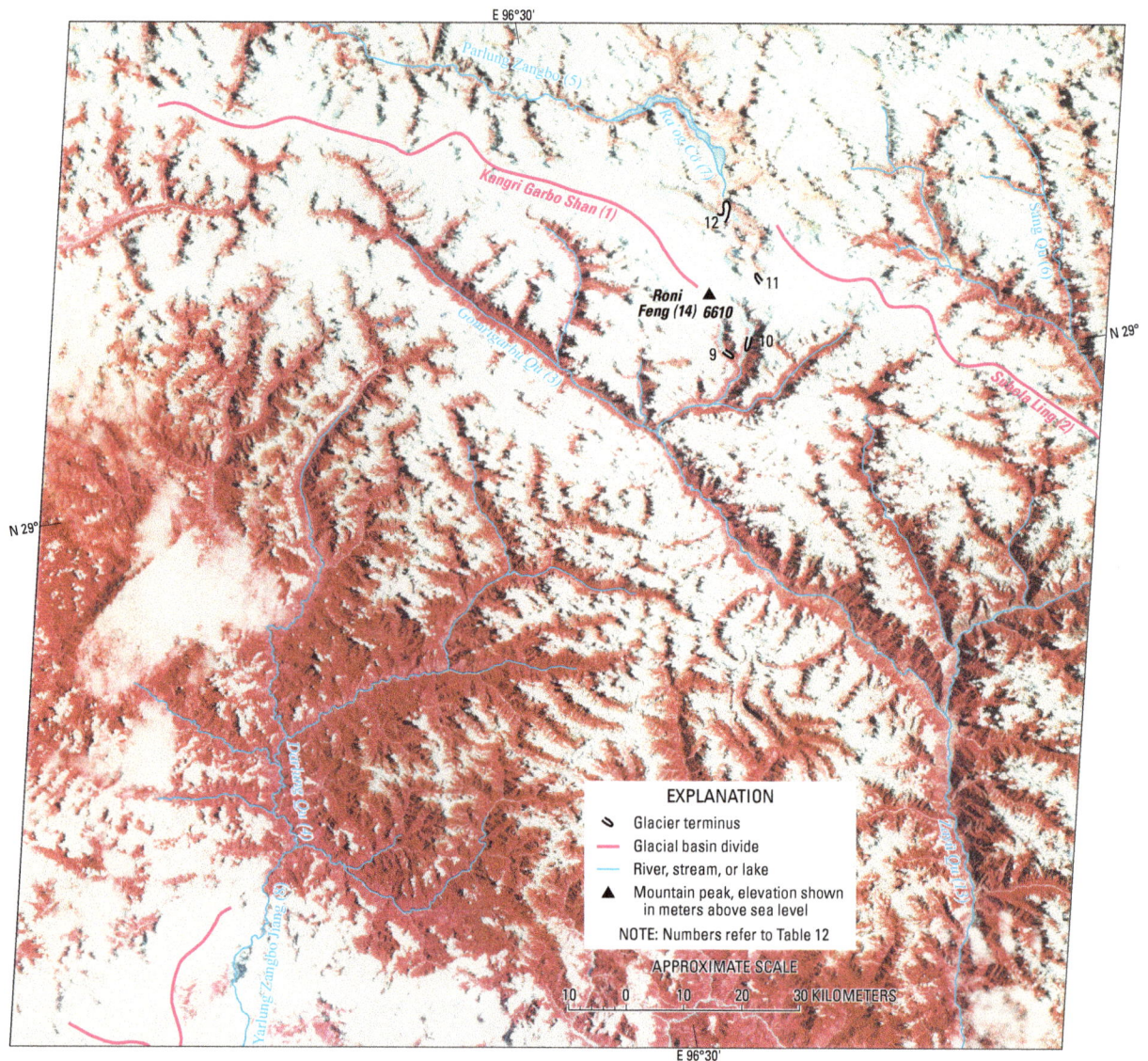

Figure 9.—Annotated Landsat 2 MSS false-color composite image of the eastern Nyainqêntanglha Shan region. Drainage divides between glacierized basins shown in red; selected rivers shown in blue; elevations of selected peaks shown in meters; selected positions of glacier termini shown in black; numbers indicate place-names of geographic features that are keyed to table 12. Landsat 2 MSS image (2925–03020, bands 4, 5, 7; 4 August 1977; Path 144–Row 40) from U.S. Geological Survey, EROS Data Center, Sioux Falls, S. Dak.

TABLE 12.—*Selected place-names of glaciers and other geographic features of the eastern Nyainqêntanglha Shan region*

[See figure 9 for number locations of the place-names on a Landsat image of the region]

Place-name of glacier or other geographic feature	Latitude	Longitude
Kangri Garbo Shan (1)	29°25'N.	96°26'E.
Subola Ling (2)	29°04'N.	97°10'E.
Gomrigarbu Qu (3)	28°53'N.	96°48'E.
Dalung Qu (4)	28°30'N.	95°50'E.
Parlung Zangbo (5)	29°37'N.	96°18'E.
Sang Qu (6)	29°05'N.	97°22'E.
Ra'og Co (7)	29°30'N.	96°45'E.
Yarlung Zangbo Jiang (8)	28°00'N.	95°42'E.
Roni Bingchuan (9)	29°08'N.	96°45'E.
Abzha Bingchuan (10)	29°07'N.	96°47'E.
Zoqupu Bingchuan (11)	29°10'N.	96°52'E.
Yagnung Bingchuan (12)	29°19'N.	96°42'E.
Zayü Qu (13)	28°21'N.	97°00'E.
Roni Feng (14)	29°09'N.	96°45'E.

Major maritime glaciers have developed in this area, and are characterized by temperate ice, intensive ablation, and high mass-exchange rate. It is estimated that average annual precipitation near the snowline is 2.5–3.0 m and average annual temperature is –2 to –4 °C (Li and Zheng, 1981). Annual ablation is the highest in China, 6.5–8.0 m of water equivalent. The largest glacier in the area *Kyaggen Bingchuan* (30°31'N., 94°42'E.) is located west of figure 9. It is 35.3 km long and has an area of 206.7 km². The snowline is at 4,900 m and the terminus at 3,900 m. It has a very broad accumulation area of about 165 km² (Li and others, 1986). The second largest glacier in the area is shown on figure 9 and is the *Yagnung Bingchuan* (12); it is 32.5 km long, with the snowline at 5,040 m, the terminus at 3,960 m, and an area of 191.45 km² (Shi and others, 2005). The 60 km² ablation area has three slightly contorted medial moraines, suggesting that it may have a surge history.

Because of the frequent cloud cover, aerial observations of these glaciers are rare. Landsat images from 1973 and 1977 show that the *Yagnung Bingchuan* (12) advanced +230 m over that period. In the early 1930s, the *Abzha* (10) and *Roni* (9) *Bingchuan* were joined (Kingdon-Ward, 1934). The 4 August 1977 Landsat MSS image (fig. 9) shows a separation between these two glaciers of 1,350 m, indicating major recession of the *Roni Bingchuan*. *Roni Feng* (14) (6,610 m) is at the head of the *Abzha Bingchuan* (10); the snowline elevation is 4,600 m and its terminus lies at 2,550 m. Observations by a local resident suggest that the *Abzha Bingchuan* has receded –700 m in the last 50 years, and observations from 1973–1977 showed a –193 m recession (Li and Zheng, 1981). Field work in 1976, however, indicated that the glacier was thickening, and by 1978 the *Abzha Bingchuan* had advanced +200 m. A small glacier, the *Zoqupu Bingchuan* (11), near the *Abzha Bingchuan*, advanced +350 m from 1973 to 1977.

Central Himalaya Region

A Landsat 2 MSS false-color composite image (fig. 10, table 13) includes the area of Qomolangma Feng (Mount Everest) (1) (8,848 m), in the southeast corner, and Xixabangma Feng (5) (8,012 m), in the southwest corner; both are in the central part of the Himalaya. Other major peaks are *Lhotse Feng* (2) (8,501 m), *Makalu Shan* (3) (8,470 m), and Cho Oyu (4) (8,153 m), all located near Mount Everest. Topographically, the northern slope tends to be gentle and the southern slope steep. Climatically, the northern slope of Himalaya is cold and dry, and the southern slope is warm and moist, indicating a general decrease, from south to north, in precipitation and temperature.

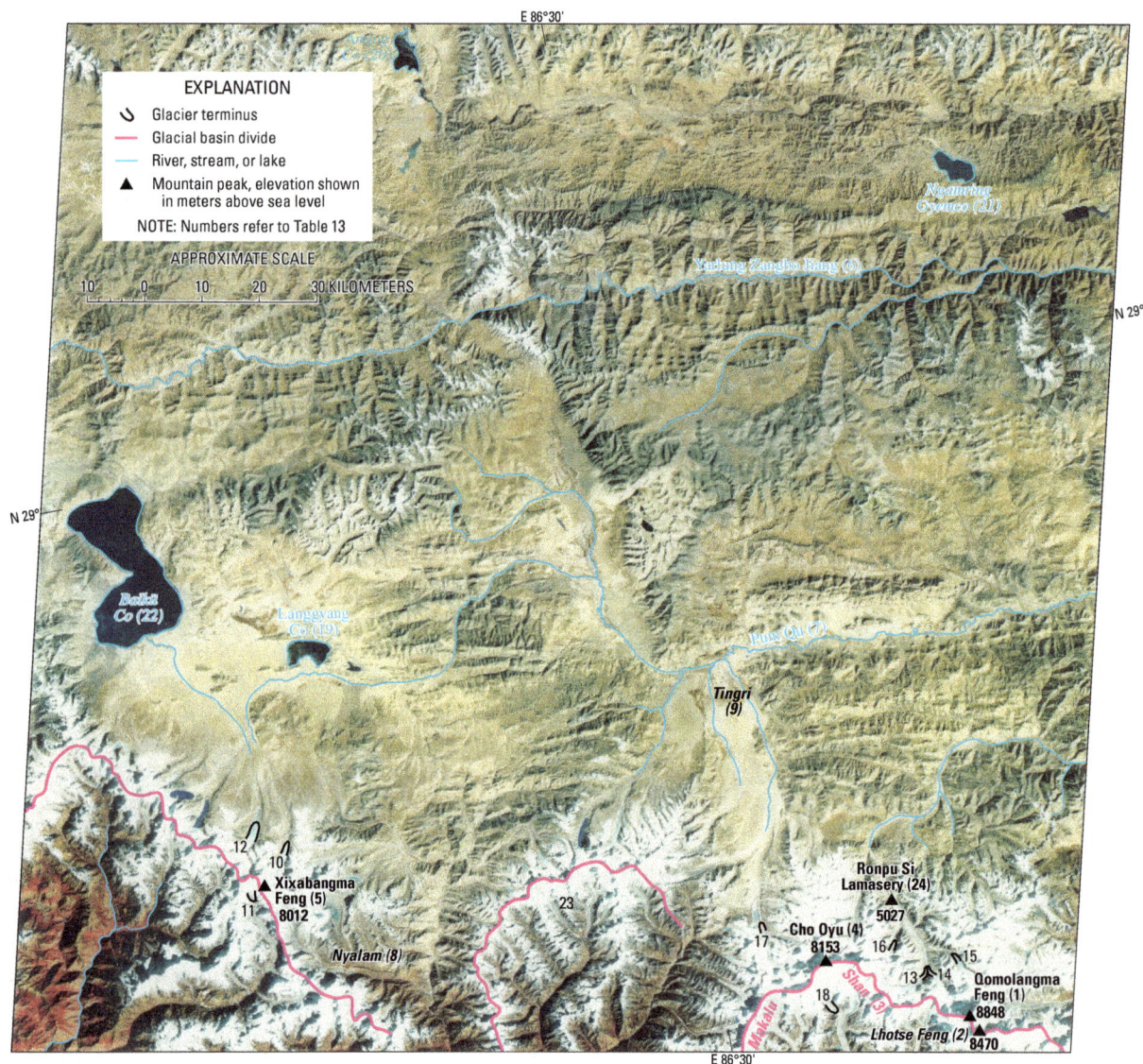

Figure 10.—*Annotated Landsat 2 MSS false-color composite image of the central Himalaya. Drainage divides between glacierized basins shown in red; selected rivers shown in blue; elevations of selected peaks shown in meters; selected positions of glacier termini shown in black; numbers indicate place-names of geographic features that are keyed to table 13. Landsat 2 MSS image (2662–03542, bands 4, 5, 7; 14 November 1976; Path 151–Row 40) from U.S. Geological Survey, EROS Data Center, Sioux Falls, S. Dak.*

[See figure 10 for number locations of the place-names on a Landsat image of the region]

Place-name of glacier or other geographic feature	Latitude	Longitude
Qomolangma Feng (1)	27°58'N.	86°54'E.
Lhotse Feng (2)	27°56'N.	86°56'E.
Makalu Shan (3)	27°56'N.	87°04'E.
Cho Oyu (peak) (4)	28°05'N.	86°42'E.
Xixabangma Feng (5)	28°21'N.	85°47'E.
Yarlung Zangbo Jiang (6)	29°13'N.	86°31'E.
Pum Qu (7)	28°34'N.	86°35'E.
Nyalam (8)	28°10'N.	85°58'E.
Tingri (9)	28°34'N.	86°35'E.
Yebokangjiale (Dasapu) Bingchuan (10)	28°24'N.	85°47'E.
West Kangboqen Bingchuan (11)	28°18'N.	85°41'E.
North Kangboqen Bingchuan (12)	28°25'N.	85°42'E.
West Rongpu Bingchuan (13)	28°02'N.	86°50'E.
Central Rongpu Bingchuan (14)	28°03'N.	86°52'E.
East Rongpu Bingchuan (15)	28°03'N.	86°55'E.
Juda Bingchuan (16)	28°05'N.	86°50'E.
Gyachungkang Bingchuan (17)	28°09'N.	86°36'E.
Gyarag Bingchuan (18)	28°01'N.	86°42'E.
Langgyang Co (19)	28°43'N.	85°53'E.
Amjog Co (20)	29°37'N.	86°13'E.
Ngamring Gyemco (21)	29°18'N.	87°11'E.
Baikü Co (22)	28°53'N..	85°38'E.
Rongxar Region (23)	28°15'N.	86°15'E.
Rongpu Si Lamasery (24)	28°09'N.	86°49'E.

Precipitation in both areas is concentrated in summer. For example, annual precipitation at *Nyalam* (8) (3,800 m), southeast of Xixabangma Feng (5), is 422 mm; at the Rongpu Si Lamasery (24) (5,027 m), on the north side of Qomolangma Feng, annual precipitation is 335 mm, with 84 percent concentrated July through September; and further north at Tingri (9) (4,300 m), annual precipitation is 243 mm, with 97 percent occurring from July through September. Moreover, because of the dry climate, low latitude, and high elevation, this region has the most abundant solar radiation in China. Annual global radiation may reach 199.9 Kcal cm^{-2} at the Rongpu Si Lamasery and annual radiation balance is 80–90 Kcal cm^{-2} (Zeng and Kou, 1975).

The major drainage basins of this region are the Yarlung Zangbo Jiang (6) (the upper Brahmaputra River), and the Pum Qu (7) (the upper Arun River); the Pum Qu joins the Sun Kosi southeast of Qomolangma Feng (1). The lakes are all internal (closed) drainage basins: the Langgyang Co (19), *Amjog Co* (20), *Ngamring Gyemco* (21), and *Baikü Co* (22). The glacier area shown on the Landsat 2 MSS image (fig. 10) amounts to about 2,000 km^2 and is concentrated in the areas of Qomolangma Feng (1), Rongxar (23), and Xixabangma

Feng (5). Glacier termini range from 4,400 to 5,200 m, and snowline elevations from 5,500 to 6,200 m (fig. 1). Because of the intense solar radiation, ice pyramids are well developed on many of the glacier termini.

Glaciers in the area of Xixabangma Feng (5) are retreating, and the glaciers of the southern slopes in this region are retreating two or three times faster than those on the northern slope. For example, the lowest limit of clear ice of the *Yebokangjiale (Dasupu) Bingchuan* (10), on the northern slope of Xixabangma Feng (5), retreated –160 m from 1974 to 1976, while for the same period the *West* (11) and *North Kangboqen* (12) *Bingchuan* retreated –460 m and –350 m, respectively.

In the Qomolangma Feng area, the clean ice of the *West Rongpu Bingchuan* (13) advanced +300 m from 1966 to 1976. During the same period, the *Central Rongpu Bingchuan* (14) and *East Rongpu Bingchuan* (15) retreated –60 m and –200 m, respectively. In the same area, the *Juda Bingchuan* (16) advanced +150 m from 1969 to 1976; the *Gyachungkang Bingchuan* (17) retreated –1,340 m during the period from 1969 to 1976; and the *Gyarag Bingchuan* (18) retreated –870 m from 1957 to 1976.

Optimum Landsat 1, 2, and 3 Images of the Glacierized Regions of China

The preceding sections discuss the use of selected Landsat images to describe glacierized areas of China. Table 14 and figure 11 list optimum Landsat 1, 2, and 3 images that cover all the glacierized areas of China.

TABLE 14.—*Optimum Landsat 1, 2, and 3 images of the glaciers of China*

[Code column indicates usability for glacier studies; see figure 11 for explanation]

Path–Row	Nominal scene center latitude and longitude		Landsat identification number	Date	Code	Cloud cover (percent)	Remarks
ALTAY SHAN							
152–27	47°20′N.	91°19′E.	2152027007718890	07 Jul 77	◑	10	
153–27	47°20′N.	89°53′E.	2153027007722590	13 Aug 77	◑	10	
154–26	48°44′N.	89°04′E.	2154026007719090	09 Jul 77	◐	15	
154–27	47°20′N.	88°27′E.	2154027007719090	09 Jul 77	◐	50	
155–26	48°44′N.	87°38′E.	2155026007720990	28 Jul 77	◐	50	
156–26	48°44′N.	86°12′E.	2156026007719290	11 Jul 77	◐	25	
TIAN SHAN							
152–30	43°05′N.	89°38′E.	2152030007726090	17 Sep 77	◑	10	
153–29	44°30′N.	88°44′E.	2153029007722590	13 Aug 77	●	0	
153–30	43°05′N.	88°12′E.	2153030007722590	13 Aug 77	●	0	
154–30	43°05′N.	86°46′E.	2154030007625090	06 Sep 76	●	0	
154–30	43°05′N.	86°46′E.	2154030007726290	19 Sep 77	●	0	Image used for fig. 2
155–30	43°05′N.	85°20′E.	1155030007224590	01 Sep 72	◐	50	
155–30	43°05′N.	85°20′E.	2155030007628790	13 Oct 76	◐	15	
155–31	41°40′N.	84°48′E.	2155031007722790	15 Aug 77	◑	10	
156–29	44°30′N.	84°26′E.	1156029007226490	20 Sep 72	●	0	
156–29	44°30′N.	84°26′E.	2156029007719290	11 Jul 77	◐	25	
156–30	43°05′N.	83°54′E.	2156030007228290	08 Oct 72	●	5	

TABLE 14.—*Optimum Landsat 1, 2, and 3 images of the glaciers of China*—Continued

[Code column indicates usability for glacier studies; see figure 11 for explanation]

Path–Row	Nominal scene center latitude and longitude		Landsat identification number	Date	Code	Cloud cover (percent)	Remarks
156–30	43°05′N.	83°54′E.	2156030007726490	21 Sep 77	◑	70	
157–29	44°30′N.	83°00′E.	2157029007724790	04 Sep 77	●	0	
157–30	43°05′N.	82°27′E.	2157030007722990	17 Aug 77	●	5	
157–31	41°40′N.	81°56′E.	2157031007722990	17 Aug 77	●	0	
158–29	44°30′N.	81°34′E.	2158029007723090	18 Aug 77	◐	15	
158–30	43°05′N.	81°01′E.	2158030007723090	18 Aug 77	●	0	
158–31	41°40′N.	80°30′E.	2158031007723090	18 Aug 77	●	5	Image used for fig. 3
159–29	44°30′N.	80°08′E.	2159029007522490	12 Aug 75	◕	10	
QILIAN SHAN							
141–34	37°24′N.	103°25′E.	2141034007527890	05 Oct 75	●	5	
142–34	37°24′N.	101°59′E.	2142034007517190	20 Jun 75	◑	40	
143–33	38°49′N.	101°01′E.	2143033007719790	16 Jul 77	●	0	
143–34	37°24′N.	100°33′E.	2143034007719790	16 Jul 77	◕	10	
144–33	38°49′N.	99°35′E.	2144033007618690	04 Jul 76	◑	50	
144–34	37°24′N.	99°07′E.	2144034007718090	29 Jun 77	◕	10	
145–33	38°49′N.	98°09′E.	1145033007330190	28 Oct 73	◕	10	
146–33	38°49′N.	96°43′E.	2146033007711090	20 Apr 77	●	0	Image used for fig. 4
146–34	37°24′N.	96°15′E.	2146034007711090	20 Apr 77	●	0	
147–33	38°49′N.	95°17′E.	2147033007712990	09 May 77	●	0	
PAMIR							
161–33	38°49′N.	75°12′E.	1161033007225190	07 Sep 72	◕	10	
KARAKORUM SHAN							
159–35	35°58′N.	77°08′E.	3159035007819990	18 Jul 78	◕	10	Image used for fig. 5
160–34	37°24′N.	76°10′E.	2160034007723290	20 Aug 77	●	0	
160–35	35°58′N.	75°42′E.	2160035007721490	02 Aug 77	●	5	
KUNLUN SHAN							
143–35	35°58′N.	100°05′E.	2143035007719790	16 Jul 77	◑	30	
143–36	34°32′N.	99°39′E.	2143036007719790	16 Jul 77	◑	30	
144–35	35°58′N.	98°39′E.	2144035007723490	22 Aug 77	◑	70	
145–35	35°58′N.	97°13′E.	2145035007710990	19 Apr 77	◕	10	
146–35	35°58′N.	95°47′E.	2146035007611190	25 Apr 76	●	5	
147–35	35°58′N.	94°21′E.	2147035007727390	30 Sep 77	◑	20	
148–35	35°58′N.	92°55′E.	2148035007713090	10 May 77	◑	20	
149–34	37°24′N.	91°57′E.	1149034007323390	21 Aug 73	●	0	
149–35	35°58′N.	91°29′E.	1149035007319790	16 Jul 73	◑	15	
150–34	37°24′N.	90°30′E.	1150034007227690	02 Oct 72	●	0	
150–35	35°58′N.	90°03′E.	2150035007635490	19 Dec 76	●	0	
151–34	37°24′N.	89°04′E.	2151034007715190	31 May 77	●	0	
152–34	37°24′N.	87°38′E.	1152034007325490	11 Sep 73	●	0	
152–35	35°58′N.	87°11′E.	1152035007235090	15 Dec 72	●	0	

Path–Row	Nominal scene center latitude and longitude		Landsat identification number	Date	Code	Cloud cover (percent)	Remarks
153–34	37°24′N.	86°12′E.	2153034007726190	18 Sep 77	●	5	
153–35	35°58′N.	85°45′E.	2153035007724390	31 Aug 77	●	5	
154–34	37°24′N.	84°46′E.	2154034007724490	01 Sep 77	●	0	
154–35	35°58′N.	84°19′E.	2154035007724490	01 Sep 77	●	0	
155–35	35°58′N.	82°53′E.	2155035007726390	20 Sep 77	◐	30	
155–36	34°32′N.	82°26′E.	2155036007706590	06 Mar 77	●	0	
156–35	35°58′N.	81°27′E.	2156035007632490	19 Nov 76	◐	30	Image used for fig. 6
156–35	35°58′N.	81°27′E	2156035007721090	29 Jul 77	◐	20	
156–36	34°32′N.	81°00′E.	2156036007704890	17 Feb 77	◑	10	
157–35	35°58′N.	80°00′E.	2157035007632590	20 Nov 76	●	5	
158–35	35°58′N.	78°34′E.	2158035007719490	13 Jul 77	◑	10	
159–34	37°24′N.	77°36′E.	1159034007326190	18 Sep 73	●	0	
TANGGULA SHAN							
145–38	31°41′N.	95°55′E.	2145038007721790	05 Aug 77	◐	40	
146–38	31°41′N.	94°29′E.	3146038008129690	23 Oct 81	●	5	Partial scene
147–37	33°07′N.	93°28′E.	2147037007705790	26 Feb 77	●	5	
148–37	33°07′N.	92°02′E.	2148037007635290	17 Dec 76	●	5	Geladaindong
149–36	34°32′N.	91°03′E.	2149036007631790	12 Nov 76	●	0	
149–37	33°07′N.	90°36′E.	2149037007705990	28 Feb 77	◑	10	Geladaindong. Image used for fig. 7
150–36	34°32′N.	89°36′E.	2150036007635490	19 Dec 76	●	0	
150–37	33°07′N.	89°10′E.	2150037007635490	19 Dec 76	●	0	
NYAINQÊNTANGLHA SHAN							
144–39	30°15′N.	96°56′E.	2144039007634890	13 Dec 76	●	0	
144–39	30°15′N.	96°56′E.	3144039008129490	21 Oct 81	●	0	Partial scene
144–40	28°49′N.	96°31′E.	2144040007634890	13 Dec 76	●	0	
144–40	28°49′N.	96°31′E.	2144040007721690	04 Aug 77	◐	50	Image used for fig. 9
145–39	30°15′N.	95°30′E.	2145039007634990	14 Dec 76	◐	30	
146–39	30°15′N.	94°04′E.	2146039007635090	15 Dec 76	●	5	
147–38	31°41′N.	93°03′E.	2147038007635190	16 Dec 76	●	0	
147–39	30°15′N.	92°38′E.	2147039007635190	16 Dec 76	●	0	
148–39	30°15′N.	91°12′E.	2148039007635290	17 Dec 76	●	0	
149–39	30°15′N.	89°46′E.	2149039007635390	18 Dec 76	●	0	Image used for fig. 8
GANGDISÊ SHAN							
151–39	30°15′N.	86°54′E.	2151039007633790	02 Dec 76	●	0	
152–39	30°15′N.	85°28′E.	2152039007633890	03 Dec 76	●	0	
153–38	31°41′N.	84°27′E.	2153038007632190	16 Nov 76	●	0	
153–39	30°15′N.	84°01′E.	2153039007700990	09 Jan 77	●	0	
154–38	31°41′N.	83°01′E.	2154038007632290	17 Nov 76	●	0	
154–39	30°15′N.	82°35′E.	2154039007632290	17 Nov 76	◑	10	
155–38	31°41′N.	81°35′E.	2155038007628790	13 Oct 76	●	5	

TABLE 14.—*Optimum Landsat 1, 2, and 3 images of the glaciers of China*—Continued

[Code column indicates usability for glacier studies; see figure 11 for explanation]

Path–Row	Nominal scene center latitude and longitude		Landsat identification number	Date	Code	Cloud cover (percent)	Remarks
155–39	30°15′N.	81°09′E.	2155039007634190	06 Dec 76	●	5	
156–37	33°07′N.	80°34′E.	2156037007632490	19 Nov 76	●	5	
156–38	31°41′N.	80°09′E.	2156038007632490	19 Nov 76	●	0	
HIMALAYA							
144–40	28°49′N.	96°31′E.	2144040007631290	07 Nov 76	◕	10	
144–40	28°49′N.	96°31′E.	2144040007721690	04 Aug 77	●	0	
145–39	30°15′N.	95°30′E.	2145039007634990	14 Dec 76	◑	30	
145–40	28°49′N.	95°95′E.	1145040007335590	21 Dec 73	◕	10	
146–40	28°49′N.	93°39′E	2146040007635090	15 Dec 76	◕	10	
147–40	28°49′N.	92°13′E.	2147040007635190	16 Dec 76	●	5	
147–41	27°23′N.	91°49′E.	2147041007700390	03 Jan 77	◑	20	
148–40	28°49′N.	90°47′E.	2148040007635290	17 Dec 76	●	5	
148–41	27°23′N.	90°23′E.	2148041007635290	17 Dec 76	◕	10	
149–41	27°23′N.	88°57′E.	2149041007635390	18 Dec 76	◑	15	
150–40	28°49′N.	87°55′E.	2150040007635490	19 Dec 76	●	0	
150–41	27°23′N.	87°31′E.	2150041007635490	19 Dec 76	●	5	
151–40	28°49′N.	86°29′E.	2151040007631990	14 Nov 76	●	0	Image used for fig. 10
151–40	28°49′N.	86°29′E.	2151040007635590	20 Dec 76	●	5	
151–41	27°23′N.	86°05′E.	2151041007635590	20 Dec 76	●	10	
152–40	28°49′N.	85°03′E.	2152040007633890	03 Dec 76	●	5	
153–39	30°15′N.	84°01′E.	2153039007700990	09 Jan 77	●	0	
153–40	28°49′N.	83°37′E.	2153040007700990	09 Jan 77	◕	10	
154–39	30°15′N.	82°35′E.	2154039007632290	17 Nov 76	◕	10	
155–39	30°15′N.	81°09′E.	2155039007634190	06 Dec 76	●	5	
156–38	31°41′N.	80°09′E.	2156038007632490	19 Nov 76	●	0	
156–39	30°15′N.	79°43′E.	2156039007632490	19 Nov 76	●	0	
157–38	31°41′N.	78°42′E.	2157038007632590	20 Nov 76	●	0	
HENGDUAN SHAN							
140–39	30°15′N.	102°40′E.	1140039007400390	03 Jan 74	◕	10	
140–40	28°49′N.	102°16′E.	1140040007402190	21 Jan 74	●	0	
140–41	27°23′N.	101°52′E.	1140041007400390	03 Jan 74	●	0	
141–39	30°15′N.	101°14′E.	1141039007400490	04 Jan 74	●	0	
141–40	28°49′N.	100°50′E.	11410400074000490	04 Jan 74	●	0	
141–41	27°23′N.	100°26′E.	11410410074000490	04 Jan 74	◕	10	
142–39	30°15′N.	99°48′E.	11420390074000590	05 Jan 74	●	0	
142–40	28°49′N.	99°24′E.	11420400074000590	05 Jan 74	●	0	
142–41	27°23′N.	99°00′E.	11420410074000590	05 Jan 74	●	0	
143–39	30°15′N.	98°22′E.	2143039007631190	06 Nov 76	●	0	
143–40	28°49′N.	97°57′E.	2143040007631190	06 Nov 76	●	0	

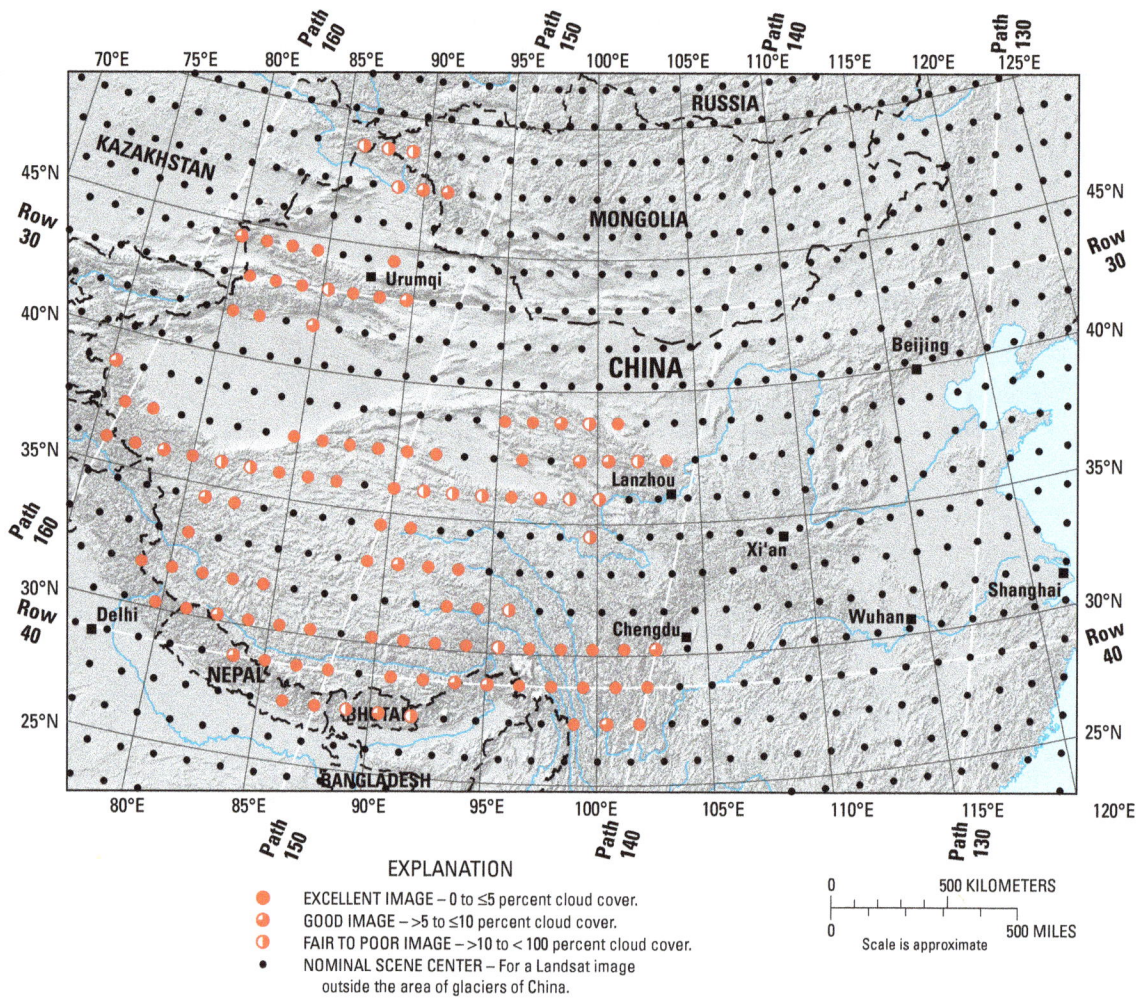

Figure 11.—Index map of optimum Landsat 1, 2, and 3 images of the glacierized areas of China.

References Cited

Alpine Ice and Snow Utilization Team, 1959, Report on modern glaciers expedition of Qilian Shan: Beijing, Science Press, 15 p. (In Chinese)

Bai Zhongyuan and Xie Weirong, 1965, The heat balance in the wide ice surface of the Ürümqi He No. 1 Glacier in ablation period; Glaciological and hydrological study on the Ürümqi He, Tian Shan: Beijing, Science Press. (In Chinese)

Bakov, E.K. 1975, The surging change of a large dendritic glacier in Tianshan, *in* Glaciological Research in Tianshan 73–79 (In Russian).

Du Ronghuan and Zhang Shucheng, 1981, Characteristics of glacial mudflows in southeastern Qinghai-Xizang Plateau, *in* Geological and ecological studies of Qinghai-Xizang (Tibet) Plateau: Beijing, Science Press, v. 2, p. 1837–1842.

Hewitt, K., 1979, Letter to the Chinese Embassy in Ottawa, Canada.

Iwata, S., and Jiao Keqin, 1993, Fluctuations of the Zepu Glacier in Late Holocene epoch, the eastern Nyainqêntanglha Mountains, Tibet Plateau, *in* Glaciological climate and environment on Qinghai-Xizang Plateau: Beijing, Science Press, p. 130–146.

Jiao Keqin and others, 1988, Glaciers of the Qiantang Gaoyuan; Glacier inventory of China 7: Beijing, Science Press. (In Chinese)

Kingdon-Ward, F., 1934, The Himalaya East of Tsangpo: Geographical Journal, v. 84, no. 5, p. 369–397.

Kou Youguan, Zeng Qunzhu, Xie Weirong, and Xie Yingqin, 1974, On the solar radiation in the region of Qomolangma Feng; Report of Scientific Expedition in the Region of Qomolangma Feng, 1966–1968, Meteorology and Solar Radiation: Beijing, Science Press (In Chinese).

Li Jijun, 1975, Recent investigation on glaciers in southeast Xizang: Journal Lanzhou University, no. 2, June 1975 (in Chinese).

Li Jijun and Zheng Benxing, 1981, The monsoon maritime glaciers in the southeastern part of Xizang, *in* Geological and ecological studies of Qinghai-Xizang (Tibet) Plateau: Beijing, Science Press, v. 2, p. 1599–1610.

Li Jijun, Zheng Benxing, Yang Xijun, and others, 1986, Glaciers in Xizang: Beijing, Science Press, 283 p. (In Chinese)

Liu Chaohai and others, 1982, Glaciers of the Altay Shan (including Muztag Ling); Glacier inventory of China 2: Beijing, Science Press. (In Chinese)

Liu Chaohai and others, 1987, Glaciers of the Tian Shan; Glacier inventory of China 3: Beijing, Science Press. (In Chinese)

Mason, K., 1930, The glaciers of the Karakoram and neighborhood: Records of the Geological Survey of India, v. 63, pt. 2, p. 214–278.

Mi Desheng, Liu Chaohai, and others, 1988 (1999 revised), Glaciers of the Pamir; Glacier inventory of China 4: Beijing, Science Press. (in Chinese)

Mi Desheng, Xie Zichu, and others, 2002, The Ganga drainage basin and Indus drainage basin; Glacier inventory of China 11: Xi'an, Xi'an Cartographic Publishing House, 572 p. (In Chinese)

Nakawo, M., Ageta, Y., and Han, J., 1990, Climatic information from the Chongce Ice Cap, West Kunlun, China: Annals of Glaciology, v. 14, p. 205–207.

Pu Jianchen and others, 1994, Glaciers of the Chang Jiang (Yangtze) drainage area; Glacier inventory of China 8: Lanzhou, Gansu Culture Publishing House. (In Chinese)

Pu Jianchen, and others, 2001, Glaciers of the Lancang Jiang and Nu Jiang Valley; Glacier inventory of China 9: Xi'an, Xi'an Cartographic Publishing House. (In Chinese)

Qin Dahe, chief ed., 1999, Map of glacier resources in the Himalayas; First Edition: Beijing, China, Science Press, scale 1:500,000, 7 sheets. [Compilation of map based on analysis of 1975–1978 Landsat MSS band 7 and false-color (bands 4, 5, 7) images and some aerial photographs]

Shipton, Eric, 1938, The Shaksgam Expedition, 1937: Geographical Journal, v. 91, no. 3, p. 313–339.

Shi Yafeng, 1961, On utilization of alpine snow and ice for enriching agricultural products in northwestern China: Kexue Tungbao, no. 1, 1961. (In Chinese)

Shi Yafeng, editor in chief, 2008, Concise glacier inventory of China: Shanghai, Popular Science Press, 205 p.

Shi Yafeng, Liu Chaohai, Wang Zhungtai, Liu Shiyin, Yi Baisheng, and others, 2005, Concise glacier inventory of China. Chinese ed.: Shanghai, Popular Science Press, p. 1–194.

Shi Yafeng and Liu Shiyin, 2000, Estimation of the response of glaciers in China to the global warming in the 21st century: Chinese Science Bulletin, v. 45, no. 7, p. 668–672.

Shi Yafeng and Wang Zhungtai, 1979, The Muzart glacial valley passway and Sino-west communication in history: Journal of Glaciology and Cryopedology, v. 2, no. 2, p. 22–26. (In Chinese)

Shi Yafeng and Xie Zichu, 1964, The basic characteristics of modern glaciers in China: Acta Geographica Sinica, v. 30, no. 3.

Su Zhen and Wang Zhichao, 1985, On the variation of glaciers in the Tomur Feng area, *in* Glaciology and meteorology in the Tomur Feng area: Ürümqi, Xinjiang People's Press, p. 85–94. (In Chinese)

Su Zhen, Zhang Wenjing, and Ding Liangfu, 1985, On the condition, amount of distribution, and morphological type of modern glaciers in the Tomur Feng area, *in* Glaciology and meteorology in the Tomur Feng area: Ürümqi, Xinjiang People's Press, p. 34–43. (In Chinese)

Thompson, L.G., Mosley-Thompson, E., Davis, M., Bolzan, J., Gundestrup, N., Yao, T., Wu, X., and Xie, Z., 1989, Holocene-Late Wisconsin Pleistocene climatic ice core records from Qinghai-Tibetan Plateau: Science, v. 246, no. 4929, p. 474–477.

Thompson, L.G., Yao, T., Davis, M.E., Henderson, K.A., Mosley-Thompson, E., Lin, P.N., Beer, J., Synal, H.-A., Cole-Dai, J., and Bolzan, J.F., 1997, Tropical climate instability. The last glacial cycle from a Qinghai-Tibetan ice core: Science, v. 276, no. 5320, p. 1821–1827.

U.S. Defense Mapping Agency, 1990, Gazetteer of the People's Republic of China; March 1990 (2d ed.): Washington, D.C., Defense Mapping Agency; Names approved by the United States Board on Geographic Names; v. I (A–H), p. 1–719; v. II (I–R), p. 720–1,331; v. III (S–Z), p. 1,332–2,074. [Editors' note: The introductory pages to v. I (p. i–xvi) contain a map of the provinces and provincial capitals of China, an index map to the 1:250,000-scale Joint Operations Graphic (Ground) Series 1501 maps of China, sources for the approximately 96,000 citations to geographic place-names and features in China, a glossary of generic terms (generally Mandarin Chinese rendered in Pinyin spellings; Mongolian, Tibetan, and Uygur terms are followed by an (M), (T), or (U)) and other relevant information.]

von Wissman, H., 1959, Die heutige vergletscherung und Shneegrenze in Hochasien [The present-day glacierization and snowline in High Asia]: Abhandlungen der Akademie der Mathematisch-Naturrwissenschaften und der Literatur (Mainz): Jahrgang 1959, no. 14, p. 1101–1407. (In German)

Wang Zongtai and others, 1981, Glaciers of the Qilian Shan (including Altun Shan); Glacier inventory of China 1: Beijing, Science Press. (In Chinese)

Xie Zichu, 1980, Mass balance of glaciers and its relationship with characteristics of glaciers: Journal of Glaciology and Cryopedology, v. 2, no. 4, p. 1–10. (In Chinese with English abstract)

Xie Zichu and Ge Guangwen, 1965, The accumulation, ablation and mass balance on the Ürümqi He No. 1 glacier, Tianshan, glaciological and hydrological study on the Ürümqi He, Tian Shan: Beijing, Science Press. (In Chinese)

Xie Zichu and Kotlyakov, V.M., eds., 1994, Glaciers and environment in the Qinghai-Xizang (Tibet) Plateau (I)— The Gongga Mountain: Beijing, China, Science Press, Reports on the Sino-Russian Joint Glaciological Expedition, Lanzhou Institute of Glaciology and Geocryology, Chinese Academy of Sciences; Institute of Geography, Russian Academy of Sciences, 201 p.

Yang Huian and others, 1989, Glaciers of the Karakorum Shan; Glacier inventory of China 5: Beijing, Science Press. (In Chinese)

Yang Huian and others, 1992–94, Glaciers of the Kunlun Shan; Glacier inventory of China 6: Beijing, Science Press. (In Chinese)

Yang Zhenniang, 1980, The characteristics of glacial hydrology in the mountains of West China (manuscript).

Yao Tandong and Thompson, L.G., 1992, Trends and features of climatic changes in the past 5000 years recorded by the Dunde ice core: Annals of Glaciology, v. 16, p. 21–24.

Yao Tandong, Thompson, L.G., Jiao Keqin, Mosley-Thompson, Ellen, and Yang Zhihong, 1995, Recent warming as recorded in the Qinghai-Tibetan cryosphere: Annals of Glaciology, v. 21, p. 196–200.

Yao Tandong, Thompson, L.G., Qin Dahe, Tian Lide, Jiao Keqin, Yang Zhihong, and Xie Chao, 1996, Variations in temperature and precipitation change in the past 2 ka on the Xizang (Tibetan) Plateau-Guliya ice core record: Science in China (Series D) v. 39, no. 4, p. 425–433.

Zeng Qunzhu and Kou Youguan, 1975, The heat balance on the Rongbu Glacier during ablation period; Report of scientific expedition in the region of Qomolangma Feng, 1966–1968: Glaciology and Geomorphology, Beijing, Science Press, p. 52–54. (In Chinese)

Zeng Qunzhu, Kou Youguan, Xie Weirong, and Xiao Shu, 1982, Study of the radiation balance on Qinghai-Xizang Plateau; The collected works of heat conditions of developing glacier and permafrost: Beijing, Science Press. (In Chinese)

Zhang Xiangsong, 1980 Recent variations of the Yinsugaiti Glacier and its adjacent glaciers in the Karakorum Mountains: Journal of Glaciology and Cryopedology, v. 2, no. 3. (In Chinese)

Zhang Xiangsong and others, 1990, Study on the glacier lake outburst flood of the Yarkant River, Karakorum Mountains: Beijing, Science Press. (In Chinese)

Glaciers of Asia—

GLACIERS OF AFGHANISTAN

By John F. Shroder, Jr., *and* Michael P. Bishop

SATELLITE IMAGE ATLAS OF GLACIERS OF THE WORLD

Edited by RICHARD S. WILLIAMS, JR., *and* JANE G. FERRIGNO

U.S. GEOLOGICAL SURVEY PROFESSIONAL PAPER 1386–F–3

CONTENTS

Abstract -- **167**

Introduction -- **167**

    Figure   1. Index map of Afghanistan showing the location of major
        geographic features mentioned in text ----------------------------------- 168

Occurrence of Glaciers -- **169**

    Figure   2. Map of Afghanistan showing the three main drainage basins -------------- **170**

    Table   1. Glacierized drainage basins in Afghanistan ----------------------------- **171**

    Figure   3. Photograph showing terminus of debris-covered *Qādzī Deh
        Glacier* joined by *Ye Safed Glacier* ---------------------------------- **173**

    Figure   4. Photograph showing argillite-covered glacier ice on the
        Qādzī Deh Glacier below Noshaq peak -------------------------------- **173**

    Figure   5. Landsat 3 return beam vidicon satellite image showing glacierized
        areas extending from Tajikistan and Ab-i-Panj river in the
        Wākhān valley to Tirich Mir peak in Pakistan ---------------------------- **174**

Mapping of Glaciers --- **175**

    Table   2. Glacier information contained in scientific and mountain-
        climbing publications on Afghanistan -------------------------------- **176**

    Satellite Imagery in the 1970s --- **177**

        Photographic Prints **177**

    Figure   6. Index map of Landsat 1, 2, and 3 satellite imagery, drainage basins,
        and large-scale maps of glacierized areas of Afghanistan----------------- **178**

    Figure   7. Map of optimum Landsat 1, 2, and 3 images of the glaciers of
        Afghanistan --- **179**

    Table   3. Optimum Landsat 1, 2, and 3 images of the glaciers of
        Afghanistan -- **180–181**

        Digital Analyses **181**

    Satellite Imagery since the 1970s -- **182**

Case Studies of Glaciers--- **182**

    Kūh-e Bābā **183**

    Figure   8. Topographic map of *Fūlādī* valley and central *Kūh-e Bābā* range---------- **184**

    Figure   9. View northwest from eastern end moraine toward large moraine
        loop above the village of Burghasūn ----------------------------------- **185**

    Figure 10. Photograph of *Fūlādī Glacier* taken in July 1978 ----------------------- **185**

    Figure 11. Stereoscopic pair of Landsat 1 images of the *Kūh-e Bābā* area------------- **186**

    Figure 12. Stereoscopic pair of ASTER images of *Kūh-e Bābā* area -------------------- **186**

    Mīr Samīr **187**

    Figure 13. Topographic map of *Mīr Samīr* (Mērsmer) area in the central
        Hindu Kush--- **187**

    Kūh-e Bandakā **188**

    Figure 14. Stereoscopic pair of ASTER images of *Mīr Samīr* area---------------------- **189**

    Figure 15. Stereoscopic pair of Landsat 1 and 2 images of the Bandakā massif------- **189**

    Figure 16. Planimetric map of the Bandakā massif ------------------------------- **190**

    Keshnīkhān Glacier **191**

    Wākhān Pamir Glaciers.. **191**

    Figure 17. Planimetric map of *Keshnīkhān Glacier* ----------------------------- **192**

    Figure 18. Landsat 1 MSS image of *Keshnīkhān Glacier* area taken 12 July 1973 ---- **193**

    Figure 19. Stereoscopic pair of ASTER images of *Keshnīkhān Glacier* ---------------- **193**

    Figure 20. Landsat 3 RBV image of the central Wākhān Pamir -------------------- **194**

    Figure 21. Topographic map of south-central part of the Wākhān Pamir ----------- **195**

Conclusions -- **196**

Acknowledgments --- **196**

References Cited --- **197**

GLACIERS OF ASIA—

SELECTED GLACIERS OF AFGHANISTAN

By John F. Shroder, Jr.,[1] *and* Michael P. Bishop[1]

Abstract

The Hindu Kush mountains of Afghanistan range in elevation from more than 7,000 meters in the east, where most of the glaciers occur, to less than 500 m in the arid west. More than 3,000 small glaciers, with an estimated area of 2,700 km², provide vital water resources to the region, especially for irrigation. The glaciers are concentrated in the highest parts of the three main drainage basins in the country: (1) the Afghan-Iranian plateau endorheic basin, (2) the Indus basin, and (3) the glacier-dominant Turkistan endorheic basin. Most glaciers occur on north-facing slopes that are shaded by mountain peaks, and on east and southeast slopes that are shaded by monsoon clouds. Frequent snow avalanches, coupled with widespread stagnation or retreat, have produced numerous debris-covered ice and rock glaciers. Limited verification of analyses from older Landsat and some ASTER imagery is provided by Russian and United States 1:100,000- and 1:50,000-scale topographic maps, compiled from vertical aerial photographs acquired during the late 1950s, and by several large-scale glacier maps made in the 1960s and 1970s. Transparent image overlays, false-color enlargements, a few digital analyses, and stereoscopic satellite images were used in a limited analytical fashion to analyze typical glaciers in the region, including the small *Fūlādī Glacier*[2] in the *Kūh-e-Bābā*, the glaciologically assessed ice masses of *Mir Samir*, the high *Sakhi Glacier* on *Kūh-e-Bandakā*, the extensively analyzed *Keshnīkhān Glacier* and the long, debris-covered *Qādzī Deh Glacier* in the western Wākhān area, and the precisely mapped *Zemestan* and *Northern* and *Southern Issīk Glaciers* in the Afghan Pamirs.

Introduction

The high elevation and remote mountains of Afghanistan (fig. 1) have been traversed for millennia by caravans, explorers, and scientific expeditions; yet little detailed information is readily available regarding the terrain, ice, and snow in these mountains. Study of glaciers in such difficult and inaccessible country necessitates the use of satellite imagery. Fortunately, high-quality imagery, topographic maps, and ground photographs are available for limited verification of satellite analyses in some areas. This report provides glacier information based on these data sources, and also includes a general review of the limited literature covering Afghanistan's glaciers.

The glaciers of Afghanistan provide an important, but variable (5–25 percent) part of the meltwater used for critical irrigation through the summer dry season (Levedeva and Larin, 1991); however, the amount of meltwater provided is expected to decline as much as 14 percent if temperatures increase as little as 1 °C because of global warming (Levedeva, 1997). Shroder (1980, 1989a) initiated a glacier inventory to provide better information for water-resource analysis and development. The unstable military situation in the late

---

[1]Department of Geography and Geology, University of Nebraska at Omaha, Omaha, NE 68182

[2]The geographic place-names used in this section conform to the usage authorized for foreign names by the U.S. Board on Geographic Names as listed on the GEOnet Names Server (GNS) website: *http://earth-info.nga.mil/gns/html/index.html*. The names not verified or not listed on the website are shown in italics.

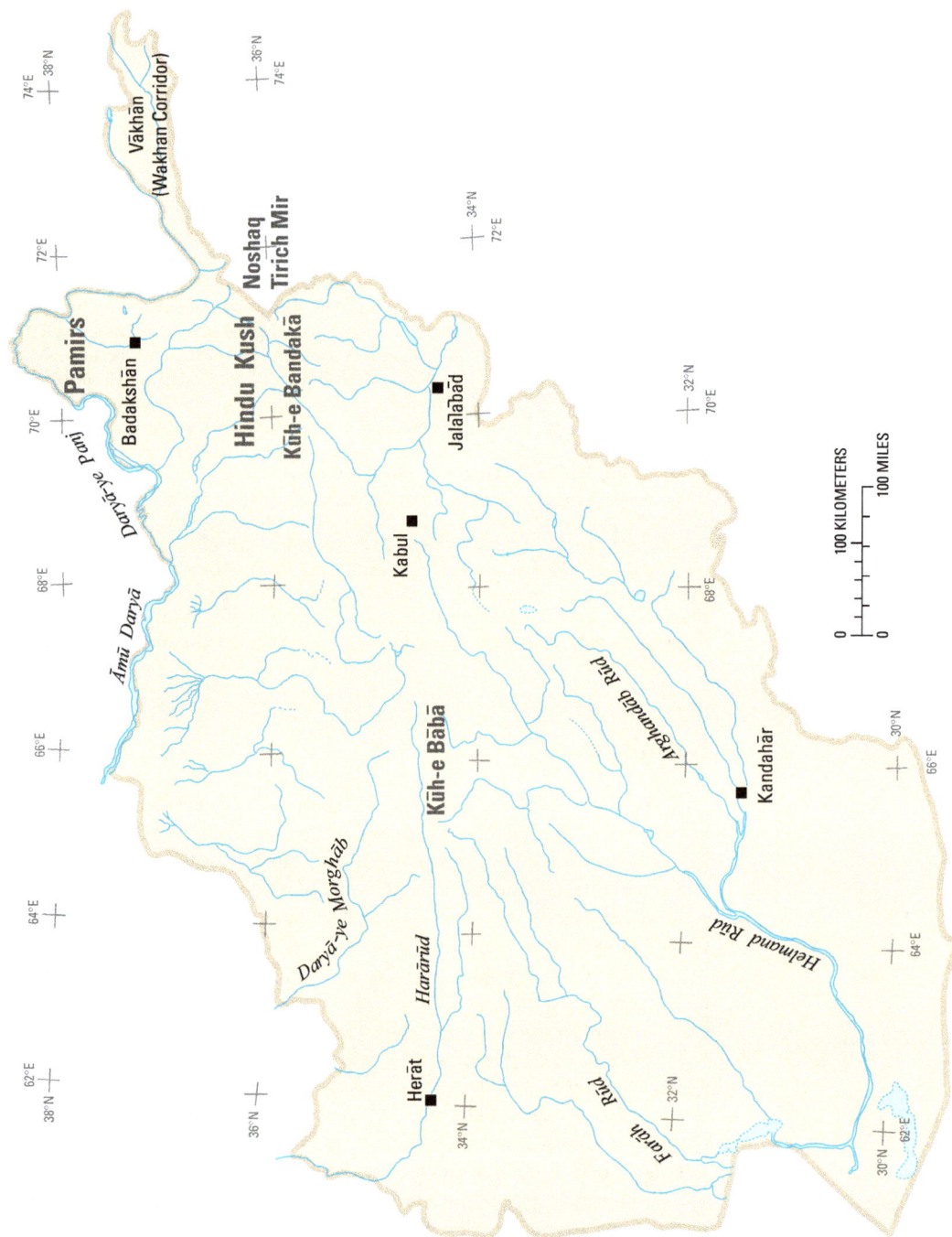

Figure 1.—Index map of Afghanistan showing the location of major geographic features mentioned in text — Hindu Kush, the Pamirs, Wākhān corridor, Badakshān, Daryā-ye Panj (river), Tirich Mir, Kūh-e-Bandakā and Kūh-e Bābā.

1970s disrupted the field-based effort, however, and forced the use of satellite imagery to accomplish the inventory. Early results were given to the U.S. Geological Survey in the early 1980s but they were not published. Then the ensuing two decades of ongoing war and religious and political upheaval, and the subsequent presence of western military forces further disrupted efforts to study glaciers in Afghanistan. The addition of seven years of catastrophic drought, starting in the late 1990s, and the related expansion of trafficking in drought-tolerant opium poppies, coupled with post-war reconstruction efforts, gave renewed impetus for water-resource assessments, a glacier inventory, and an analysis of glacier-related hazards. Water resources and the potential problems from ongoing climate change are viewed as critical to future political stability, with the National Aeronautics and Space Administration (NASA)-supported Global Land Ice Measurements from Space (GLIMS) Project [*http://www.glims.org*] and the U.S. Geological Survey tasked to provide additional baseline information for better predictive and reconstruction capabilities in Afghanistan (Shroder, 2004).

Occurrence of Glaciers

The mountains of Afghanistan generally decline in elevation from more than 7,000 meters (m) in the east to less than 500 m on the Iranian border. Several thousand small glaciers occur in the rugged alpine topography of the central and northeast Hindu Kush and in the Pamirs of eastern Afghanistan, with a total area estimated at approximately 2,700 km^2. In the central part of the country, three major basins meet: (1) in the Turkistan endorheic basin (does not drain to the sea), drainage is north or west into the Aral Sea basin between Uzbekistan and Kazakhstan; (2) in the Afghan-Iran plateau endorheic basin, drainage is south or west, and (3) in the Indus (exorheic) basin, drainage is to the east or south into Pakistan and thence to the Indian Ocean (fig. 2). Delineation and letter/numeral designation of the drainage basins of Afghanistan was originally done according to the revised guidelines of the Temporary Technical Secretariat (Müller, 1978) of the World Glacier Inventory (Shroder, 1980). The three main basins are subdivided into 77 smaller ones, of which 25 are glacierized (contain glaciers) (fig. 2) (Shroder, 1980). Nineteen of the glacierized basins have more than 75 percent of the total ice mass of Afghanistan (table 1) and drain into the Turkistan endorheic system, whereas the Afghan-Iran plateau basin has only seven small glaciers in one small drainage area.

A variety of snowline or equilibrium line altitudes (ELA) was determined in Afghanistan during the 1960s and 1970s. General altitudes in subtropical high Asia was first distinguished by von Wissman (1960); Grötzbach and Rathjens (1969) refined this work for much of the central and western Hindu Kush. Grötzbach and Hillebrandt (1964) and Gilbert and others (1969) contributed some detailed work in the central Hindu Kush; Desio (1975) extended these analyses to the north into the Pamirs of the Badakhshān region. In general, however, the glaciers of northern Badakhshān are the most poorly known in Afghanistan. Braslau (1972) and Patzelt (1978) provided additional ELA data for the Wākhān corridor. Porter (1985) first used Landsat imagery to determine the extent of late Pleistocene glaciers in Afghanistan, but not modern ice.

During the war years of the 1980s, snow cover mapping of Afghanistan's mountains was undertaken by Soviet scientists (Tsarev and others, 1986; Kravtsova, 1990) by using satellite imagery and field measurements to assess seasonal snow distribution and depth, dynamics of the snow boundary, seasonal snow-lines, and dates of formation, melting, and duration of snow cover.

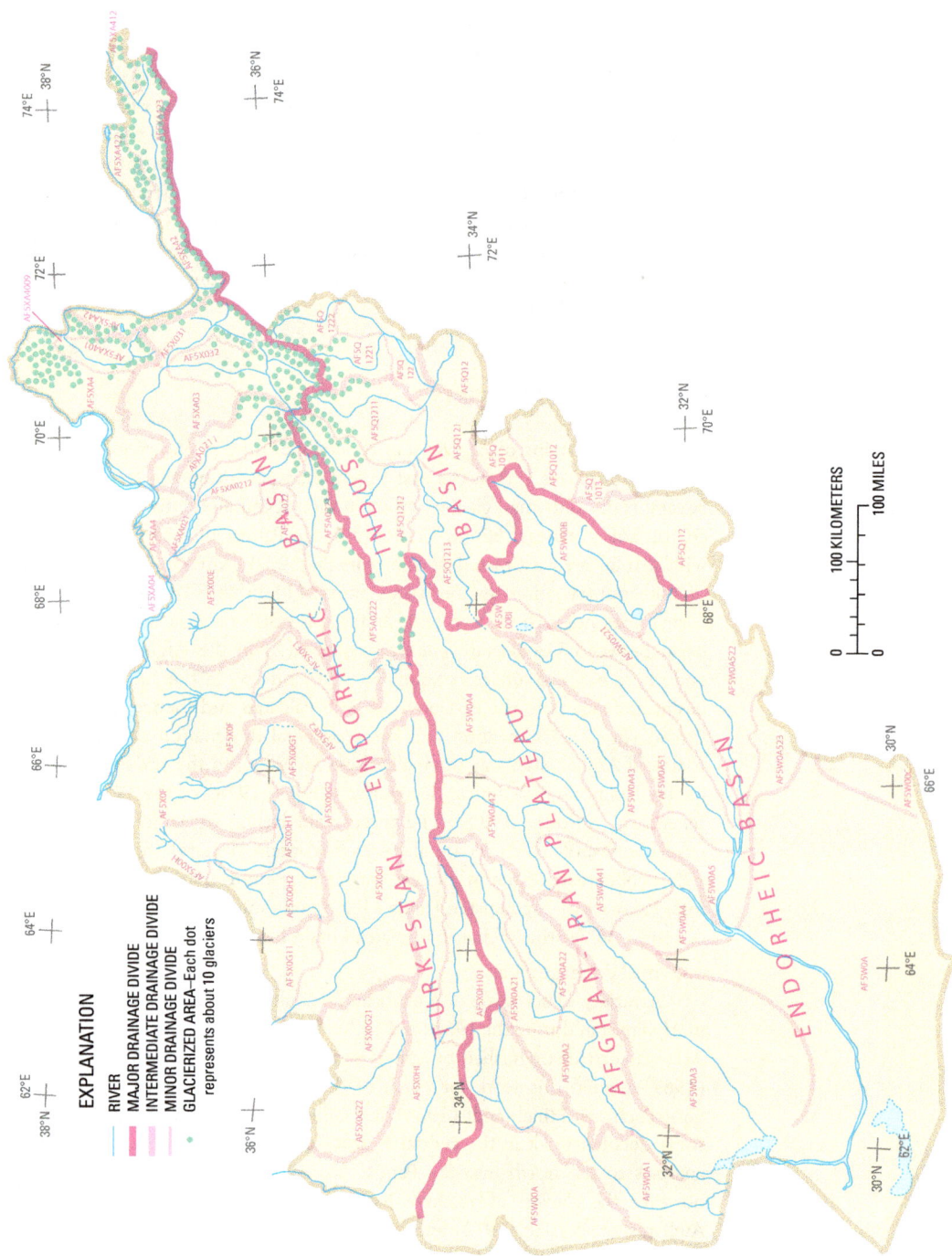

Figure 2.—Map of Afghanistan showing the three main drainage basins, the detailed delineation of 77 smaller drainage basins according to Temporary Technical Secretariat (TTS) for World Glacier Inventory (Müller and others, 1977; Müller, 1978) guidelines, and the generalized glacierized areas. Each dot represents about 10 glaciers (see table 1).

EXPLANATION

RIVER

MAJOR DRAINAGE DIVIDE

INTERMEDIATE DRAINAGE DIVIDE

MINOR DRAINAGE DIVIDE

GLACIERIZED AREA—Each dot represents about 10 glaciers

TABLE 1.—*Glacierized drainage basins in Afghanistan*

[Code letters and numbers identify individual drainage basins delineated on the basis of Temporary
Technical Secretariat (TTS) for World Glacier Inventory guidelines (Müller, 1978; Shroder, 1980).
The number of glaciers in parentheses are minimum counts made from inadequate maps,
satellite imagery, or from both sources. Other numbers are more precise counts made from
1:100,000-scale maps. All total counts include a few ice-cored rock glaciers or
debris-covered glaciers in which little or no surface ice is exposed]

Basins	Code	Number of glaciers
TURKESTAN ENDORHERIC DRAINAGES AF5X		
Vardūj	A031	(199)
Monjān	A0321	(220)
Upper Kowkcheh	A032	(57)
Yakhshendar (Anjumān)	A0322	(141)
Darang	A4003	6
Kulab	A4005	104
Sabz	A4006	54
Jaway	A4007	113
Maynay	A4008	91
Yarhk	A4009	65
Shīveh	A401	(137)
Chakmaktin Kol	A412	(80)
Upper Ab-Panj	A42	(275)
Pāmīr	A422	174
Wākhān	A423	(381)
Khānābād (Taliqan)	A0211	(183)
Chal Bangi	A0212	(60)
Andarāb	A0221	80
Surkh Ab	A0222	73
INDUS RIVER BASIN DRAINAGES AF5Q		
ʿAlīshang - ʿAlīngār	1211	178
Ghowr Band - *Panshir*	1212	262
Pīch	1221	(100)
Katīgal	1222	(103)
Upper Kabul	121	(13)
AFGHAN-IRANIAN PLATEAU DRAINAGES AF5W		
Upper Helmand	A4	7

Maximum snow storage accumulation was mapped, and indices of snow-pack instability were compiled (Tsarev, 1988). After the Soviet departure from the country in 1989, efforts continued during the following decade to assess and map at a small scale the climatic characteristics of the snow cover and avalanche regimes in Afghanistan (Kravtsova and Tsarev, 1997). Morphology, climate, mass exchange, and runoff of glaciers in Afghanistan were also studied (Lebedeva and Larin, 1991), and predictions were made regarding glacier conditions and runoff under the influence of global warming in the early 21st century (Lebedeva, 1997). For the World Atlas of Snow and Ice Resources (Kotlyakov, 1997; Kotlyakov and Lebedeva, 1998), ice of the Hindu Kush was mapped, at the very generalized scale of 1:5,000,000, in terms of such parameters as ice water-volume equivalent, temperature, precipitation, glacier regime, river runoff, avalanche activity, and late Pleistocene glaciation.

Snowline contours for glaciers of the central Hindu Kush and the Wākhān corridor range in elevation from 4,600 m to 5,200 m, with the highest values at the northeast and southwest ends of the ranges, where precipitation is less.

The lowest snowline values occur on the north-facing slopes, and presumably also to the north into the high Badakhshān Pamirs. The low ELA contours also form a prominent dip at the Sālang pass area, north of Kabul in the central Hindu Kush, where precipitation is highest in the country. This increase in the amount of snow is the result of the concentration of precipitation in the narrowest part of the range through which westerly storms pass, and also because monsoon precipitation sometimes influences the area (Sivall, 1977).

Glaciers in Afghanistan are generally small because the climatic snowline is above many peaks, and glaciers form only where ablation is retarded in shaded areas (Grötzbach and Rathjens, 1969). This produces a strong northerly glacierization in much of the central Hindu Kush; toward the Wākhān corridor in the northeast this tendency is less strong. Breckle and Frey (1976b) noticed relatively strong glacierization facing the east and southeast near the Pakistan border, where summer monsoon weather from India produces clouds that provide shade and slow melt. This part of Afghanistan is the site of the intertropical convergence zone (ITCZ) in summer, and is regularly influenced by monsoon precipitation (Sivall, 1977; Shroder, 1989b). The cloudiness to the southeast, especially over the glaciers in Pakistan, obscures much of the satellite imagery taken in late summer near the end of the ablation season.
Most of the glaciers of Afghanistan are mountain or valley glaciers that occur in cirques or simple basins. The higher peak areas have some compound basins. Nourishment is generally by avalanche, with the result that ice fronts are extensively debris-covered. Although there has been little study of glacier terminus activity, the large amount of debris cover gives the appearance of widespread stagnation, equilibrium, or retreat. In general, retreat is thought to be dominant throughout much of the Himalayan chain (Mayewski and Jeschke, 1979) and the Hindu Kush (Shroder, 1980).

Numerous rock glaciers (Grötzbach, 1965; Shroder and Giardino, 1978) occur in Afghanistan; the most common are tongue-shaped and spatulate types, although many lobate valley-wall types also occur. Many have glaciers at their heads and water-filled kettle depressions lower down. Internal ice movement produces a steep rubble front at the angle of repose that is characteristic of rock glaciers. Outbursts of water from some of these rock glaciers or from moraines have been reported and these may be related to water-pressure induced movement (Scott and Cheverst, 1968). Grötzbach (oral commun., 1978) observed ice lenses, interspersed with open-matrix rock fragments in which water could be trapped, in some rock glaciers in the Hindu Kush. Many observations elsewhere indicate a relation between movement of some rock glaciers and internal water (Shroder and Giardino, 1978). Analyses of rock glaciers in Afghanistan with satellite imagery have produced few results. This is largely because, even with ground-based observations, there is as yet little agreement on the differences between rock glaciers, ice-cored moraines, some types of landslides, or boulder streams (Giardino and others, 1987). In addition, many of these landforms show only minor topographic variation, which makes differentiation with satellite imagery especially difficult.

The longest and largest glaciers in Afghanistan are located in the narrowest part of the entrance to the Wākhān corridor, along the north-facing left bank of the Daryā-ye Panj. In that area, the mountains on the Afghanistan-Pakistan border are more than 6,000 m high and at least 15 of the north-flowing glaciers are more than 10 km long. Several of these are also the largest in area and, although precise measures are not yet possible, the largest of these glaciers are about 75 to 100 km² in surface area. For example, from Noshaq peak (7,492 m), a subsidiary peak of the great Tirich Mir in Pakistan, *Qādzī Deh Glacier* flows more than 14 km down to an elevation of about 3,580 m (fig. 3). More than half of the lower part of this glacier is covered extensively with black slate and argillite fragments (fig. 4) derived from the Wakhan Formation (Buchroithner,

Figure 3.—(at left) Terminus of debris-covered Qādzī Deh Glacier *in background; joined by* Ye Safed Glacier *entering on the far right (east). The varicolored* Rakhe Kuchek Glacier *in the foreground is covered in part by light-colored, probably granitic debris, and in part by dark argillite from the peak* Kharposhte Yakhi (5,698 m) *from which the photograph was taken. See figure 5 for location. Photograph by B. Ehmann.*

Figure 4.—(below) Argillite-covered glacier ice at an elevation of about 5,075 m on the Qādzī Deh Glacier *below Noshaq peak (7,492 m). See figure 5 for location. Photograph by B. Ehmann.*

1978), or the Wakhan Slates (Desio, 1975). This large quantity of supraglacier debris is provided by profuse avalanches across the weak and friable rocks that make climbing here so hazardous (B. Ehmann, verbal communication, after the Noshaq expedition, 1978). This supraglacier debris cover produces the anomalous, nearly black glacier surfaces that are so prominent on the satellite imagery (fig. 5). The lack of contrast between buried ice and surrounding slopes adds to the uncertainty of determining the true size of some of these glaciers.

In addition to the authors mentioned previously, other observations and studies of Afghanistan glaciers have been made (table 2). Detailed field study was done only by Gilbert and others (1969), Grötzbach and Rathjens (1969), and Patzelt (1978). Minor field data were provided by Grötzbach (1964), Braslau (1972), Breckle and Frey (1976b), and Desio (1975). First, von Wissman (1960), then Maksimov and Perugina (1975) worked mainly from maps, and Shroder (1980) used maps, satellite imagery, and field observations to initiate a glacier inventory following Temporary Technical Secretariat for World Glacier Inventory guidelines (Müller and others, 1977; Müller, 1978). The needs of this inventory led directly to computer processing of satellite digital data of the *Keshnīkhān Glacier*, at the mouth of the Wākhān corridor (Braslau and Bussom, 1978a, b). Numerous other minor observations of glaciers and useful photographs and planimetric maps have been produced by various scientists and mountaineers (table 2).

Mapping of Glaciers

Detailed maps do not exist for most of the glaciers in Afghanistan because of poor accessibility and an unstable military and political climate. Fortunately, reasonably high-quality topographic maps exist for the entire country, although they are difficult to obtain. These maps were produced in the late 1950s, using stereographic, vertical aerial photography that was flown to produce the maps (Glicken, 1960; Anonymous, 1960). The southern two-thirds of the country was officially mapped by the United States and the northern one-third by the former Soviet Union (Reiner, 1966), although both countries subsequently made their own complete maps of the entire country clandestinely. Maps of the country were produced at scales of 1:50,000, 1:100,000, 1:200,000, and 1:250,000. Nearly complete sets of all scales are housed in the Afghanistan collection at the University of Nebraska at Omaha, and also in the scholarly map collections of various British, German, and Russian institutions. Although several problems of incompatibility exist between the American and Russian maps (Shroder, 1980), glaciers are shown reasonably well at 1:100,000-scale, as they existed in 1958–59, when the aerial photographs were acquired. Considerably greater detail is presented on the 1:50,000 series, the old Soviet versions of which are now more generally available. The original vertical aerial photographs were restricted but are now reported to be available in Kabul, Afghanistan. For a short time in the 1960s, photomosaics and a few stereoscopic pairs were available, and a few photomosaics of the glacierized areas are held by scientists today (Breckle and Frey, 1976b).

In spite of the political changes in recent years, obtaining maps in Afghanistan remains difficult. Consequently, some agencies have implemented more transparent means of accessing topographic data (Shroder, 2004). The Afghanistan Information Management Service (AIMS) and the Afghanistan Research and Evaluation Unit were set up with backing from the United Nations and the U.S. Agency for International Development (USAID) to help with this initiative. AIMS is actively producing high-quality, GIS-based maps at many scales, and is generating Landsat-based terrain maps with a variety of GIS overlays. Most significantly for the glacier inventory, all of the old Soviet-era topographic maps at 1:50,000-scale have been scanned, overlain with Roman alphabet conversions of the former Cyrillic characters, and made available for reconstruction and assessment efforts.

Figure 5.—*(at left) Landsat 3 return beam vidicon (RBV) satellite image showing glacierized areas extending from Tajikistan and Ab-i-Panj river in the Wākhān valley in the north to Tirich Mir peak (7,706 m) in Pakistan on the extreme south. (See figure 6 for location.) The large glacier flowing from center to right is the Darband and Udren system. The large and complex glacier system in the lower right third of the image is the upper and lower Tirich glaciers. All these large glacier systems originate on the Pakistan side of the Tirich Mir massif. International border is approximate. The Landsat 3 image (30553–05112, Subscene B; 9 September 1979; Path 163–Row 35) is from the U.S. Geological Survey, EROS Data Center, Sioux Falls, S. Dak. 57198.*

TABLE 2.—*Glacier information contained in scientific and mountain-climbing (mountaineering) publications on Afghanistan*

Report	Planimetric maps	Useful photographs or topographic diagrams	Minor observations of glaciers	Detailed analyses of glaciers	Detailed maps of glaciers	Digital satellite maps	Moraines and geomorphology
Austrian scientific expedition in the Wakhan (1970, 1972)					X		
Bartolami (1972, 1973)	X	X					
Biel and Wala, (1972, 1973)	X	X	X				
Braslau (1972)		X		X			X
Braslau and Bussom (1978a)						X	
Braslau and Bussom (1978b)		X				X	
Breckle and Frey (1976a)		X					
Breckle and Frey (1976b)	X	X	X				
Chwascinski (1966)	X	X					
Desio (1962)							X
Desio (1975)	X	X	X				X
Diemberger (1966)	X	X	X				
Dozier (1970)	X	X	X				
Dunsheath and Baillie (1961)	X	X					
Exploration Pamir 75 (1978a, b)					X		
Frey (1967, 1968)	X	X					
Giger (1972, 1973)	X						
Gilbert and others (1969)	X	X		X	X		X
Grötzbach (1964)	X		X				
Grötzbach (1965)	X	X	X				
Grötzbach and Hillebrandt (1964)	X		X				
Grötzbach and Rathjens (1969)	X		X	X			X
Heiss (1972, 1973)	X	X					
Kick (1969)			X				
Klaer (1969)			X				X
Lister (1969)			X				
Maksimov and Perugina (1975)				X			X
Naumann (1974)	X						
Newby (1958)	X	X					
North (1970)	X	X	X				
Patzelt (1978)	X	X	X				X
Paytubi (1972, 1973)	X						
Rathjens (1978)			X				X
Schweizer (1969)			X				
Scott and Cheverst (1968)			X				
Scott (1972, 1973)			X				
Shroder (1980)				X			
Shroder and Guardino (1978)			X				X
Shroder and others (1978)							
Sidorov (1960)							
von Wissman (1960)				X			
Wala (1971)	X						
Wala (1973)	X	X		X			
Wala (1976a, b; 1977)	X	X					

Many of the glacierized areas occur in northeastern Afghanistan, along the high mountain borders with Pakistan and China. The Russians had trouble taking aerial photographs close to these borders and clouds obscured much of the landscape as well, so only planimetric maps produced by mountain climbers (mountaineers) were easily available for some of these border areas for many years. The comprehensive work of Wala (1971; 1973; 1976a, b; 1977) was the best source of information for many years, but several other scientists and mountaineers also contributed (table 2). Mapping of some glaciers in these areas was accomplished with imagery from the Landsat 3 return beam vidicon (RBV) sensor, because of its 30-m picture element (pixel) resolution and optical clarity, although even on these images many of the debris-covered small glaciers are difficult to discern. [Editors' note: Some of the Landsat 3 RBV images are archived by the author. The U.S. Geological Survey's EROS Data Center (EDC) no longer provides copies of Landsat 3 RBV images. EDC has permanently sequestered these historically important images, so they are no longer available for analysis in either digital or film formats.]

Only three high-quality, large-scale maps exist that show glaciers well in parts of Afghanistan. These are the excellent maps, by Kostka and his students (The Austrian Scientific Expedition in the Wakhan 1970, 1972; Exploration Pamir 75, 1978a, b), of the *Keshnīkhān Glacier* group near the west end of the Wākhān corridor as it was in 1970, and of the Afghan Pamir group in the central Wākhān as it was in 1975 (fig. 6). These maps show detailed moraine configurations, transient snowlines, accurate topography, and other important glacier landforms. The great detail of these maps at scales of 1:25,000 and 1:50,000 makes them excellent sources for verification of analyses of Landsat and other satellite imagery. The glaciological study by Gilbert and others (1969) of the *Mīr Samīr* glaciers is also useful, although the maps are far less precise than the Austrian examples.

Satellite Imagery in the 1970s

The lack of historical information, together with ongoing military and insurgent activities, increases the value of remote-sensing studies of the glaciers of Afghanistan. Satellite imagery represents the best available tool for constant and selective monitoring of glaciers (fig. 7, table 3). Of the two basic types of imagery available of Afghanistan during the 1970s, the greater resolution of the Landsat 3 RBV imagery (30-m pixels) offered more information than the Landsat 1, 2, and 3 multispectral scanner (MSS) (79-m pixels) imagery. However, few RBV images were acquired, and those are no longer available [see Editors' note in previous section].

Photographic Prints

Standard black-and-white and false-color composite prints of Landsat 1, 2, and 3 MSS imagery at a scale of 1:250,000 were used photogrammetrically in several preliminary studies for this project. P. Hearty first compared small-scale imagery of glaciers in northern Badakhshān with the Russian-made topographic maps and noted considerable differences. The apparent general retreat of glaciers was suspected to be partly an artifact of problems with scale and the absence of confirmatory field observations. T. Shipley photographically enlarged film positives of false-color composite images to a scale of about 1:175,000 and established mutual registration points for overlay with various topographic maps and the glacier maps of Kostka (Exploration Pamir 75, 1978a, b). Shipley thought he detected a slight advance of glacier fronts between the 1972 imagery and Kostka's field survey in 1975.

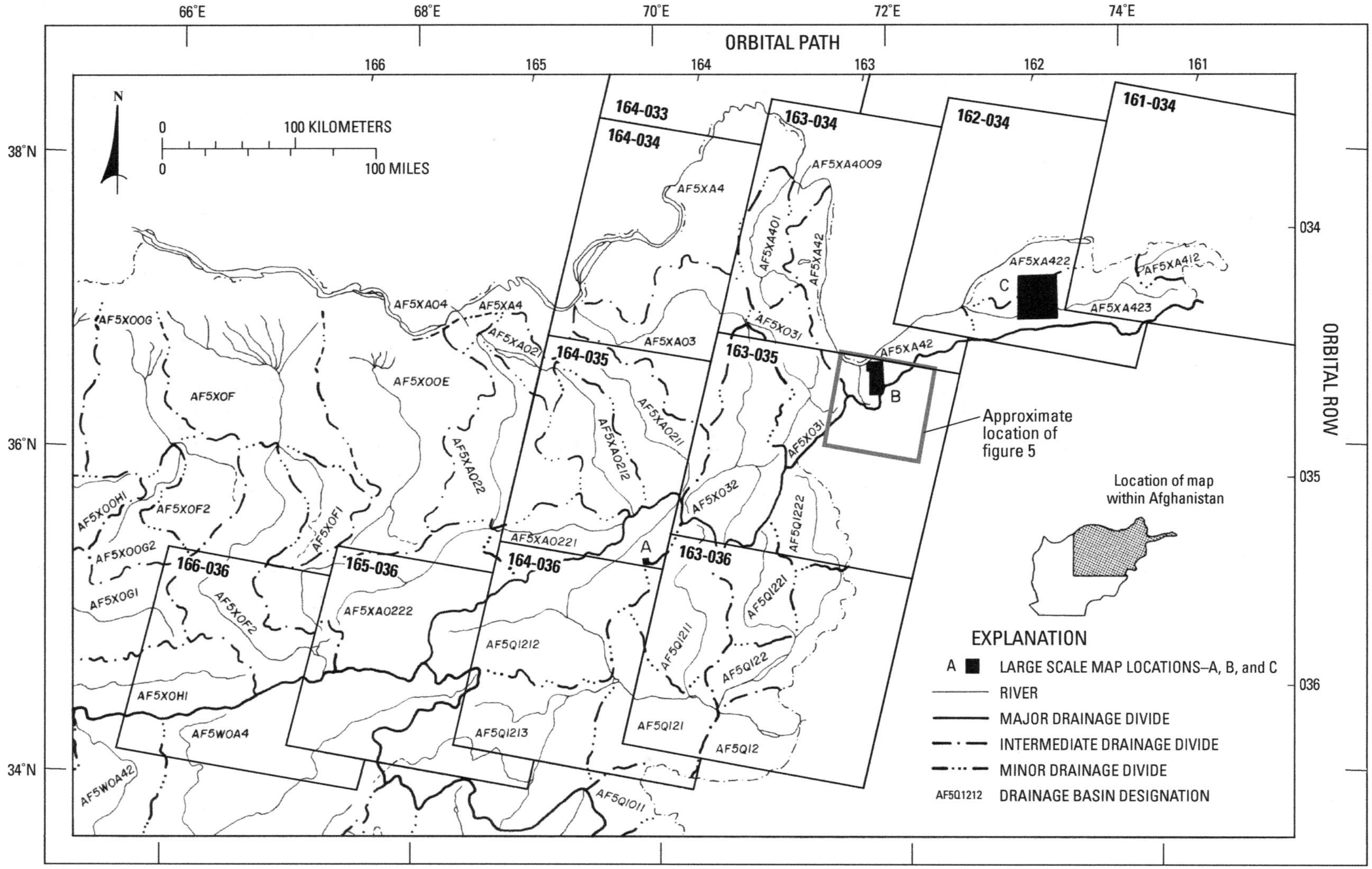

Figure 6.—Index map of Landsat 1, 2, and 3 satellite imagery, drainage basins, and large-scale maps of glacierized areas of Afghanistan. Satellite images are listed by Path–Row numbers and keyed to table 3. Large-scale maps are listed by letters keyed to references: A – Gilbert and others, 1969; B – Austrian Scientific Expedition in the Wakhan 1970, 1972; C – Exploration Pamir 75, 1978a, b.

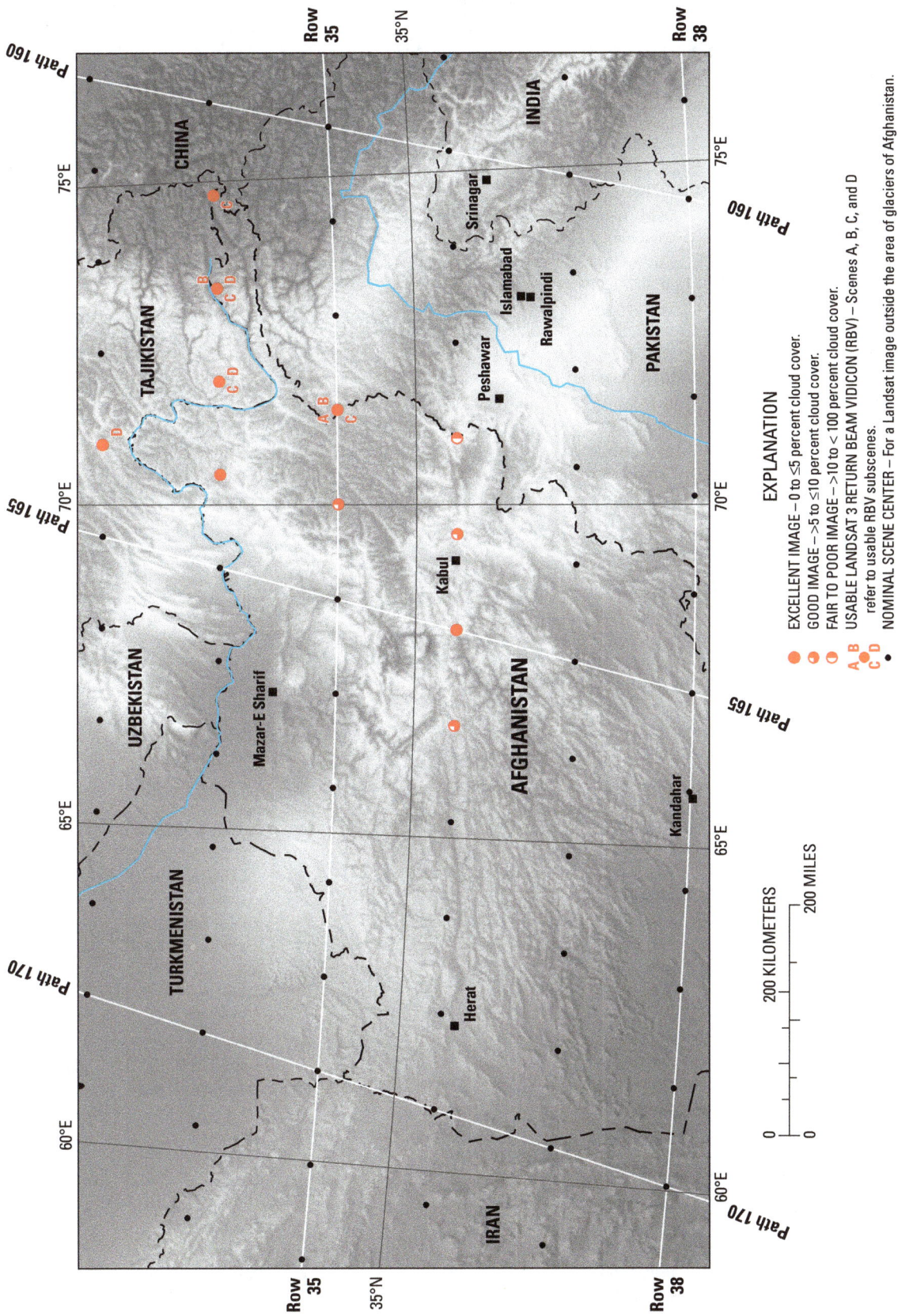

Figure 7.—Map of optimum Landsat 1, 2, and 3 images of the glaciers of Afghanistan.

EXPLANATION

EXCELLENT IMAGE – 0 to ≤5 percent cloud cover.

GOOD IMAGE – >5 to ≤10 percent cloud cover.

FAIR TO POOR IMAGE – >10 to < 100 percent cloud cover.

USABLE LANDSAT 3 RETURN BEAM VIDICON (RBV) – Scenes A, B, C, and D refer to usable RBV subscenes.

NOMINAL SCENE CENTER – For a Landsat image outside the area of glaciers of Afghanistan.

TABLE 3.—*Optimum Landsat 1, 2, and 3 images of the glaciers of Afghanistan*

[Code column indicates usability for glacier studies; see figure 7 for explanation]

Path-Row	Actual scene center (latitude and logitude)		Landsat identification number (9-10 digits) or entity identification number (16 digits)[1]	Date	Solar elevation angle (in degrees)	Code	Remarks
161–34	37°31′N.	74°41′E.	2564–04522 (2161034007622190)	08 Aug 76	53	◑	No band 4
161–34	37°24′N.	74°34′E.	2924–04370 (2161034007721590)	03 Aug 77	51	◑	Good in Afghanistan. Clouds to north
161–34	37°06′N.	74°00′E.	30515–04595 Subscene C	02 Aug 79	55	●	Landsat 3 RBV image, archived by author
162–34	37°21′N.	73°13′E.	1047–05171 (1162034007225290)	08 Sep 72	50	◑	Few clouds
162–34	37°28′N.	73°12′E.	2565–04580 (2162034007622290)	09 Aug 76	53	●	Excellent in Afghanistan
162–34	37°24′N.	73°08′E.	2601–04570 (2162034007625890)	14 Sep 76	45	◑	Few clouds
162–34	37°35′N.	73°44′E.	30516–05053 Subscene B	03 Aug 79	55	◑	Landsat 3 RBV image, archived by author
162–34	37°01′N.	72°33′E.	30516–05053 Subscene C	03 Aug 79	55	◑	Landsat 3 RBV image, archived by author
162–34	36°52′N.	73°30′E.	30516–05053 Subscene D	03 Aug 79	55	◑	Landsat 3 RBV image, archived by author
163–34	37°32′N.	71°38′E.	1354–05231 (1163034007319390)	12 Jul 73	60	◑	Preceding winter snow
163–34	37°30′N.	71°47′E.	2566–05034 (2163034007622390)	10 Aug 76	53	●	
163–34	37°31′N.	71°46′E.	2584–05031 (2163034007624190)	28 Aug 76	49	●	
163–34	37°12′N.	71°08′E.	30553–05105 Subscene C	09 Sep 79	48	●	Landsat 3 RBV image, archived by author
163–34	37°04′N.	72°04′E.	30553–05105 Subscene D	09 Sep 79	48	●	Landsat 3 RBV image, archived by author
163–35	36°05′N.	71°11′E.	1354–05233 (1163035007319390)	12 Jul 73	61	●	Excellent in Afghanistan
163–35	34°04′N.	71°19′E.	2566–05041 (2163035007622390)	10 Aug 76	53	●	Excellent in Afghanistan
163–35	36°07′N.	71°19′E.	2980–04454 (2163035007727190)	28 Sep 77	39	◑	Shadows too large in cirques
163–35	36°29′N.	70°55′E.	30553–05112 Subscene A	09 Sep 79	48	●	Landsat 3 RBV image, archived by author
163–35	36°21′N.	71°50′E.	30553–05112 Subscene B	09 Sep 79	48	◑	Few clouds. Landsat 3 RBV image, archived by author
163–35	35°46′N.	70°42′E.	30553–05112 Subscene C	09 Sep 79	48	●	Landsat 3 RBV image, archived by author
163–36	34°39′N.	70°44′E.	1354–05240 (1163036007319390)	12 Jul 73	61	◐	Clouds and old snow
163–36	34°38′N.	70°53′E.	2566–05043 (2163036007622390)	10 Aug 76	53	◐	Cloud cover
163–36	34°39′N.	70°53′E.	2584–05040 (2163036007624190)	28 Aug 76	50	◐	Cloud cover
163–36	34°33′N.	70°49′E.	2926–04991 (2163036007721790)	05 Aug 77	51	◐	Cloud cover
164–33	38°53′N.	70°53′E.	1049–05281 (1164033007225490)	10 Sep 72	48	●	Excellent in Afghanistan

[Code column indicates usability for glacier studies; see figure 7 for explanation]

Path-Row	Actual scene center (latitude and logitude)	Landsat identification number (9-10 digits) or entity identification number (16 digits)[1]	Date	Solar elevation angle (in degrees)	Code	Remarks
164–33	38°49′N. 70°47′E.	2531–05100 (2164033007618890)	06 Jul 76	57	◐	Preceding winter snow
164–33	38°48′N. 70°45′E.	2909–04544 (2164033007720090)	19 Jul 77	53	◐	Good in Afghanistan
164–33	38°30′N. 71°07′E.	30554–05161 Subscene D	10 Sep 79	47	●	Landsat 3 RBV image, archived by author
164–34	37°28′N. 70°24′E.	1049–05283 (1164034007225490)	10 Sep 72	49	●	
164–34	37°31′N. 70°21′E.	2567–05092 (2164034007622490)	11 Aug 76	52	●	
164–34	37°30′N. 70°18′E.	2585–05085 (2164034007624290)	29 Aug 76	49	●	
164–34	37°23′N. 70°17′E.	2909–04550 (2164034007720090)	19 Jul 77	53	◐	Preceding winter snow
164–34	37°33′N. 70°20′E.	2963–04520 (2164034007725490)	11 Sep 77	43	●	
164–35	36°02′N. 69°56′E.	1049–05290 (1164035007225490)	10 Sep 72	50	◐	Few clouds
164–35	36°02′N. 70°01′E.	1067–05290 (1164035007227290)	28 Sep 72	45	◑	Marginal. New snow
164–35	36°07′N. 69°53′E.	2963–04523 (2164035007725490)	11 Sep 77	44	◑	Good in band 7 only
164–36	34°36′N. 69°29′E.	1049–05292 (1164036007225490)	10 Sep 72	51	◐	Few clouds
164–36	34°36′N. 69°34′E.	1067–05292 (1164036007227290)	28 Sep 72	46	◑	Marginal. New snow
164–36	34°40′N. 69°26′E.	2963–04525 (2164036007725490)	11 Sep 77	44	◑	Marginal. Clouds, band 7 only
165–36	34°39′N. 67°55′E.	1410–05343 (1165036007324990)	06 Sep 73	51	●	
165–36	34°39′N. 67°59′E.	2982–04573 (2165036007727390)	30 Sep 77	40	●	
166–36	34°35′N. 66°42′E.	1069–05405 (1166036007227490)	30 Sep 72	45	◐	Few clouds

Digital Analyses

Several different digital analyses of the various glacierized areas in Afghanistan were made in the 1970s, using computer-compatible tapes (CCTs). Following these initial analyses (Shroder and others, 1978), a full research program to perform a satellite glacier inventory of Afghanistan was initiated. To further develop the methodology, several areas outside Afghanistan were selected by Rundquist (1978), in consultation with Fritz Müller, a Swiss glaciologist. Braslau and Bussom (1978a, b) worked on the *Keshnīkhān Glacier* because of the initial experimentation there, and because of Kostka's excellent base map (The Austrian Scientific Expedition in the Wakhan 1970, 1972). In their analyses, the *Keshnīkhān Glacier* was classified according to 11 symbols, each representing some number of pixel radiance values.

Radiance thresholds for each of the four bands were constantly adjusted until the computer output and the Kostka map bore a close resemblance to one another. At one point, in response to collegial criticism of the technique, Bussom manually rectified a computer printout to remove the skewed distortion of the image in order to demonstrate the close geometric registration possible between the two. Finally, a random sample of about 14 percent of the pixels was tested, using multivariate linear discriminate analysis to derive the percentage of pixels correctly classified within each band. About 90 percent accuracy was typically achieved. Results of the *Keshnīkhān Glacier* analysis was significant because the glacier is similar to the majority of glaciers in Afghanistan, and the successful computer mapping indicates the feasibility of monitoring other nearby glaciers where field-based observations have not been made or where adequate maps do not exist.

The preliminary success with digital analyses of CCTs were misleading in some cases, however, and a number of problems emerged:

1. The techniques and equipment were expensive, time consuming, and prone to operator error when the level of complexity eclipsed realistic interpretations.

2. The original hand-colored thresholding techniques of 25 years ago were inexpensive and made the technology easily available, but were too time-consuming to be of much real use, and are now outmoded technology.

3. The Landsat CCTs commonly deteriorated in storage (degaussed) over a few years if not periodically copied, and several of the Afghanistan tapes purchased in 1977–78 were unusable by the mid-1980s.

Nevertheless, thresholding done by skilled operators using the equipment available at many universities (Rundquist, 1978; Rundquist and Samson, 1980; Rundquist and others, 1980) indicated a potential for using satellite data for carrying out glacier inventories by applying more modern technologies.

Satellite Imagery since the 1970s

A plethora of new satellite images of Afghanistan have become available since the 1970s; these include SPOT (Satellite Pour l'Observation de la Terre), TM (Landsat Thematic Mapper), and ASTER (Advanced Spaceborne, Thermal Emission and Reflection Radiometer), some of which were also used in this study. As the USGS-designated GLIMS Regional Center for Southwest Asia (Afghanistan and Pakistan), the University of Nebraska at Omaha is undertaking regional glacier inventories of the Hindu Kush and western Himalaya, along with glacier-change detection with these satellite-image resources. The acquisition of terrabytes of high-resolution (15-m) stereographic data is enabling production of digital elevation models (DEMs) of these mountain regions with the ASTER system in the GLIMS project, and is allowing unprecedented glacier assessments (Bishop and others, 2000; Kieffer and others, 2000).

Case Studies of Glaciers

Five areas of Afghanistan were selected as most representative of the variations in size, character, altitude, and radiance values of the glaciers. Small glaciers of the *Kūh-e-Bābā* are the westernmost in the country. The *Mīr Samīr* region has a number of small glaciers that were studied glaciologically and mapped planimetrically by a British team in 1965 (Gilbert and others, 1969). *Kūh-e-Bandakā* has glaciers of small and intermediate size, with a general character typical of the central Hindu Kush. Glaciers at the entrance to the Wākhān corridor are either small, as in the case of the *Keshnīkhān*, or large, and atypically dark colored, as in the case of the *Qādzī Deh Glacier*. Glaciers in the Wākhān Pamirs are noteworthy because of their considerable volume, in spite of the area having the lowest annual precipitation in the high mountains of Afghanistan.

Kūh-e-Bābā

Most of the glaciers of the more habitable parts of Afghanistan are small remnants of the larger Pleistocene ice masses. They are difficult to detect on the small-scale satellite imagery and are easily confused with clouds and rocks of high reflectivity. Also, the pixel resolution of the older imagery was commonly insufficient to detect what could be significant changes in small glaciers. However, these smaller ice masses are as important to the local economies as are the larger glaciers. They are essential sources of late summer irrigation water, and considerable human effort is required at high altitudes each spring to maintain the makeshift diversion dams and ditches to exploit this water. At lower elevations, entire villages turn out to repair the irrigation structures along the major snow- and glacier-fed larger rivers.

A total of 18 small glaciers occur in the *Kūh-e-Bābā* range of central Afghanistan, averaging about 0.5 km^2 in area, and all have northern exposures (Shroder and Giardino, 1978). The range of the lowest elevation of exposed glacial ice is 4,075–4,657 m, with an average of 4,365 m. All of the glaciers terminate in large, tongue-shaped boulder deposits that can be variously classified as debris-covered glaciers, ice-cored moraines, or ice-cored rock glaciers, depending upon certain fine distinctions made between relative activity or inactivity, steep fronts or gentle fronts, and general surface morphologic differences. The lower altitude range of these rocky termini is 3,850–4,600 m, with an average of 4,232 m. In the late Pleistocene, these glaciers are believed to have terminated at elevations as low as 2,600 m near Burghasūn, in the Fūlādī valley above Bāmyān (figs. 8 and 9).

During July 1978, *Fūlādī Glacier* was visited by the senior author (fig. 10), and it had reduced in surface area by about 20 to 25 percent since the 1959 aerial photographs. Mean annual precipitation at *Fūlādī Glacier* is about 600 mm, and the climatic snowline is about 5,100 m (Grötzbach and Rathjens, 1969). This snowline is well above the tops of all the surrounding peaks except Fūlādī peak (5,150 m), so *Fūlādī Glacier* only survives in the shadow of the cirque walls on the northern exposures. It receives maximum exposure to Sun during midsummer. A negative mass balance on this glacier is evident based on the location of the transient snowline high in the cirque close to the bergschrund, the concave-up cross profile of the ice surface, and the extensive till cover on the margins of the ice.

Low-altitude aerial reconnaissance of the *Kūh-e-Bābā* in 1977 and topographic maps revealed 35 boulder deposits that do not have surficial ice exposed at their heads. Three have large kettle-like lakes on them, and many have major springs, all indicating the presence of buried ice. Twenty nine of these rock-glacier-like features face north, four face east, and one each faces south and west. The lower elevation range is 3,650–4,425 m, with an average of 4,078 m.

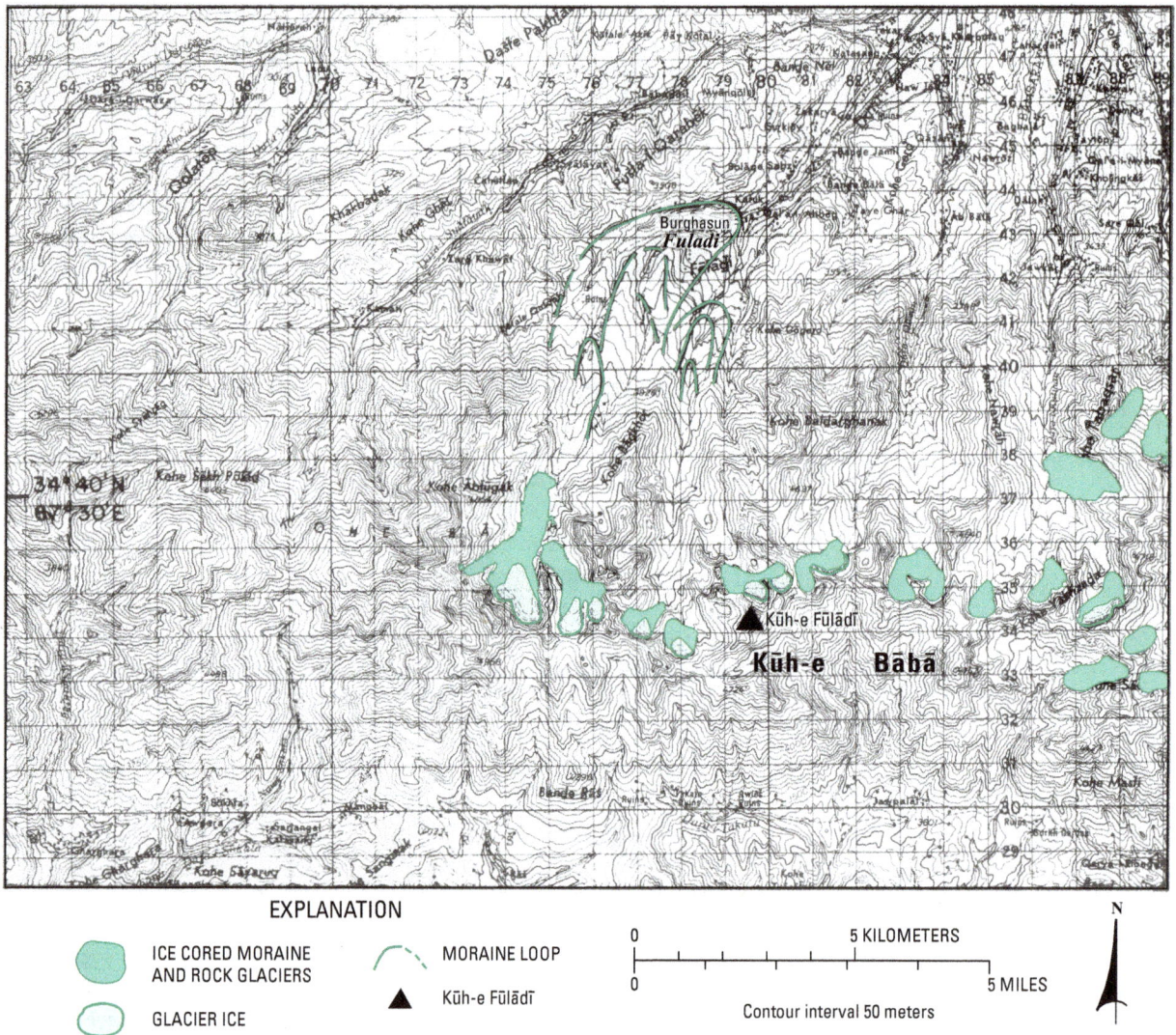

EXPLANATION

ICE CORED MORAINE AND ROCK GLACIERS		MORAINE LOOP	
GLACIER ICE		▲ Kūh-e Fūlādī	

0   5 KILOMETERS

0   5 MILES

Contour interval 50 meters

N

Figure 8.—*Topographic map of* Fūlādī *valley and central* Kūh-e Bābā *range.* Fūlādī Glacier *is 0.6 km directly north of* Fūlādī *peak. The largest moraines emanate mainly from the western valley between* Ābtowgak *and* Bādkhūr *peaks; the eastern valley between* Bādkhūr *and* Bāldārghanak *peaks produced the smaller set of moraine loops. Geographic place-names on map may be variant.*

Analysis and monitoring with early satellite imagery of these small ice glaciers (fig. 11) and buried-ice glaciers or rock glaciers was difficult, and results were ambiguous for the following reasons:

- Retreat of the *Fūlādī Glacier* in 20 years was only a few partial MSS pixels in size and thus difficult to detect or compare with newer, higher resolution ASTER imagery.

- Strong shadows in the cirques obscured the ice or mimicked small water bodies in certain places.

- Debris cover commonly masked the buried ice.

- Tills, talus, and fractured source bedrock all appeared to have similar radiance values and thus were difficult to distinguish.

- Clouds and rocks of high reflectivity mimicked ice and snow.

Figure 9.— View northwest from eastern end moraine toward large moraine loop above the village of Burghasūn. Compare with figure 8 for location, about 9 km north of Fūlādī peak. Photograph by J.F. Shroder, Jr.

Figure 10.— Fūlādī Glacier *in July 1978. At this time, the transient snow line was close to or at the base of the steep firn slopes and the bergschrund below the large rock tower on the shoulder of Fūlādī peak. See figure 8 for location. The rocks of this area are granitic intrusions of early Tertiary age. Photograph by J.F. Shroder, Jr.*

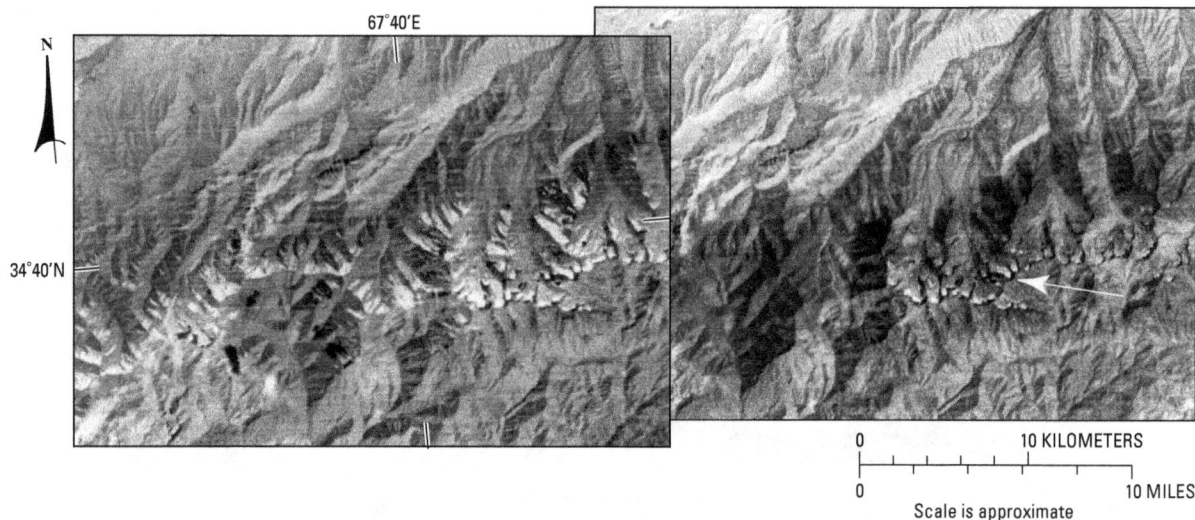

Figure 11.—*Stereoscopic pair of Landsat 1 images of the Kūh-e Bābā area. Arrow designates Fūlādī Glacier. Compare with figure 12, and with figure 8 for location. Landsat 1 MSS images (west, Landsat 1 MSS image 1069–05405, band 5; 30 September 1972; Path 166–Row 36; east, Landsat 1 MSS image 1410–05343, band 5; 6 September 1973; Path 165–Row 36) are from the U.S. Geological Survey, EROS Data Center, Sioux Falls, S. Dak. 57198. [Editors' note: Use stereoscope for viewing.]*

Nevertheless, satellite imagery of an area like the Fūlādī valley can be useful, especially in stereoscopy. In comparison with the topographic map (fig. 8), the stereoscopic views of figure 12 enable better determinations of ice, moraine, lakes, and other features merely because of their topographic location. The older French satellites (for example, SPOT) and the newer ASTER imagery have provided a regular source of stereoscopic imagery that facilitates analysis of high mountain areas and glaciers. Careful comparison of the 1973 MSS imagery (fig. 11) with the ASTER stereopair (fig. 12) reveals that *Fūlādī Glacier* has continued to downwaste so that it is highly incised inside its curvilinear terminal moraine. A small lake inside the moraine in 1973 was not observed in 1978 or 2003.

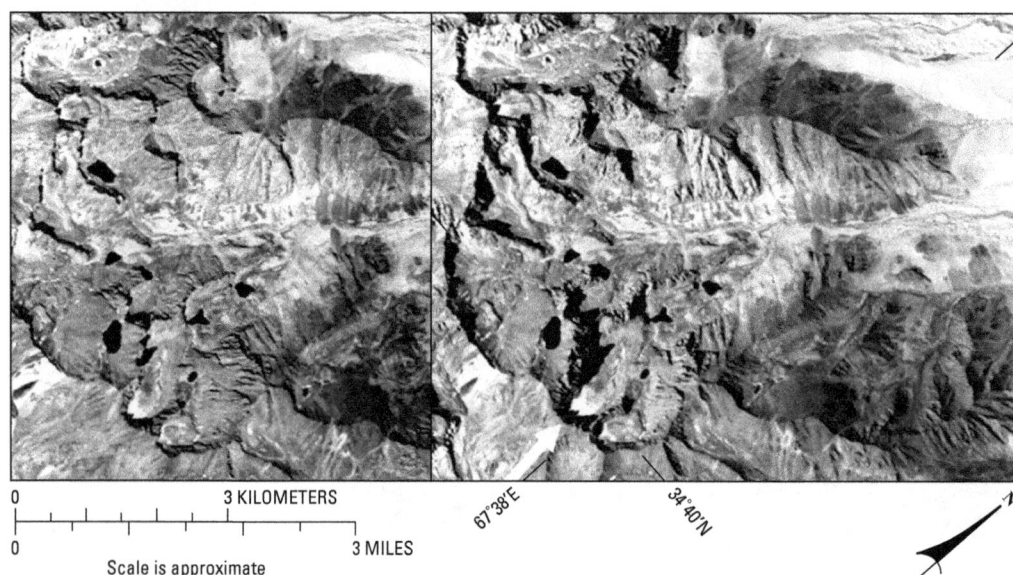

Figure 12.—*Stereoscopic pair of ASTER images of Kūh-e Bābā area. White arrow points to Fūlādī Glacier. North is to the right. Compare with figure 11. ASTER Scene AST_L1B-003_10022003061744_10162003100406; 2 October 2003. [Editors' note: Use stereoscope for viewing.]*

Mīr Samīr

The *East Glacier* (*Yakhchaal-i-Sherq*) and *West Glacier* (*Yakhchaal-i-Gharb*), below the north face of *Mīr Samīr* peak (5,809 m) in the central Hindu Kush (fig. 13), were first assessed glaciologically by a British team in 1965 (Gilbert and others, 1969). They studied snow accumulation, ablation, meltwater discharge, lichenometry, micrometeorological heat balance, and other glaciologic characteristics. Their simple planimetric maps of ice distribution and

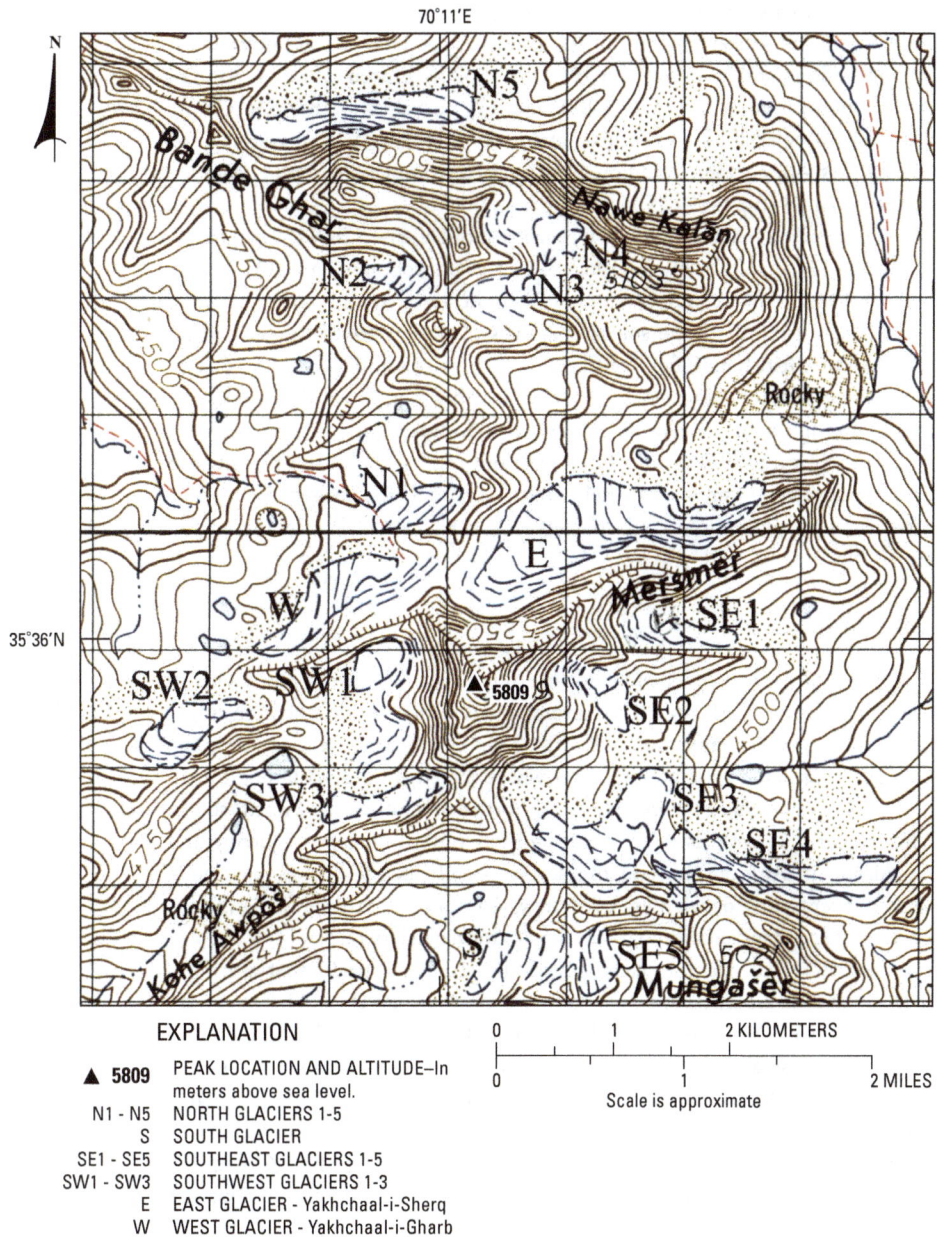

EXPLANATION

▲ **5809**   PEAK LOCATION AND ALTITUDE—In meters above sea level.
N1 - N5   NORTH GLACIERS 1-5
S   SOUTH GLACIER
SE1 - SE5   SOUTHEAST GLACIERS 1-5
SW1 - SW3   SOUTHWEST GLACIERS 1-3
E   EAST GLACIER - Yakhchaal-i-Sherq
W   WEST GLACIER - Yakhchaal-i-Gharb

Figure 13.—*Topographic map of Mīr Samīr (Mērsmēr) area in the central Hindu Kush (source: U.S. Department of Defense (DOD) U.S. Army Topographic command; compiled in 1970 from Afghanistan, 1:100,000, Ministry of Mines, Fairchild Aerial Surveys, Sheets 505A & 505C, 1967 [surveyed 1957–59, reliability fair]. Map not field checked.*

lichen zones on the moraine allow a comparison with satellite imagery for change detection. Gilbert and others (1969) estimated that the *East Glacier* measured about 1 km² and the *West Glacier* measured about 0.5 km². These small glaciers could not be assessed with much precision with the early Landsat imagery but are clearly shown on the new 15-m resolution ASTER imagery. The five new images, obtained in the past two years between July and November, show the *Mīr Samīr* glaciers quite clearly. The ASTER image of 12 November 2002 is noteworthy because no new snow appears to have fallen so late in the melt season before winter amelioration began again (fig. 14). It is also possible that any earlier light snowfall could have already melted away by the time this scene was taken (Kravtsova, 1990). Two small linear proglacial lakes at the western margin of the main part of *West Glacier*, both about 60–100 m long and 75 m apart in 1965, by 2002 had fused into one rounded lake about 30 m in diameter. The 2002 imagery also showed that *East Glacier* has a new proglacial lake, measuring about 60 m long and 15 m wide, which did not exist 40 years ago. Additional lakes have also appeared around *Southeast Glaciers* 3 and 4 (fig. 14).

Comparison between the U.S. Department of Defense (DOD) topographic map (fig. 13), made from the 1957–59 aerial photographs, and the British map of 1965 (Gilbert and others, 1969), indicates that the two maps were likely based upon the same source materials (Glicken, 1960; Reiner, 1966), although the British modified contours and glacier size and location according to their field observations. A number of possibly significant differences are visible through time. The *North Glacier* (N1), between *East* and *West Glaciers*, is prominent on the DOD map (1957-59) but minuscule on the British map (1965) and is only a minor snowfield on the 2002 ASTER imagery. Numerous other similar diminutions and changes in glaciers and lakes in the region are visible

Kūh-e-Bandakā

The *Kūh-e-Bandakā* (Bandakhor in some usages) group of mountains of the central Hindu Kush consists of several nearly parallel peaks and high ridges of Precambrian metamorphic rocks (fig. 15). These mountains are basically fault slivers isolated by strike-slip shear where the edge of the colliding Indian and Asian plates changes direction from northeast-southwest to north-south. This massif is directly astride the earthquake zone of highest frequency and magnitude in Afghanistan, and the several small- and few large-magnitude seismic events that occur there every few years may contribute a significant portion of snow avalanche and debris load to local glaciers.

The *Bandakā* massif is characterized by six major peaks that are more than 6,000 m in elevation and many others almost as high (fig. 16). The north-facing slopes of the main peak of *Bandakā* (6,843 m) are small and therefore its glaciers are smaller than those on the lower peak, *Bandakā Sakhī* (6,414 m), 4 km to the north. *Sakhī* peak has a large cirque and steep snowfield which is the source of avalanches on the north side that generate the *Sakhī Glacier* — the largest ice mass in the area. Comparison of the 1959 map (fig. 16) with the satellite imagery of the early 1970s (fig. 15) shows about 0.5 km of apparent retreat of the exposed ice front.

Meltwater from the *Sakhī Glacier* flows into the *Jay* river, which passes through several moraine areas and into two 1- to 2-km-long, flat-floored and boulder-covered, intermittent lake beds (Scott and Cheverst, 1968) before joining the main *Darreh-ye Sakhī* (river), which comes from another valley to the west.

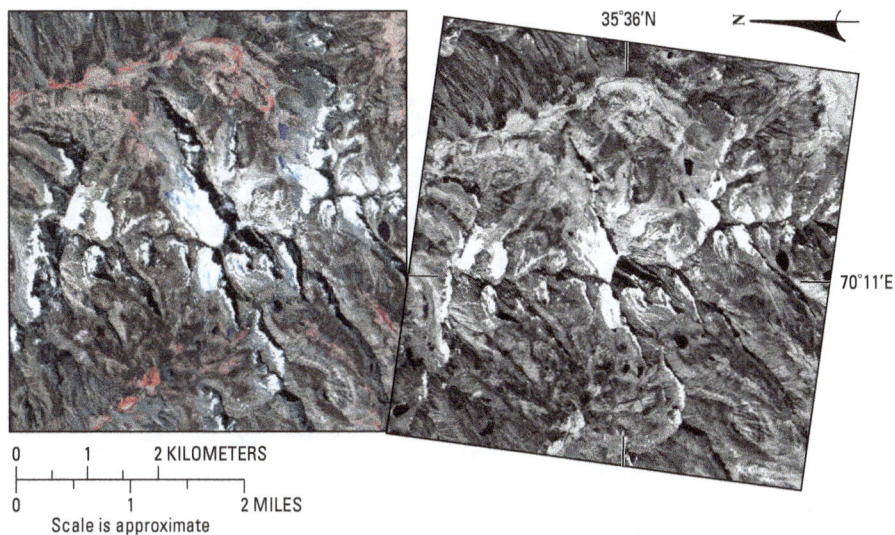

Figure 14.—Stereoscopic pair of ASTER images of Mīr Samīr area. North is to the left. The image portrays the same area as figure 13. ASTER scene AST_L1B_003_ 08172004061706_083020041222 36; 17 August 2004. [Editors' note: Use stereoscope for viewing.]

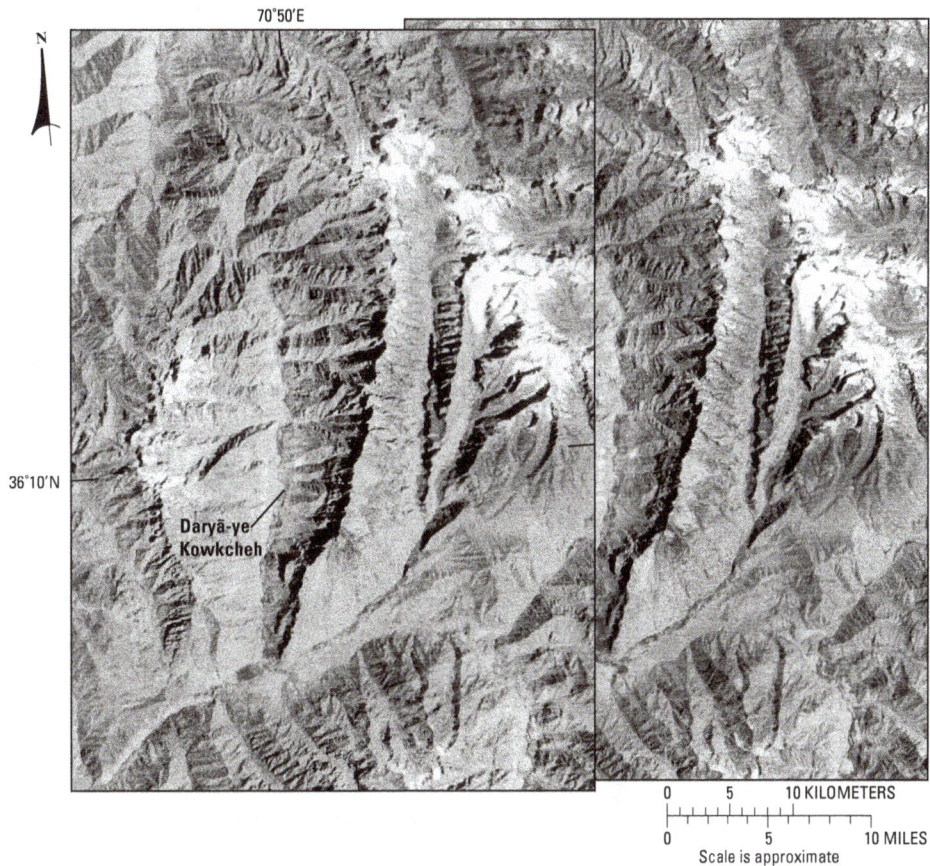

Figure 15.—Stereoscopic pair of Landsat 1 and 2 images of the Bandakā massif. The westernmost glacierized range on the left bank of the Daryā-ye Kowkcheh (Kokcha river) has peaks up to 5,862 m in elevation. Features to the east of the river may be compared with the map of figure 16. Landsat 1 and 2 MSS images (west, Landsat 1 MSS image 1049–05290, band 5; 10 August 1972; Path 164–Row 35; east, Landsat 2 MSS image 2566–05041, band 5; 10 August 1976; Path 163–Row 35) are from the U.S. Geological Survey, EROS Data Center, Sioux Falls, S. Dak. 57198. [Editors' note: Use stereoscope for viewing.]

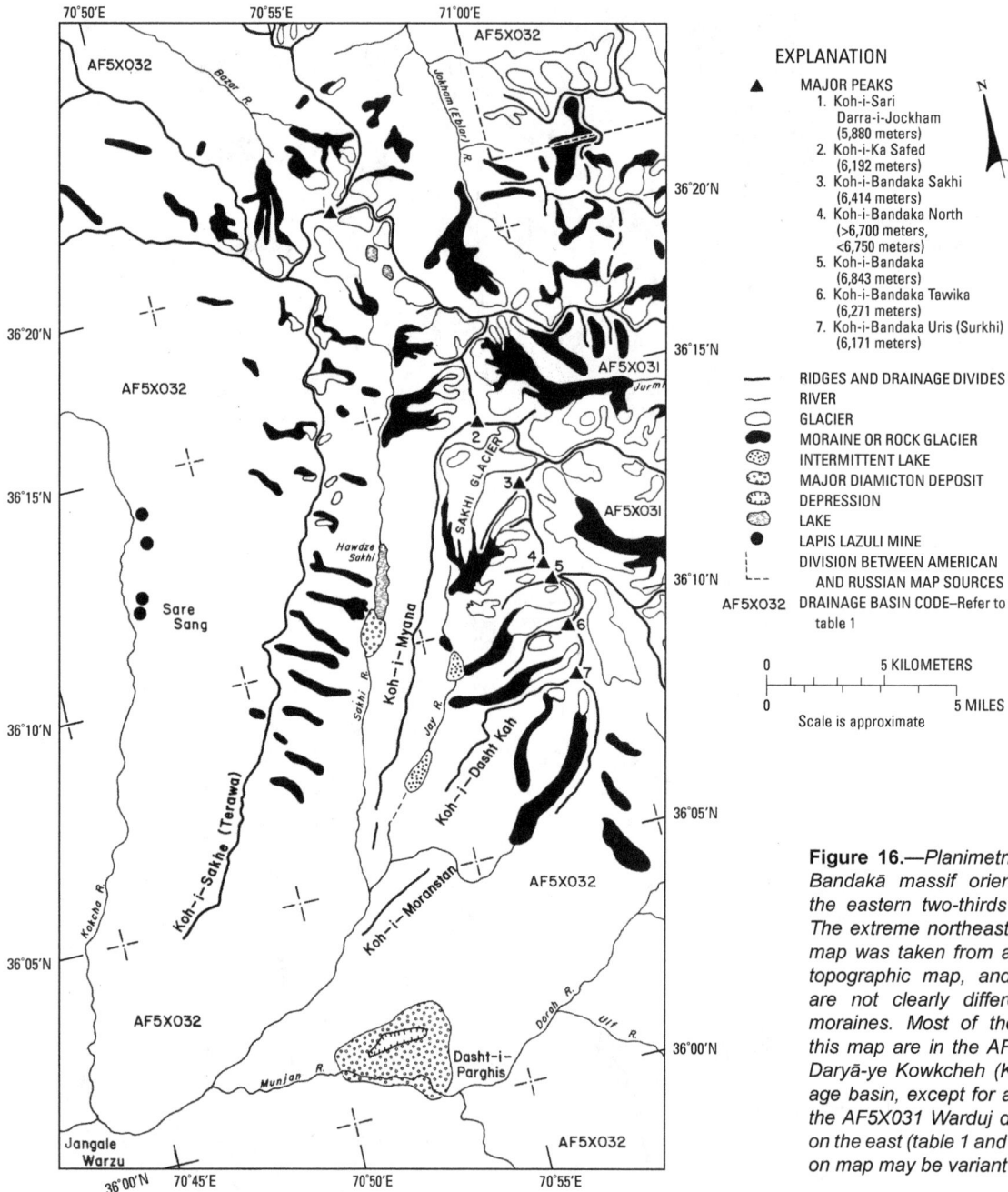

EXPLANATION

▲ MAJOR PEAKS
1. Koh-i-Sari
Darra-i-Jockham
(5,880 meters)
2. Koh-i-Ka Safed
(6,192 meters)
3. Koh-i-Bandaka Sakhi
(6,414 meters)
4. Koh-i-Bandaka North
(>6,700 meters,
<6,750 meters)
5. Koh-i-Bandaka
(6,843 meters)
6. Koh-i-Bandaka Tawika
(6,271 meters)
7. Koh-i-Bandaka Uris (Surkhi)
(6,171 meters)

—— RIDGES AND DRAINAGE DIVIDES
—— RIVER
⬭ GLACIER
⬤ MORAINE OR ROCK GLACIER
⬚ INTERMITTENT LAKE
⬚ MAJOR DIAMICTON DEPOSIT
⬚ DEPRESSION
⬚ LAKE
● LAPIS LAZULI MINE
╌╌ DIVISION BETWEEN AMERICAN
AND RUSSIAN MAP SOURCES
AF5X032 DRAINAGE BASIN CODE—Refer to
table 1

0   5 KILOMETERS
0   5 MILES
Scale is approximate

Figure 16.—*Planimetric map of the Bandakā massif oriented to cover the eastern two-thirds of figure 15. The extreme northeast corner of the map was taken from a Soviet-made topographic map, and the glaciers are not clearly differentiated from moraines. Most of the glaciers on this map are in the AF5X032 Upper Daryā-ye Kowkcheh (Kokcha) drainage basin, except for a small part of the AF5X031 Warduj drainage basin on the east (table 1 and fig. 2). Names on map may be variant names.*

The *Sakhī* river has its origin in glaciers and glacial lakes at the head of a south-facing, 4-km-wide, compound cirque basin below the peak, *Kūh-e-Sari Darreh-ye-Jow Khām* (5,880 m). *Hawdze Sakhī* (figs. 15 and 16) is a 3-km-long lake impounded behind a probable moraine dam; the *Sakhī* river passes through the moraine to re-emerge 1.5 km below.

In addition to several medium-sized glaciers around the *Bandakā* massif, there are also a large number of ice-cored moraines and rock glaciers in all the mountain ranges in the area. These features are covered with rock fragments that have nearly the same radiance values as the surrounding slopes. The result is that differentiation on the satellite images can be difficult unless collateral information is available, such as stereoscopic overlap (fig. 15), high-quality topographic maps, or digital analyses in selected cases.

The Dasht Parghish diamicton deposit, at an elevation of 2,750–3,250 m, sits astride the Monjān river and is prominent on maps and satellite imagery (fig. 12). Several large depressions up to 1.5 km long are evident, and suggest that the deposit may be a remnant end moraine produced by a major Pleistocene glacier from the Monjān or Dorāh valleys. Alternatively, the deposit may be a fault- or landslide-generated mass, inasmuch as it is located directly across a major strike-slip fault zone in an area of high seismic activity.

Keshnīkhān Glacier

The *Keshnīkhān Glacier*, at the western end of the Wākhān corridor, was chosen for inclusion here because it has been extensively studied (Braslau, 1972; The Austrian Scientific Expedition in the Wakhan 1970, 1972; Braslau and Bussom, 1978a, b; and Shroder and others, 1978), and because it is similar to the majority of glaciers in Afghanistan. It is a relatively small, steep ice mass, 15 km^2 in area and 4 km long (fig. 17). Braslau (1972) was impressed with the continuous belt of morainic material from 2,600 m elevation near the Daryā-ye Panj river valley, all the way up to the ice front — a vertical range of almost 2,000 m. There is, however, a major break in slope in this long moraine at about 3,600 m elevation, the place chosen by the Russian cartographers to delineate the terminus of the glacier ice/moraine mass, although they set the elevation at about 3,460 m. The actual stagnant terminus occurred at about 4,400 m in 1970 and the transient snowline was at 5,000 m. The actual equilibrium line altitude (ELA) at that time was estimated to be about 5,100–5,200 m. Based on elevation, a small ablation zone, and other general glacial characteristics, the glacier was judged to be receding (Braslau, 1972).

The original Landsat MSS imagery of 1973 was assessed in detail by Braslau and Bussom (1978a, b), as noted above in the section on "Digital analyses." Therefore, the 12 July 1973 image (fig. 18) is included for comparison to the ASTER image of 31 August 2000 (fig. 19). Because of the additional month and a half of melt season between the two scenes, the variance between the obvious transient snowlines is not viewed as significant of long-term trends. Nonetheless, a comparison of the approximate ELA on the 1970 image (fig. 17) with the ELA on the 2000 ASTER image (fig. 19) shows the likely uphill movement of the ELA on the left (western) side and a new emergence of medial moraine-producing rocks on the right (east).

Wākhān Pamir Glaciers

The part of the Pamirs in the middle of the Wākhān panhandle of Afghanistan has nine peaks in the central part of the range that are more than 6,000 m high. The most areally extensive glaciers of Afghanistan occur there because of the large size and general high altitude of the massif, in spite of a mean annual precipitation of less than 100 mm (Lalande and others, 1974). Glacial meltwater from the Wākhān Pamir Mountains feeds the Pāmīr, Wākhān, and Daryā-ye Panj tributaries which flow into the major Amu Darya river of Tajikistan and Uzbekistan, and its Garagum (Karakum) canal outtake into Turkmenistan. The core of the range is light-colored granitic rocks of Late Cretaceous-Tertiary age, with peripheral rocks of the Wakhan Formation (Buchroithner, 1978). Most of the large moraines therefore are composed of light-colored granitic rocks (fig. 20).

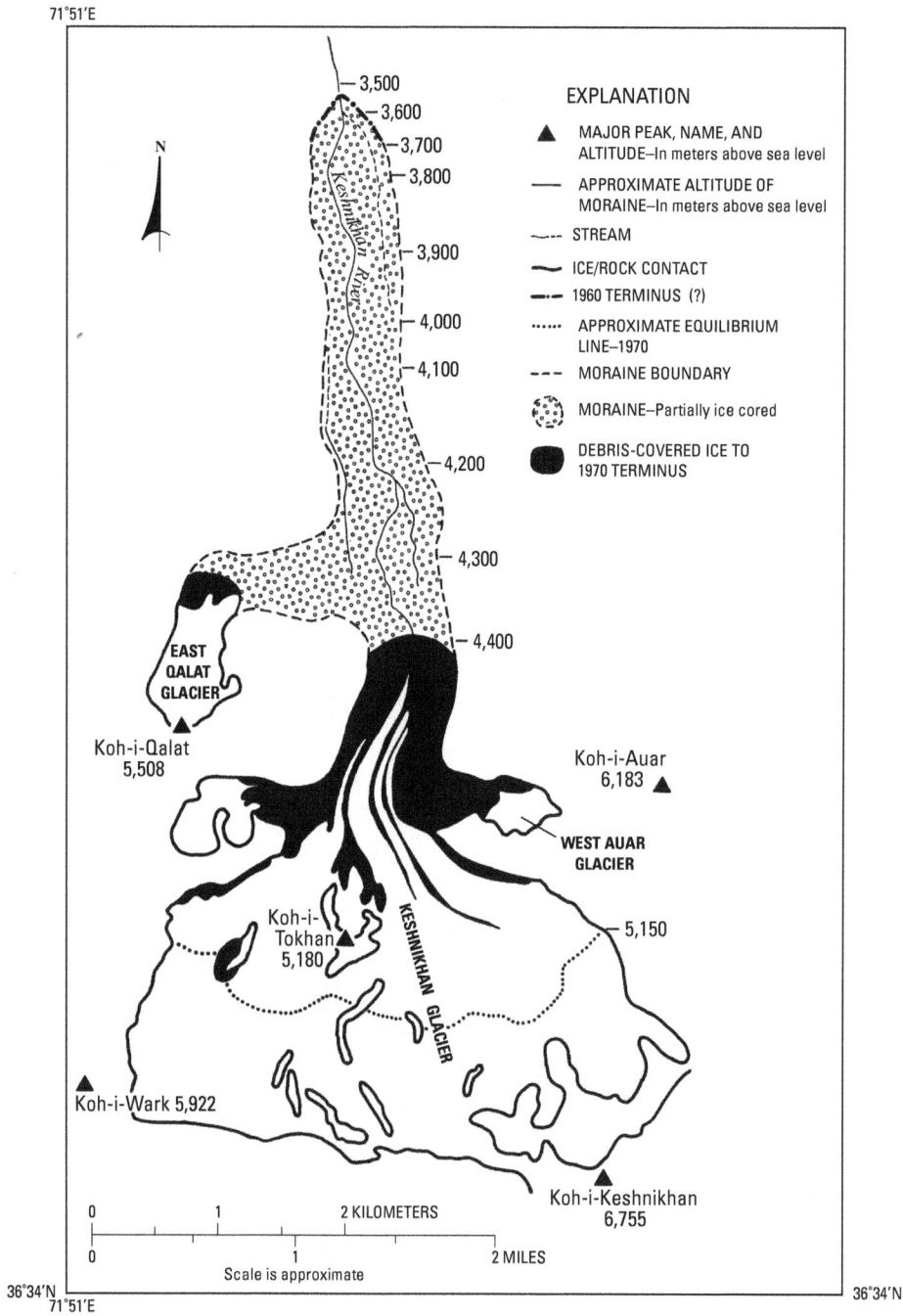

Figure 17.—*Planimetric map of* Keshnīkhān *Glacier adapted from Austrian Scientific Expedition in the Wakhan 1970 (1972). The approximate terminus of 1970 was delineated by Braslau (oral commun., 1977) and was only a few hundred meters below exposed ice. Compare with figure 5 for location. Names on map may be variant names.*

71°52'E

N

36°38'N

0 1 2 KILOMETERS

0 1 2 MILES
Scale is approximate

Figure 18.—*Landsat MSS image of Keshnīkhān Glacier area (approximate lat 36°35'N., long 71°52'E.) taken 12 July 1973. The Ab-i-Panj river is in the upper left. Compare with figures 5, 17, and 19.*

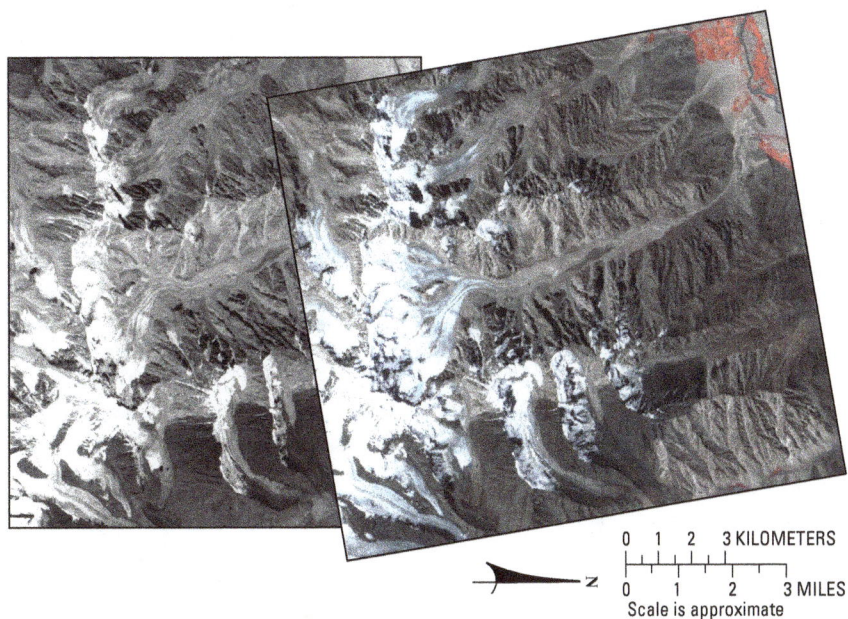

0 1 2 3 KILOMETERS

N

0 1 2 3 MILES
Scale is approximate

Figure 19.—*Stereoscopic pair of ASTER images of Keshnīkhān Glacier (approximate lat 36°35'N., long 71°52'E.) and environs. Compare with figures 5, 17, and 18. North is to the right. ASTER scene AST_L1B-003_08312000062611_ 09182003202055; 31 August 2000. [Editors' note: Use stereoscope for viewing.]*

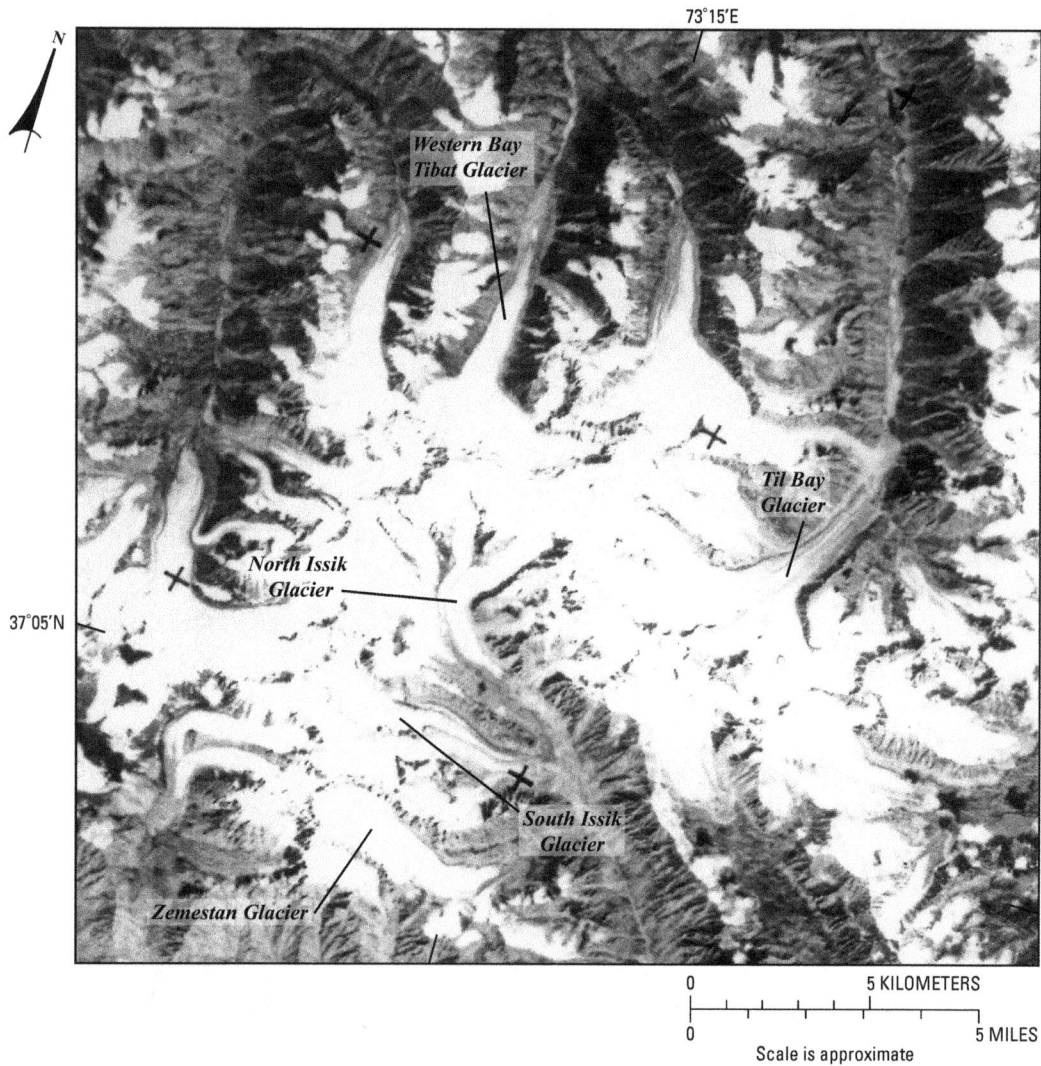

Figure 20.—Landsat 3 RBV image of the central Wākhān Pamir. Compare with figure 21. The Landsat 3 RBV image (30534–05053, subscene D; 21 August 1979; Path 162–Row 34) is from the U.S. Geological Survey, EROS Data Center, Sioux Falls, S. Dak. 57198.

The three main glaciers of interest occur on the south side of the range at the head of the Issīk valley. The *Northern* and *Southern Issīk Glaciers* and *Zemestan Glacier* are surrounded on three sides by six of the highest peaks in the area (more than 6,000 m in elevation). Maps of the glacier termini made by Russian topographers from 1960 vertical aerial photographs appear to be based on an average between clean ice, debris-covered ice, and ice-free moraine; but these are actually an inaccurate representation of the real fronts, as indicated by the detailed mapping by Kostka and coworkers (Exploration Pamir 75, 1978a, b) (fig. 21). The work of Patzelt (1978) and his support team on the *Southern Issīk Glacier* shows two lateral moraines produced by short-lived prior advances — one in the 19th century (ca. 1850), and one in the early 20th century (ca. 1920). The ice has since retreated from both moraines.

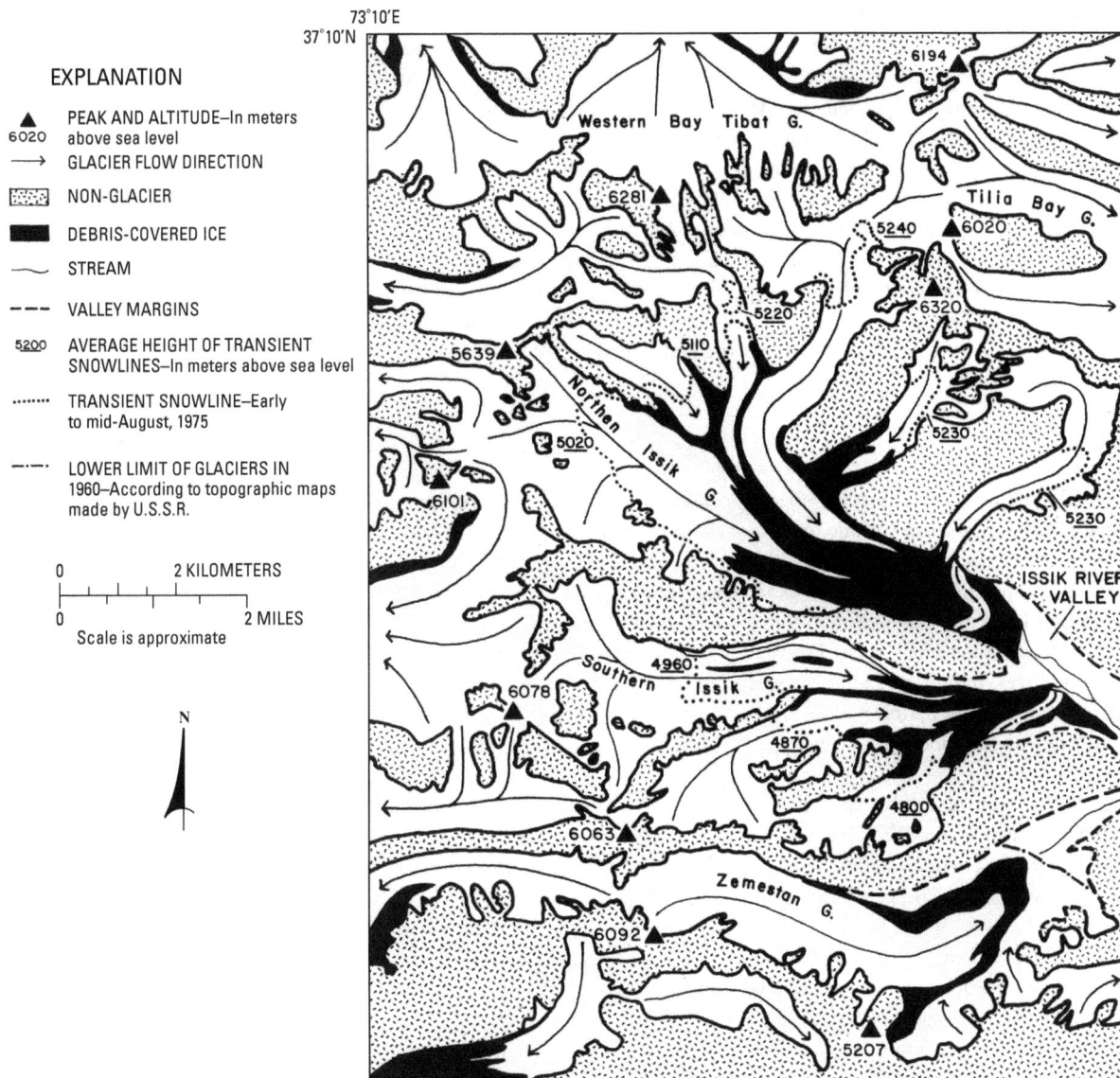

EXPLANATION

▲ 6020 PEAK AND ALTITUDE–In meters above sea level

→ GLACIER FLOW DIRECTION

░ NON-GLACIER

■ DEBRIS-COVERED ICE

〰 STREAM

--- VALLEY MARGINS

5200 AVERAGE HEIGHT OF TRANSIENT SNOWLINES–In meters above sea level

.... TRANSIENT SNOWLINE–Early to mid-August, 1975

–··– LOWER LIMIT OF GLACIERS IN 1960–According to topographic maps made by U.S.S.R.

0   2 KILOMETERS

0   2 MILES
Scale is approximate

N

Figure 21.—*Topographic map of south-central part of the Wākhān Pamir. After Exploration Pamir 75 (1978b). The map covers part of the area of figure 20.*

The younger moraine closely corresponds to the glacier front as mapped by the Russian topographers, but this is probably just fortuitous as none of their other determinations coincide. Several other glaciers in the area also reached their maxima within the last century and have not retreated since, apparently due to the slow response time of the ice masses rather than to any climatic control. Lichens and desert varnish have not developed on these moraines and no other older moraines occur below them, so a clear differentiation can be made based on color and radiance between glacierized and unglaciated terrain.

The transient snowlines at various sites on the three glaciers in August 1975 ranged from 4,800–5,420 m, with a mean of about 5,100 m. The lower snowlines occurred in north-facing cirques and the higher occurred on southern aspects (fig. 21). In general, the glaciers of the Issīk valley seem to be retreating or downwasting somewhat, as is so common throughout Afghanistan.

The glaciers of the Issīk valley were analyzed by several remote sensing students using a variety of transparency-overlay and digital techniques. The students reported various observations of glacier advance or retreat. But these academic studies illustrate the potential pitfalls of using satellite imagery without proper controls. Apparent glacier movements were created by almost undetectable differences in scale and unrectified geometric differences between imagery and map overlays, by differences in Russian and Austrian maps, by problems in distinguishing parts of some glaciers from snow fields, clouds, or old moraines, and by seasonal variability. Ultimately, visual comparisons at the same scales of early imagery (for example, Landsat 1 MSS image 1047–05171; 08 September 1972; Path 162–Row 34) with later RBV material (fig. 20) showed little significant change in the Issīk valley.

Conclusions

Glaciers in Afghanistan are generally small and debris covered and have been retreating or down wasting for a long time; however, only limited recent change in the glaciers of the higher, colder northeast has been observed in our limited comparisons of the earlier MSS to later ASTER imagery. Some new meltwater lakes may pose a threat from glacier lake outburst floods (GLOFs), but in general, their small size and considerable incisement inside high moraine walls reduces hazard potential. Nevertheless, without a thorough assessment using high resolution ASTER imagery and new digital elevation models (DEMs), considerable uncertainty exists about the possibility for future GLOFs.

Difficulties posed by the rough terrain and the ongoing insurgent activities have limited available information and access to the glaciers of Afghanistan; therefore, satellite imagery is extremely valuable in conducting glacier inventories. The generally small size and dark debris cover of many of the ice masses make interpretation of small-scale imagery difficult, but stereoscopic and higher resolution imagery provides the additional information needed in such cases. Renewed inventory and assessment of the glaciers of Afghanistan as a part of the Global Land Ice Measurements from Space (GLIMS) Project is continuing to allow detailed investigation of the characteristics of the glaciers of the country; in this project, the authors maintain the USGS- and NASA-supported Regional Center for Southwest Asia.

Acknowledgments

Acknowledgment is made to the late Fritz Müller, Swiss glaciologist and Director, Temporary Technical Secretariat for World Glacier Inventory, Department of Geography, Swiss Federal Institute of Technology (ETH), Zürich, Switzerland, who first encouraged the senior author to pursue this work over thirty years ago. Numerous colleagues and students at the University of Nebraska at Omaha contributed to this effort, especially J. Olsenholler. B. Ehmann contributed photographs and information about glaciers of the Noshaq peak area taken during the 1978 expedition. In Afghanistan, J. Summers and his fine staff of the Afghan-American Educational Commission assisted during the 1970s. Many thanks go to all the Afghan friends who aided this project. We are especially grateful to Lutfullah Safi, who gave unstintingly of himself through many decades of field work. Finally, we must thank and remember our field counterpart, Alam Malabar Nuristani.

References Cited

Anonymous, 1960, Making a map of Afghanistan: Prepared for Fairchild Aerial Surveys, Inc., for American Society of Photogrammetry, Afghanistan, July, August, September, p. 1–10.

The Austrian scientific expedition in the Wakhan 1970, 1972, Hoher Hindukusch Koh-e-Keshnikhan map: Vienna, Kartographische Anstalt Freytag-Berndt und Artaria, 1:25,000 scale (also in Gratzl, K., ed., 1972 Hindukusch-Österreichische Forschungsexpedition in den Wakhan 1970: Graz, Austria, Akademische Druck-u. Verlagsanstalt).

Bartolami, Lino, 1972–73, Expedition Jurm, 1972: The Himalayan Journal, v. 32, p. 56–66.

Biel, Stanislaw, and Wala, Jerzy, 1972–73, The Polish Hindu Kush and Pamir expedition, 1971: The Himalayan Journal, v. 32, p. 39–55.

Bishop, M.P., Kargel, J.S., Kieffer, H.H., MacKinnon, D.J., Raup, B.H., and Shroder, J.F., Jr., 2000, Remote-sensing science and technology for studying glacier processes in high Asia: Annals of Glaciology, v. 31, p. 164–170.

Braslau, David, 1972, The glaciers of Keshnikhan, in Gratzl, K., ed., Hindukusch-Österreichische Forschungsexpedition in den Wakhan 1970: Graz, Austria, Akademische Druck-u. Verlagsanstalt, p. 112–116.

Braslau, David, and Bussom, D.E., 1978a, A glacier inventory method using Landsat MSS CCT, in Rundquist, D.C., ed., The use of Landsat digital information for assessing glacier inventory parameters; an evaluation: Final report, project sponsored by Temporary Technical Secretariat for World Glacier Inventory, International Commission of Snow and Ice, and United Nations Educational, Scientific and Cultural Organization, p. 23–40.

Braslau, David, and Bussom, D.E., 1978b, Landsat sensing of glaciers with application to mass-balance and runoff, in Colbeck, S.C., and Ray, Manoranjan, eds., Proceedings of a meeting on modeling of snow cover runoff, 26–28 September 1978, Hanover, New Hampshire: U.S. Army Cold Regions Research and Engineering Laboratory, CRREL SR 79–36, p. 77–82.

Breckle, S.W., and Frey, W., 1976a, Die hochsten Berge im Zentralen Hindukusch [The highest mountains in the central Hindu Kush]: Afghanistan Journal, v. 3, no. 3, p. 91–94.

Breckle, S.W., and Frey, W., 1976b, Beobachtungen zur heutigen Vergletscherung der Hauptkette des Zentralen Hindukusch [Observations of the present-day glacierization of the principal mountain ranges in the central Hindu Kush]: Afghanistan Journal, v. 3, no. 3, p. 95–100.

Buchroithner, M.F., 1978, Zur geologie des Afghanischen Pamir [Geology of the Pamir Mountains in Afghanistan], in Senarclens de Grancy, Roger, and Kostka, Robert, eds., Grosser Pamir: Graz, Austria, Akademische Druck-u. Verlagsanstalt, p. 85–118.

Chwascinski, Boleslaw, 1966, The exploration of the Hindu Kush: Alpine Journal (Alpine Club, London), v. 71, no. 313, p. 199–214.

Desio, Ardito, 1962, Espansioni glaciali quaternarie nel territorio di Faizabad (Afghanistan) [Advance of glaciers during the Quaternary in the territory of Faizabad (Afghanistan)]: Roma, Rendiconti Accademia Nazionale dei Lincei, ser. 8, v. 32, no. 3, p. 281–285.

Desio, Ardito, 1975, Notes on the Pleistocene of central Badakhshan, in Desio, Ardito, ed., Geology of central Badakhshan (north-east Afghanistan) and surrounding countries, Italian expedition to the Karakorum (K2) and Hindu Kush, Scientific Reports, v. 3, Geology-petrology: Leiden, E.J., Brill, p. 339–409.

Diemberger, Adolf, 1966, Development of mountaineering in the Hindu Kush: The Himalayan Journal, v. 27, p. 3–41.

Dozier, Jack, 1970, South watershed of the Afghan Hindu Kush, 1967–70: The Himalayan Journal, v. 30, p. 301–304.

Dunsheath, Joyce, and Baillie, Eleanor, 1961, Afghan quest; the story of their Abinger Afghanistan expedition 1960: London, George G. Harrap & Co. Ltd., 236 p.

Exploration Pamir 75, 1978a, Darrah-e Issik-e Bala glacier map: Vienna, Freytag-Berndt und Artaria, 1:25,000 scale (also in Senarclens de Grancy, Roger, and Kostka, Robert, eds., Grosser Pamir: Graz, Austria, Akademische Druck-u. Verlagsanstalt).

Exploration Pamir 75, 1978b, Koh-e Pamir topographic map: Vienna, Freytag-Berndt und Artaria, 1:50,000 scale (also in Senarclens de Grancy, Roger, and Kostka, Robert, eds., Grosser Pamir: Graz, Austria, Akademische Druck-u. Verlagsanstalt).

Frey, Wolfgang, 1967–68, Between Anjuman and Munjan valleys: The Himalayan Journal, v. 28, p. 98–101.

Giardino, J.R., Jr., Shroder, J.F., Jr., and Vitek, J.D., eds., 1987, Rock glaciers: Boston, Allen & Unwin, 355 p.

Giger, W., 1972–73, Ismurgh Valley-Wakhan corridor: The Himalayan Journal, v. 32, p. 67–70.

Gilbert, O., Jamieson, D., Lister, H., Pendlington, A., 1969, Regime of an Afghan glacier: Journal of Glaciology, v. 8, no. 52, p. 51–65.

Glicken, Milton, 1960, Making a map of Afghanistan: Photogrammetric Engineering, v. 26, no. 5, p. 743–745.

Grötzbach, Erwin, 1964, Vorläufiger Bericht über die Münchner Hindukusch-Kundfahrt 1963 [Preliminary report on the Munich Hindu Kush Scientific Expedition 1963]: Die Erde, v. 95, no. 4, p. 291–298.

Grötzbach, Erwin, 1965, Beobachtungen an Blockströmen im afghanischen Hindukusch und in den Ostalpen [Observations of rock glaciers in the Hindu Kush of Afghanistan and in the eastern Alps]: Mitteilungen der Geographischen Gesellschaft in München, v. 50, p. 175–201.

Grötzbach, Erwin, and Hillebrandt, Axel, 1964, Die rezente und eiszeitliche Vergletscherung im mittleren Khwája Muhammad Gebirge [Recent and Pleistocene glacierization in the mid-Khwája Muhammed Mountains], in Münchner Hindukusch-Kundfahrt 1963: München, Akademische Sektion München, p. 26–31.

Grötzbach, Erwin, and Rathjens, Carl, 1969, Die heutige und die jungpleistozäne Vergletscherung des afghanischen Hindukusch [The present-day and late Pleistocene glacierization of the Hindu Kush of Afghanistan]: Zeitschrift für Geomorphologie, Supplementband 8, p. 58–75.

Heiss, Lorenz, 1972–73, The Berchtesgadener Hindukush expedition 1972: The Himalayan Journal, v. 32, p. 74–76.

Kick, W., 1969, Comments on "Regime of an Afghan glacier": Journal of Glaciology, v. 8, no. 54, p. 493.

Kieffer, Hugh, Kargel, J.S., Barry, R., Bindschadler, R., and others, 2000, Satellite measurements of glaciers and ice sheets: EOS (Transactions, American Geophysical Union), v. 81, no. 24, p. 265, 270–271.

Klaer, Wendelin, 1969, Glazialmorphologische probleme in den Hochgebirgen Vorderasiens [A glacial morphology problem in the high mountains of Western Asia]: Erdkunde, v. 23, no. 3, p. 192–200.

Kotlyakov, V.M., 1997, Atlas snezhno-ledovykh resursov mira [World atlas of snow and ice resources]: Moscow, Rossiiskaia Akedemiia Nauk, v. 3.

Kotlyakov, V.M., and Lebedeva, I.M., 1998, Melting and evaporation of glacier systems in the Hindu Kush — Himalaya region and their possible changes as a result of global warming, in Chalise, S.R., Herrmann, A., Khanal, N.R., and others, eds., Ecohydrology of High Mountain Areas: p. 367–375.

Kravtsova, V.I., 1990, Snow cover mapping of Afghanistan's mountains with space imagery, in Mapping sciences and remote sensing: V.H. Winston & Son, v. 27, no. 4, p. 295–302 [originally published in Russian in Moscow in Materialy Glyatsiologicheskikh Issledovaniy (MGI), 1989, Pub. 67, p. 44–49].

Kravtsova, V.I., and Tsarev, B.K., 1997, Snezhnyi pokrov i Laviny Afganistana [Snow cover and avalanches of Afghanistan]: Middle Asia Science Research Hydro-Meteorological Institute of V.A. Baguev (SANIGMI) Tashkent, 136 p.

Lalande, P., Herman, N.M., and Zillhardt, J., 1974, Cartes climatiques de l'Afghanistan [Climate maps of Afghanistan]: Kaboul, L'Institut de Météorologie, Publication 4, v. 1, text, 47 p.; and v. 2, maps.

Lebedeva, I.M., 1997, Izmenenie lednikovogo stoka vek Gindukusha pri global'nom poteplenii klimata [Change of the glacial runoff of the Hindu Kush rivers under the global climate warming]: Materialy Glyatsiologicheskikh Issledovaniy (MGI) [Data of Glaciological Studies], Pub. 83, p. 65–72.

Lebedeva, I.M., and Larin, A.D., 1991, Lednikovaia sistema Afghanistana: Morfologiia, klimat, massoobmen, stok [The glacier system of Afghanistan: Morphology, climate, mass-exchange and runoff]: Materialy Glyatsiologicheskikh Issledovaniy (MGI) [Data of Glaciological Studies], Pub. 72, p. 76–87.

Lister, H., 1969, Reply to Dr. W. Kick's comments on "Regime of an Afghan glacier": Journal of Glaciology, v. 8, no. 54, p. 493–494.

Maksimov, Ye.V., and Perugina, N.N., 1975, Opyt gliatsiomorfometricheskogo analiza kart Gindukusha [Glaciomorphological analysis of a Hindu Kush map]: Vsesoiuznoe geograficheskoe Obshchestva Izvestiia, v. 107, no. 3, p. 238–243.

Mayewski, P.A., and Jeschke, P.A., 1979, Himalayan and trans-Himalayan glacier fluctuations since AD 1812: Arctic and Alpine Research, v. 11, no. 3, p. 267–287.

Müller, Fritz, 1978, Instructions for the compilation and assemblage of data for a world glacier inventory. Supplement: Identification/glacier number: Temporary Technical Secretariat for World Glacier Inventory. International Commission on Snow and Ice, International Association of Hydrological Sciences, United Nations Environment Programme, United Nations Educational, Scientific, and Cultural Organization, 7 p., plus apps.

Müller, Fritz, Caflisch, T., and Müller, G., 1977, Instructions for the compilation and assemblage of data for a world glacier inventory. Supplement: Identification/glacier number: Temporary Technical Secretariat for World Glacier Inventory. International Commission on Snow and Ice, International Association of Hydrological Sciences, United Nations Environment Programme, United Nations Educational, Scientific, and Cultural Organization, 19 p., plus apps.

Naumann, C.M., 1974, Pamir und Wakhan: Afghanistan Journal, v. 1, no. 4, p. 91–104.

Newby, Eric, 1958, A short walk in the Hindu Kush: London, Secker and Warburg Ltd., 247 p.

North, R.A., 1970, Scottish Hindu Kush expeditions, 1965–70: The Himalayan Journal, v. 30, p. 282–300.

Patzelt, Gernot, 1978, Gletscherkundliche Untersuchlungen im "Grossen Pamir" [Glacier research in the "Greater Pamir"], in Senarclens de Grancy, Roger, and Kostka, Robert, eds., Grosser Pamir: Graz, Austria, Akademische Druck-u. Verlagsanstalt, p. 131–149.

Paytubi, Jose, 1972–73, Urgunt-e-Bala: The Himalayan Journal, v. 32, p. 71–73.

Porter, S.C., 1985, Extent of Late-Pleistocene glaciers in Afghanistan based on interpretation of Landsat imagery, in Climate and geology of Kashmir and central Asia: The last four million years: Agrawal, D.P., Kusumgar, S., and Krishnamurthy, R.K., eds., Current Trends in Geology: New Delhi, India, Today & Tomorrow's Printers and Publishers, v. VI, p. 191–195.

Rathjens, Carl, 1978, Klimatische Bedingungen der Solifluktionsstufe im sommertrockenen Hochgebirge, am Beispiel des afghanischen Hindukusch [Climatic limitations on the degree of solifluction in the dry summer period in the high mountains, with reference to the Hindu Kush of Afghanistan]: Zeitschrift für Geomorphologie, New Folge, Supplement Band 30, p. 132–142.

Reiner, Ernst, 1966, Die kartographie in Afghanistan [Cartography in Afghanistan]: Kartographische Nachrichten, v. 16, no. 4, p. 137–145.

Rundquist, D.C., ed., 1978, The use of Landsat digital information for assessing glacier inventory parameters; an evaluation: Final report, project sponsored by the Temporary Technical Secretariat for World Glacier Inventory of the International Commission of Snow and Ice and the United Nations Educational, Scientific and Cultural Organization, 148 p.

Rundquist, D.C., and Samson, S.A., 1980, A Landsat digital examination of Khumbu glacier, Nepal: Remote Sensing Quarterly, v. 2, no. 1, p. 4–15.

Rundquist, D.C., Collins, S.G., Barnes, R.B., Bussom, D.E., Samson, S.A., and Peake, J.S., 1980, The use of Landsat digital information for assessing glacier inventory parameters, in World Glacier Inventory Workshop, Riederalp, Switzerland, 17–22 September 1978, Proceedings: International Association of Hydrological Sciences-Internationale des Sciences Hydrologiques, IAHS-AISH Publication 126, p. 321–331.

Schweizer, Gunther, 1969, Büsserschnee in Vorderasien [Snow penitentes in Western Asia]: Erdkunde, v. 23, no. 3, p. 200–205.

Scott, D.K., and Cheverst, W., 1968, The Midlands Hindu Kush Expedition, 1967, Report: Nottingham, U.K., Chilwell Ltd., Nottingham Climber's Club, 60 p.

Scott, R., 1972–73, The British Central Hindukush Expedition, 1971: The Himalayan Journal, v. 32, p. 77–78.

Shroder, J.F., Jr., 1980, Special problems of glacier inventory in Afghanistan, in World Glacier Inventory Workshop, Riederalp, Switzerland, 17–22 September 1978, Proceedings: International Association of Hydrological Sciences-Internationale des Sciences Hydrologiques, IAHS-AISH Publication No. 126, p. 149–154.

Shroder, J.F., Jr., 1989a, Glacierized areas of Afghanistan, *in* Haeberli, W., Bosch, H., Scherler, K., and others, eds., World glacier inventory, status 1988: Teufen, Switzerland, IAHS (ICSI)–UNEP–UNESCO, p. C39–C40, C346–C353.

Shroder, J.F., Jr., 1989b, Slope failure — Extent and economic significance in Afghanistan and Pakistan, *in* Brabb, E.E., and Harrod, B.L., eds., Landslides — Extent and economic significance: Rotterdam, A.A. Balkema, p. 325–341.

Shroder, J.F., Jr., 2004, Afghanistan redux — Better late than never?: Geotimes, October, v. 49, no. 10, p. 34–37.

Shroder, J.F., Jr., DiMarzio, C.M., Bussom, D.E., and Braslau, David, 1978, Remote sensing of Afghanistan: Afghanistan Journal, v. 5, no. 4, p. 123–128.

Shroder, J.F., Jr., and Giardino, J.R., 1978, Progress on rock glacier research: Nebraska Academy of Sciences and Affiliated Societies Transactions, v. 6, p. 51–54.

Sidorov, L.F., 1960, Early glaciation of the Pamirs: Doklady, Earth Sciences Section, Academy of Sciences U.S.S.R., v. 127, p. 732–733.

Sivall, T.R., 1977, Synoptic-climatological study of the Asian summer monsoon in Afghanistan: Geografiska Analer, v. 59A, nos. 1–2, p. 67–87.

Tsarev, B.K., 1988, Makromasshtabnye zakonomernosti dinamiki snezhnogo pokrova gindukusha [Macro-scale regularities of snow-cover dynamics in the Hindu Kush]: Materialy Glyatsiologicheskikh Issledovaniy (MGI) [Data of Glaciological Studies], Pub. 62, p. 136–139.

Tsarev, B.K., Getler, M.I., Pyatova, R.B., 1986, Nekotorye kharakteristiki rezhima ustoichivogo snezhnogo pokrova v gorakh gindukusha [Some properties of stable snow cover regime in the Hindu Kush Mountains]: Materialy Glyatsiologicheskikh Issledovaniy (MGI) [Data of Glaciological Studies], Pub. 56, p. 73–78.

von Wissman, Herman, 1960, Die heutige Vergletscherung und Schneegrenze in Hochasien [The present-day glacierization and snowline in High Asia]: Abhandlungen der Mathematisch-Naturwissenschaftlichen Klasse, Jg. ,1959, no. 14, p. 1101–1407.

Wala, Jerzy, 1971, Skizzenlandkarte des Umkreises von Noshaq-Kohe Tez [Sketch map of the Noshoq Koh-e Tez region]: Krakow, Poland, Klub Wysokogorski, blue-print copy, 6 p.

Wala, Jerzy, 1973, Hindu Kush; the regional division: Krakow, Poland, Klub Wysokogorski, blue-print copy, 45 p.

Wala, Jerzy, 1976a, Kohe Pamire Wakhan; orographical sketch map: Krakow, Poland, Klub Wysokogorsky, blue-print copy, 5 p.

Wala, Jerzy, 1976b, Kohe Wakhan; the six-thousanders in the Kohe Wakhan mountain group in the Pamir: Krakow, Poland, Klub Wysokogorsky, blue-print copy, 11 p.

Wala, Jerzy, 1977, Hindu Kush, mountain group; Hendukushe Agram: Krakow, Poland, Klub Wysogorski, 29 p.

Glaciers of Asia—

GLACIERS OF PAKISTAN

By John F. Shroder, Jr., *and* Michael P. Bishop

SATELLITE IMAGE ATLAS OF GLACIERS OF THE WORLD

Edited by RICHARD S. WILLIAMS, JR., *and* JANE G. FERRIGNO

U.S. GEOLOGICAL SURVEY PROFESSIONAL PAPER 1386–F–4

CONTENTS

Abstract--**201**

Introduction--**201**

Background---**203**

    Figure   1. Index map of the glacierized region of Pakistan showing major
                  geographic features, approximate Path–Row locations of
                  Landsat 1, 2, and 3 imagery, and locations of glaciers selected
                  for this study--- **204**

Remote Sensing of Glaciers---**207**

    Figure   2. Optimum Landsat 1, 2, and 3 images of the glaciers of Pakistan ----------- **208**

    Table   1. Optimum Landsat 1, 2, 3 MSS and RBV images of the glaciers
                of Pakistan--- **209**

Glacier Fluctuations and Hazards ---**210**

Field Studies of Selected Glaciers ---**211**

    Tirich Glacier--**211**

    Gorshai Glacier --**211**

        Figure   3. Landsat 1 MSS false-color composite image taken on 12 July
                  1973, and ASTER image taken on 28 September 2001 of
                  Tirich Mir glaciers--- **212**

        Figure   4. Photograph taken in July 1999 of the disconnection between
                  Upper Tirich Glacier and Lower Tirich Glacier----------------------- **213**

        Figure   5. Sketch map of *Gorshai Glacier,* near *Matiltan,* Swat Kohistan -------------- **213**

        Figure   6. Landsat 2 MSS false-color composite image taken on
                  6 June 1978, and ASTER image taken on 13 October 2003 of
                  Gorshai Glacier -- **214**

    Glaciers of Nanga Parbat---**215**

        Figure   7. Part of a Landsat 3 RBV image showing Nanga Parbat massif ------------- **215**

        Figure   8. Landsat 2 MSS false-color composite image taken on 2 August
                  1977, and ASTER image taken on 29 October 2003, showing
                  Rāikot Glacier and *Sachen Glacier*-------------------------------- **216**

    Sachen Glacier ---**217**

        Figure   9. Photograph showing *Sachen Glacier* on the northeast slopes of
                  Nanga Parbat above *Astor* valley----------------------------------- **217**

    Rāikot Glacier --**218**

        Figure 10. Photograph taken in July 1991 of the *Rāikot Glacier* terminus,
                  and photo taken in June 1996, 2 years after the breakout flood
                  of early spring 1994--- **219**

        Figure 11. Photograph showing largest ice portal in the upper west margin
                  of *Rāikot Glacier* from the 1994 outburst flood ----------------------- **220**

    Batūra Glacier --**220**

        Figure 12. Photograph taken in August 1997 showing the connection
                  between *Chongra-Rāikot Glacier* and the main *Rāikot Glacier*----------- **221**

        Figure 13. Landsat 3 RBV image mosaic showing glaciers in the Hunza
                  region of the Karakoram Range-------------------------------------- **222**

        Figure 14. Photograph showing lower part of Batūra Glacier from *Yunz,* site
                  of a former diffluent valley from Batūra to *Pasu Glacier*-------------------- **223**

        Figure 15. Photograph taken June 1984 showing terminus of Batūra Glacier--------- **223**

        Figure 16. Landsat 3 MSS false-color composite image of Batūra Glacier taken
                  on 15 July 1979, and ASTER image taken on 30 April 2001 ------------- **224**

    Biafo Glacier ---**225**

        Figure 17. Landsat 2 MSS false-color composite image of Biafo Glacier taken
                  on 2 August 1977, showing linear medial and lateral moraines---------- **226**

        Figure 18. Sketch maps and ASTER image of terminus, ice portal, and
                  meltwater stream locations of Biafo Glacier -------------------------- **227**

FIGURE 19. Sketch map of Biafo Glacier comparing conditions of the overall glacier in 1908, 1939, and 1979 -- **228**

FIGURE 20. Landsat 3 RBV image mosaic showing glaciers in the K2 area of the Karakoram Range -- **229**

Glaciers Having Unusual or Irregular Flow -- **229**

Criteria for Recognizing Different Types of Unusual or Rapid Glacier Flow or Retreat --- **230**

TABLE 2. Glaciers in northern Pakistan that were directly observed to retreat, advance, or surge sometime in the past --------------------------- **231**

Movement Criteria 1 and 2: Convoluted Medial or Lateral Moraines **232**

TABLE 3. Glaciers of northern Pakistan selected for remote-sensing analysis-- **233–240**

FIGURE 21. Landsat 3 RBV image of Siachen, *Teram Shehr, north, central, and south Rimo, North Terong-Selkar Chorten, South Terong,* and Bilafond Glaciers--- **241**

Movement Criteria 3: Tributary Ice Overriding or Displacing Main Glacier Ice **241**

FIGURE 22. Landsat 3 MSS false-color composite image taken on 15 July 1979, and ASTER image taken on 30 September 2001 showing the terminus of Hispar Glacier-- **242**

FIGURE 23. Photograph showing terminus of the Hispar Glacier in late July 1984-- **243**

FIGURE 24. Photograph of the valley of Yengutz Har Glacier above Hispar village-- **245**

FIGURE 25. Landsat 3 MSS image taken on 18 July 1978, and ASTER image taken on 14 August 2004, showing terminus of Baltoro Glacier and *Liligo Glacier* -- **246**

Movement Criteria 4: Marked Depression of Ice Surface Below Lateral Moraines or Tributary Glaciers **246**

Movement Criteria 5: Lateral and Medial Moraines That are Slightly Sinuous, Lobate, or Offset by Crevasses or Ogives **246**

Movement Criteria 6: Recent Major Melting or Retreat of Icefront, Leaving Light-Colored Scars **247**

FIGURE 26. Landsat 2 MSS false-color composite and ASTER satellite images of the *Balt Bare Glacier* and its breakout flood of 1974 as it appeared in 1977, and the appearance of the same features in 2003 -- **247**

FIGURE 27. Photograph of the *Balt Bare* fan in July 1992 near the village of *Shash Kat* -- **248**

Movement Criteria 7: Extensive Area of Debris-Covered Downwasting or Stagnant Ice **248**

Movement Criteria 8: Linear Moraines and Accordant Tributaries **249**

Mountain Geomorphology --**249**

Conclusions ---**251**

References Cited --**252**

GLACIERS OF ASIA—

GLACIERS OF PAKISTAN[1]

By John F. Shroder, Jr.,[2] *and* Michael P. Bishop[2]

Abstract

Glaciers of northern Pakistan are some of the largest and longest mid-latitude glaciers on Earth, with an estimated area of 15,000 km². Field-based and space-based glacier studies in this region are necessary to elucidate their role in providing melt-water for irrigation, hazard potential, their role in erosion and geodynamics, and their sensitivity to climate forcing. Repeated field surveys in the 1980s, 1990s, and 2000s of several glaciers in the Hindu Kush and Swat, Kohistan region, the Nanga Parbat Himalaya, and the Karakoram Himalaya, provided reference data and verification for information extracted from satellite imagery. A number of these field assessments (*Balt Bare*, Batūra, Biafo, *Gorshai, Rāikot, Sachen*, Tirich Mir, and Yengutz Har) are reported here as case studies, thought to be representative of the diversity of glaciers and their fluctuations in the western Himalaya. A number of change-detection studies using the older Landsat satellite imagery, in some cases compared to newer ASTER imagery, provide useful comparisons.

Emphasis was also directed toward glaciers with unusual advance, retreat, or surge histories. Eight criteria were developed for recognition of different types of flow as displayed on the older imagery and then applied to 169 glaciers; many have multiple criteria. Five glaciers had strongly convoluted medial or lateral moraines; 13 glaciers had medial or lateral moraines convoluted or offset by extensive crevasses or ogives; 7 had prominent tributary ice overriding or displacing main glacier ice; 14 had marked depression of ice surface below lateral moraines or tributary glaciers; 76 had lateral and medial moraines that were slightly sinuous, lobate, or offset by crevasses or ogives; 9 had recent major melting or retreat of ice front that left light-colored deglacierized terrain behind; 106 had extensive areas of debris-covered downwasting or stagnant ice; and 47 glaciers had linear moraines and accordant tributary glaciers. Some of these conditions will have changed in the ensuing two decades since this study was done originally. Finally, the issues of complex interdependencies and scale dependencies of climate forcings, glacier erosion, and mountain geodynamics in the Himalaya of Pakistan and elsewhere are areas of active research that are currently receiving much attention by a variety of workers. Study of the role of present-day glacierization and past glaciation of the western Himalaya in these efforts is greatly augmented by the availability of some of the historical satellite imagery presented here.

Introduction

Concerns over greenhouse-gas forcing and warmer temperatures have spurred research into climate forcing and associated Earth-System responses. Considerable scientific debate concerns climate forcing and landscape response, because complex geodynamics regulate feedback mechanisms that couple climatic, tectonic and surface processes (Molnar and England, 1990; Ruddiman, 1997; Bush, 2000; Zeitler, Koons, and others, 2001; Zeitler, Meltzer, and others, 2001; Bishop and others, 2002). A significant component

---

[1]Editors' Footnote: This manuscript was originally written in the early 1980s to describe the glaciers of Pakistan in the late 1970s and early 1980s, the "benchmark" time period for the Satellite Image Atlas of Glaciers of the World, U.S. Geological Survey Professional Paper 1386A–K. Because there were delays in publishing, the authors updated their manuscript by adding references to more recent work while intentionally retaining the original benchmark information.

[2]Department of Geography and Geology, University of Nebraska at Omaha, Omaha, NE 68182

in the coupling of Earth's systems involves the cryosphere, as glacier-related feedback mechanisms govern atmospheric, hydrospheric and lithospheric response (Bush, 2000; Shroder and Bishop, 2000; Meier and Wahr, 2002). Specifically, snow and ice mass distributions partially regulate atmospheric properties (Henderson-Sellers and Pitman, 1992; Kaser, 2001), sea level variations (Meier, 1984; Haeberli and others, 1998; Lambeck and Chappell, 2001; Meier and Wahr, 2002), surface and regional hydrology (Schaper and others, 1999; Mattson, 2000), erosion (Hallet and others, 1996; Harbor and Warburton, 1992, 1993), and topographic evolution (Molnar and England, 1990; Brozovik and others, 1997; Bishop and others, 2002). Consequently, scientists have recognized the significance of understanding glacier fluctuations and their potential as direct and indirect indicators of climate change (Kotlyakov and others, 1991; Seltzer, 1993; Haeberli and Beniston, 1998; Maisch, 2000). In addition, the international scientific community now recognizes the need to assess glacier fluctuations at a global scale to elucidate the complex scale-dependent interactions involving climate forcing and glacier response (Haeberli and others, 1998; Meier and Dyurgerov, 2002). Satellite data products from NASA's Earth Observation System (EOS) (Hall and others, 2005) assist in the identification and characterization of those regions that are changing most rapidly and that have the most significant impact on sea level, water resources, economics, and geopolitics (Haeberli, 1998).

Mountain environments are known for their complexity and sensitivity to climate change (Beniston, 1994; Meier and Dyurgerov, 2002). Numerous mountain systems have been identified as "critical regions" and include Alaska, Patagonia and the Himalaya (Haeberli and others, 1998; Meier and Dyurgerov, 2002). Within the Himalaya, alpine glaciers are thought to be very sensitive to climate forcing due to their range in altitude and variability in debris cover (Nakawo and others, 1997). Furthermore, such high-altitude geodynamic systems are thought to be the direct result of climate forcing (Molnar and England, 1990; Bishop and others, 2002), although climate versus tectonic causation is still being debated (Raymo and others, 1988; Raymo and Ruddiman, 1992). To resolve these glaciological arguments, a fundamental understanding of the feedbacks between climate forcing and glacier response is needed (Dyurgerov and Meier, 2000). This requires detailed information about glacier distribution and ice volumes, mass-balance gradients, regional mass-balance trends, and landscape factors that significantly control ablation.

The western Himalaya is an ideal location for studying the causal interrelationships between climate, present-day glacierization, and past glaciation because the magnitude of such geodynamics is relatively high — the region has experienced dramatic climatic change, and it contains significant ice volumes (Tsvetkov and others, 1998; Bishop and others, 2002). From a practical point of view, the rapidly changing glaciological, geomorphological, and hydrological conditions of this region present a "looming crisis" to the world in terms of a decreasing water supply, increased hazard potential, and geopolitical destabilization.

Scientific progress in understanding the glaciers in Pakistan, however, has been slow because of the complex topography, paucity of field measurements, and limitations associated with information extraction from satellite imagery. Information is limited or nonexistent regarding: (1) enumeration and distribution of glaciers; (2) glacier mass-balance gradients and regional trends; (3) the contribution of glacial meltwater to the observed rise in sea level; and (4) natural hazards and the imminent threat of landslides, ice and moraine dams, and outburst flooding caused by rapid glacier fluctuations.

McClung and Armstrong (1993) have indicated that detailed studies of a few well-monitored glaciers do not permit characterization of regional mass-balance trend, the advance/retreat behavior of glaciers, or global extrapolation.

To our knowledge, the alpine glaciers in Pakistan have not been adequately studied or understood in terms of their role in mountain geodynamics, or their response and sensitivity to climate forcing, to adequately characterize existing conditions and future trends. An integrated approach to studying these glaciers, developing base-line information, and improving our understanding of climate forcing and glacier fluctuations requires the use of new remote sensing and geographic information systems (GIS) technologies (Haeberli and others, 1998; Bishop and others, 2004; Bishop and Shroder, 2004).

Much of the work for this paper was originally done over two and a half decades ago using only the Landsat Multispectral Scanner (MSS) imagery (Shroder and Bishop, 2005). Subsequent delays in publication necessitated some updating. In the intervening years since the initial research was undertaken, numerous expeditions and analyses have added to the knowledge store, but only limited such information is presented here. Furthermore, advanced numerical assessments are not included in this paper because they are presented elsewhere (Bishop and others, 1995, 1998a, 1999, 2000, 2001, 2002, 2004). Instead, this paper presents baseline information to facilitate subsequent comparative studies of glacier change. Our original work using small-scale Landsat imagery was augmented by field studies in the 1980s and 1990s, and by the use of Advanced Spaceborne Thermal Emission and Reflection Radiometer (ASTER) imagery. This work is now part of the international Global Land-Ice Measurements from Space (GLIMS) project, designed to produce quantitative information about the world's glaciers through remote sensing and field observations (Bishop and others, 2000; Kieffer and others, 2000; Bishop and others, 2004). In this paper, we first provide background information about the region and the important geoscience issues there, and discuss the role of remote sensing using the reflective region of the electro-magnetic spectrum in glaciological studies. Selected glacier fluctuations and glacier hazards across the region are then discussed, with an emphasis on those glaciers on which field and remote sensing studies have been conducted. The larger glaciers in Pakistan were assessed first with 1970s MSS and RBV imagery to identify observable flow variations such as the highly deformed medial moraines characteristic of surges, as well as other features. Selected ASTER scenes were then used to look at particularly interesting features first observed on the older imagery. Finally, the important role of alpine glaciers in mountain geodynamics is dicussed, and results are summarized.

Background

Pakistan has some of the world's highest and most spectacular mountains. A total of 13 of the world's 30 tallest peaks are located there, including K2 (8,611 m), the second highest peak in the world, Nanga Parbat (8,125 m), the ninth highest peak, and Tirich Mir (7,690 m) in the Hindu Kush. Because of the numerous high mountains, and abundant precipitation characteristic of a monsoon climate, the mountains of northern Pakistan, including the Hindu Kush, Hindu Raj, Kohistan ranges, Nanga Parbat massif, and Karakoram Himalaya[3], host some of the largest and longest mid-latitude glaciers on Earth (fig. 1). The glacierized area in northern Pakistan is estimated to cover 15,000 km², and as much as 37 percent of the Karakoram region is covered by glaciers.

---

[3]U.S. Government publications require that official geographic place-names for foreign countries be used to the greatest extent possible. In this section the use of geographic place-names is based on the U.S. Board on Geographic Names website: *http://earth-info.nga.mil/gns/html/index.html*. The names not listed on the website are shown in italics. However, in the case of italicized names, the authors have been careful to use the most accurate transliteration of local names or generally accepted local names.

EXPLANATION

160/37   APPROXIMATE PATH / ROW LOCATION OF LANDSAT 1, 2, AND 3 IMAGERY–Number
is located in the north-east corner of the area it represents

92   GLACIER PLOTTED BY NUMBER–Listed in table 3

0   100 KILOMETERS

0   100 MILES

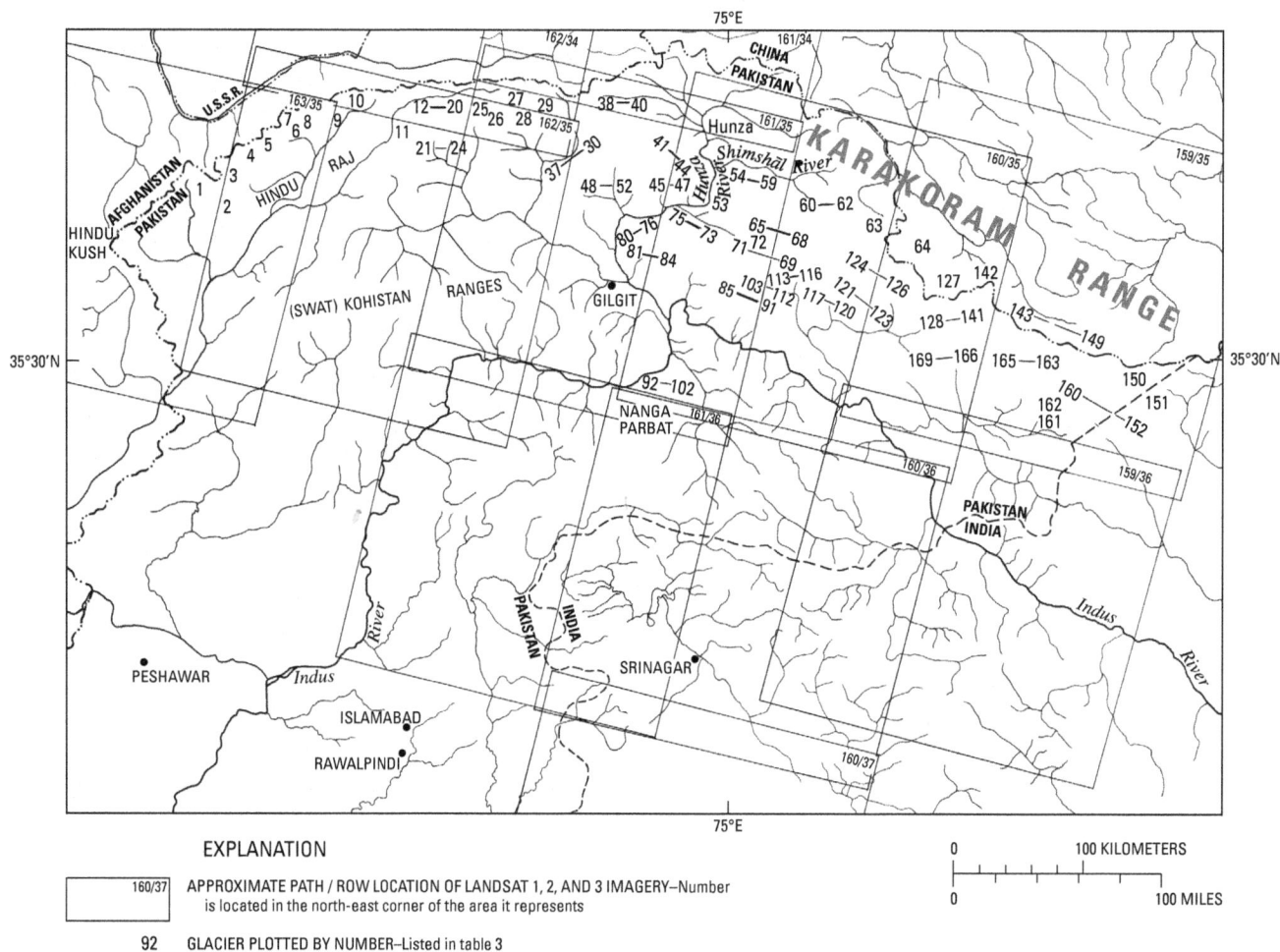

Figure 1.—Index map of the glacierized region of Pakistan showing major geographic features, approximate Path–Row locations of Landsat 1, 2, and 3 imagery, and locations of glaciers selected for this study. Optimum Landsat images are listed in table 1 and plotted in figure 2. Glaciers selected for this study are plotted by number and listed in table 3. Small glaciers in the Hindu Kush, Hindu Raj, and Kohistan ranges were not included. Glaciers 41–169 are those of the main Karakoram Himalaya.

The glaciers are highly variable in terms of their morphogenic and functional character. Some are mainly summer accumulation types (southwest monsoon), while others also receive mass from the westerlies. The partitioning of accumulation from both sources is thought to have dramatically changed from the past, with evidence suggesting that there was greater northward penetration of the monsoon during the Holocene (Phillips and others, 2000). Glaciers in the Himalaya have been classified according to their dominant nourishment type (Visser and Visser-Hooft, 1938; Washburn, 1939) as: (1) basin-reservoir (Firnmulden) types that are predominantly in cirques; (2) firn plateau types; (3) incised reservoir (Firnkesel) types in deep, narrow valleys with multiple catchments fed mainly by snow and ice avalanches; and (4) avalanche (Lawinen) types that lack tributary reservoirs. In practice, the latter two types are basically the same (Maull, 1938; Washburn, 1939). The huge, steep peaks of the Himalaya produce a characteristic dominance of avalanche source materials, including the copious rock debris that dominates the lower ablation regions of many glaciers in the region. The glaciers are highly variable in size, and the region has cirque, ribbon, hanging, alpine valley, and large compound valley glaciers. An example of the later is the Baltoro Glacier [128] at K2, which is fed by about 30 tributaries and covers about 1,300 km^2 — almost 10 percent of the country's total glacier area. The glaciers are commonly also heavily debris-covered, with highly variable debris thicknesses and cover (Bishop and others, 1995). Given the complexity of topography, including landscape relief, slope angles and basin orientation, many glaciers exhibit highly diverse morphologies and ice-flow velocities (Shroder and others, 1999). Many have advanced to the opposite valley wall, blocking the meltwater runoff from up-valley glaciers, creating large alpine valley lake impoundments that later release catastrophically.

Various early reports of unusual features, catastrophic advances, and glacier-related floods (jökulhlaups) (Hayden, 1907; Mason, 1930, 1935; Hewitt, 1969) piqued an interest in studying these glaciers more closely, but the difficult terrain and uncertain politics allowed only intermittent observation. By the 1970s, an increasing interest in environmental changes led to several syntheses of previous work (Mercer, 1963, 1975; Horvath, 1975; Mayewski and Jeschke, 1979; Mayewski and others, 1980).

The construction of the Karakoram Highway (KKH) between Pakistan and China in 1978 provided more access to this region, and the opportunity for extensive fieldwork. This resulted in some detailed studies of glaciers in the Hunza area by the Chinese-sponsored Batūra Glacier [141] Investigation Group (BGIG; 1979, 1980), the British-sponsored International Karakoram Project (IKP) in 1980 (Miller, 1984), and the Canadian Snow and Ice Hydrology Project in the 1980s (Hewitt, 1988, 1990). Some remote sensing and glacial geomorphological studies were conducted by Shroder (1980, 1989), Bishop and others, (1995; 1998a, b; 1999), Bishop and Shroder (2000), and Shroder and Bishop (2000), although other groups from Austria, Britain, Germany, Italy, Russia, Switzerland, Nepal, and other countries including most recently Pakistan, have also made significant contributions. For example, Russian scientists used satellite imagery to evaluate glacier distribution climate regimes, equilibrium lines, and glacier surge types and locations in the western Himalaya (Kravtsova and Labatina, 1974; Desinov and others, 1982; Kravtsova, 1982; Nosenko, 1991; Lebedeva, 1993, 1995; Lebedeva and Kislov, 1993; Knizhnikov and others, 1997; Osipova and Tsvetkov, 1998, 2002, 2003). They compiled a small scale atlas of snow and ice resources over the western Himalaya and Karakoram that presented plentiful data on glaciers, glacial processes, snowmelt and runoff (Kotlyakov, 1997). Their classification of surging glaciers, similar to that used herein, was presented as: (1) identified repeated surges, with morphological features of instability, (2) some indications of pulses, (3) indications of pulses not defined, and (4) some data about massive changes in the ice zone.

In essence, however, the huge region still lacks reliable quantitative estimates of fundamental glacier parameters such as glacier distribution and ice volume, ablation rates, flow rates, ELA determinations, and has but a few detailed mass balance studies. Our field investigations and remote sensing analyses indicate that some glaciers are retreating and downwasting. Not currently known is what the regional mass-balance trend is, although we suspect a general negative mass-balance trend based upon our data and work on mass balance in other areas (Aizen and others, 1997; Fujita and others, 1997; Cao, 1998; Cogley and Adams, 1998; Bhutiyani, 1999; Meier and others, 2003; Berthier and others, 2007). Furthermore, the exact number of alpine glaciers, and their size distribution, regional ice volume, modern-day and historical spatial-distribution patterns, and sensitivity of these heavily debris-covered glaciers, is not nearly as well known in the western as in the eastern Himalaya (Mool, Bajracharya, and Joshi, 2001; Mool, Wangda, and others, 2001). Recently, however, the Government of Pakistan has finally begun to focus on glacial assessments, using Landsat-7 Enhanced Thematic Mapper Plus (ETM+) data to study the glaciers with new appreciation of the effects of global warming (Roohi and others, 2005; Roohi, 2007). Study results are critical to the understanding of climate variability and ice-volume fluctuations, and to predicting the environmental consequences for Pakistan.

From a scientific perspective, a variety of glaciological parameters must be accurately estimated. Climate forcing has had a significant impact on this region in a relatively short period of time (Shroder and Bishop, 2000; Bishop and others, 2002). Research, however, has yet to definitively determine whether atmospheric warming will produce a negative or positive regional mass-balance trend. In addition, it is difficult to effectively characterize local variations in glacier mass balance, because the glaciers respond to a variety of climatic and geologic forcing factors (Molnar and England, 1990; Haeberli and Beniston, 1998; Dyurgerov and Meier, 2000; Shroder and Bishop, 2000). To complicate matters, the local and mesoscale topographic variations modify the regional climatic patterns and tectonic forcing (isostatic and tectonic uplift), which in turn govern climate and glacier dynamics (Hubbard, 1997; Shroder and Bishop, 2000). Consequently, glacier mass-balance gradients and the spatial distribution in the magnitude of mass balance may be highly variable, such that detailed studies of individual glaciers do not produce representative results for the region. These complex feedback mechanisms have not been adequately studied because they require sophisticated geographic information science (GISc) analyses, remote sensing, and numerical modeling of physical processes to produce information that can be integrated with climate, surface process, and tectonic physical models (Bishop and Shroder, 2000, Bishop and others, 2002).

From a resources perspective, ice masses in the Hindu Kush, Karakoram, and Nanga Parbat Himalaya constitute potential meltwater for the country, although the ice masses in India also contribute to Pakistan's water supply. As such, a far better understanding of the total volume and condition of these long- and short-term storages is critical to the management and prediction of future resource availability in both countries. Furthermore, the condition of the glacier ice mass provides important information on many factors that have significant ramifications in climate-change scenarios; these factors include glacier advance, retreat, backwasting, downwasting, surge, avalanche, debris accumulation, new meltwater lakes, and ice- and moraine-dammed lakes. Changing conditions of the regional ice mass can significantly affect irrigation, agriculture, tourism, hydroelectricity, drinking water, catastrophic floods, and cross-border conflict.

Glaciation and deglaciation in the region generate numerous hazards that can have catastrophic effects, by disrupting meltwater resources through damming and diversion, producing instabilities from mountain wall undercutting and debuttressing, and removing ice and rock supports as glaciers downwaste. A host of landforms long interpreted as Pleistocene recessional moraines have recently been reinterpreted as massive rockwall collapses which followed deglaciation and removal of wall support in a Last Glacial Maximum (LGM) (Hewitt, 1988, 1998, 1999). Furthermore, as discussed below, our prior work and that of others have now shown that the LGM in many places in the region is actually Holocene in age – much more recent and therefore more similar in conditions to the present day (Phillips and others, 2000; Shroder and Bishop, 2000).

Glacier chronologies have been attempted in the mountains of high Asia for more than a century, but only in the past two decades has any real progress been made in the western Himalaya (Derbyshire and others, 1984; Shroder and others, 1989; Gillespie and Molnar, 1995; Phillips and others, 2000; Shroder and Bishop, 2000; Owen and others, 2002). Accurate dates are required for assessing the influence of the Asian monsoon.

These recent chronologies show that a localized last glacial maximum also occurred early in the last glacial cycle. In the Swat Himalaya and the Zanskar valley, this occurred during oxygen isotope stage 5a [~91,000 to 79,000 years before the present (BP)], while in the Hunza valley, Nanga Parbat, and the Middle Indus valley, it occurred during oxygen isotope stage 3(~59,000 to 24,000 yrs BP) (Phillips and others, 2000; Richards and others, 2000; Taylor and Mitchell, 2000; Owen and others, 2002). Ice-core records from the Guliya ice cap support the view that interstadial conditions existed in western Tibet during oxygen isotope stages 3, 5a, and 5c (~111,000 to 99,000 yrs BP) (Thompson and others, 1997). The last glacial maximum in the Garhwal Himalaya occurred during oxygen isotope stage 4 (~74,000 to 59,000 yrs BP), although the large error bars suggest that it could have occurred during oxygen isotope stage 3. Extensive early glaciations have been recognized in the Khumbu and Lahul Himalaya, but these have not been numerically dated (Owen and others, 2000; Richards and others, 2000). Limited advances in the Hunza valley and middle Indus valley occurred during the global LGM in oxygen isotope stage 2 (~24,000 to 12,000 yrs BP), and advances in the Hunza valley and Lahul Himalaya occurred during the Late Glacial Interstadial (Owen and others, 2000, 2002; Richards and others, 2000). There is no evidence of glaciation during the Younger Dryas (~12,000 yrs BP) in any area of the Himalayas. In addition, as mentioned above, there is abundant evidence for an early-middle Holocene glacial advance in the Hunza valley, Nanga Parbat, Lahul Himalaya, and Khumbu Himal.

These data show that glaciation across the Himalaya appears to be broadly synchronous and that the timing of glaciation throughout the Himalaya occurred during times of insolation maxima, when the monsoon was strengthened and its influence extended further north to supply snow at high altitudes. Glaciers would have exhibited positive mass balance, allowing them to advance. Because the spatial extent of these earlier advances is not well known, it is difficult to compare them to modern glacial extents to estimate average rates of change.

Remote Sensing of Glaciers

Given the challenging logistics, geopolitical restrictions, and other uncertainties in Pakistan, many of the glaciers of Pakistan must be assessed via remote sensing. With the advent of new sensors and geographic information technologies (for example, GPS, GIS), scientists are able to estimate some glaciological parameters such as ice velocity fields and equilibrium line altitudes, and can investigate mountains and glaciers in new ways that address important scientific issues and problems (Bishop and Shroder, 2004; Bishop and others, 2004).

Remote sensing analysis of glaciers traditionally focused on simple image interpretation and the identification, characterization, and mapping of glaciers, although quantitative approaches such as image transformations and pattern recognition were used in trying to differentiate snow and glacier facies. This was characteristic of glacier studies in the 1970s (Shroder and others, 1978), including the work done originally for this paper two decades ago, which utilized Landsat 1, 2, and 3 MSS imagery. The relatively poor geometric accuracy and coarse spatial, spectral, and radiometric resolution greatly limited the use of the imagery for more than identification of the general location of moderate to large glaciers, production of general outlines of glacial extent, the identification of some glaciers that surged, and some change detection in comparisons with modern imagery. Although this information represents important base-line information (fig. 2, table 1), there was error and uncertainty associated with the information extracted.

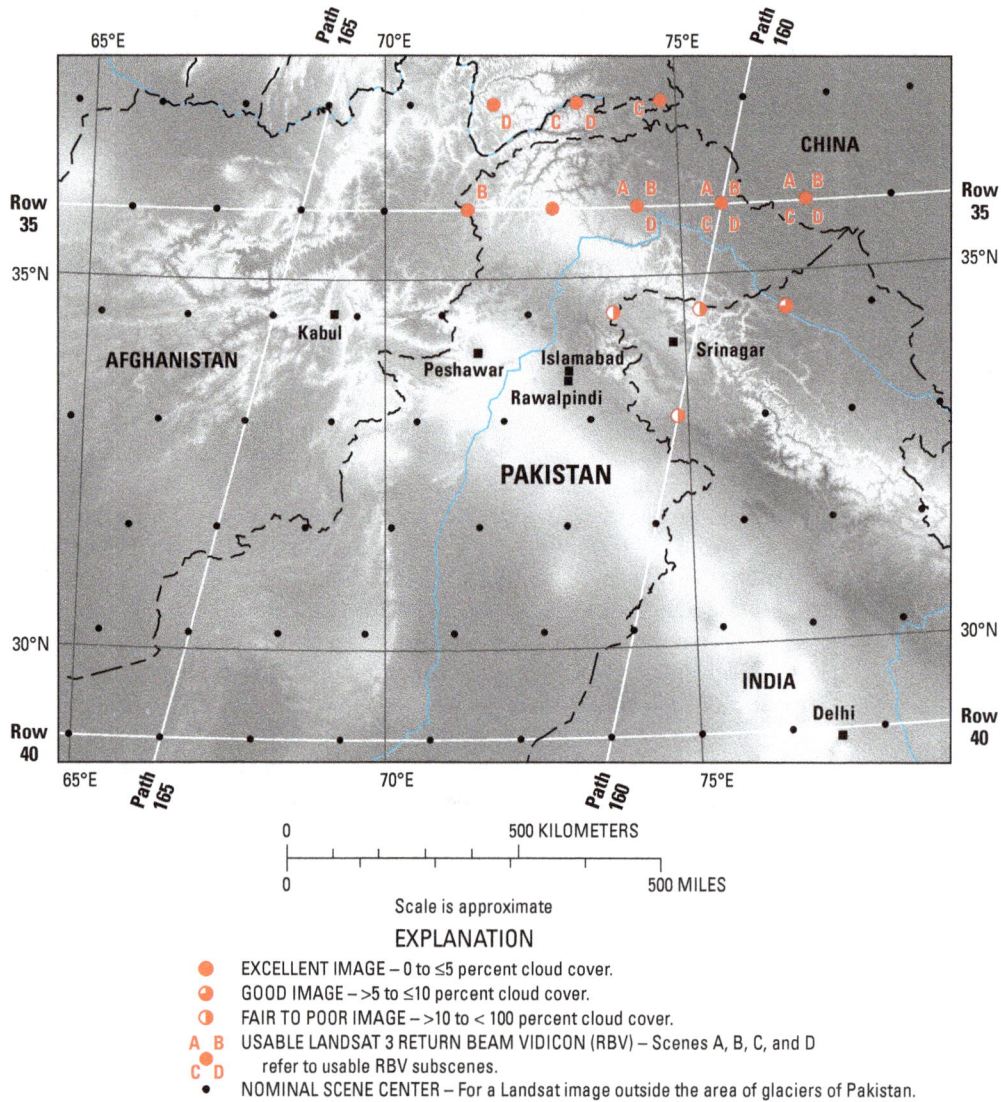

EXPLANATION

- EXCELLENT IMAGE – 0 to ≤5 percent cloud cover.
- GOOD IMAGE – >5 to ≤10 percent cloud cover.
- FAIR TO POOR IMAGE – >10 to < 100 percent cloud cover.
- A B USABLE LANDSAT 3 RETURN BEAM VIDICON (RBV) – Scenes A, B, C, and D
- C D   refer to usable RBV subscenes.
- NOMINAL SCENE CENTER – For a Landsat image outside the area of glaciers of Pakistan.

Figure 2.—*Optimum Landsat 1, 2, and 3 images of the glaciers of Pakistan.*

Since then, investigators have been studying glaciers using moderate to high resolution data that exhibit greater radiometric and geometric fidelity. This permits an extraction of additional information for assessing glacier conditions. For example, Aniya and others (1996) conducted an inventory of outlet glaciers of the southern Patagonia Icefield. Eleven parameters related to glacier morphology were extracted, and the results were comparable to results generated from topographic maps. Other researchers have investigated the potential of satellite imagery for mapping alpine snow (Dozier and Marks, 1987; Haefner and others, 1997), and for assessing and mapping supraglacial features (Bishop and others, 1995, 1998a, 1999) and glacier ice and snow facies (Hall and others, 1988; Williams and others, 1991). This work is commonly concerned with accurate delineation of glaciers and facies, and comparing satellite-derived surface reflectance with *in-situ* measurements. Results from these and other studies have indicated that high-resolution spectral data and topographic information are neccessary for delineating

TABLE 1.—Optimum Landsat 1, 2, 3 MSS and RBV images of the glaciers of Pakistan

[Abbreviations: MSS, multi spectral scanner; RBV, return beam vidicon; m, meter; μm, micrometer; km, kilometer; cm, centimeter; EROS, Earth Resources Observation and Science]

[Code column indicates usability for glacier studies; see figure 2 for explanation. —, unknown

Path-Row	Nominal scene center latitude and longitude	Landsat identification number (10 digits) or entity identification number (16 digits)[1]	Date	Code	Cloud cover (percent)	Remarks
159–35	35°58′N. 77°08′E.	3159035007819990	18 Jul 78	◑	10	
159–35	35°58′N. 77°08′E.	30531–04484 Subscenes A, B, C, D	18 Aug 79	●	0–20	Landsat 3 RBV archived by the authors[2]. Images used for figs. 20 and 21
159–36	34°32′N. 76°42′E.	3159036007819990	18 Jul 78	◑	10	Image used for fig. 25
160–35	35°58′N. 75°42′E.	2160035007721490	02 Aug 77	●	0	Image used for figs. 8A, 17, and 26A
160–35	35°58′N. 75°42′E.	2160035007723290	20 Aug 77	◑	10	
160–35	35°58′N. 75°42′E.	30532–04542 Subscenes A, B, C, D	19 Aug 79	●	5–40	Landsat 3 RBV archived by the authors[2]. Images used for figs. 13 and 20
160–36	34°32′N. 75°16′E.	2160036007723290	20 Aug 77	◐	20	
160–37	33°07′N. 74°49′E.	2160037007723290	20 Aug 77	◐	50	
161–34	37°24′N. 74°44′E.	2161034007622190	08 Aug 76	●	5	Some technical problems
161–34	37°24′N. 74°44′E.	30515-04595 Subscene C	02 Aug 79	—	—	Landsat 3 RBV archived by the authors[2]
161–35	35°58′N. 74°16′E.	2161035007622190	08 Aug 76	●	5	
161–35	35°58′N. 74°16′E.	2161035007726990	26 Sep 77	●	0	Some technical problems
161–35	35°58′N. 74°16′E.	3161035007919690	15 Jul 79	●	0	Image used for figs. 16A, and 22A
161–35	35°58′N. 74°16′E.	30515-05001 Subscenes A, B, D	02 Aug 79	●	0–20	Landsat 3 RBV archived by the authors[2]. Images used for figs. 7 and 13
161–36	34°32′N. 73°50′E.	2161036007821090	29 Jul 78	◐	0	Some technical problems
162–34	37°24′N. 73°18′E.	2162034007622290	09 Aug 76	●	0	
162–34	37°24′N. 73°18′E.	30516–05053 Subscenes C, D	03 Aug 79	—	—	Landsat 3 RBV archived by the authors[2]
162–35	35°58′N. 72°50′E.	2162035007723490	22 Aug 77	◐	20	
162–35	35°58′N. 72°50′E.	2162035007815790	06 Jun 78	●	0	Image used for fig. 6A
163–34	37°24′N. 71°52′E.	30553–05105 Subscene D	09 Sep 79	●	0	Landsat 3 RBV archived by the authors[2]
163–35	35°58′N. 71°24′E.	1163035007319390	12 Jul 73	◑	10	Image used for fig. 3A
163–35	35°58′N. 71°24′E.	2163035007622390	10 Aug 76	●	5	
163–35	35°58′N. 71°24′E.	30553–05112 Subscene B	09 Sep 79	—	—	Landsat 3 RBV archived by the authors[2]

[1]Landsat images were originally assigned a unique identification number that incorporated the satellite number, the number of days since satellite launch and the time the image was acquired. All archived Landsat images are now assigned an entity (or order) identification number based on the satellite number, the path and row of the scene, and the year and Julian day of the scene acquisition.

[2] Landsat 3 RBV images of the Earth were acquired by two RBV cameras during the operation of the Landsat 3 spacecraft (launched on 5 March 1978 and deactivated on 31 March 1983). Landsat 3 RBV images have a pixel resolution of ~30 m, were acquired in a single panchromatic band (0.505–0.750 μm; spanning part of the visible and part of the near-infrared bands of the electromagnetic spectrum) by two side-by-side slightly overlapping RBV cameras. The images covered an area 98 × 98 km (4 RBV images coincide with a single MSS frame). RBV images were recorded on film media (18.5 cm × 18.5 cm) and archived as positive and negative film transparencies at the U.S. Geological Survey, EROS Data Center, Sioux Falls, S. Dak. 57198. Landsat 1 images (3 bands, 79-m pixels) and Landsat 3 RBV images (single band, 30-m pixels) are archived in film format in deep storage at the EROS Data Center. Landsat RBV images are no longer available to scientists or to the public and can only be obtained from private archives.

debris-covered glaciers, mapping glacier facies, and estimating albedo and mass balance (Hall and others, 1989; Bishop and others, 2001). Mass balance estimates can be generated by estimating the equilibrium line altitude (ELA) during the later portion of the ablation season, or using topographic information to do so (Hall and others, 1989).

The debris-covered glaciers in Pakistan and other countries are exceptionally difficult to study because they exhibit extreme spectral and topographic variability (Bishop and others, 2001; Kääb and others, 2002; Paul and others, 2002). Snow and ice facies, glacial till, and different mineral and rock types produce highly variable reflectance, making glacier mapping very difficult in the solar reflective regions of the electromagnetic spectrum. For relatively large parts of many glaciers, the supraglacial debris cannot be spectrally differentiated from the surrounding rocks and ground moraine. In these cases, it is necessary to use topographic information, because the morphometric properties of the glacier can be used for delineating the terminus and identifying supraglacial landform features (Bishop and others, 1995; Bishop and others, 1998a; Bishop and others, 2001; Paul and others, 2004).

Given our current rate of collecting glacier information in Pakistan, it is expected that far too many glaciers will significantly change or be entirely gone before we can measure and understand them adequately (Roger Barry and Wilfried Haeberli, oral comm., March 2003). This time-sensitive issue requires us to acquire global and regional coverage of glaciers via satellite imagery before they disappear. Change-detection studies using earlier baseline information generated with Landsat MSS, TM, SPOT and other data should be compared with information obtained from fieldwork and modern sensors such as ASTER. Consequently, we present several case studies here that characterize selected Pakistan glaciers and their fluctuations.

When feasible, fieldwork has been conducted to augment and validate remote sensing and GIS investigations. The field data has greatly enhanced glacier study results, because some parameters can only be reliably measured in the field.

Glacier Fluctuations and Hazards

Multi-temporal imagery and field studies document the dynamic nature of glaciers and their continual adjustment to climatic, glaciological, and geological conditions. Change-detection studies are useful for assessing basic morphological conditions (for example, advance and retreat) caused by glacier flow dynamics and mass balance variations. In addition, satellite imagery documents numerous processes and provides insights into feedback mechanisms that govern glacier fluctuations. Furthermore, the spatial context of glaciers and their spatial topological relationships to other topographic features provides valuable information about hazard potential.

Extracting such information from satellite imagery has its technical challenges and limitations caused by sensor problems and difficulties related to image texture, scale, etc. Nevertheless, higher resolution imagery can document surface features and processes, and glacier-related fluctuations, including: (1) mass movements onto glaciers; (2) spatial variation of ablation; (3) surface moisture variations; (4) influence of adjacent-terrain irradiance on ablation; (5) terminus position changes; (6) supraglacial-lake development; (7) ice-velocity variations; (8) white-ice-stream position changes; (9) glacial-lake development; and others. Commonly, however, fieldwork is required to collect information that is not easily extracted from imagery, such as the location of ice portals, features related to ablation, moulins, and other surface characteristics. When integrated with image information, a more complete picture of glacial fluctuations and sensitivity emerges.

Field Studies of Selected Glaciers

A number of glaciers in the Pakistan Himalaya were studied in the field to determine their regional variability. For each glacier, general characteristics were identified, a change detection was performed, surges and other hazards were assessed, and baseline data was developed. Background information provided in the case studies that follow can be used to establish benchmark glacier sites (Kaser and Fountain, 2001) for long-term monitoring. Information gleaned from this long-term monitoring and from analysis of ASTER or other high-resolution stereoscopic imagery allows Pakistan managers to more reliably determine the availability of water resources for irrigation and hydroelectric power, and more accurately identify potential glacier-related hazards.

Tirich Glacier

The Tirich Mir massif (7,690 m) in the Chitral area of northwestern Pakistan represents the tallest mountain in the Hindu Kush. Large glaciers radiate from it, but relatively little is known about their dynamics and mass-balance histories. Previous research indicates that the North and South Barum Glaciers on the southeastern side of the massif were confluent 50 years ago (Bateson and Bateson, 1952), and satellite imagery reveals the presence of high altitude (~4,000 m) glacier erosional and depositional surfaces. We suspect that these surfaces were formed from monsoon-enhanced glaciation between 70 and 20 K yrs BP, because similar surfaces exist at that altitude throughout the Karakoram and Nanga Parbat Himalaya, and have been dated at 60–70 K yrs BP (Phillips and others, 2000, Owen and others, 2000). Although exposure age dates for the surfaces at Tirich Mir have yet to be determined, they clearly developed during more extensive glacial conditions, and document deglaciation and valley formation. This is evident on the north side of the massif, where the large compound Tirich Glacier flows down-valley in a northeasterly direction.

Examination of Landsat MSS imagery of the Tirich Glacier reveals numerous tributary glaciers that form the large Upper Tirich Glacier [1] (fig. 3A). A variety of debris cover — ice, black slate, and granite/granodiorite — is also evident in the imagery. Upon comparison of MSS imagery with ASTER stereo pairs, a flow discontinuity was revealed that divides the Upper from the Lower Tirich Glacier [2] (fig. 3B) and has existed for more that 20 years. Presumably, this ice flow discontinuity was caused by a decrease in regional mass balance and an associated change in surface hydrology. Ground photography of this discontinuity depicts the deposition of a ground moraine and its transport by an active river originating from the terminus of a tributary glacier on the south side of the Upper Tirich Glacier (fig. 4). In addition, visual comparative analysis of satellite imagery and field observations indicates the development of large supraglacial lakes on the Upper Tirich Glacier. This observational evidence suggests that the glacier is downwasting rather than exhibiting significant retreat.

Gorshai Glacier

Gorshai Glacier, about 6 km southeast of *Matiltan* in Swat Kohistan, heads in a north-facing cirque at about 4,725 m, beneath an unnamed peak 5,254 m in elevation; this peak is south of the main Kohistan, Hindu Raj, and Hindu Kush ranges. The glacier flows 2.8 km NNW in a series of ice falls before passing, at 3,660 m (12,000 ft) in elevation, into a zone of debris cover with prominent 2-km long lateral moraines which extend down to about

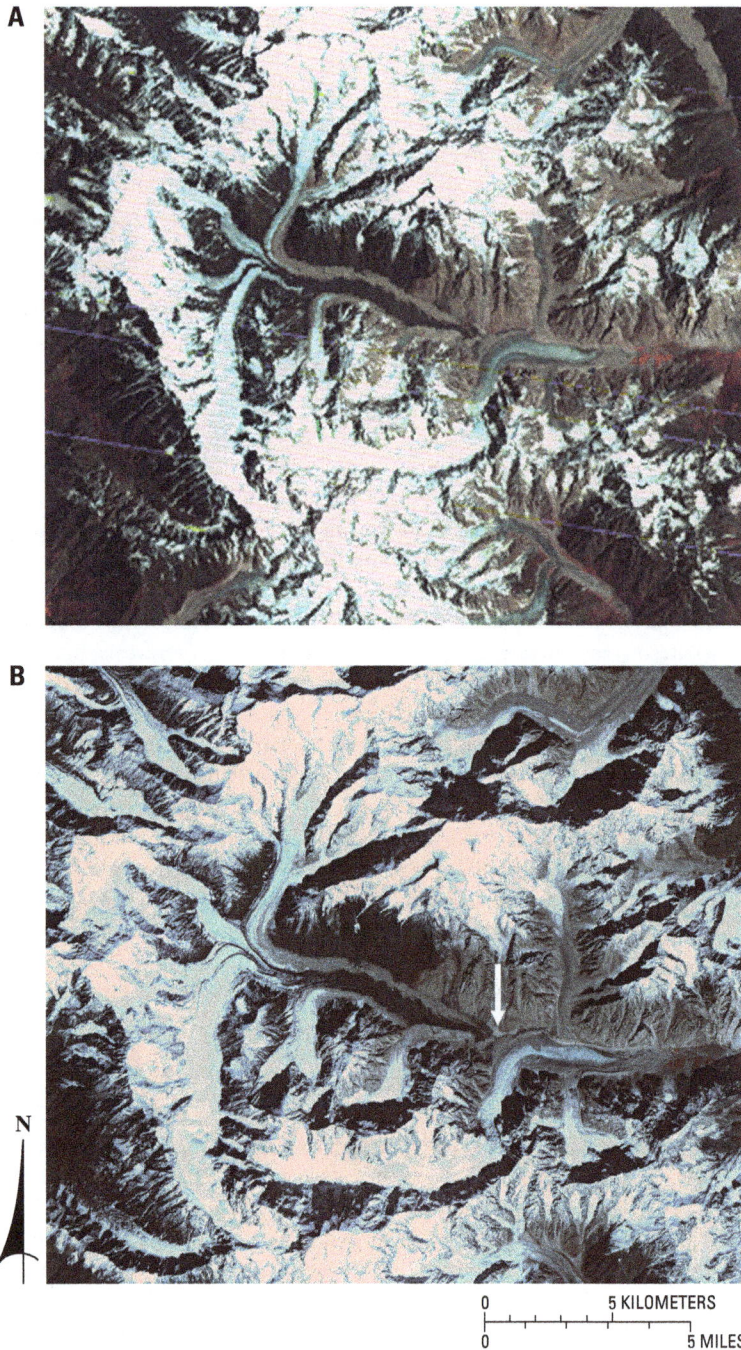

A

B

N

Figure 3.—A, Landsat 1 MSS false-color composite image taken on 12 July 1973, and **B**, ASTER image taken on 28 September 2001 of Tirich Mir glaciers (approximate lat 36°19'N., long 71°53'E.). Both scenes show that the Upper Tirich Glacier [left center] is disconnected from the Lower Tirich Glacier [right center], which is also shown in the ground photograph of this feature in figure 4. The location and direction of figure 4 are shown by the arrow. Many of the glaciers in the later ASTER scene appear somewhat larger than in the older MSS scene, but this may be only an artifact due to new snow in the late September scene. Landsat 1 MSS image no. 1163035007319390, bands 4, 5, 7; 12 July 1973; Path 163–Row 35; and ASTER image no. AST-L1B-00309282001061101112920030800213; 28 September 2001, are from the U.S. Geological Survey, EROS Data Center, Sioux Falls, South Dakota 57198.

| 0 | | | | | 5 KILOMETERS |
| 0 | | | | | 5 MILES |

3,000 m (~9,800 ft) in elevation (figs. 5, 6). Porter (1970, p. 1,441) noted general snowlines at elevations of 4,100–4,400 m for the Swat area. Massive snow avalanches fall from the peak above the glacier and travel down its entire length to a zone 2–3 km past the end of the lowest Neoglacial moraine. Slopes surrounding the lower moraine and avalanche chute below are heavily vegetated with grasses, birch, and conifers. The rich vegetation is caused by greater precipitation south of the main mountain ridge, in areas more exposed

Figure 4.—Photograph taken in July 1999 of the disconnection between Upper Tirich Glacier on the right and the ogives of the Lower Tirich Glacier on the left. Photograph by M.P. Bishop.

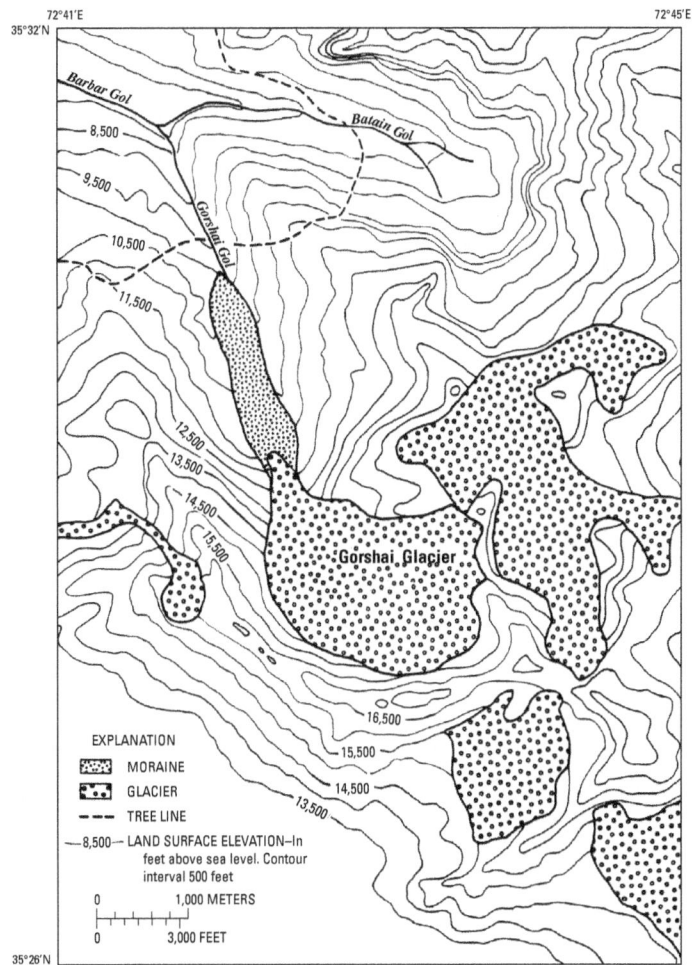

Figure 5.—Gorshai Glacier, near Matiltan, Swat Kohistan. This glacier is typical of hundreds of unmapped glaciers in the Hindu Kush, Hindu Raj, and Kohistan ranges; however, it was unusual in this study because the otherwise classified large-scale aerial photographs and topographic maps of it were made available by the Government of Pakistan for this study.

EXPLANATION

MORAINE

GLACIER

TREE LINE

—8,500— LAND SURFACE ELEVATION—In feet above sea level. Contour interval 500 feet

Figure 6.—Gorshai Glacier (arrows) (approximate lat 35°29'N., long 72°43'E.) **A**, Landsat 2 MSS false-color composite image taken on 6 June 1978, and **B**, ASTER image taken on 13 October 2003 of the same area. The small size and lower pixel resolution (79 m) of the MSS imagery precludes much change detection between the two scenes. Such small and more easily accessible glaciers can, however, make good benchmark glaciers for long-term monitoring, although crevasse hazards may be a problem with this glacier. Landsat 2 MSS image no. 2162035007815790, bands 4, 5, 7; 6 June 1978; Path 162–Row 35, and ASTER image no. AST_LIB_003 10132003055911_10272003095214; 13 October 2003, are from U.S. Geological Survey, EROS Data Center, Sioux Falls, South Dakota 57198.

to monsoon and other westerly precipitation. The result is a strong contrast between glacial features and surrounding terrain on Landsat MSS false-color composite imagery, such as the Landsat 2 MSS image used for figure 6A. In general, the elevations of the mountains and the aridity increase and the vegetation decreases toward the north, with the result that contrast between glaciers, moraine, and the surrounding areas declines from Swat toward the interior of the Hindu Kush, Hindu Raj, and the Karakoram Himalaya.

Glaciers of Nanga Parbat

The Nanga Parbat massif or Diāmir (8,125 m), the ninth highest mountain in the world, is located in the Indus River drainage basin and is significant in terms of its high magnitude erosion and unique mountain geodynamics (Zeitler, Koons, and others, 2001; Zeitler, Meltzer, and others, 2001). The mountain exhibits 69 glaciers that extend down to approximately 3,000 m and cover a total area of 302 km². The mountain receives less than 120 mm a^{-1} in precipitation below 2,500 m, to more than 8,000 mm a^{-1} above 4,500 m; the variation greatly influences glacier type and dynamics (Kick, 1980).

The glaciers of Nanga Parbat were first described by European explorers in the mid-1800s (Kick, 1975). The first detailed measurements, however, were made by a German expedition in 1934 (Finsterwalder and others, 1935; Finsterwalder, 1937). The German expedition produced a highly accurate 1:50,000-scale map of Nanga Parbat that is one of the most detailed maps in the western Himalaya. Additional glaciological and glacial geomorphological studies have been conducted by Pillewizer (1956), Loewe (1959, 1961), Gardner (1986), Scott (1992), Gardner and Jones (1993), Bishop and others (1999), Shroder and Bishop (2000), and Shroder and others (2000). The glaciers surrounding Nanga Parbat (figs. 7, 8) are directly exposed to high monsoon-generated precipitation and extremes of relief on cirque headwalls (3,000–4,000 m). This contributes to massive ice falls and snow avalanches that create extensive debris cover. Some of the glaciers on the steep slopes are moving rapidly in a tumbling "Blockschollen" fashion, in

Figure 7.—Part of a Landsat 3 RBV image showing Nanga Parbat massif. Many of the glaciers surrounding the massif are covered with supraglacier debris and therefore rather dark colored. Only Rākiot Glacier [92] on the north side of the massif has much white snow-and-ice exposed. Sachen Glacier [95] on the northeast side was investigated in detail for this study. Rākiot [92] and Chungpar Glaciers [96] have the greatest rates of retreat since 1930. Landsat 3 RBV image no. 30515–05001, subscene D; 2 August 1979; Path 161–Row 35, was from the U.S. Geological Survey, EROS Data Center, Sioux Falls, South Dakota 57198, but is now only archived by the authors.

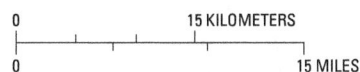

74°30'E

35°30'N

0   15 KILOMETERS

0   15 MILES

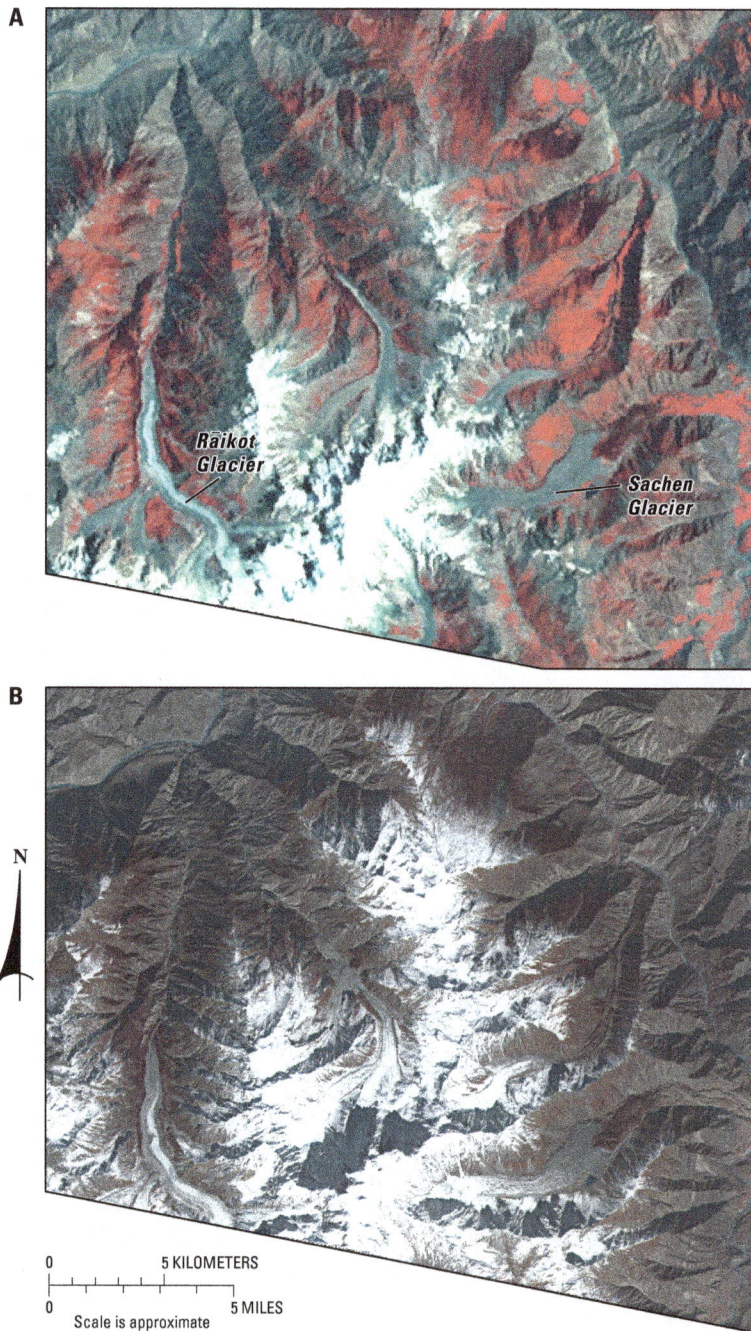

A

Rāikot
Glacier

Sachen
Glacier

B

N

0   5 KILOMETERS
0   5 MILES
Scale is approximate

Figure 8.—**A**, *Landsat 2 MSS false-color composite image taken on 2 August 1977, and* **B**, *ASTER image taken on 29 October 2003, showing* Rāikot Glacier *[92] (approximate lat 35°20'N., long 74°35'E.) on the lower left and* Sachen Glacier *[95] (approximate lat 35°20'N., long 74°45'E.) on the lower right. Landsat 2 MSS image no. 2160035007721490, bands 4, 5, 7; 2 August 1977; Path 160–Row 35; and ASTER image no. AST_LIA_ 00310292003055903_ 11122003124818; 29 October 2003, are from the U.S. Geological Survey, EROS Data Center, Sioux Falls, S. Dak. 57198.*

which the entire volume of the glacier is moving as a plug broken up into large ice-clods or blocks that jostle about somewhat independently (Mason, 1930; Finsterwalder and Pillewizer, 1939). The glaciers of Nanga Parbat descend from areas of fairly low accumulation to termini at about 3,000–3,800 m elevation — not into the nearby hot and arid Indus valley at 1,500 m. Elsewhere in the Karakoram, many glaciers descend from zones of large accumulation and ablate in the low, hot valleys, transporting large loads of ice and debris to lower elevations.

In general, the glaciers of Nanga Parbat have experienced a rather minor retreat since AD 1850. The two glaciers with greatest rates of retreat since 1930, *Rāikot* [*Rakhiot*] [92] and *Chungpar* [96], have had ice velocities that have increased two or more times since retreat began (Mayewski and Jeschke, 1979). At the same time, the *Sachen Glacier* [95] experienced almost no change, probably because it is a special class of glacier, a "dam glacier," that flows on its own bed of debris and raises the level of debris at the glacier bottom (Kick (1962, p. 228–229)). The lower portion of the glacier may become stagnant through excessive loading with rock debris, thereby allowing the upper portion to shear over it (Visser and Visser-Hooft, 1938; as reported by Washburn, 1939).

Sachen Glacier

Sachen Glacier [95] (Shroder and others, 2000) occurs on the northeast side of Nanga Parbat (figs. 8, 9). Its accumulation area is composed of several small firn fields and ice falls on the upper slopes of *Chongra* peak (6,891 m). Accumulation also occurs due to snow avalanches from surrounding ridges. The glacier is characterized by long, wide ablation valleys between the valley walls and large lateral moraines along both sides of the glacier. In at least four places associated with the lateral moraine, rock-glacier-like features occur, with

Figure 9.—Sachen Glacier *[95] on the northeast slopes of Nanga Parbat above* Astor *valley. The digitate terminus has five lobes that have been active at different times in the past. The presently active terminus is in the lower center of the 19 July 1984 photograph.* Lake Sango Sar *is impounded behind the inactive rock-glacier-like lobe on left. Photograph by J.F. Shroder, Jr.*

the characteristic morphology of a steep front at the angle of repose, a sharp angle between front and top, and transverse ridges and furrows. These rock glaciers either issue from the moraine at right angles to it, or come out of a tributary glacier and flow into the ablation valley.

The digitate front of *Sachen Glacier* has not backwasted significantly in 50 years, although we observed up to 25 m of recent downwasting on its northernmost front. The southern group of fronts (fig. 9) comprise two rock-glacier-like masses with steep fronts at an angle of repose that impounded lake *Sango Sar*, and two fronts of debris-covered, kettle-pocked ice. The most active front is in the center of figure 9. In 1934, this front had two arcuate moraine-covered ice lobes across its upper surface (Finsterwalder, 1937, fig 10 and topographic map), but by 1984 they had disappeared and high up on the front of the mass a large meltwater stream was flowing out.

Rāikot Glacier

The *Rāikot* (formerly *Rakhiot*) *Glacier* [92], on the north side of Nanga Parbat, is one of the best studied in the Nanga Parbat Himalaya because of the ease of access from the Karakoram Highway (KKH) at the base of the mountain. The glacier is one of the largest on the mountain, with a length of approximately 14 km, an elevation range of 3,150 to almost 8,000 m, and a total connected area of 32 km^2 (Gardner and Jones, 1993).

The *Rāikot Glacier* includes a firn basin surrounded by steep slopes above 5,500 m; it is the accumulation area and is frequently fed by snow and ice avalanches. Below the firn basin, a steep ice fall occurs between 5,500 and 4,100 m, and below that the glacier becomes increasingly debris-covered toward the glacier terminus at about 3,150 m (figs. 7, 8). Based upon measurements made by us and the Pakistan-Canada Snow and Ice Hydrology project (Gardner and Jones, 1993, and Finsterwalder, 1937), maximum ice velocities are about 900 m a^{-1} near the base of the ice fall 8 km from the terminus, 302 m a^{-1} at 2 km from the terminus, 171 m a^{-1} at 1 km from the terminus, and 94 m a^{-1} at 0.5 km from the terminus. Between 1934 and 1954, the glacier terminus thinned and retreated approximately 450 m, whereas between 1954 and 1985, it thickened and advanced approximately 250 m (Gardner, 1986). Repeat photography from 1985, 1989, 1991, 1993, 1995, 1996, and 1997 shows that the advance culminated in a thickening and steepening of the front and sides, coupled with an outburst flood in early spring, 1994, that eroded much lateral, terminal, and supraglacial debris (figs. 10, 11).

The *Rāikot Glacier* is covered with a thick mantle of supraglacial debris over most of its ablation zone (figs. 8, 10). The glacier has downwasted considerably since its most recent advance, and is therefore bordered by very steep and high lateral moraines. Most of the supraglacial debris originates from these 50- to 70-m high lateral moraines, mainly as a result of landsliding and rockfall during rainstorms. Snow and rock avalanches also transport debris onto the glacier surface in the upper portion of the basin. Our debris-cover depth measurements taken in 1996 and 1997 indicate that glacier surface topography and debris depth is highly variable, although the supraglacial debris is thickest at the margins and terminus of the glacier, where it reaches depths of up to 4–5 m, and is thinner near the glacier center, where depths are typically 10–20 cm. Areas of shallow debris depths can be mapped via satellite imagery, because reflection in the near-infrared part of the spectrum is enhanced by ablation and melt-water production that occur in these areas.

Glacier flow dynamics and hydrology also play an important role in governing the variability of supraglacial debris depths and sediment transport. Fieldwork in 1995 led to the discovery of catastrophic glacier outburst floods

Figure 10.—(at right) **A**, *Photograph taken in July 1991 of the* Rāikot Glacier *[92] terminus (looking south). View showing generally gentle sides and plentiful lateral and terminal morainal debris. The main melt-water ice portal is in the center, with a smaller subsidiary ice portal issuing on the right (arrows). Photograph by J.F. Shroder, Jr.* **B**, *photo taken in June 1996 of the* Rāikot Glacier *terminus 2 years after the breakout flood of early spring 1994. The terminus had begun to form its steep sides from a renewed advance about 1993. The dark scar on the upper right lateral margin of the glacier (west glacier side) was the main meltwater flood ice portal in 1994. Note the removal of a considerable quantity of ice-marginal debris in the interim. Photograph by M.P. Bishop.*

A

B

Figure 11.—*Largest (47 m wide, 23 m deep) ice portal in the upper west margin of* Rāikot Glacier [92] *from the 1994 outburst flood. Three other flood outlet portals were also noted, two others on the west side, and another on the top of the terminus. The outburst flood from the four ice portals in early spring 1994 allowed the flood water to remove large quantities of ice-marginal and supraglacial debris. August 1995 photograph by J.F. Shroder, Jr.*

from four ice portals in the early spring of 1994 that originated on the surface of the glacier terminus and on the western lateral margin (Shroder and others, 2000). Presumably, internal blockage of water flow through englacial conduits led to the build-up of hydrostatic pressure, eventually causing the catastrophic outburst floods from these ice portals (figs. 10, 11). This led to the removal of significant quantities of supraglacial debris cover and the transport of sediment along the western side of the glacier, thereby increasing ablation along the exposed ice margin and over portions of the terminus.

Satellite imagery reveals that this relatively small glacier contains a limited number of supraglacial lakes, in comparison to the numerous, sizeable lakes observed on larger glaciers in Pakistan. Multitemporal imagery documents minor terminus position fluctuations, although a temporally consistent trend, of either advance or retreat, has not been observed. Comparison of SPOT panchromatic and ASTER imagery reveals that the terminus position of the glacier has not significantly changed, presumably because of summer monsoon accumulation. However, at a higher elevation two tributary glaciers, *Ganalo* and the *Chongra-Rāikot Glaciers*, were connected in the recent past; but we have observed that there is currently a disconnect between these tributary glaciers and with the *Rāikot Glacier* (figs. 8, 12).

Batūra Glacier

The large Batūra Glacier [41] to the north in the upper Hunza region is prominent on satellite imagery (fig. 13) and has been assessed in detail, in part because of its accessibility from the KKH. Batūra Glacier flows from west to east below the north side of the peaks of Batūra Muztagh (7,795 m) from 6,200 m into the upper Hunza valley at 2,516 m. The glacier is notorious for

Figure 12.—Photograph taken in August 1997 showing the connection (arrow) between Chongra-Rāikot Glacier *and the main* Rāikot Glacier *[92]. The two glaciers have been observed subsequently in the field and on recent imagery and are now disconnected. Photograph by M.P. Bishop.*

large avalanches (Edwards, 1960; Finsterwalder, 1960; Shi and Wang, 1980) that contribute greatly to the mass balance, as well as to the plentiful debris load. The snowline occurs at about 5,000 m, and the annual 0 °C isotherm is near 4,200 m (BGIG, 1979). The glacier is therefore cold in its upper reaches, and temperate in its middle and lower reaches, where two-thirds of the main glacier is covered with debris. Only a narrow (about 700 m) strip of white ice occurs along the south side of the glacier to within about 4 km of the terminus (fig. 14).

Chinese glaciologists (BGIG, 1979) found evidence that the Batūra Glacier was not a surging glacier, but predicted that its terminus would advance in the 1980s to threaten the KKH and then retreat in the 1990s. The predicted advance never occurred and instead the frontal ice cliff has downwasted and the portal above the main meltwater channel has backwasted (Shroder and others, 1984). The debris-covered ice is not easily seen on the imagery, but vegetation on the outwash and kettle ponds on the terminus provide good contrast (figs. 15, 16B).

The glacier consists of five main ice flows and more than 20 smaller tributary glaciers (fig. 16). The basin consists of heavy ice cover and snow on the higher north-facing slopes, with less extensive tributary glaciers on the north side. Extensive ablation valley complexes exist on the north side, and glacial meltwater flow and breakout floods from supraglacial ponds have been documented in these valleys.

The dominant uppermost surface of the glacier is characterized by featureless cirques that pass downward through areas of extentional flow over ice falls where the glacier breaks up into seracs. These areas are commonly below the equilibrium line, and thin medial moraines first appear from beneath the firn and pass downward into the crevasses. The western ice flow

Figure 13.—*Landsat 3 RBV image mosaic showing glaciers in the Hunza region of the Karakoram Range. The main valleys here are the Hunza valley with the Batūra [41], Pasu [42], and Ghulkin [43] Glaciers, the Shimshal valley with the Momhil [57], Mulungūtti [58], Yāzghil [59], Yukshin Gardan [60], Khurdopīn [61], and Vijerab [62] Glaciers, and the Hispar valley with Hispar [72], Lak-Khiang [65], Pumārkish [66], Jutmau [67], Khanbusa [68], Haigatum [69], Makrong [70], Garumbar [71], and Yengutz Har [73] Glaciers. The images provide sufficient overlap and can, therefore, be used as a stereoscopic pair. Landsat 3 RBV image nos. 30515–05001, subscene B; 2 August 1979, Path 161–Row 35 (left) and 30532–04542, subscene A; 19 August 1979; Path 160–Row 35 (right) were from the U.S. Geological Survey, EROS Data Center, Sioux Falls, South Dakota 57198, but are now only archived by the authors.*

and its serac field flows downward as a thin white ice stream, and joins an ice flow from the north, which exhibits ogives due to compressive flow from higher elevations. The white ice stream that extends to lower elevations is subject to greater ablation, and the ice undergoes structural changes from ogives to reticulated ice hills. All of these features are clearly visible on SPOT and ASTER imagery.

The remainder of the glacial surface is heavily debris covered and exhibits a variety of supraglacial features. Field measurements and observations show that the debris is granodiorite on the north side of the glacier and black slate on the south side. Satellite imagery also indicates spectral differentiation of granodiorite that is related to weathering. The northern side of the glacier exhibits weathered granodiorite, suggesting that the less weathered debris on the south side results from more active ice flow and sediment redistribution. In general, debris thickness increases toward the terminus and the edges of the glacier.

Figure 14.—Lower part of Batūra Glacier [41] from Yunz, site of a former diffluent valley from Batūra to Pasu Glacier [42]. The white ice surface in the background is particularly noticeable on the imagery (figs. 13, 16) and a useful marker for monitoring future changes. Photograph taken June 1984 by J.F. Shroder, Jr.

Figure 15.—Photograph taken June 1984 showing terminus of Batūra Glacier [41]. The moraine-covered steep slopes directly behind the vegetation in the middle ground were a prominent ice cliff in the late 1970s when investigated by Chinese scientists. The cliff is not obvious on the 1979 imagery, but the vegetation and water-filled kettles produce a strong signature (figs. 13, 16). Photograph by J.F. Shroder, Jr.

A

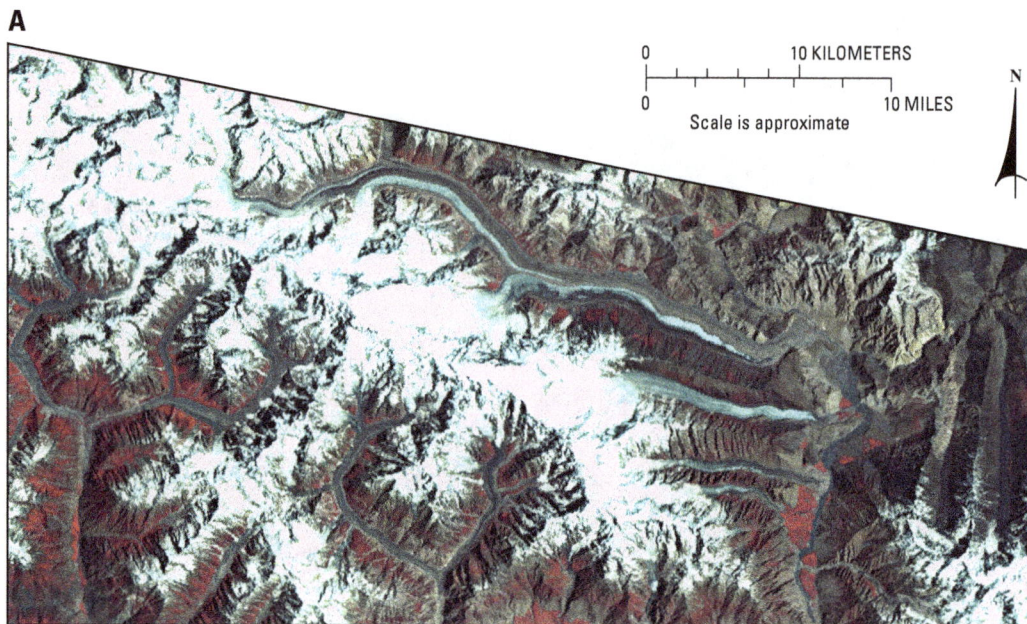

B

Figure 16.—A, *Landsat 3 MSS false-color composite image of Batūra Glacier [41] (approximate lat 36°32'N., long 74°40'E.) taken on 15 July 1979, and* B, *ASTER image taken on 30 April 2001. Landsat 3 MSS image no. 3161035007919690, bands 4, 5, 7; 15 July 1979; Path 161–Row 35; and ASTER image no. AST_L1B__003_0 4302001060850_04172002082052_vnir_sub; 30 April 2001 are from the U.S. Geological Survey, EROS Data Center, Sioux Falls, South Dakota 57918.*

The Batūra Glacier exhibits a variety of features and characteristics that suggest negative mass balance. From the 19th century until the middle 1940s, the glacier terminus approached the Hunza River (Goudie and others, 1984). Since then, that region has undergone ice flow stagnation and downwasting such that hummocky ground moraine covers a large portion of the proglacial environment (fig. 15). Similarly, the glacier surface has significantly downwasted below the lateral moraines.

Satellite image analysis and change-detection studies support these interpretations, and provide additional evidence. For example, Bishop and others (1995, 1998) conducted a change-detection study and found that the major white-ice stream has been systematically retreating since 1974. Examination of recent ASTER imagery indicates the continuation of this trend (fig. 16B). Differential downwasting of the glacier surface is expected to produce this result; englacial debris can accumulate at the surface and the higher ablation over the exposed ice can generate topographic variation, resulting in the redistribution of nearby supraglacial debris.

Biafo Glacier

At approximately 853 km² in size, 628 km² of which is permanent snow and ice, Biafo Glacier [121] is one of the largest and longest (68 km if upper-most tributaries are included) of the Karakoram glaciers. It differs from many others in the Himalaya because it is mainly nourished by direct snowfall rather than by avalanching (Hewitt and others, 1989). Biafo Glacier's measured annual snow accumulation rate of approximately 0.6 km³, thickness (perhaps 1.4 km at the equilibrium line), and flow rates of 0.8 m d⁻¹ in summer equate to an annual ice flux through the equilibrium line of 0.7 km³ a⁻¹; this matches stake ablation rates equating to 0.7 km³ a⁻¹. The approximate concordance of the three measurements by Hewitt and others (1989) indicates that the ablation zone of Biafo Glacier, whose area covers about 0.09 percent of the whole upper Indus basin, produces approximately 0.09 percent of the total runoff.

The Biafo Glacier has largely linear moraines (fig. 17), and a reported history of considerable fluctuation of its terminus (Mason, 1930; Auden, 1935; Hewitt, 1969; Mayewski and Jeschke, 1979; Hewitt and others, 1989) (fig. 18). Some of this variability may be seasonal, but there has been a general retreat of 0.5 –1 km in the last 120 years. Perhaps of greater significance, however, are the changes on the upper parts of the glacier (fig. 19). During the Shaksgam expedition, Conway's "Snow Lake" consisted of "mostly bare ice" when it was traversed on 20 August 1937 (Shipton, 1938, p. 325). Close-up and panoramic photographs of the same area by Workman and Workman (1911) taken on 16 August 1908 and by Shipton (1940) on 19 August 1939 show extensive snow and white firn fields over clean ice. Shipton (1940, p. 414) noted that travel on the Biafo Glacier was remarkably easy on 19-21 August 1939 "as the smooth white ice extends from its upper reaches almost to the snout." The expedition map (Mott, 1950) also shows extensive clean ice or firn covering much of the upper two-thirds of the glacier. Only thin lateral moraines occurred along the margins, including a narrow strip about 700–800 m wide on the north side of the *Sim Gang* part of "Snow Lake." About forty years later, in 1977 and 1979, (figs. 17, 19, and 20), only a thin band of white ice occurred in the lower third of the glacier, and much of the upper part was dark and debris covered. Even the *Sim Gang* area, at an elevation of about 5,000 m, has a debris-covered surface up to 2 km wide and 8 km long. People who have trekked across this region in recent years have reported that progress in summer months can be most difficult because much of the *Sim Gang* firn has become water saturated, in contrast to prior years of easy travel (Searle, oral comm. to JFS, 1988).

Figure 17.—*Landsat 2 MSS false-color composite image of Biafo Glacier [121] (approximate lat 35°51'N., long 75°45'E.) taken on 2 August 1977, showing linear medial and lateral moraines, together with a complex equilibrium zone in the ice-fall area as well as in the Sim Gang ("Snow Lake") area (see fig. 19 for locations). The south-facing high peaks on the north side of the east-west trending Sim Gang tributary to Biafo provide topographic control of reflected irradiance that strongly affects local ablation, and the location of the transient snowlines. Landsat 2 MSS image no. 2160035007721490, bands 4, 5, 7; 2 August 1977; Path 160–Row 35 is from the U.S. Geological Survey, EROS Data Center, Sioux Falls, South Dakota 57198.*

The ablation region of Biafo Glacier was once characterized by ice standing high above its lateral moraines and the bounding lateral troughs, or 'ablation valleys', but Hewitt and others (1989) have calculated that the glacier has lost at least 2 km³ of its ice mass between 1910 and 1960. Net annual ice losses due to the wastage of the glacier in the 20th century are estimated to be 0.4–0.5 m a⁻¹, which represents 12–15 percent of the annual water yield from ice melt.

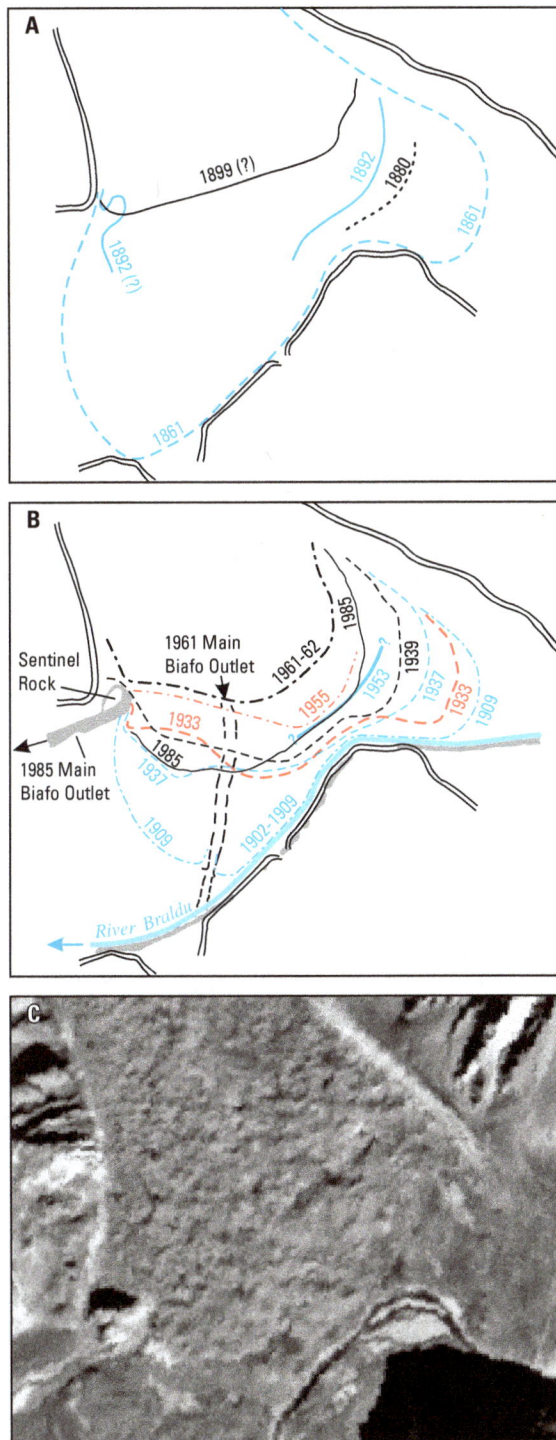

Figure 18.—Terminus, ice portal, and meltwater stream locations of Biafo Glacier [121] (approximate lat 35°51'N., long. 75°45'E.) in the **A**, 19th century map sources; **B**, 20th century map sources, and **C**, 21st century (2004) ASTER image (A and B after Hewitt and others, 1989). Note the existence of the ice portal and melt-water stream discharge in the same position in 2004 as in 1985, as well as the new lake at the terminus. Aster image no. AST_LIA_ 003_08142004054614_08282004141219, 14 August 2004, is from the U.S. Geological Survey, EROS Data Center, Sioux Falls, South Dakota 57198.

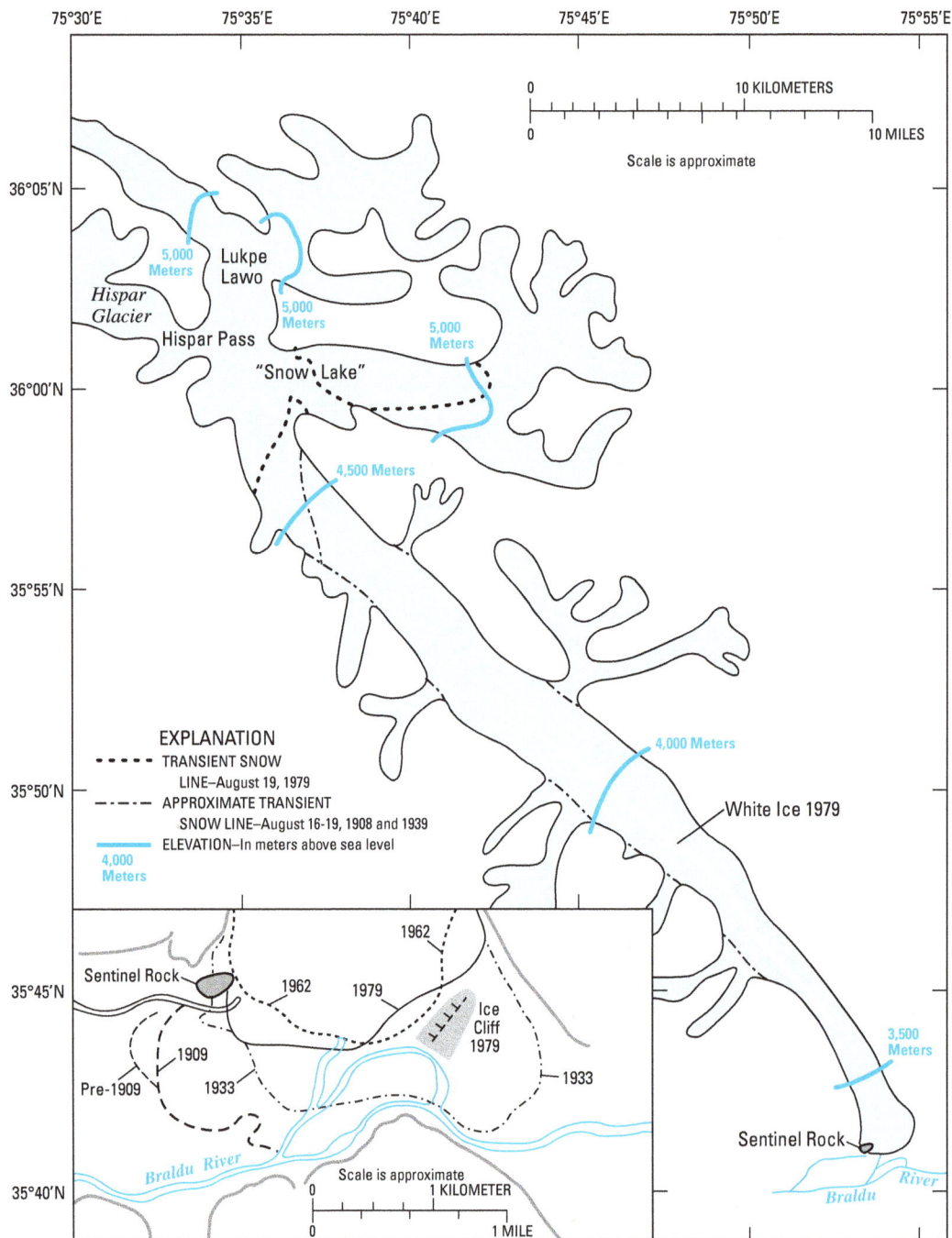

Figure 19.—*Sketch map of Biafo Glacier [121] comparing conditions of the overall glacier in August 1908, 1939, and 1979. The transient snow lines of 1908 and 1939 are estimates based on oblique photographs by Workman and Workman (1911) and by Shipton (1940), and on the map by Mott (1950). Mott's map also provided the base for this figure. The terminus positions are from Auden (1935) and Hewitt (1969). All 1979 information is from Landsat 3 RBV image no. 30532–04542, subscene D, 19 August 1979, Path 160, row 35 (fig. 20), with stereo overlap of the "Snow Lake" head area using Landsat 3 RBV image no. 30532–04542, subscene B, 19 August 1979, Path 160–Row 35, and the terminus area using Landsat 3 RBV image no. 30531–04484, subscene C, 18 August 1979, Path 159– Row 35 (fig. 20).*

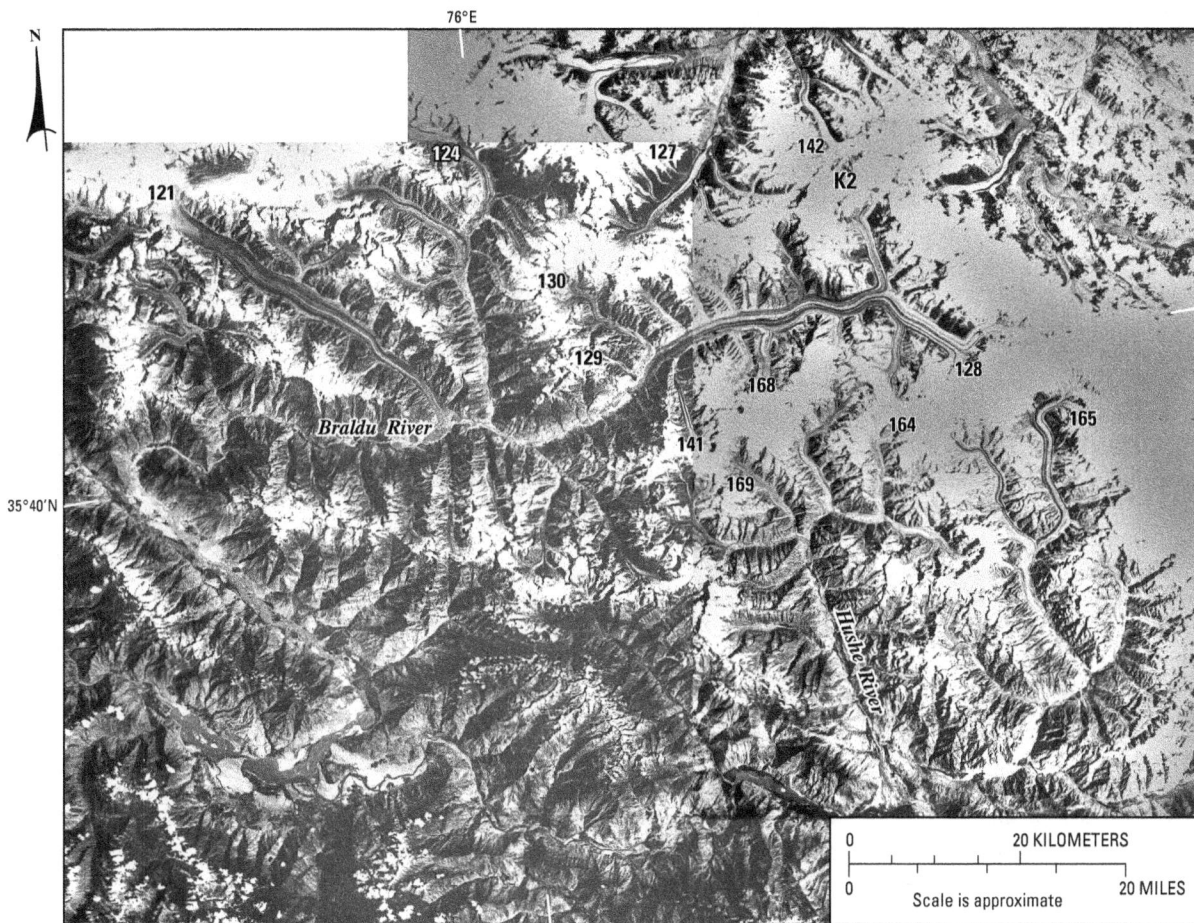

Figure 20.—*Landsat 3 RBV image mosaic showing glaciers in the K2 area of the Karakoram Range. Within the mosaic are part of Biafo [121], Choktoi-Panmah [124], Baltoro [128], Sarpo Laggo [127], Kondus [165], and Kaberi [164] Glaciers. K2 (Qogir Feng) is at the upper right. The images provide sufficient overlap that they can be used as a stereoscopic pair. The Landsat 3 RBV image nos. 30532–04542, subscene D; 19 August 1979; Path 160–Row 35 (left) and 30531–04484, subscene C; 18 August 1979; Path 159–Row 35 (right) were from the U.S. Geological Survey, EROS Data Center, Sioux Falls, South Dakota 57198, but are now only archived by the authors.*

Glaciers Having Unusual or Irregular Flow

The Karakoram and Alaska-Yukon regions account for about 90 percent of the known surging glacier events (Hewitt, 1969). These events are thought to be caused by the buildup of high water pressure in the basal passageway system or in deformable bed sediments (Kamb and others, 1985; Hambrey, 1994; Menzies, 2002). Wang and others (1984) described a breakout flood from the surging *Balt Bare Glacier* in the Karakoram that destroyed part of the KKH. The exceptionally hot (summer temperatures commonly greater than 35 °C) ablation zones in the deep arid valleys of the Karakoram Himalaya contributed meltwater that may have been a factor in surging there.

Unusual or irregular glacier flow was investigated using MSS imagery on western Himalaya glaciers which had histories of advance, retreat, or surge (table 2). Most of these glaciers show no obvious surficial evidence of such movement. The smaller glaciers generally showed the least evidence of movement, in part because the pixel resolution of the MSS imagery did not allow discrimination of diagnostic features on smaller glaciers. Also, many glaciers have a history of rapid movement followed by stagnation and downwasting that produces an extensive debris cover, kettles, and other similar thermokarst phenomena. These surface features are common in the western Himalaya, and limit the potential to differentiate glaciers with irregular rapid movement on MSS imagery.

Using features visible on the older Landsat small-scale imagery, we developed a set of criteria to categorize as many different types of supraglacial features as possible. A few criteria had already been developed to differentiate between surging and nonsurging glaciers using satellite-image interpretation (Krimmel and others, 1976; Meier, 1976; Post and others, 1976; Holdsworth and others, 2002). Kotlyakov (1980) listed surge criteria used in the Soviet Union; Liestøl (1993, table 4, p. E137) listed the names of 86 glaciers in Svalbard which surged between 1838 and 1990; and Elson (1980) listed some general criteria, not all of which are visible on MSS imagery. Elson (1980) also noted that glacial features on Landsat images should be listed as "resolved" if they were identified without prior knowledge, and "detected" if previous information was used to locate them. Major "resolved" surge features include convoluted medial moraines (definitive evidence of surge-type glaciers), unusual pits on the surface of the accumulation area, a stagnant lower part, chaotic crevasses, sheared-off tributary glaciers, the thrust of glacier tongues over other glaciers, large horizontal displacements, and a bulging terminal profile. Goudie and others (1984, p. 413) studied "so-called 'surging' glaciers" in the Karakoram, and noted that major advances and retreats seem to simply reflect the way in which glaciers can respond rapidly as environmental conditions dictate. In other words, they appear to be saying that a sharp distinction between surge-type and non-surge-type glaciers might not exist. Based on these ideas, and using the satellite imagery, we developed a list of resolvable criteria for recognizing the various glacier types in northern Pakistan. The criteria tend to be gradually transitional from one to another, so that a clear distinction between surge-type and non-surge-type glaciers in northern Pakistan may not be possible, if in fact this distinction really exists at all.

Criteria for Recognizing Different Types of Unusual or Rapid Glacier Flow or Retreat

Criteria were developed primarily for use with the older Landsat 3 RBV and MSS imagery, and the numbers of glaciers identified as being of particular types have changed in the intervening decades. The criteria and the numbers of glaciers exhibiting these criteria are:

1. Strongly convoluted medial or lateral moraines (5 glaciers).

2. Medial or lateral moraines convoluted or offset by extensive crevasses or ogives (13 glaciers).

3. Tributary ice overriding or displacing main glacier ice (7 glaciers).

4. Marked depression of ice surface below lateral moraines or tributary glaciers (14 glaciers).

5. Lateral and medial moraines slightly sinuous, or lobate, or offset by crevasses or ogives (76 glaciers).

6. Recent major melting or retreat of ice front leaving light-colored deglacierized terrain (9 glaciers).

TABLE 2.—*Glaciers in northern Pakistan that were directly observed to retreat, advance, or surge sometime in the past*

[Glaciers were selected to test whether or not prior movement or retreat could be detected on Landsat imagery. Glaciers are located in figure 1. Exceptional advances are from Hewitt (1969) and Miller (1984)]

Glacier number	Glacier name and location	Latitude	Longitude	Advance(s)	Features visible on satellite images[1]
34	*Karambar* (Gilgit-*Iskhoman* valley)	36°38′N.	74°10′E.	1930	Slight ogive or crevasse pattern on right (north) lateral moraine; medial moraine not convoluted; end moraine extensively downwasted. Images used for analysis were (2) and (3) as a stereoscopic pair
41	Batūra (Hunza valley)	36°32′N.	74°40′E.	1913–25; 1974–79	Ogives; slightly undulose medial moraine (figs. 14, 15, 16). Images used for analysis were (4) and (5) as a stereoscopic pair
42	*Pasu* (Hunza valley)	36°28′N.	74°45′E.	c.1910–1930	Ogives; ice surface deeply below lateral moraines; snout backwasted. Images used for analysis were (4) and (5) as stereoscopic pair
43	*Ghulkin* (Hunza valley)	36°25′N.	74°50′E.	1913–25, 1966–78	Ogives; ice surface below lateral moraine; snout digitate and far advanced to furthest end moraine above Hunza river. Images used for analysis were (4) and (5) as a stereoscopic pair
45–47	*Hasanabad* and tributaries (Hunza valley)	36°21′N.	74°34′E.	1903–06 surge	Extensive dirty ice; no evidence of rapid movement; two tributary glaciers joined and flow together about 1 km from terminus. Images used for analysis were (1) and (4)
54	*Balt Bare* (Hunza valley)	36°20′N.	74°56′E.	1976 surge	No evidence of rapid movement; light-colored scars of 1974 debris-flow from glacier visible in meltwater channel. Images used for analysis were (4) and (5) as a stereoscopic pair
71	*Garumbar* (Hispar valley)	36°07′N.	75°01′E.	1892–1925?	Prominent recently deglaciated gorge for 4 km to Hispar Glacier with strong downwasted and ice-collapse features nearly to snow-line (fig. 13). Images used for analysis were (4) and (5) as a stereoscopic pair
73	Yengutz Har (Hispar valley)	36°07′N.	74°58′E.	1901–03 surge	Small with dirty ice; no evidence of rapid movement; sizeable alluvial fan at canyon mouth below glacier at Hispar village (fig. 13). Images used for analysis were (4) and (5) as a stereoscopic pair
75	*Hopar* (Hispar valley)	36°10′N.	74°57′E.	1929–30	Dirty ice; no evidence of rapid movement. Images used for analysis were (4) and (5) as a stereoscopic pair
77	Minapin (Hunza valley)	36°11′N.	74°34′E.	1892–93	Moraines not convoluted. Images used for analysis were (1) and (4)
90	*Kutiah* (Stak valley)	35°48′N.	75°00′E.	1953 surge	Some dirty ice; no evidence of rapid movement. Image used for analysis was (1)
108	*Niamul Gans* (tributary of Chogo Lungma, Shigar valley)	35°48′N.	75°15′E.	1902–03	Dark and obscure. Image used for analysis was (6)

[1]Landsat 2 MSS and 3 RBV images used for analysis were (1) 2161035007726990, (2) 30515–04595–C, (3) 30515–05001–A, (4) 30515–05001–B, (5) 30532–04542–A and (6) 30532–04542–C. See table 1 for more information about the images.

7. Extensive area of debris-covered downwasting ice or stagnant ice (106 glaciers).

8. Linear moraines and accordant tributaries (47 glaciers).

Because of difficulty distinguishing different types of motion observed in the field on the MSS imagery, a general study was conducted on northern Pakistan glaciers for which good data existed. Many of the glaciers of the Ghujerab Mountains (Chapchingal and Karun Kuh Groups) north of the main Karakoram, and the small glaciers of the Hindu Kush, Hindu Raj, and Kohistan, were not considered. A few glaciers on the Chinese side were included because of their close proximity to Pakistan and/or because they exhibited some unusual features. The larger glaciers were analyzed preferentially because of the pixel resolution of the MSS imagery, limited ASTER imagery, and because the larger ice masses had more prior information available. Although some smaller glaciers on the periphery of the main Karakoram were included because they had specific characteristics observable on the imagery, most less significant, smaller glaciers in the central Karakoram were excluded. In some cases, tributaries to larger glaciers were treated separately; other tributaries that were unnamed or not particularly different were omitted.

A total of 161 of the larger glaciers of northern Pakistan and 8 in adjacent China were analyzed in this study (listed by number and name in table 3). The longest of these, the Siachen Glacier [155] (Bhutiyana, 1999) in the far northeast corner of Pakistan, is about 67 km long (see fig. 21). The mean length of all glaciers studied is about 15 km. Approximately 75 of the glaciers are between 10–19 km long, 25 between 20–29 km, 4 each from 30–39 km and 40–49 km, 3 between 50–59 km, and only one is 60–70 km in length.

Movement Criteria 1 and 2: Convoluted Medial or Lateral Moraines

A total of 18 glaciers have moraines that are convoluted or strongly convoluted, and these range across northern Pakistan from the mountains on the eastern border to those on the western border. The largest and most prominent examples are in the valleys of Shimshal (fig. 13), Hispar (fig. 13) and Braldu, although several also occur across the border in China, north of the K2 (Qogir Feng) mountain area (fig. 20). In a less prominent but interesting example, the *Selkar Chorten* and *North Terong Glaciers* [153], near the terminus of the Siachen Glacier [155] (fig. 21), have linear moraines and flow together accordingly before turning at right angles into the South Terong river valley. The lowermost 4 km of the terminus was strongly convoluted, suggesting perhaps that meltwater from the *South Terong Glacier* [152] could have been partly responsible. The *South Terong Glacier* was mapped (1938–1946) as connected with *North Terong-Selkar Chorten Glacier* [153], but they were later separated by several kilometers of valley train and meltwater channels (fig. 21).

The glaciers of the Shimshal valley (fig. 13) are well known because of their proclivity to periodically dam the main river (Mason, 1930). The Mulungūtti [58] and Yāzghil [59] Glaciers still extend to the river, essentially the same as they have done since the first maps and descriptions were made in 1892 (Mason, 1930). The Khurdopīn Glacier [61] seems, however, to have undergone some changes due to unusual flow in this same time frame. The glacier is notorious for causing glacier-outburst floods (jökulhlaups), as it periodically blocks meltwater from the Vījerāb Glacier [62]. Mason (1930, p. 248) reported that in about the lower 5 km the "moraine is represented by two dark median lines." Either this is a mistake or some unusual movement has occurred since that time, because high up the glacier there is one linear medial moraine and one wide lateral moraine that continue down the ice to the terminus — a zone of strongly convoluted moraines almost 10 km long (fig. 13, table 3).

TABLE 3.—*Glaciers of northern Pakistan selected for remote-sensing analysis*

[Glacier numbers are cited in the text and plotted on figure 1. Selection based on the following criteria: (1) size large enough to provide resolvable evidence of unusual movement or other resolvable distinguishing landforms; (2) availability of maps or imagery detailed enough to allow discrimination of features. Movement criteria: (1) strongly convoluted medial or lateral moraines (5 glaciers); (2) medial or lateral moraines convoluted or offset by extensive crevasses or ogives (13 glaciers); (3) tributary ice overriding or displacing main glacier ice (7 glaciers); (4) marked depression of ice surface below lateral moraines or tributary glaciers (14 glaciers); (5) lateral and medial moraines slightly sinuous, or lobate, or offset by crevasses or ogives (76 glaciers); (6) recent major melting or retreat of ice front leaving light-colored deglacierized terrain behind (9 glaciers); (7) extensive area of debris-covered downwasting ice or stagnant ice (106 glaciers); and (8) linear moraines and accordant tributaries (47 glaciers). Glacier maps were obtained from many different sources and our field work. In some cases, hyphenated names include tributary names, in others the name itself is a hyphenated phrase. Abbreviations: km, kilometer; ?, glaciers with uncertain names or different spellings]

Glacier number	Glacier name (general location)	Latitude and longitude	Length (km)	Movement criteria								Satellite images[1]
				1	2	3	4	5	6	7	8	
1	Upper Tirich (Chitrāl)	36°20′N. 75°50′E.	20					X				RBV30553–05112–B; 09 Sep 79
2	Lower Tirich (Chitrāl)	36°19′N. 70°57′E.	8								X	RBV30553–05112–B; 09 Sep 79
3	*Darban-Udren* (Chitrāl)	36°27′N. 72°00′E.	24							X	X	RBV30553–05112–B; 09 Sep 79
4	Hoski-o Shayoz (NW Chitrāl)	36°31′N. 72°10′E.	16				X					RBV30553–05105–D; 09 Sep 79 Stereo30516–05053–C; 03 Aug 79
5	*Shaghordak* (*Hurusko Kuh*, NW Chitrāl)	36°38′N. 72°14′E.	7		X			X				RBV30553–05105–D; 09 Sep 79 Stereo30516–05053–C; 03 Aug 79
6	Chikār (N Chitrāl)	36°38′N. 72°19′E.	9			X		X				RBV30553–05015–D; 09 Sep 79 Stereo30516–05053–C; 03 Aug 79
7	Kotgāz (N Chitrāl)	36°43′N. 72°17′E.	18				X	X				RBV30553–05105–D; 09 Sep 79 Stereo30516–05053–C; 03 Aug 79
8	Chhutidum (N Chitrāl)	36°45′N. 72°25′E.	10		X			X				RBV30553–05105–D; 09 Sep 79 Stereo30516–05053–C; 03 Aug 79
9	Noroghikun (N Chitrāl)	36°44′N. 72°32′E.	15		X			X				RBV30553–05105–D; 09 Sep 79 Stereo30516–05053–C; 03 Aug 79
10	Kach (N Chitrāl)	36°47′N. 72°41′E.	10							X	X	RBV30516–05053–C; 03 Aug 79
11	Madit (*Yarkun*, Hindu Raj)	36°42′N. 73°00′E.	8				X	X				RBV30516–05053–D; 03 Aug 79
12	Risht (*Yarkun*, Hindu Raj)	36°43′N. 73°01′E.	9				X	X				RBV30516–05053–D; 03 Aug 79
13	Shetor (*Yarkun*, Hindu Raj)	36°43′N. 73°05′E.	14				X	X				RBV30516–05053–D; 03 Aug 79
14	Ponārilio (*Yarkun*, Hindu Raj)	36°45′N. 73°09′E.	10							X	X	RBV30516–05053–D; 03 Aug 79
15	Kotalkash (*Yarkun*, Hindu Raj)	36°45′N. 73°12′E.	13			X					X	RBV30516–05053–D; 03 Aug 79
16	Koyo (*Yarkun*, Hindu Raj)	36°45′N. 73°14′E.	5							X	X	RBV30516–05053–D; 03 Aug 79
17	Pechus (*Yarkun*, Hindu Raj)	36°45′N. 73°16′E.	10			X					X	RBV30516–05053–D; 03 Aug 79
18	Chhatiboi (*Yarkun*, Hindu Raj)	36°45′N. 73°19′E.	12				X					RBV30516–05053–D; 03 Aug 79
19	Chikār-Darkot (*Yarkun*, Hindu Raj)	36°46′N. 73°23′E.	10				X	X				RBV30516–05053–D; 03 Aug 79
20	Gazin-Bārbīn (*Yarkun*, Hindu Raj)	36°36′N. 73°02′E.	9					X				RBV30516–05053–D; 03 Aug 79
21	*Mushk Bar* (Thui Gol, Gilgit)	36°33′N. 73°06′E.	10					X				RBV30516–05053–D; 03 Aug 79
22	*Kerun Bar* (Thui Gol, Gilgit)	36°39′N. 73°14′E.	8					X				RBV30516–05053–D; 03 Aug 79
23	*West Gamu Bar* (Thui Gol, Gilgit)	36°37′N. 73°19′E.	7				X					RBV30516–05053–D; 03 Aug 79

[Glacier numbers are cited in the text and plotted on figure 1. Selection based on the following criteria: (1) size large enough to provide resolvable evidence of unusual movement or other resolvable distinguishing landforms; (2) availability of maps or imagery detailed enough to allow discrimination of features. Movement criteria: (1) strongly convoluted medial or lateral moraines (5 glaciers); (2) medial or lateral moraines convoluted or offset by extensive crevasses or ogives (13 glaciers); (3) tributary ice overriding or displacing main glacier ice (7 glaciers); (4) marked depression of ice surface below lateral moraines or tributary glaciers (14 glaciers); (5) lateral and medial moraines slightly sinuous, or lobate, or offset by crevasses or ogives (76 glaciers); (6) recent major melting or retreat of ice front leaving light-colored deglacierized terrain behind (9 glaciers); (7) extensive area of debris-covered downwasting ice or stagnant ice (106 glaciers); and (8) linear moraines and accordant tributaries (47 glaciers). Glacier maps were obtained from many different sources and our field work. In some cases, hyphenated names include tributary names, in others the name itself is a hyphenated phrase. Abbreviations: km, kilometer; ?, glaciers with uncertain names or different spellings]

Glacier number	Glacier name (general location)	Latitude and longitude	Length (km)	Movement criteria								Satellite images[1]
				1	2	3	4	5	6	7	8	
24	*East Ghamu Bar* (Darkot Bar, Gilgit)	36°36′N. 73°22′E.	9					X		X		RBV30516–05053–D; 03 Aug 79
25	Zindikharam (*Yarkun*, Hindu Raj)	36°46′N. 73°27′E.	8		X		X			X		RBV30516–05053–D; 03 Aug 79 Stereo30515–04595–C; 02 Aug 79
26	Chikzar (*Yarkun*, Hindu Raj)	36°45′N. 73°31′E.	7				X					RBV30516–05053–D; 03 Aug 79 Stereo30515–04595–C; 02 Aug 79
27	Chiãntar (*Yarkun*, Hindu Raj)	36°47′N. 73°45′E.	28								X	RBV30516–05053–D; 03 Aug 79 Stereo30515–04595–C; 02 Aug 79
28	Garmush (Chiãntar tributary)	36°45′N. 73°37′E.	14								X	RBV30516–05053–D; 03 Aug 79 Stereo30515–04595–C; 02 Aug 79
29	*Chhateboi* or *Chashboi* (upper Karambar)	36°48′N. 73°52′E.	15		X							RBV30516–05053–D; 03 Aug 79 Stereo30515–04595–C; 02 Aug 79
30	*Sokha Robat* (upper Karambar)	36°47′N. 73°58′E.	9		X			X				RBV30516–05053–D; 03 Aug 79 Stereo30515–04595–C; 02 Aug 79 30515–05001–A; 02 Aug 79
31	*Chillinji* (upper Karambar)	36°46′N. 74°03′E.	8					X				RBV30516–05053–D; 03 Aug 79 Stereo30515–04595–C; 02 Aug 79 30515–05001–A; 02 Aug 79
32	*Wargot* (upper Karambar)	36°44′N. 73°58′E.	8								X	RBV30516–05053–D; 03 Aug 79 Stereo30515–04595–C; 02 Aug 79 30515–05001–A; 02 Aug 79
33	*Pekhin* (upper Karambar)	36°42′N. 73°55′E.	12					X				RBV30516–05053–D; 03 Aug 79 Stereo30515–04595–C; 02 Aug 79 30515–05001–A; 02 Aug 79
34	*Karambar* (upper Karambar)	36°38′N. 74°10′E.	15		X		X	X				RBV30516–05053–D; 03 Aug 79 Stereo30515–04595–C; 02 Aug 79 30515–05001–A; 02 Aug 79
35	*Bohrt* (upper Karambar)	36°33′N. 74°07′E.	10					X				RBV30515–05001–A; 02 Aug 79
36	*Bad Swat* (upper Karambar)	36°01′N. 74°04′E.	10					X				RBV30515–05001–A; 02 Aug 79
37	*Bajgaz* (upper Karambar)	36°23′N. 74°01′E.	15						X			RBV30515–05001–A; 02 Aug 79
38	*Koz Yaz* (upper Hunza)	36°48′N. 74°09′E.	16					X				RBV30515–05001–A; 02 Aug 79 30515–04595–C; 02 Aug 79
39	Yashkūk Yāz (upper Hunza)	36°43′N. 74°18′E.	20					X		X		RBV30515–05001–A; 02 Aug 79 30515–04595–C; 02 Aug 79
40	*Ku-ki-jerab* (upper Hunza)	36°45′N. 74°25′E.	17		X							RBV30515–05001–A; 02 Aug 79 30515–04595–C; 02 Aug 79
41	Batūra (Hunza)	36°32′N. 74°40′E.	55					X		X		RBV30515–05001–B; 02 Aug 79 30532–04542–A; 19 Aug 79
42	*Pasu* (Hunza)	36°28′N. 74°45′E.	26				X	X				RBV30515–05001–B; 02 Aug 79 30532–04542–A; 19 Aug 79
43	*Ghulkin* (Hunza)	36°25′N. 74°50′E.	15					X		X		RBV30515–05001–B; 02 Aug 79 30532–04542–A; 19 Aug 79

TABLE 3.—*Glaciers of northern Pakistan selected for remote-sensing analysis*—Continued

[Glacier numbers are cited in the text and plotted on figure 1. Selection based on the following criteria: (1) size large enough to provide resolvable evidence of unusual movement or other resolvable distinguishing landforms; (2) availability of maps or imagery detailed enough to allow discrimination of features. Movement criteria: (1) strongly convoluted medial or lateral moraines (5 glaciers); (2) medial or lateral moraines convoluted or offset by extensive crevasses or ogives (13 glaciers); (3) tributary ice overriding or displacing main glacier ice (7 glaciers); (4) marked depression of ice surface below lateral moraines or tributary glaciers (14 glaciers); (5) lateral and medial moraines slightly sinuous, or lobate, or offset by crevasses or ogives (76 glaciers); (6) recent major melting or retreat of ice front leaving light-colored deglacierized terrain behind (9 glaciers); (7) extensive area of debris-covered downwasting ice or stagnant ice (106 glaciers); and (8) linear moraines and accordant tributaries (47 glaciers). Glacier maps were obtained from many different sources and our field work. In some cases, hyphenated names include tributary names, in others the name itself is a hyphenated phrase. Abbreviations: km, kilometer; ?, glaciers with uncertain names or different spellings]

Glacier number	Glacier name (general location)	Latitude and longitude	Length (km)	Movement criteria								Satellite images[1]
				1	2	3	4	5	6	7	8	
44	*Gulmit* (Hunza)	36°24'N. 74°45'E.	12					X		X		RBV30515–05001–B; 02 Aug 79 30532–04542–A; 19 Aug 79
45	*Shispar* (*Hasanābād* tributary)	36°24'N. 74°36'E.	14							X		RBV30515–05001–B; 02 Aug 79
46	*Mutschual* (*Hasanābād* tributary)	36°23'N. 74°31'E.	21							X		RBV30515–05001–B; 02 Aug 79
47	*Hasanābād* (Hunza)	36°21'N. 74°34'E.	1							X		RBV30515–05001–B; 02 Aug 79
48	*Shittinbar* (Chalt, Hunza)	36°22'N. 74°24'E.	7					X		X		RBV30515–05001–B; 02 Aug 79
49	*Baltar* (Hunza)	36°27'N. 74°24'E.	17					X	X	X		RBV30515–05001–B; 02 Aug 79
50	*Sat Maro-Kukuar* (Hunza)	36°29'N. 74°15'E.	20					X	X	X		RBV30515–05001–A; 02 Aug 79
51	*Aldar Kush* (Hunza)	36°24'N. 74°12'E.	6					X		X		RBV30515–05001–A; 02 Aug 79
52	*Diantar* (Hunza)	36°24'N. 74°08'E.	8					X		X		RBV30515–05001–A; 02 Aug 79
53	*Gharesha-Trivor* (Nagir, Hunza)	36°14'N. 75°00'E.	21					X		X		RBV30515–05001–B; 02 Aug 79 Stereo30532–04542–A; 19 Aug 79
54	*Balt Bare* (Hunza)	36°20'N. 74°56'E.	6					X		X		RBV30515–05001–B; 02 Aug 79 Stereo30532–04542–A; 19 Aug 79
55	*Ghutulji* (Shimshāl)	36°24'N. 74°58'E.	7					X		X		RBV30515–05001–B; 02 Aug 79 Stereo30532–04542–A; 19 Aug 79
56	*Lupghar* (Shimshāl)	36°16'N. 75°24'E.	17					X		X		RBV30515–05001–B; 02 Aug 79 Stereo30532–04542–A; 19 Aug 79
57	Momhil (Shimshāl)	36°21'N. 75°06'E.	22					X		X		RBV30515–05001–B; 02 Aug 79 Stereo30532–04542–A; 19 Aug 79
58	*Mulunqūtti* (Shimshāl)	36°27'N. 75°13'E.	15				X	X				RBV30515–05001–B; 02 Aug 79 Stereo30532–04542–A; 19 Aug 79
59	*Yāzghil* (Shimshāl)	36°20'N. 75°20'E.	25					X				RBV30532–04542–A; 19 Aug 79
60	*Yukshin Gardan* (Shimshāl)	36°15'N. 75°25'E.	17					X				RBV30532–04542–A; 19 Aug 79
61	Khurdopīn (Shimshāl)	36°13'N. 75°30'E.	32	X							X	RBV30532–04542–A; 19 Aug 79 RBV30532–04542–B; 19 Aug 79
62	Vīrjerāb (Shimshāl)	36°15'N. 75°40'E.	40	X							X	RBV30532–04542–A; 19 Aug 79 RBV30532–04542–B; 19 Aug 79
63	Braldu	36°10'N. 75°52'E.	35		X						X	RBV30532–04542–B; 19 Aug 79
64	*Skamri* (China)	36°03'N. 76°15'E.	36	X				X		X		RBV30532–04542–B; 19 Aug 79 Stereo30531–04484–A; 18 Aug 79 30531–04484–B; 18 Aug 79
65	*Lak-Khiang* (Hispar tributary)	36°10'N. 75°08'E.	19					X		X		RBV30515–05001–B; 02 Aug 79 Stereo30532–04542–A; 19 Aug 79

[Glacier numbers are cited in the text and plotted on figure 1. Selection based on the following criteria: (1) size large enough to provide resolvable evidence of unusual movement or other resolvable distinguishing landforms; (2) availability of maps or imagery detailed enough to allow discrimination of features. Movement criteria: (1) strongly convoluted medial or lateral moraines (5 glaciers); (2) medial or lateral moraines convoluted or offset by extensive crevasses or ogives (13 glaciers); (3) tributary ice overriding or displacing main glacier ice (7 glaciers); (4) marked depression of ice surface below lateral moraines or tributary glaciers (14 glaciers); (5) lateral and medial moraines slightly sinuous, or lobate, or offset by crevasses or ogives (76 glaciers); (6) recent major melting or retreat of ice front leaving light-colored deglacierized terrain behind (9 glaciers); (7) extensive area of debris-covered downwasting ice or stagnant ice (106 glaciers); and (8) linear moraines and accordant tributaries (47 glaciers). Glacier maps were obtained from many different sources and our field work. In some cases, hyphenated names include tributary names, in others the name itself is a hyphenated phrase. Abbreviations: km, kilometer; ?, glaciers with uncertain names or different spellings]

Glacier number	Glacier name (general location)	Latitude and longitude	Length (km)	Movement criteria								Satellite images[1]
				1	2	3	4	5	6	7	8	
66	Pumārikish (Hispar tributary)	36°08'N. 75°12'E.	8					X		X		RBV30515–05001–B; 02 Aug 79 Stereo30532–04542–A; 19 Aug 79
67	*Jutmau* (Hispar tributary)	36°08'N. 75°19'E.	20							X	X	RBV30532–04542–A; 19 Aug 79
68	*Khanbasa* (Hispar tributary)	36°07'N. 75°24'E.	19							X	X	RBV30532–04542–A; 19 Aug 79
69	*Haigatum* (Hispar tributary)	36°04'N. 75°13'E.	5					X		X		RBV30515–05001–B; 02 Aug 79 Stereo30532–04542–A; 19 Aug 79 30532–04542–C; 19 Aug 79
70	*Makrong* (Hispar tributary)	36°04'N. 75°09'E.	8					X		X		RBV30515–05001–B; 02 Aug 79 Stereo30532–04542–A; 19 Aug 79 30532–04542–C; 19 Aug 79
71	*Garumbar* (Hispar)	36°07'N. 75°01'E.	12					X	X			RBV30515–05001–B; 02 Aug 79 Stereo30532–04542–A; 19 Aug 79 30532–04542–C; 19 Aug 79
72	Hispar	36°05'N. 75°15'E.	48	X		X				X		RBV30515–05001–B; 02 Aug 79 Stereo30532–04542–A; 19 Aug 79 30532–04542–C; 19 Aug 79
73	Yengutz Har (Hispar)	36°07'N. 74°58'E.	5					X		X		RBV30515–05001–B; 02 Aug 79 Stereo30532–04542–A; 19 Aug 79 30532–04542–C; 19 Aug 79
74	Miar or *Shalhubu* Sumaiyar Bar-Barpu (Nagir, Hispar)	36°10'N. 74°50'E.	23				X	X				RBV30515–05001–B; 02 Aug 79 Stereo30532–04542–A; 19 Aug 79 30532–04542–C; 19 Aug 79
75	Buāltar or *Hopar* (Nagir, Hispar)	36°10'N. 74°45'E.	20				X	X		X		RBV30515–05001–B; 02 Aug 79 Stereo30532–04542–A; 19 Aug 79 30532–04542–C; 19 Aug 79
76	*Sumaiyar* or *Silkiang* (Rakaposhi, Hunza)	36°12'N. 74°39'E.	7					X		X		RBV30515–05001–B; 02 Aug 79
77	Minapin (Rakaposhi, Hunza)	36°11'N. 74°34'E.	10					X				RBV30515–05001–B; 02 Aug 79
78	Pisan (Rakaposhi, Hunza)	36°11'N. 74°30'E.	9					X				RBV30515–05001–B; 02 Aug 79
79	*Ghulmet* (Rakaposhi, Hunza)	36°11'N. 74°28'E.	6								X	RBV30515–05001–B; 02 Aug 79
80	*Jaglot?* (Rakaposhi, Hunza)	36°05'N. 74°24'E.	6							X		RBV30515–05001–B; 02 Aug 79
81	*Surgin* (Rakaposhi, Gilgit)	36°06'N. 74°29'E.	7							X		RBV30515–05001–B; 02 Aug 79
82	*Hinarche* (Rakaposhi, Gilgit)	36°05'N. 74°34'E.	14					X		X		RBV30515–05001–B; 02 Aug 79
83	*Yuna* (Rakaposhi, Gilgit)	36°04'N. 74°39'E.	6							X		RBV30515–05001–B; 02 Aug 79
84	*Burcha* (Rakaposhi, Gilgit)	36°03'N. 74°40'E.	15							X		RBV30515–05001–B; 02 Aug 79
85	*Saltli* (Haramosh)	36°02'N. 74°45'E.	4							X		RBV30515–05001–B; 02 Aug 79

[Glacier numbers are cited in the text and plotted on figure 1. Selection based on the following criteria: (1) size large enough to provide resolvable evidence of unusual movement or other resolvable distinguishing landforms; (2) availability of maps or imagery detailed enough to allow discrimination of features. Movement criteria: (1) strongly convoluted medial or lateral moraines (5 glaciers); (2) medial or lateral moraines convoluted or offset by extensive crevasses or ogives (13 glaciers); (3) tributary ice overriding or displacing main glacier ice (7 glaciers); (4) marked depression of ice surface below lateral moraines or tributary glaciers (14 glaciers); (5) lateral and medial moraines slightly sinuous, or lobate, or offset by crevasses or ogives (76 glaciers); (6) recent major melting or retreat of ice front leaving light-colored deglacierized terrain behind (9 glaciers); (7) extensive area of debris-covered downwasting ice or stagnant ice (106 glaciers); and (8) linear moraines and accordant tributaries (47 glaciers). Glacier maps were obtained from many different sources and our field work. In some cases, hyphenated names include tributary names, in others the name itself is a hyphenated phrase. Abbreviations: km, kilometer; ?, glaciers with uncertain names or different spellings]

Glacier number	Glacier name (general location)	Latitude and longitude	Length (km)	Movement criteria								Satellite images[1]
				1	2	3	4	5	6	7	8	
86	*Kaltaro* (Haramosh)	36°06′N. 74°47′E.	10							X		RBV30515–05001–B; 02 Aug 79
87	*Baskai* (Haramosh)	35°56′N. 74°52′E.	9							X		RBV30515–05001–B; 02 Aug 79
88	*Mani* (Haramosh)	35°53′N. 74°53′E.	12							X		RBV30515–05001–B; 02 Aug 79 Stereo30515–05001–D; 02 Aug 79
89	*Ishkapal* (Haramosh)	35°49′N. 74°50′E.	7							X		RBV30515–05001–B; 02 Aug 79 Stereo30515–05001–D; 02 Aug 79
90	*Kutiah* (Haramosh)	35°48′N. 75°00′E.	16				X	X		X		RBV30515–05001–B; 02 Aug 79 Stereo30515–05001–D; 02 Aug 79
91	*Stak* (Haramosh)	35°48′N. 75°04′E.	9					X		X		RBV30515–05001–B; 02 Aug 79 Stereo30515–05001–D; 02 Aug 79
92	*Rākhiot* (Nanga Parbat)	35°20′N. 74°35′E.	13				X	X				RBV30515–05001–D; 02 Aug 79 Stereo30532–04542–C; 19 Aug 79
93	*Buldar* (Nanga Parbat)	35°22′N. 74°42′E.	7							X		RBV30515–05001–D; 02 Aug 79
94	*Lotang* (Nanga Parbat)	35°22′N. 74°45′E.	5							X		RBV30515–05001–D; 02 Aug 79
95	*Sachen* (Nanga Parbat)	35°20′N. 74°45′E.	9							X		RBV30515–05001–D; 02 Aug 79
96	*Tsongra-Chungpar* (Nanga Parbat)	35°16′N. 74°42′E.	10							X		RBV30515–05001–D; 02 Aug 79
97	*Bizin* (Nanga Parbat)	35°12′N. 74°38′E.	11							X		RBV30515–05001–D; 02 Aug 79
98	*Tap* (Nanga Parbat)	35°13′N. 74°35′E.	5							X		RBV30515–05001–D; 02 Aug 79
99	*Sheigri* (Nanga Parbat)	35°11′N. 74°34′E.	6							X		RBV30515–05001–D; 02 Aug 79
100	*Toshain* or *Rupal* (Nanga Parbat)	35°10′N. 74°31′E.	15								X	RBV30515–05001–D; 02 Aug 79
101	*Diamir* (Nanga Parbat)	35°16′N. 74°32′E.	15							X		RBV30515–05001–D; 02 Aug 79
102	*Patro* (Nanga Parbat)	35°18′N. 74°31′E.	4							X		RBV30515–05001–D; 02 Aug 79
103	Chogo Lungma (Haramosh, *Basna*)	36°00′N. 75°00′E.	40							X	X	RBV30515–05001–B; 02 Aug 79 Stereo30532–04542–C; 19 Aug 79
104	Haramosh (Chogo Lungma tributary)	35°57′N. 75°00′E.	16				X					RBV30515–05001–B; 02 Aug 79 Stereo30532–04542–C; 19 Aug 79
105	*West Kupultung-Kung?* (Chogo Lungma tributary)	35°55′N. 75°05′E.	15				X				X	RBV30515–05001–B; 02 Aug 79 Stereo30532–04542–C; 19 Aug 79
106	*East Kupultung-Kung?* (Chogo Lungma tributary)	35°54′N. 75°08′E.	13				X					RBV30515–05001–B; 02 Aug 79 Stereo30532–04542–C; 19 Aug 79
107	*Marpah-Remendok* (Chogo Lungma valley)	35°53′N. 75°11′E.	15				X	X		X		RBV30515–05001–B; 02 Aug 79 Stereo30532–04542–C; 19 Aug 79

[Glacier numbers are cited in the text and plotted on figure 1. Selection based on the following criteria: (1) size large enough to provide resolvable evidence of unusual movement or other resolvable distinguishing landforms; (2) availability of maps or imagery detailed enough to allow discrimination of features. Movement criteria: (1) strongly convoluted medial or lateral moraines (5 glaciers); (2) medial or lateral moraines convoluted or offset by extensive crevasses or ogives (13 glaciers); (3) tributary ice overriding or displacing main glacier ice (7 glaciers); (4) marked depression of ice surface below lateral moraines or tributary glaciers (14 glaciers); (5) lateral and medial moraines slightly sinuous, or lobate, or offset by crevasses or ogives (76 glaciers); (6) recent major melting or retreat of ice front leaving light-colored deglacierized terrain behind (9 glaciers); (7) extensive area of debris-covered downwasting ice or stagnant ice (106 glaciers); and (8) linear moraines and accordant tributaries (47 glaciers). Glacier maps were obtained from many different sources and our field work. In some cases, hyphenated names include tributary names, in others the name itself is a hyphenated phrase. Abbreviations: km, kilometer; ?, glaciers with uncertain names or different spellings]

Glacier number	Glacier name (general location)	Latitude and longitude	Length (km)	Movement criteria 1	2	3	4	5	6	7	8	Satellite images[1]
108	*Niamul Gans* (Chogo Lungma valley)	35°48′N. 75°15′E.	14						X	X		RBV30532–04542–C; 19 Aug 79
109	*Tippur Gans* (upper Basna valley)	35°51′N. 75°20′E.	7							X		RBV30532–04542–C; 19 Aug 79
110	*Sgari-bian* (Chogo Lungma tributary)	35°04′N. 75°00′E.	9								X	RBV30515–05001–B; 02 Aug 79 30532–04542–C; 19 Aug 79
111	*Bolucho* (Chogo Lungma tributary)	36°08′N. 75°00′E.	5					X	X			RBV30515–05001–B; 02 Aug 79 30532–04542–C; 19 Aug 79
112	*Arincu* (Chogo Lungma tributary)	36°00′N. 75°11′E.	8					X				RBV30515–05001–B; 02 Aug 79 30532–04542–C; 19 Aug 79
113	*Kero Lungma* (upper Basna valley)	36°00′N. 75°15′E.	15					X		X		RBV30532–04542–A; 19 Aug 79 Stereo30532–04542–C; 19 Aug 79
114	*Hucho Alchori* (upper Basna valley)	35°59′N. 75°23′E.	13					X		X		RBV30532–04542–A; 19 Aug 79 Stereo30532–04542–C; 19 Aug 79
115	*Niaro Gans* (upper Basna valley)	35°55′N. 75°17′E.	5							X		RBV30532–04542–C; 19 Aug 79
116	*Solu-Sokha* (upper Basna valley)	35°59′N. 75°30′E.	15					X		X		RBV30532–04542–C; 19 Aug 79
117	*Sosbun* (Braldu)	35°52′N. 75°34′E.	16					X		X		RBV30532–04542–D; 19 Aug 79
118	*Tsilbu-Hoh Lungma* (Braldu)	35°50′N. 75°33′E.	9							X		RBV30532–04542–D; 19 Aug 79
119	*Chongahanmong* (Braldu)	35°48′N. 75°33′E.	5					X		X		RBV30532–04542–D; 19 Aug 79
120	*West Tongo?* (Braldu)	35°48′N. 75°39′E.	5					X		X		RBV30532–04542–D; 19 Aug 79
121	Biafo (Braldu)	35°50′N. 75°45′E.	54					X		X	X	RBV30531–04484–C; 18 Aug 79 Stereo30532–04542–B; 19 Aug 79 30532–04542–D; 19 Aug 79
122	*Sim Gang* (Biafo tributary)	36°00′N. 75° 35′E.	15					X		X		RBV30531–04484–C; 18 Aug 79 Stereo30532–04542–B; 19 Aug 79 30532–04542–D; 19 Aug 79
123	*Uzun Blakk Baintha Lukpar* (Biafo tributary)	35°53′N. 75°45′E.	10					X		X		RBV30531–04484–C; 18 Aug 79 Stereo30532–04542–B; 19 Aug 79 30532–04542–D; 19 Aug 79
124	*Choktoi-Panmah* (Dumordo-Braldu)	35°53′N. 75°58′E.	30	X	X	X	X	X	X	X		RBV30531–04484–C; 18 Aug 79 Stereo30532–04542–B; 19 Aug 79 30532–04542–D; 19 Aug 79
125	*Drenmang* (Panmah tributary)	35°59N. 76°03.E.	8			X		X				RBV30531–04484–C; 18 Aug 79 Stereo30532–04542–B; 19 Aug 79 30532–04542–D; 19 Aug 79
126	*Chiring* (Panmah tributary)	35°55′N. 76°04′E.	16			X		X	X			RBV30531–04484–C; 18 Aug 79 Stereo30532–04542–B; 19 Aug 79 30532–04542–D; 19 Aug 79

[Glacier numbers are cited in the text and plotted on figure 1. Selection based on the following criteria: (1) size large enough to provide resolvable evidence of unusual movement or other resolvable distinguishing landforms; (2) availability of maps or imagery detailed enough to allow discrimination of features. Movement criteria: (1) strongly convoluted medial or lateral moraines (5 glaciers); (2) medial or lateral moraines convoluted or offset by extensive crevasses or ogives (13 glaciers); (3) tributary ice overriding or displacing main glacier ice (7 glaciers); (4) marked depression of ice surface below lateral moraines or tributary glaciers (14 glaciers); (5) lateral and medial moraines slightly sinuous, or lobate, or offset by crevasses or ogives (76 glaciers); (6) recent major melting or retreat of ice front leaving light-colored deglacierized terrain behind (9 glaciers); (7) extensive area of debris-covered downwasting ice or stagnant ice (106 glaciers); and (8) linear moraines and accordant tributaries (47 glaciers). Glacier maps were obtained from many different sources and our field work. In some cases, hyphenated names include tributary names, in others the name itself is a hyphenated phrase. Abbreviations: km, kilometer; ?, glaciers with uncertain names or different spellings]

Glacier number	Glacier name (general location)	Latitude and longitude	Length (km)	Movement criteria								Satellite images[1]
				1	2	3	4	5	6	7	8	
127	*Sarpo Lago* (Baltoro Mutagh)	36°00'N. 76°22'E.	29		X			X		X		RBV30531–04484–A; 18 Aug 79 Stereo30532–04542–B; 19 Aug 79 30532–04542–D; 19 Aug 79
128	Baltoro (Biaho Lungma, Braldu)	35°46'N. 76°15'E.	52							X	X	RBV30531–04484–C; 18 Aug 79 Stereo30532–04542–D; 19 Aug 79
129	*Uli Biaho* (Baltoro tributary)	35°13'N. 76°09'E.	8		X							RBV30531–04484–C; 18 Aug 79 Stereo30532–04542–D; 19 Aug 79
130	*Trango* (Baltoro tributary)	35°45'N. 76°10'E.	16		X			X				RBV30531–04484–C; 18 Aug 79 Stereo30532–04542–D; 19 Aug 79
131	*Dunge* (Baltoro tributary)	35°45'N. 76°14'E.	10				X					RBV30531–04484–C; 18 Aug 79 Stereo30532–04542–D; 19 Aug 79
132	*Biaie* (Baltoro tributary)	35°46'N. 76°16'E.	6								X	RBV30531–04484–C; 18 Aug 79 Stereo30532–04542–D; 19 Aug 79
133	*Mutagh* (Baltoro tributary)	35°47'N. 76°18'E.	11				X					RBV30531–04484–C; 18 Aug 79 Stereo30532–04542–D; 19 Aug 79
134	Biange-*Younghusband* (Baltoro tributary)	35°48'N. 76°24'E.	12								X	RBV30531–04484–C; 18 Aug 79 Stereo30532–04542–D; 19 Aug 79
135	K2-Godwin Austen (Baltoro tributary)	35°50'N. 76°31'E.	18								X	RBV30531–04484–C; 18 Aug 79 Stereo30532–04542–B; 19 Aug 79
136	*South Gasherbrum* (Baltoro tributary)	35°52'N. 76°40'E.	6								X	RBV30531–04484–C; 18 Aug 79
137	*Upper Baltoro* (Baltoro tributary)	35°43'N. 76°31'E.	14								X	RBV30531–04484–C; 18 Aug 79
138	*Vigne* (Baltoro tributary)	35°43'N. 76°26'E.	5								X	RBV30531–04484–C; 18 Aug 79 Stereo30532–04542–B; 19 Aug 79
139	*Yermanendu* (Baltoro tributary)	35°38'N. 76°17'E.	10								X	RBV30531–04484–C; 18 Aug 79 Stereo30532–04542–B; 19 Aug 79
140	Mundu (Baltoro tributary)	35°40'N. 76°17'E.	8								X	RBV30531–04484–C; 18 Aug 79 Stereo30532–04542–B; 19 Aug 79
141	*Liligo* (Baltoro tributary)	35°40'N. 76°13'E.	6						X			RBV30531–04484–C; 18 Aug 79 Stereo30532–04542–B; 19 Aug 79
142	K2 North (Baltoro Mutagh, China)	36°00'N. 76°28'E.	16		X			X				RBV30531–04484–C; 18 Aug 79 Stereo30532–04542–B; 19 Aug 79
143	*Skyang Lungpa* (Baltoro Mutagh, China)	35°54'N. 76°40'E.	11								X	RBV30531–04484–C; 18 Aug 79 Stereo30532–04542–B; 19 Aug 79
144	*Gasherbrum-Chagharlung* (Baltoro Mutagh, China)	35°51'N. 76°42'E.	15				X					RBV30531–04484–C; 18 Aug 79 Stereo30532–04542–B; 19 Aug 79
145	*Urdok* (Baltoro Mutagh, China)	35°47'N. 76°45'E.	24				X	X				RBV30531–04484–C; 18 Aug 79 Stereo30532–04542–B; 19 Aug 79
146	*Staghar* (Siachen Mutagh, China)	35°45'N. 76°48'E.	18								X	RBV30531–04484–C; 18 Aug 79 Stereo30532–04542–B; 19 Aug 79
147	*Singhi* (Siachen Mutagh, China)	35°41'N. 77°00'E.	24								X	RBV30531–04484–D; 18 Aug 79
148	*Kyagar* (Siachen Mutagh, China)	35°38'N. 77°11'E.	19								X	RBV30531–04484–D; 18 Aug 79

[Glacier numbers are cited in the text and plotted on figure 1. Selection based on the following criteria: (1) size large enough to provide resolvable evidence of unusual movement or other resolvable distinguishing landforms; (2) availability of maps or imagery detailed enough to allow discrimination of features. Movement criteria: (1) strongly convoluted medial or lateral moraines (5 glaciers); (2) medial or lateral moraines convoluted or offset by extensive crevasses or ogives (13 glaciers); (3) tributary ice overriding or displacing main glacier ice (7 glaciers); (4) marked depression of ice surface below lateral moraines or tributary glaciers (14 glaciers); (5) lateral and medial moraines slightly sinuous, or lobate, or offset by crevasses or ogives (76 glaciers); (6) recent major melting or retreat of ice front leaving light-colored deglacierized terrain behind (9 glaciers); (7) extensive area of debris-covered downwasting ice or stagnant ice (106 glaciers); and (8) linear moraines and accordant tributaries (47 glaciers). Glacier maps were obtained from many different sources and our field work. In some cases, hyphenated names include tributary names, in others the name itself is a hyphenated phrase. Abbreviations: km, kilometer; ?, glaciers with uncertain names or different spellings]

Glacier number	Glacier name (general location)	Latitude and longitude	Length (km)	Movement criteria								Satellite images[1]
				1	2	3	4	5	6	7	8	
149	*North Rimo* (*Rimo Mutagh*, China)	35°29'N. 77°30'E.	20		X							RBV30531–04484–D; 18 Aug 79
150	*Central Rimo* (*Rimo Mutagh*)	35°27'N. 77°30'E.	40					X		X		RBV30531–04484–D; 18 Aug 79
151	*South Rimo* (*Rimo Mutagh*)	35°20'N. 77°30'E.	22					X		X		RBV30531–04484–D; 18 Aug 79
152	*South Terong* (upper Nubra valley)	35°12'N. 77°24'E.	21							X	X	RBV30531–04484–D; 18 Aug 79
153	*Selkar Chorten-North Terong* (upper Nubra valley)	35°15'N. 77°20'E.	24	X						X	X	RBV30531–04484–D; 18 Aug 79
154	*Dzingrulma* (upper Nubra valley)	35°10'N. 77°10'E.	5						X			RBV30531–04484–D; 18 Aug 79
155	Siachen (upper Nubra valley)	35°30'N. 77°00'E.	67							X		RBV30531–04484–D; 18 Aug 79
156	*Chumick Saltoro I?* (Siachen tributary)	35°13'N. 77°08'E.	11							X	X	RBV30531–04484–D; 18 Aug 79
157	*Chumick Saltoro II?* (Siachen tributary)	35°21'N. 77°07'E.	15							X		RBV30531–04484–D; 18 Aug 79
158	Lolofong (Siachen tributary)	35° 26'N. 77°00'E.	12							X		RBV30531–04484–D; 18 Aug 79
159	*Saltoro* (Siachen tributary)	35°30'N. 76°56'E.	18							X		RBV30531–04484–D; 18 Aug 79
160	*Teram Shehr* (Siachen tributary)	35°30'N. 77°05'E.	28							X		RBV30531–04484–D; 18 Aug 79
161	*Gyong* (upper Saltoro valley)	35°07'N. 77°00'E.	12						X			RBV30531–04484–D; 18 Aug 79
162	Chūmik-Bilafond (upper Saltoro valley)	35°15'N. 76°52'E.	21							X	X	RBV30531–04484–D; 18 Aug 79
163	*Sherpi Kang* (*Kondus*, Saltoro valley)	35°22'N. 76°47'E.	19					X	X	X		RBV30531–04484–D; 18 Aug 79
164	Kaberi (*Kondus*, Saltoro valley)	35°30'N. 76°38'E.	23					X		X	X	RBV30531–04484–C; 18 Aug 79
165	*Kondus* (*Kondus*, Saltoro valley)	35°30'N. 76°41'E.	25					X		X	X	RBV30531–04484–C; 18 Aug 79
166	*Chogolisa* (Hushe valley)	35°30'N. 76°30'E.	21					X		X		RBV30531–04484–C; 18 Aug 79
167	*Chundugaro* (Hushe valley)	35°32'N. 76°24'E.	13							X	X	RBV30531–04484–C; 18 Aug 79 Stereo30532–04542–D; 19 Aug 79
168	Masherbrum (Hushe valley)	35°34'N. 76°19'E.	12							X		RBV30531–04484–C; 18 Aug 79 Stereo30532–04542–D; 19 Aug 79
169	*Aling* (Hushe valley)	35°29'N. 76°15'E.	12			X	X			X		RBV30531–04484–C; 18 Aug 79 Stereo30532–04542–D; 19 Aug 79

[1] Landsat 3 return beam vidicon (RBV) subscenes; see footnote 2 in table 1.

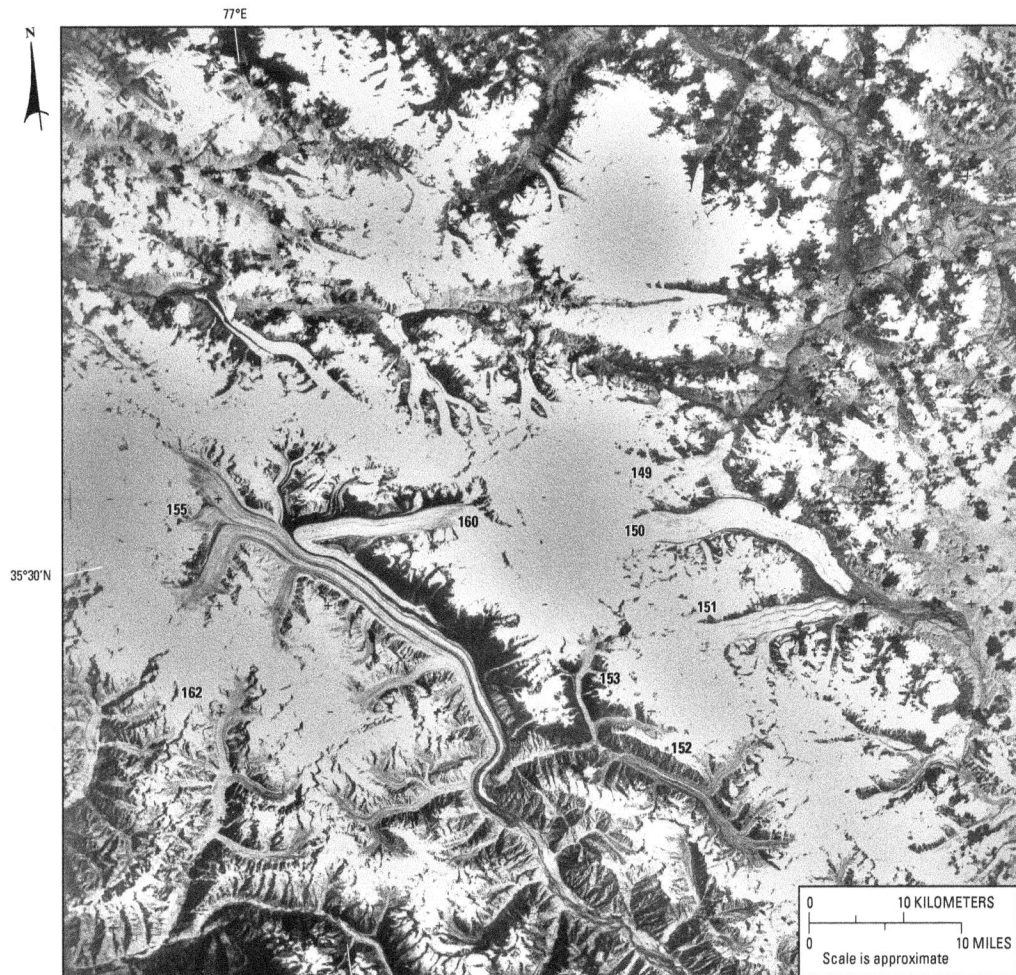

Figure 21.—*Landsat 3 RBV image of Siachen [155], Teram Shehr [160], north [149], central [150], and south Rimo [151], North Terong-Selkar Chorten [153], South Terong [152], and Bilafond [162] Glaciers. Especially prominent are the abrupt bend of the* Teram Shehr Glacier *to join the Siachen Glacier, and the seeming failure of the diffluent north Rimo Glacier to join the clean white ice of the central Rimo Glacier. Also obvious are plentiful supraglacier meltwater lakes on central Rimo Glacier and slightly sinuous medial moraines on south Rimo Glacier. The Landsat 3 RBV image no. 30531–04484, subscene D; 18 August 1979; Path 159–Row 35, is from the U.S. Geological Survey, EROS Data Center, Sioux Falls, South Dakota 57198, but is now only archived by the authors.*

Movement Criteria 3: Tributary Ice Overriding or Displacing Main Glacier Ice

Tributary ice overrode or displaced the main glacier ice in seven obvious examples, notably on Hispar [72] (figs. 13, 22), *Panmah* [124] (fig. 20), and Baltoro [128] (fig. 20) Glaciers. Odell (1937), and Visser and Visser-Hooft (1938; as reported by Washburn, 1939), noted that one glacier overriding another might be caused by differences in elevation of their floors and by their differences in volume, density, degree of compactness, texture, temperature, meltwater, and rate of flow. Because the Hispar Glacier has both convoluted and overridden or displaced moraine, and because of reasonable accessibility, it was selected for field verification, and was visited in 1984.

A

N 0   5 KILOMETERS

0   5 MILES
Scale is approximate

B

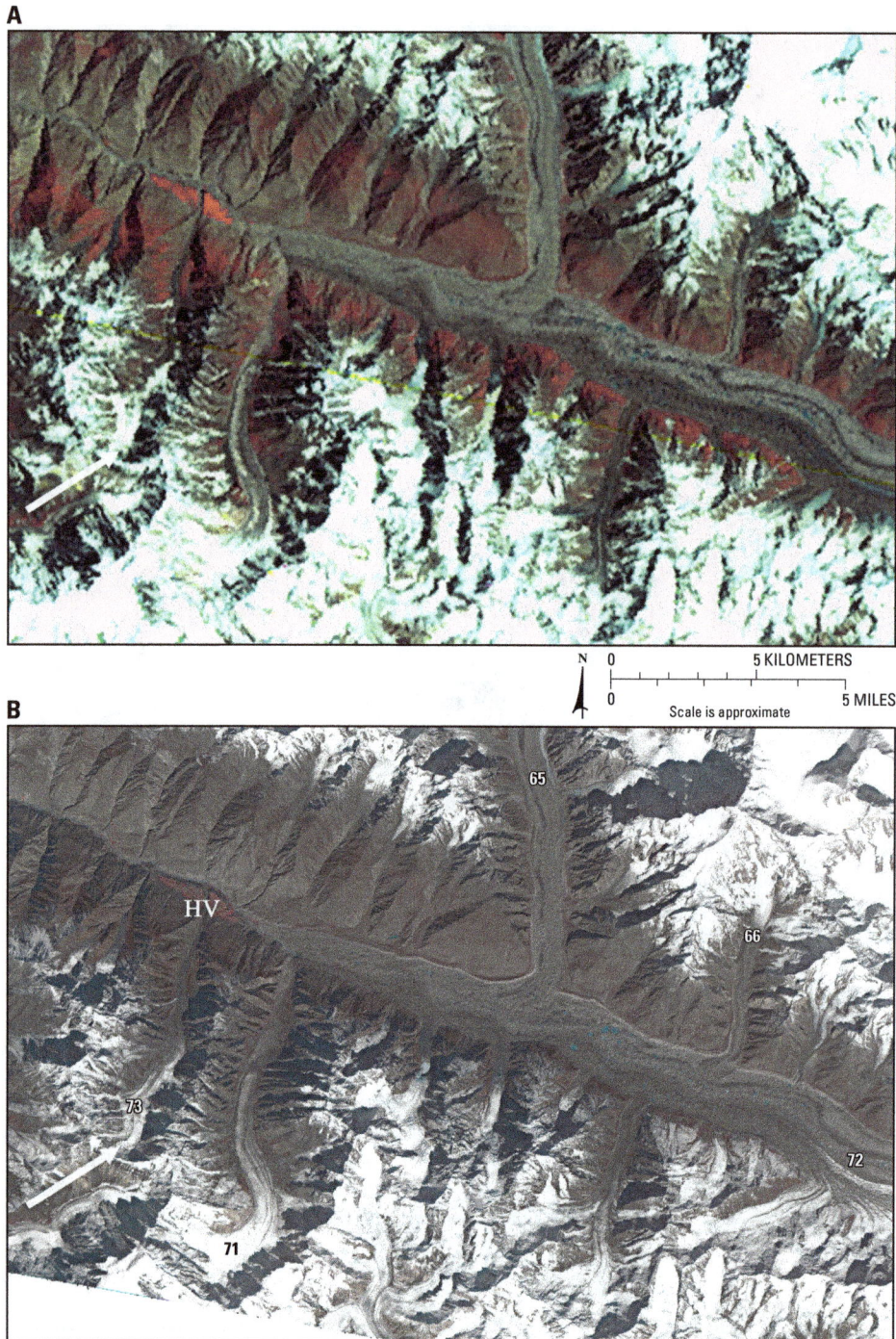

Figure 22.—A, *Landsat 3 MSS false-color composite image taken on 15 July 1979, and* **B**, *ASTER image taken on 30 September 2001, showing the terminus of Hispar Glacier [72] (approximate lat 36°05'N., long 75°15'E.) above which are various flow discontinuities expressed as overridden or displaced ice from tributary glaciers, such as* Lak-Khiang *[65] and* Pumārikish *[66] Glaciers. HV is Hispar Village. The white arrows indicate Yengutz Har Glacier [73] that advanced rapidly between 1892 and 1901. Historic debris flows from this glacier created the large fan where Hispar Village is located. Landsat 3 MSS image no. 3161035007919690, bands 4, 5, 7; 15 July 1979; Path 161– Row 35, and ASTER image no. AST_L1B-003_09302001055833_04172002120543 are from the U.S. Geological Survey, EROS Data Center, Sioux Falls, South Dakota 57198.*

The east-to-west flowing Hispar Glacier was described by Workman and Workman (1911) as being forced against its south wall by its four large tributaries, (*Lak-Khiang* [65], Pumārikish [66], *Jutmaru* [67], and *Khanbasa* [68] Glaciers and several smaller tributaries that flow in from the north (fig. 13). The terminus of the Hispar Glacier has reportedly changed little in more than 70 years of record during the early to middle 20th century (Mercer, 1963); general downwasting of just a few tens of meters and backwasting of less than a kilometer has been observed since the photographs of Hayden (1907) in 1906 and Workman and Workman (1911) in 1908 (fig. 23). The *Lak-Khiang Glacier*, furthest downstream of the tributaries entering from the north, still enters the Hispar valley with a "huge west lateral moraine sweeping around in a splendid curve" (Workman and Workman, 1911, p. 51 and photo), and we observed that the tributary glacier has displaced the entire lower Hispar Glacier to its terminus, 8 km below. The lighter colored Lak moraine is dominantly granite and marble, whereas the darker, narrower, and more compressed moraine on the southern margin is composed of weathered, iron-stained schist and gneiss. Up ice from the *Lak Glacier* tributary is an area about 10 km long, that has extensive kettles and convoluted moraines from the *Jutmaru* (*Jutmau*) *Glacier* [67] tributary on the north side. Some of these convolutions seem to be also influenced by other smaller side glaciers — Pumārikish [*Pumari Chhish*] Glacier [66] and *Khatumburumbun Glacier*.

Figure 23.—*Terminus of the Hispar Glacier [72] in late July 1984. The dark ice cliff above the river has a strong signature on the satellite imagery (figs. 13, 22). The small dark ice cliff directly right (south) of the river is a former ice portal of the Hispar river, now cut off by retreat of the new portal to its present position further back into the glacier terminus. A collapse zone above the subglacial Hispar river is barely visible to the right (south) side of the terminus. Photograph by J.F. Shroder, Jr.*

Searle (1991) and Wake and Searle (1993) noted that the Pumārikish Glacier [66] (figs. 13, 22) is the smallest of the four major tributaries on the north side of the Hispar Glacier, and has produced the most impressive surge in the Karakoram in recent years. In 1988, the surface of the glacier had risen somewhat, but in 1989, the glacier had surged about 1.5 km out and over the Hispar Glacier. The ice had grown dramatically higher, by more than 50 m, with ice towers and seracs looming 20 m or more above the lateral moraines. Searle (1991) was at a loss to explain the surge, beyond speculation of an overall rock uplift of the source region, and did not recognize the far more likely possibility of increased meltwater.

Workman and Workman (1911) reported that about 25 km above the Hispar terminus, white ice occurred near the merger of the small *Haigatum Glacier* [69], a tributary on the south side. Shipton's map of 1939 also shows this (the preparation and planning for the map is described in Shipton, 1938), but the satellite imagery of 1979 shows only a thin strip of light debris. Instead, white ice is prominent 15 km further up the glacier. Additional evidence for downwasting is the position of the Hispar Glacier, several tens of meters below and now partly cut off from the Pumārikish Glacier and *Khatumburumbun Glacier* tributaries.

Finally, the *Garumbar Glacier's* [71] snout, on the south side of the Hispar valley near the terminus, was about 2.5 km from the Hispar when Workman and Workman (1911) were there in 1908. Mason (1930) reported that it had joined the Hispar Glacier by 1925, but in 1984 it was about 4 km away. Its meltwater river passed beneath the Hispar Glacier and joined the main subglacier river in a zone of general roof collapse that is 100–200 m deep between the *Garumbar Glacier* and the main terminus. The Hispar river abandoned one ice portal on the south side several years ago and moved up ice about 100 m toward the Garumbar valley, close to its present position (fig. 23). The ice portal of the Hispar river was a black cliff of dirty ice that is prominent on the imagery (figs. 13, 22).

In order to make field observations of a small glacier with a known history of rapid flow, we visited the Yengutz Har Glacier [73] (figs. 13, 22) in Hispar valley. In 1892, Conway (1894) noted that the glacier was about 3.5 km from the Hispar river, but by 1901 it had advanced to a position only about 900 m from the river, where it buried several water mills. It was adjacent to Hispar village through at least 1906–1908, when Hayden (1907) and Workman and Workman (1910, 1911) were there. In 1984, it had retreated to close to its position in 1892 (fig. 24). Neither evidence from the original surge, nor signs of the subsequent melting and retreat, are apparent on the satellite imagery (figs. 13, 22).

On the Baltoro Glacier [128] (figs. 20, 25), the debris-covered and kettle-pocked *Trango Glacier* [130] tributary has overridden and displaced the main glacier, and was overridden and displaced in turn by the *Uli Biaho Glacier* [129]. Similarly, the lowermost of the southern tributaries to the *Aling Glacier* [169], at the northwest head of Hushe valley below Masherbrum, has overridden the downwasting or stagnant debris-covered ice of the main glacier. This tributary glacier has lobate moraines and a depressed ice surface.

Several glaciers have tributaries with convoluted moraines that terminate against another ice stream. For example, the *Kondus Glacier* [165] is tributary to the Kaberi Glacier [164], which has an extensive area of debris-covered ice over its lower third (fig. 20). The *Kondus Glacier* apparently has piled up against the Kaberi Glacier, causing the moraine of the *Kondus Glacier* to back up in a sinuous pattern. Similarly, the *Panmah Glacier* [124] has convoluted moraines pushed up against the *Choktoi Glacier*, which itself is somewhat convoluted and has a terminus of extensive debris-covered ice (fig. 20). A case might be made that both the *Panmah* and *Kondus Glaciers* surged forward to produce

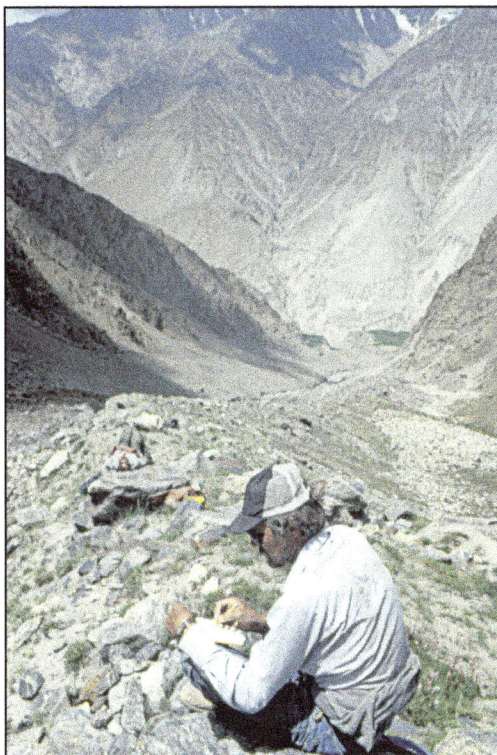

Figure 24.—The valley of Yengutz Har Glacier [73] above Hispar village; view from right lateral moraine with glacier terminus out of picture to left. This valley was filled with ice down to the fields in the valley bottom below, after a surge in about 1901. July 1984 photograph by J.F. Shroder, Jr.

convoluted moraines, overriding or displacing the other glaciers. Subsequent stagnation of the *Panmah* or *Kondus Glaciers* would then allow the Kaberi Glacier or *Choktoi Glacier* to reassert flow dominance and cut off the lower convoluted part of the other glaciers. Close scrutiny of the confluence areas does not show an abrupt truncation of the *Kondus Glacier*, nor is there evidence of interference with the *Choktoi Glacier* by the *Panmah Glacier*. In another case, the clean ice stream of the north Rimo Glacier [149] (fig. 21) was diffluent; the south ice lobe flowed directly up against the clean ice of the central Rimo Glacier [150]. At the confluence, the ice foliation of the north Rimo Glacier lobe was extensively convoluted and did not penetrate through the lateral moraine of the central Rimo Glacier (fig. 21). This moraine between the central and north Rimo Glaciers dominated these glaciers in about 1914, but was "smothered in ice" in 1930 (Dainelli, 1932, p. 398; Mercer, 1963, p. 21). Thus, it appears that tributary glaciers, in some cases, may override or displace the main glacier, and then be overridden, displaced, or completely incorporated themselves. In other cases, the tributary glaciers may run up against another glacier and the forward motion of the tributary glacier is slowed. Reasons for these differences are not clear, but may be controlled in part by spatially and temporally different velocity gradients.

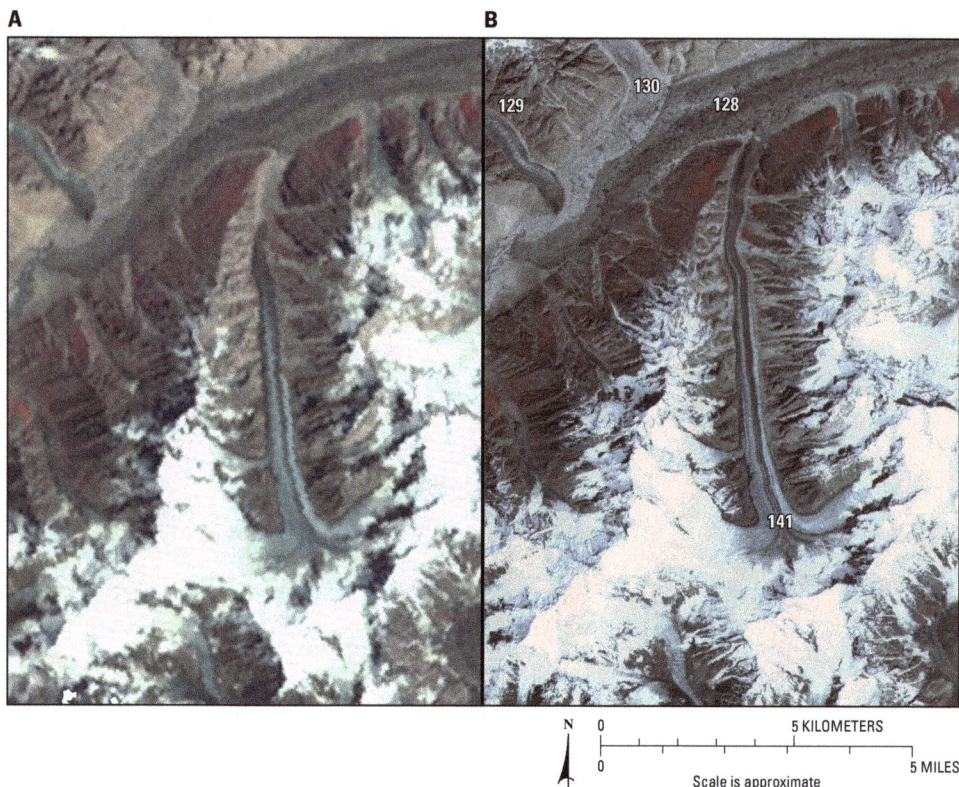

A

B

130
129 128

141

N
0   5 KILOMETERS

0   5 MILES
Scale is approximate

Figure 25.—**A**, *Landsat 3 MSS image taken on 18 July 1978, and* **B**, *ASTER image taken on 14 August 2004, showing terminus of* Baltoro Glacier *[128] in upper left and* Liligo Glacier *[141] (lat 35°40'N., long 76°13'E.) in center middle. The light-colored, granitic-debris covered* Trango Glacier *[130] overrides and displaces much of the northwestern portion of the Baltoro terminus, and is itself overridden by the dark-colored metasediment debris cover of the* Uli Biaho Glacier *[129]. The* Liligo Glacier *shows a rapid advance or surge in the interim between the two scenes. At present the terminus of* Liligo Glacier *exists in close proximity to the Baltoro terminus, with a small proglacial lake between the two. Landsat 3 MSS image no. 3159036007819990, bands 4, 5, 7; 18 July 1978; Path 159–Row 36, and ASTER image no. AST_L1A_003_08142004054614_08282004141219, 14 August 2004, are from the U.S. Geological Survey, EROS Data Center, Sioux Falls, South Dakota 57198.*

Movement Criteria 4: Marked Depression of Ice Surface Below Lateral Moraines or Tributary Glaciers

This criterion indicates past changes in mass balance or possible surges. Only 14 glaciers were observed with what appeared on the RBV stereo imagery to be unusually depressed or down-wasted surfaces below prominent lateral moraines or tributary glaciers. Based on extensive field observations and subsequent imagery, we now know that if more such RBV stereo imagery had been available, then the number would have been much larger.

Movement Criteria 5: Lateral and Medial Moraines That Are Slightly Sinuous, Lobate, or Offset by Crevasses or Ogives

More than 76 glaciers have slightly sinuous or lobate moraines that primarily reflect nonlinear flow regimes or shear and variable melting near crevasses and ogives. In one prominent example, the *Teram Shehr Glacier* [160] (fig. 21) turns 140° from its due west flow to flow southeast as it joins the Siachen Glacier [155]. The sharp bend is marked by numerous crevasses and large melt ponds on the tensional outside margin of the bend (fig. 21).

Movement Criteria 6: Recent Major Melting or Retreat of Icefront, Leaving Light-Colored Scars

Several glaciers showed recent major melting or retreat of the ice front, and nine of them left light-colored deglacierized areas behind. The *Liligo Glacier* [141], on the south side of Baltoro Glacier [128] near its terminus, has been observed for more than a century by a succession of climbing expeditions to the peak of K2 (Pecci and Smiraglia, 2000). The *Liligo Glacier* was about 0.5 km from the Baltoro Glacier in 1909 (De Filippi, 1912, p. 186–187), 0.75 km away in 1953 (Desio and others, 1961), and 1.75 km away in the 1970s, with a prominent light-colored deglacierized zone between the two. Then in the late 1990s the *Liligo Glacier* advanced partly over the Baltoro, and in 2004 it was still in close proximity to the Baltoro terminus (figs. 20 and 25); in late summer, there was a large lake between the two that drained away into a crevasse. Pecci and Smiraglia (2000) suspected a surge of the *Liligo Glacier*, but noted that the absence of seemingly surge-definitive convoluted medial moraines precluded them from making a definite determination.

In another important example, prior to about 1974 the terminus of the *Balt Bare Glacier* [54] in the upper Hunza Valley had been at about 3,800 m altitude at the base of the main ice fall. In the spring of that year, an extensive mud-rock (rapid, wet debris) flow burst out from the glacier, reaching a maximum discharge of approximately 63 m^3 sec^{-1} and pouring out more than 5×10^6 m^3 of material into the river valley (fig. 26). The flood destroyed part of the KKH, a 120-m long bridge, as well as numerous dwellings and fields, and killed one pedestrian (fig. 27). For a while, the Hunza River was backed up into a lake, behind the debris dam. Then, primarily in 1976, the glacier surged forward more than 2 km (Wang and others, 1984). Interestingly, the 1–3 km-sized fan upon which the nearby *Shash Kat* (*Shishket*) village rests is one of the most well-developed fans in the area; it occurs at the base of a small drainage basin, suggesting that other similar depositional events may have occurred there in the past.

A   **B**

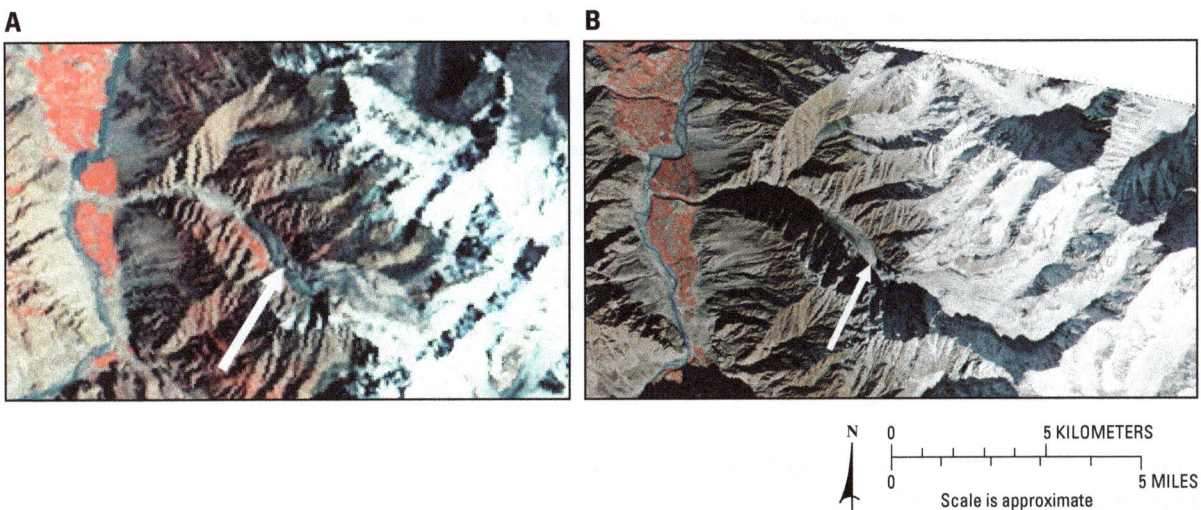

Figure 26.—*Landsat 2 MSS false-color composite and ASTER satellite images of the* Balt Bare Glacier [54] *(approximate lat 36°20'N., long 74°56'E.) and its breakout flood of 1974 as it appeared in 1977, and the appearance of the same features in 2003.* **A.** *Landsat 2 MSS false-color image obtained on 2 August 1977 immediately after the surge. The arrow indicates the original pre-surge terminus at the base of the ice fall, below which occurs the dark, debris-covered surge mass of glacial ice, the scar of the flood across the fan, as well as the impoundment of the Hunza River.* **B.** *The ASTER image of 2003 shows that the* Balt Bare Glacier *has maintained its surge length, the flood scar has been revegetated, and the river has aggraded its channel. Landsat 2 MSS image no. 2160035007721490, bands 4, 5, 7; 2 August 1977; Path 160–Row 35, and ASTER image no. AST_LIA_00310292003055903_11122003124818, 29 October 2003, are from the U.S. Geological Survey, EROS Data Center, Sioux Falls, South Dakota 57198.*

Figure 27.—*Photograph of the* Balt Bare *fan in July 1992 near the village of* Shash Kat *taken from the Karakoram Highway (KKH) on the west side of the Hunza River valley. The flood of 1974 destroyed the original KKH, part of which can be seen preserved on the river bank on the right side of the photograph. The large size of the fan at the mouth of the small* Balt Bare *valley may indicate past surge-sedimentation events as well. Photograph by J.F. Shroder, Jr*

In addition, the previously mentioned Yengutz Har Glacier [73], which advanced rapidly in the early 1900s and subsequently melted back about 3.5 km, presently appears on the MSS and ASTER imagery only as an insignificant small dark mass with an equally obscure retreat zone (fig. 22). However, it appears to be the source of the large fan on which Hispar village was constructed, in the otherwise rather inhospitable Hispar Valley. Perhaps repeated surges or related outburst floods from such small valleys produce sufficiently large debris fans to enable agriculture in otherwise difficult mountain environments.

Movement Criteria 7: Extensive Area of Debris-Covered Downwasting or Stagnant Ice

Debris-covered termini are exceptionally common in this area and reflect the plentiful mass-wasting and snow-avalanche origin of much of the debris. In some cases, the termini are covered with a chaotic mass of debris simply because so much is in transport; in other cases, the debris is the result of a period of rapid movement followed by slowing or stagnation. It is generally impossible to differentiate between types of debris masses unless collateral information is available. Also, using only imagery, it is impossible to determine if the debris covers clean ice or debris-laden ice. The Hispar [72] and

Yengutz Har [73] Glaciers, for example, have extensive dark-debris cover with debris-laden ice beneath (figs. 13, 22). The K2 North Glacier [142] also is depicted on the imagery as having a dark debris-covered terminus (fig. 20), but on Chinese photographs (Chi and Ren, 1980) the terminus appears as clean white ice overlain by a thin debris cover.

Movement Criteria 8: Linear Moraines and Accordant Tributaries

Finally, large numbers of glaciers in northern Pakistan have linear lateral and medial moraines and accordant tributaries that show dominantly linear flow. Many of these were not designated as exhibiting these characteristics in table 3, because other features seemed more important.

Mountain Geomorphology

Globally significant interactions between climate, surface processes, and tectonics have recently been proposed to explain mountain building and climate change (Raymo and others, 1988; Molnar and England, 1990; Koons 1995; Avouac and Burov, 1996). Much research has focused on climatic versus tectonic forcing (Raymo and others, 1988; Molnar and England, 1990) and the dominant surface processes responsible for relief production and topographic evolution (Montgomery, 1994; Burbank and others, 1996; Brozovik and others, 1997; Whipple and Tucker, 1999).

Research indicates that climate-driven surface processes are capable of reducing lithospheric mass, thereby inducing isostatic uplift or accelerating tectonic uplift, collectively resulting in relief production and complex topography (Gilchrist and others, 1994; Montgomery, 1994; Shroder and Bishop, 2000). The linkages between global climate and topography, however, are controversial, and the operational scale-dependencies of surface processes that govern the geomorphometry of topography and geodynamics are not well understood (Shroder and Bishop, 1998; Bishop and others, 1998a, b; 2002; Bishop and Shroder, 2000).

Multidisciplinary research has provided strong evidence of climate-induced geomorphic change in high-mountain environments (Brozovik and others, 1997; Bishop and others, 1998b; Bush, 2002; Phillips and others, 2000), although there is considerable controversy regarding the role of glaciation (Harbor and Warburton, 1992; Whipple and Tucker, 1999; Whipple and others, 1999). Complex feedback mechanisms within and between climatic, glacial, and lithospheric systems are suspected, although it is difficult to study these interactions because of the need to account for the scale-dependencies of numerous processes and polygenetic topographic evolution in modeling efforts (Koons, 1995; Bishop and others, 2002).

The influence of glaciation on the unloading history of mountain massifs, however, is still relatively unknown. We know that many major valleys in Pakistan, including the Indus, have undergone extensive and periodic glaciation (Shroder and others, 1989). Similarly, glacier-related, catastrophic geomorphic events are also responsible for the denudational unloading. Therefore, because of climate fluctuations and geomorphic temporal over-printing, modern-day river or glacier incision rates alone cannot be applied over millions of years to characterize denudational unloading.

The influence of glacial erosion on the relief structure of the mountains in Pakistan is an important issue. Bishop and others (2002, 2003) concur with Brozovik and others (1997) that, for intermediate altitudes, glaciation can limit relief up to some altitude approaching the equilibrium-line altitude (ELA).

Bishop and others (2003) also suggested that glaciation can produce greater mesoscale relief than river incision because glaciation, in concert with uplift, may be primarily responsible for relief production at high altitudes. Remote sensing of glaciers is expected to provide new insights into the role of glaciation in denudation and relief production, as spatial information on glacierization and estimates of glaciological parameters are needed for erosional and landscape-evolution modeling.

Glacier mapping supports glacial reconstructions and the assessment of glacial geomorphic features. Such studies in Pakistan have revealed the presence of high-altitude glacier erosion surfaces throughout the Hindu Kush, Karakoram, and Nanga Parbat Himalaya. Cosmogenic exposure-age dating of these surfaces and moraines in various locations have revealed glacier advances that likely represent monsoon-enhanced glaciation caused by orbital forcing (Phillips and others, 2000; Owen and others, 2002). Paleoclimatic conditions were significantly different from the climatic conditions of today, which strongly suggests that the landscape is not in topographic equilibrium (for example, it has unequal erosion and uplift rates). This explains the highly differential denudation rates reported by Shroder and Bishop (2000).

The issue of glacier erosion and relief production is at the heart of the climate-forcing hypothesis. It is notoriously difficult to estimate the magnitude of glacial erosion, because observations and measurements cannot be adequately obtained, and glacial processes such as plucking cannot be easily measured. This makes it difficult to account for various processes in models and constrain simulations using observational data. Nevertheless, remote sensing studies facilitate modeling efforts that provide new insights into the role of glaciation in landscape evolution. Preliminary one-dimensional, glacier-erosion simulations, with input parameter estimates obtained from ASTER imagery and digital elevation models (DEMs), depict nonlinear variation in glacial incision rates caused by valley evolution. These simulations also indicate that the spatially averaged rate of incision for a glacier with 300 m of ice thickness is likely to exceed 5 mm a^{-1}. These findings support empirical work indicating that glaciers in Pakistan can produce significant relief over time (Bishop and others, 2003). Collectively, results from remote sensing and geomorphological research suggest that glaciers may play a more significant role in landscape evolution than previously thought.

Another new insight into the role of glaciers in landscape evolution has come from assessing glacier ice velocity fields over mountain ranges. Kääb (2005) addressed the issue of satellite-derived DEM generation and derivation of ice velocity estimates in the Bhutan Himalaya. His results demonstrate the significant spatial variability in glacier flow, such that the glacier velocities in the terminus zones on the south side of the mountain range exhibit significantly lower velocities (10–20 m a^{-1}) than on the north side (20–200 m a^{-1}). We suspect similar, if not more spatial variation in ice flow characteristics in Pakistan. This has implications for glacier erosion, as it is generally assumed that glacial abrasion scales with basal sliding velocity (MacGregor and others, 2000; Tomkin and Braun, 2002). Although relief production via glacial erosion involves numerous factors and processes such as glacial plucking, glacial meltwater production and subglacial fluvial erosion, and cold-based ice at higher altitudes with no basal slip or erosion, we might expect higher glacier denudation rates in those areas in Pakistan that have fast-moving glaciers (other controlling factors being constant). We lack, however, adequate knowledge about numerous climate and lithospheric feedback mechanisms that influence glacier erosion, although glacier erosion and landscape evolution studies reveal that glacier processes are governed by a variety of internal and external factors. The spatial variability in flow velocities suggests the potential for significant variations in glacier erosion with altitude and geographic location, as well as with time during glacial/interglacial cycles. This may place limits on the magnitude of glacial erosion and relief production as a function of local topographic conditions, cold-based ice (at higher altitudes), and radiative and atmospheric forcing (Bishop and others 2001, 2003).

To understand the role of glaciers in mountain geodynamics, topographic information, glacier boundaries, ice-velocity fields, and land-cover information must be incorporated into glacier flow and mass balance models which consider various glacial dynamics (for example, MacGregor and others, 2000). In this way, the spatial variability of glacier geometry and ice flow are taken into account, and paleo-simulations can be constrained by glacier surface topography and variations in width and surface velocity. Similarly, supraglacial conditions can be assessed from satellite imagery (Bishop and others, 1995, Bishop and others, 2004). Using these tools, earth scientists can better investigate the complex interdependencies associated with climatic, tectonic, and surface processes that are responsible for polygenetic topographic evolution.

Conclusions

Glaciers of the Hindu Kush, Nanga Parbat, and the Karakoram Himalaya of Pakistan have diverse characteristics resulting from complex topography, spatial and temporal differences in snow and rock-debris supply, differences of flow, and variability of ablation. Remote-sensing assessments using older Landsat MSS, RBV, and newer ASTER imagery provide a new means of characterizing various aspects of glaciers and detecting perimeter changes.

In general, glaciers on the south sides of the ranges or in the deeper valleys tend to be more debris covered than those in higher locations or further into the interior. This may be because there is greater exposure to monsoon-generated deep snows and erosive avalanches on the south sides, greater debris generation from slopes in deep valleys of exceptional relief, and greater ablation in the lower altitude valleys tributary to the Indus River. Flow behavior and flow history appear to have no regional control, but instead vary throughout the ranges. From a regional perspective, many glaciers of northern Pakistan are in a general state of down wasting and back wasting. Rapid advances or surges are reasonably common and may relate to meltwater increases. Although initially catastrophic and devastating, surge events in several places have been accompanied by strong water-deposited sediment that adds to potentially arable land for later use by mountain villagers.

More sophisticated analysis and modeling of glaciers utilizing ASTER imagery and DEMs, such as those in the ongoing GLIMS research effort, promise to greatly improve our understanding of glacial fluctuation in the western Himalaya and Hindu Kush. These studies will result in the first quantitative assessments of regional mass balance and elucidation of the role of glaciation in relief production and landscape evolution in Pakistan.

References Cited

Aizen, V.B., Aizen, E.M., Dozier, J., Melack, J.M., Sexton, D.D., and Nesterov, V.N., 1997, Glacial regime of the highest Tien Shan Mountains, Pobeda-Khan Tengry Massif: Journal of Glaciology, v. 33, no. 145, p. 503–512.

Aniya, M., Sato, H., Naruse, R., Skvarca, P., and Casassa, G.,1996, The use of satellite and airborne imagery to inventory outlet glaciers of the Southern Patagonia Icefield, South America: Photogrammetric Engineering and Remote Sensing, v. 62, no. 12, p. 1361–1369.

Auden, J.B., 1935, The snout of the Biafo Glacier in Baltistan: Geological Survey of India, Records, v. 68, pt. 4, p. 400–413.

Avouac, J.P., and Burov, E.B., 1996, Erosion as a driving mechanism of intracontinental mountain growth: Journal of Geophysical Research, v.101, p. 747–769.

Bateson, S., and Bateson, R., translators, 1952, Tirich Mir — The Norwegian Himalayan Expedition: London, Hodder and Stoughton, 192 p.

Batūra Glacier Investigation Group (BGIG), 1979, The Batūra Glacier in the Karakoram Mountains and its variations: Scientia Sinica, v. 22, no. 8, p. 958–974.

Batūra Glacier Investigation Group (BGIG), 1980, Professional papers on the Batūra Glacier, Karakoram Mountains: Beijing, Science Press, 271 p. (In Chinese with English abstracts)

Beniston, M., 1994, Mountain environments in changing climates: Routledge, London, 496 p.

Berthier, E., Arnaud, Y., Kumar, R., Ahmad, S., Wagnon, P., and Chevallier, P., 2007, Remote sensing estimates of glacier mass balance in the Himachal Pradesh (western Himalaya, India): Remote Sensing of Environment, v. 108, issue. 3, p. 327–338. doi:10.1016/j.rse.2006.11.017

Bhutiyana, M.R., 1999, Mass-balance studies on Siachen Glacier in the Nubra Valley, Karakoram Himalaya, India: Journal of Glaciology, v. 45, no. 149, p. 112–118.

Bishop, M.P., Barry, R.G., Bush, A.B.G., Copland, L., and others, 2004, Global land-ice measurements from space (GLIMS) — Remote sensing and GIS investigations of the Earth's cryosphere: Geocarto International, v. 19, no. 2, p. 57–84.

Bishop, M.P., Bonk, R., Kamp, U., Jr., and Shroder, J.F., Jr., 2001, Terrain analysis and data modeling for alpine glacier mapping: Polar Geography, v. 24, no. 4, p. 257–276.

Bishop, M.P., Kargel, J.S., Kieffer, H.H., MacKinnon, D.J., Raup, B.H., and Shroder, J.F., Jr., 2000, Remote-sensing science and technology for studying glacier processes in high Asia: Annals of Glaciology, v. 31, p. 164–170.

Bishop, M.P., and Shroder, J.F., Jr., 2000, Remote sensing and geomorphometric analysis for assessing topographic complexity and erosion dynamics in the Nanga Parbat Himalaya, in Khan, M.A., Treloar, P.J., Searle, M.P., and J.M. Qasim, eds., Tectonics of the Nanga Parbat Syntaxis and Western Himalaya: Geological Society of London Special Publication 170, p. 181–200.

Bishop, M.P., and Shroder, J.F., Jr., eds., 2004, Geographic Information Science and Mountain Geomorphology: Springer-Praxis, Chichester, U.K., 486 p.

Bishop, M.P., Shroder, J.F., Jr., Bonk, R., and Olsenholler, J., 2002, Geomorphic change in high mountains: A western Himalayan perspective: Global and Planetary Change, v. 32, p. 311–329.

Bishop, M.P., Shroder, J.F., Jr., and Colby, J.D., 2003, Remote sensing and geomorphometry for studying relief production in high mountains: Geomorphology, v. 55, p. 345–361.

Bishop, M.P., Shroder, J.F., Jr., and Hickman, B.L., 1999, SPOT panchromatic imagery and neural networks for information extraction in a complex mountain environment: Geocarto International, v. 14, p. 17–26.

Bishop, M.P., Shroder, J.F., Jr., Hickman, B.L., and Copland, L., 1998a, Scale-dependent analysis of satellite imagery for characterization of glacier surfaces in the Karakoram Himalaya, in Walsh, S., and Butler, D., eds., Special volume on Remote Sensing in Geomorphology: Geomorphology, v. 21, p. 217–232.

Bishop, M.P., Shroder, J.F., Jr., Sloan, V.F., Copland, L., and Colby, J.D., 1998b, Remote sensing and GIS science and technology for studying lithospheric processes in a mountain environment: Geocarto International, v. 13, p. 75–87.

Bishop, M.P., Shroder, J.F., Jr., and Ward, J.L., 1995, SPOT multispectral analysis for producing supraglacial debris-load estimates for Batūra Glacier, Pakistan: Geocarto International, v. 10, no. 4, p. 81–90.

Brozovik, N., Burbank, D.W., and Meigs, A.J., 1997, Climatic limits on landscape development in the northwestern Himalaya: Science, v. 276, no. 5312, p. 571–574.

Burbank, D., Leland, J., Fielding, E., Anderson, R.S., Brozovik, N., Reid, M.R., and Duncan, C., 1996, Bedrock incision, rock uplift and threshold hillslopes in the northwestern Himalaya: Nature, v. 379, no. 6565, p. 505–510.

Bush, A.B.G, 2000, A positive feedback mechanism for Himalayan glaciation: Quaternary International, v. 65–6, p. 3–13.

Bush, A.B.G., 2002, A comparison of simulated monsoon circulations and snow accumulation in Asia during the mid-Holocene and at the Last Glacial Maximum: Global and Planetary Change, v. 32, p. 331–347.

Cao, M.S., 1998, Detection of abrupt changes in glacier mass balance in the Tien Shan mountains: Journal of Glaciology, v. 44, no. 147, p. 352–358.

Chi Jian-mei and Ren Bing-hui, eds., 1980, Glaciers in China: Shanghai Scientific and Technical Publishers, unpaginated.

Cogley, J.G., and Adams, W.P., 1998, Mass balance of glaciers other than the ice sheets: Journal of Glaciology, v. 44, no. 147, p. 315–325.

Conway, W.M., 1894, Climbing and exploration in the Karakoram-Himalayas: New York, Appleton & Co., 709 p.

Dainelli, Giotto, 1932, Italia Pass in the eastern Karakoram: Geographical Review, v. 22, no. 3, p. 392–402.

DeFilippi, Filippo, 1912, Karakoram and western Himalaya, 1909: London, Constable and Co., 469 p.

Derbyshire, E., Li Jijun, Perrot, F.A., Xu Shuying, and Waters, R.S., 1984, Quaternary glacial history of the Hunza Valley, Karakoram mountains, Pakistan, in Miller, K.J., ed., The International Karakoram Project, v. 2, p. 456–495.

Desinov, L.V., Ivanchenkov, A.S., Kotlyakov, V.M., and Nosenko, G.A., 1982, Resul'taty eksperimenta po izucheniyu oledeniya Karakoruma s borta orbital'noi stantsiy "salyut-6" [Results of experimental studies of Karakoram glaciers from the Salyut-6 orbital station]: Materialy Glyatsiologicheskikh Issledovaniy, Pub. 42, p. 22–40.

Desio, Ardito, Marussi, Antonio, and Caputo, Michele, 1961, Glaciological research of the Italian Karakoram Expedition 1953–1955, *in* General Assembly of Helsinki, 1960: Association Internationale d'Hydrologie Scientifique (AIHS), Pub. 54, p. 224–232.

Dozier, J., and Marks, D., 1987, Snow mapping and classification from Landsat Thematic Mapper data: Annals of Glaciology, v. 9, p. 97–103.

Dyurgerov, M.B., and Meier, M.F., 2000, Twentieth century climate change — Evidence from small glaciers: Proceedings of the National Academy of Science, v. 97, no. 4, p. 1406–1411.

Edwards, J.I., 1960, The Batūra Muztagh Expedition, 1959: The Alpine Journal, v. 65, no. 300, p. 48–52.

Elson, J.A., 1980, Glacial geology, *in* Siegel, B.S., and Gillespie, A.R., eds., Remote sensing in geology: New York, John Wiley, p. 505–551.

Finsterwalder, Richard, 1937, Die Gletscher des Nanga Parbat, Glaciologische Arbeiten der Deutschen Himalaya Expedition 1934 und ihre Ergebnisse: Zeitschrift für Gletscherkunde, v. 25, p. 57–108, and topographic map at 1:100,000.

Finsterwalder, Richard, 1960, German glaciological and geological expeditions to the Batūra Mustagh and Rakaposhi Range: Journal of Glaciology, v. 3, no. 28, p. 787–788.

Finsterwalder, Richard, and Pillewizer, W., 1939, Photogrammetric studies of glaciers in high Asia: The Himalayan Journal, v. 11, p. 107–113.

Finsterwalder, R., Raechl, W., and Misch, P., 1935, The scientific work of the German Himalayan expedition to Nanga Parbat, 1934: The Himalayan Journal, v. 7, p. 44–52.

Fujita, K., Nakawo, M., Fujii, Y., and Paudyal, P., 1997, Changes in glaciers in Hidden Valley, Mukut Himal, Nepal Himalayas, from 1974 to 1994: Journal of Glaciology, v. 43, no. 145, p. 583–588.

Gardner, J.S., 1986, Recent Fluctuations of Rakhiot Glacier, Nanga Parbat, Punjab Himalaya, Pakistan: Journal of Glaciology, v. 32, no. 112, p. 527–529.

Gardner, J.S., and Jones, N.K., 1993, Sediment transport and yield at the Rāikot Glacier, Nanga Parbat, Punjab Himalaya, *in* Shroder, J.F., Jr., ed., Himalaya to the Sea: Geology, Geomorphology and the Quaternary, p. 184–197.

Gilchrist, A.R., Summerfield, M.A., and Cockburn, H.A.P., 1994, Landscape dissection, isostatic uplift, and the morphologic development of orogens: Geology, v. 22, no. 11, p. 963–966.

Gillespie, A., and Molnar, P., 1995, Asynchronous maximum advances of mountain and continental glaciers: Reviews of Geophysics, v. 33, p. 311–364.

Goudie, A.S., Jones, D.K.C., and Brunsden, Denys, 1984, Recent fluctuations in some glaciers of the western Karakoram mountains, Hunza, Pakistan, *in* Miller, K.J., ed., The International Karakorum Project, v. 2, p. 411–455.

Haeberli, W., and Beniston, M., 1998, Climate change and its impacts on glaciers and permafrost in the Alps: Ambio, v. 27, no. 4, p. 258–265.

Haeberli, W., Hoelzle, M., and Suter, S., 1998, Into the second century of worldwide glacier monitoring — Prospects and strategies: Paris, UNESCO, 227 p.

Haefner, H., Seidel, K., and Ehrler, H., 1997, Applications of snow cover mapping in high mountain regions: Physics and Chemistry of the Earth, v. 22, p. 275–278.

Hall, D.K., Chang, A.T.C., Foster, J.L., Benson, C.S., and Kovalick, W.M., 1989, Comparison of *in situ* and Landsat derived reflectance of Alaskan glaciers: Remote Sensing of the Environment, v. 28, p. 23–31.

Hall, D.K., Chang, A.T.C., and Siddalingaiah, H., 1988, Reflectances of glaciers as calculated using Landsat-5 Thematic Mapper data: Remote Sensing of the Environment, v. 25, p. 311–321.

Hall, D.K., Kaminski, M., Cavalieri, D.J., Dickinson, R.E., and others, 2005, Earth observing system snow and ice products for observation and modeling: EOS (Transactions, American Geophysical Union), v. 86, no. 7, p. 67.

Hallet, B., Hunter, L., and Bogen, J., 1996, Rates of erosion and sediment evacuation by glaciers — A review of field data and their implications: Global and Planetary Change, v. 12, p. 213–235.

Hambrey, Michael, 1994, Glacial environments: London, UCL Press, 296 p.

Harbor, J., and Warburton, J., 1992, Glaciation and denudation rates: Nature, v. 356, no. 6372, p. 751.

Harbor, J., and Warburton, J., 1993, Relative rates of glacial and nonglacial erosion in alpine environments: Arctic and Alpine Research, v. 25, p. 1–7.

Hayden, H.H., 1907, Notes on certain glaciers in northwest Kashmir: Geological Survey of India, Records, v. 35, p. 127–137.

Henderson-Sellars, A., and Pitman, A.J., 1992, Land-surface schemes for future climate models — Specification, aggregation, and heterogeneity: Journal of Geophysical Research, v. 97, no. D3, p. 2687–2696.

Hewitt, Kenneth, 1969, Glacier surges in the Karakoram Himalaya (Central Asia): Canadian Journal of Earth Sciences, v. 6, no. 4, p. 1009–1018.

Hewitt, Kenneth, 1988a, Catastrophic landslide deposits in the Karakoram Himalaya: Science, v. 242, no. 4875, p. 64–67.

Hewitt, Kenneth, 1988b, The snow and ice hydrology project — Research and training for water resource development in the Upper Indus Basin: Journal of Canada-Pakistan Cooperation, v. 2, no. 1.

Hewitt, Kenneth, 1990, Snow and ice hydrology project, Upper Indus Basin, Canada-Pakistan — Final Report for the Hydrology Research Division, Water and Power Development Authority (WAPDA), Pakistan and International Development Research Centre (IDRC), Ottawa: Waterloo, Ontario, Wilfrid Laurier University, 9 v.

Hewitt, Kenneth, 1998, Catastrophic landslides and their effect on the Upper Indus streams, Karakoram Himalaya, northern Pakistan: Geomorphology, v. 26, p. 47–80.

Hewitt, Kenneth, 1999, Quaternary moraines vs catastrophic rock avalanches in the Karakoram Himalaya, northern Pakistan: Quaternary Research, v. 51, p. 220–237.

Hewitt, Kenneth, Wake, C.P., Young, G.J., and David, C., 1989, Hydrological investigations at Biafo Glacier, Karakoram Himalaya — An important source of water for the Indus River: Annals of Glaciology, v. 13, p. 103–108.

Horvath, Eva, 1975, Glaciers of the Hindu Kush, *in* Field, W.O., ed., Mountain glaciers of the northern hemisphere: Hanover, New Hampshire, U.S. Army Cold Regions Research and Engineering Laboratory, v. 1, p. 361–370.

Holdsworth, Gerald, Howarth, P.J., and Ommanney, C.S.L., 2002, Quantitative measurements of Tweedsmuir Glacier and Lowell Glacier imagery, *in* Williams, R.S., Jr., and Ferrigno, J.G., eds., Satellite image atlas of glaciers of the world: U.S. Geological Survey Professional Paper 1386–J (North America), p. J312–J324.

Hubbard, A.L., 1997, Modelling climate, topography and paleoglacier fluctuations in the Chilean Andes: Earth Surface Processes and Landforms, v. 22, p. 79–92.

Kääb, A., 2005, Combination of SRTM3 and repeat ASTER data for deriving alpine glacier flow velocities in the Bhutan Himalaya: Remote Sensing of Environment, v. 94, p. 463–474.

Kääb, A., Paul, F., Maisch, M., Hoelzle. M., and Haeberli, W., 2002, The new remote sensing derived Swiss glacier inventory — II. First results: Annals of Glaciology, v. 34, p. 362–366.

Kamb, Barclay, Raymond, C.F., Harrison, W.D., Engelhardt, Hermann, and others, 1985, Glacier surge mechanism — 1982–1983 surge of Varigated Glacier, Alaska: Science, v. 227, no. 4686, p. 469–479.

Kaser, G., 2001, Glacier-climate interaction at low latitudes: Journal of Glaciology, v. 47, no. 157, p. 195–204.

Kaser, G., and Fountain, A., 2001, A manual for monitoring the mass balance of mountain glaciers, with particular attention to low latitude characteristics — A contribution from ICSI to the UNESCO HKH-Friend program: ICSI/UNESCO/HKH-Friend glacier mass balance manual 12.01, draft December 2001.

Kick, W., 1962, Variations of some central Asiatic glaciers, *in* Symposium of Obergurgl: Association Internationale d'Hydrologie Scientifique (AIHS), Pub. 58, p. 223–229.

Kick, W., 1975, Application of geodesy, photogrammetry, history and geography to the study of long-term mass balances of Central Asiatic glaciers, Snow and Ice Symposium (Proceedings of Moscow Symposium, August 1971): IAHS-AISH Publication No. 104, p. 150–160.

Kick, W., 1980, Material for a glacier inventory of the Indus drainage basin — The Nanga Parbat Massif, *in* World Glacier Inventory, Proceedings of the Reideralp workshop: IAHS-AISH Publication, v. 126, p. 105–109.

Kieffer, H., Kargel, J.S., Barry, R., Bindschadler, R., Bishop, M.P., and others, 2000, New eyes in the sky measure glaciers and ice sheets: EOS (Transactions, American Geophysical Union), v. 81, no. 24, p. 265, 270–271.

Knizhnikov, Yu.F., Osipova, G.B., Tsvetkov, D.G, and Kharkovets, E.G., 1997, Measurements of the movement of surging glaciers by the method of aeropseudoparallaxes (Using the Medvezhiy Glacier as an example): Materialy Glyatsiologicheskikh Issledovaniy, Pub. 81, p. 55–60.

Koons, P.O., 1995, Modeling the topographic evolution of collisional belts: Annual Review Earth and Planetary Science, v. 23, p. 375–408.

Kotlyakov, V.M., 1980, Problems and results of studies of mountain glaciers in the Soviet Union, *in* World glacier inventory, Proceedings of the Reideralp workshop: IAHS-AISH Publication 126, p. 129–137.

Kotlyakov, V.M., 1997, Atlas snezhno – ledovykh resursov mira [World atlas of snow and ice resources]: Moscow, Rossiiskaya Akademiya Nauk, 3 v., 392 p.

Kotlyakov, V.M., Serebrjanny, L.R., and Solomina, O.N., 1991, Climate change and glacier fluctuations during the last 1,000 years in the southern mountains of the USSR: Mountain Research and Development, v. 11, p. 1–12.

Kravtsova, V.I., 1982, Ispol'zovanie kosmicheskikh shimkov v kartografirovaniy gornogo oledeneniya [The use of space images in the mapping of mountain glaciers]: Materialy Glyatsiologicheskikh Issledovaniy, Pub. 42, p. 18–22.

Kravtsova, V.I., and Lebedeva, I.M., 1974, Ispol'zovanie kosmicheskikh shimkov dlya izucheniya glyatsiologicheskikh v gornykh regionov [Using satellite images for glaciological investigations of mountain regions], *in* Issledovanie prirodnoi sredny kosmicheskimisredstvami [Investigation of the Natural Environment by Space Means]: Academy of Sciences of the USSR, Committee on Natural Resources, Investigation from Spacecraft, v. 3, p. 57–68.

Krimmel, R.M., Post, Austin, and Meier, M.F., 1976, Surging and nonsurging glaciers in the Pamir Mountains, U.S.S.R., *in* Williams, R.S., Jr., and Carter, W.D., eds., ERTS–1, a new window on our planet: U.S. Geological Survey Professional Paper 929, p. 178–179.

Lambeck, K., and Chappel, J., 2001, Sea level change through the last glacial cycle: Science, v. 292, no. 5517, p. 679–686.

Lebedeva, I.M., 1993, Let'nyaya temperatura vozdukha v nival'no-glyatsial'nom poyase tsentral'no Aziatskogo gornogo massiva [Summer air temperature in the nival-glacial band of central Asian mountain massifs]: Izvestia RAN Seriya Geograficheskaya, no. 6, p. 20–39.

Lebedeva, I.M., 1995, Shirotnye faktory izmenchivosti rezhima lednikovykh sistem tsentral'no Aziatskogo gornogo massiva [Latitude-induced factors underlying the changeability of the regime of glacier systems within the Central Asia mountain massif]: Materialy Glyatsiologicheskikh Issledovaniy, Pub. 79, p. 86–94.

Lebedeva, I.M., and Kislov, A.V., 1993, Atmosfernye osadki I vysota granitsy pitaniya lednikov v subtropicheskom tsentral'no-Aziatskom regione [Atmospheric precipitation and the equilibrium line altitude on glaciers of subtropical Central Asian region]: Materialy Glyatsiologicheskikh Issledovaniy, Pub. 76, p. 60–76.

Liestøl, Olav, 1993, Glaciers of Svalbard, Norway, *in* Williams, R.S., Jr., and Ferrigno, J.G., eds., Satellite image atlas of glaciers of the world: U.S. Geological Survey Professional Paper 1386–E (Europe), p. E127–E151.

Loewe, F., 1959, Some observations of the radiation budget and of the ablation of glacier ice in the Nanga Parbat region: Pakistan Journal of Science, v. 11, no. 5, p. 227–236.

Loewe, F., 1961, Glaciers of Nanga Parbat: Pakistan Geographical Review, v. 16, p. 19–24.

MacGregor, K.R., Anderson, R.S., Anderson, S.P., and Waddington, E.D., 2000, Numerical simulations of glacial-valley longitudinal profile evolution: Geology, v. 28, no. 11, p. 1031–1034.

Maisch, M., 2000, The long term signal of climate change in the Swiss Alps: Glacier retreat since the end of the Little Ice Age and future ice decay scenarios: Geografia Fisica e Dinamica Quaternaria, v. 23, no. 2, p. 139–151.

Mason, Kenneth, 1930, The glaciers of the Karakoram and neighborhood: Geological Survey of India, Records, v. 63, pt. 2, p. 214–278.

Mason, Kenneth, 1935, The study of threatening glaciers: Geographical Journal, v. 85, no. 1, p. 24–41.

Mattson, L.E., 2000, The influence of a debris cover on the midsummer discharge of Dome Glacier, Canadian Rocky Mountains, in Nakawo, M., Raymond, C.F., and Fountain, A., eds., Debris-Covered Glaciers — Proceedings of a workshop, Seattle, Washington, 13–15 September 2000: International Association of Hydrological Sciences Publication 264, p. 25–33.

Maull, Otto, 1938, Geomorphologie: Leipzig and Vienna, Franz Deuticke, 294 p.

Mayewski, P.A., and Jeschke, P.A., 1979, Himalayan and trans-Himalayan glacier fluctuations since AD 1812: Arctic and Alpine Research, v. 11, no. 3, p. 267–287.

Mayewski, P.A., Pregent, G.P., Jeschke, P.A., and Ahmad, Naseeruddin, 1980, Himalayan and trans-Himalayan glacier fluctuations and the south Asian monsoon record: Arctic and Alpine Research, v. 12, no. 2, p. 171–182.

McClung, D.M., and Armstrong, R.L., 1993, Temperate glacier time response from field data: Journal of Glaciology, v. 39, no. 132, p. 323–326.

Meier, M.F., 1976, Monitoring the motion of surging glaciers in the Mount McKinley massif, Alaska, in Williams, R.S., Jr., and Carter, W.D., eds., ERTS-1, a new window on our planet, U.S. Geological Survey Professional Paper 929, p. 185–187.

Meier, M.F., 1984, Contribution of small glaciers to global sea level: Science, v. 226, no. 4681, p. 1418–1421.

Meier, M.F., and Dyurgerov, M.B., 2002, How Alaska affects the world: Science, v. 297, no. 5580, p. 350–351.

Meier, M.F., Dyurgerov, M.B., and McCabe, G.J., 2003, The health of glaciers — Recent changes in glacier regime: Climate Dynamics, v. 59, p. 123–135.

Meier, M.F., and Wahr, J.M., 2002, Sea level is rising — Do we know why?: Proceedings of the National Academy of Science, v. 99, no. 10, p. 6524–6526.

Menzies, John, 2002, Modern and past glacial environments: Oxford, U.K, Butterworth Heinemann, 543 p.

Mercer, J.H., 1963, Glacier variations in the Karakoram: Glaciology Notes, World Data Center, American Geographical Society, v. 14, p. 19–33.

Mercer, J.H., 1975, Glaciers of the Karakoram, in Field, W.O., ed., Mountain glaciers of the northern hemisphere: Hanover, NH, U.S. Army Cold Regions Research and Engineering Laboratory, v. 1, p. 371–409.

Miller, K.J., ed., 1984, The International Karakoram Project: Cambridge, U.K., Cambridge University Press, v. 1, 412 p., v. 2, 635 p.

Molnar, Peter, and England, P., 1990, Late Cenozoic uplift of mountain ranges and global climate change — Chicken or egg?: Nature, v. 346, no. 6279, p. 29–34.

Montgomery, D.R., 1994, Valley incision and the uplift of mountain peaks: Journal of Geophysical Research, v. 99, no. B7, p. 13,913–13,921.

Mool, P.K., Bajracharya, S.R., and Joshi, S.P., 2001, Inventory of glaciers, glacial lakes and glacial lake outburst floods, Nepal: Kathmandu, Nepal, International Center for Integrated Mountain Development, 363 p.

Mool, P.K., Wangda, D., Bajracharya, S.R., Kunzang, K., Gurung, D.R., and Joshi, S.P., 2001, Inventory of glaciers, glacial lakes and glacial lake outburst floods, Bhutan: Kathmandu, Nepal, International Center for Integrated Mountain Development, 227 p.

Mott, P.G., 1950, Karakoram survey, 1939: A new map: Geographical Journal, v. 116, nos. 1–3, p. 89–95.

Nakawo, M., Fujita, K., Ageta, U., Shankar, K., Pokhrel, P.A., and Yao Tandong, 1997, Basic studies for assessing the impacts of the global warming on the Himalayan cryosphere: Bulletin of Glacier Research, v. 15, p. 53–58.

Nosenko, G.A., 1991, Otsenka raspredeleniya rezhimia lednikov Karakoruma po materialam kosmicheskikh semok [Evaluation of the distribution and regime of Karakoram glaciers from the data of space images]: Materialy Glyatsiologicheskikh Issledovaniy, Pub. 72, p. 87–94.

Odell, N.E., 1937, The glaciers and morphology of the Franz Josef Fjord region of north-east Greenland: Geographical Journal, v. 90, no. 2, p. 111–125.

Osipova, G.B., and Tsvetkov, D.G., 1998, Opyt katalogizatsiy pul'siruyushchikh lednikov Pamira [Inventorying surging glaciers of the Pamirs]: Materialy Glyatsiologicheskikh Issledovaniy, Pub. 85, p. 223–232.

Osipova, G.B., and Tsvetkov, D.G., 2002, Issledovanie dinamiki slozhnykh lednikov po kosmicheskim snimkam [Investigations of complex glacier dynamics on the basis of space images]: Izvestia RAN, Seriya Geograficheskaya, no. 3, p. 29–38.

Osipova, G.B., and Tsvetkov, D.G., 2003, Chto daet monitoring pul'siruyushchikh lednikov? [What is the benefit of monitoring surging glaciers?]: Priroda, v. 4, p. 3–13.

Owen, L.A., Finkel, R.C., Caffee, M.W., and Gualtieri, L., 2002, Timing of multiple late Quaternary glaciations in the Hunza Valley, Karakoram Mountains, northern Pakistan: Defined by cosmogenic radionuclide dating of moraines: Geological Society of America Bulletin, v. 114, no. 5, p. 593–604.

Owen, L.A., Scott, C.H., and Derbyshire, E., 2000, The Quaternary glacial history of Nanga Parbat, Quaternary International, v. 65–66, p. 63–79.

Paul, F., Huggel C., and Kääb, A., 2004, Combining satellite multispectral image data and a digital elevation model for mapping debris-covered glaciers: Remote Sensing of Environment, v. 89, no. 4, p. 510–518.

Paul, F., Kääb, A., Maisch, M., Kellenberger, T., and Haeberli, W., 2002, The new remote sensing derived Swiss glacier inventory — I. Methods: Annals of Glaciology, v. 34, p. 355–361.

Pecci, M., and Smiraglia, C., 2000, Advance and retreat phases of the Karakorum glaciers during the 20th century — Case studies in Braldu Valley (Pakistan): Geografia Fisica e Dinamica Quaternaria, v. 23, p. 73–85.

Phillips, W.M., Sloan, V.F., and Shroder, J.F., Jr., Sharma, P., Clarke, M.L., and Rendell, H.M., 2000, Asynchronous glaciation at Nanga Parbat, northwestern Himalaya Mountains, Pakistan: Geology, v. 28, no. 5, p. 431–434.

Pillewizer, Wolfgang, 1956, Der Rakhiotgletscher am Nanga Parbat im Jahr 1954: Zeitschrift für Gletscherkunde und Glazialgeologie, v. 3, no. 2, p. 181–199.

Porter, S.C., 1970, Quaternary glacial record in Swat Kohistan, West Pakistan: Geological Society of America Bulletin, v. 81, no. 5, p. 1421–1446.

Post, Austin, Meier, M.F., and Mayo, L.R., 1976, Measuring the motion of the Lowell and Tweedsmuir surging glaciers of British Columbia, Canada, in Williams, R.S., Jr., and Carter, W.D., eds., ERTS-1, A new window on our planet: U.S. Geological Survey Professional Paper 929, p. 180–184.

Raymo, M.E., and Ruddiman, W.F., 1992, Tectonic forcing of the Late Cenozoic climate: Nature, v. 359, no. 6391, p. 117–122.

Raymo, M.E., Ruddiman, W.F., and Froelich, P.N., 1988, Influence of Late Cenozoic mountain building on geochemical cycles: Geology, v. 16, no. 7, p. 649–653.

Richards, B.W., Benn, D.I., Owen, L.A., Rhodes, E.J., and Spencer, J.Q., 2000, Timing of late Quaternary glaciations south of Mount Everest in the Khumbu Himal, Nepal: Geological Society of America Bulletin, v. 112, no. 10, p. 1621–1632.

Roohi, Rakhshan, 2007, Research on global changes in Pakistan, in Baudo, R., Tartari, G., Vuillermoz, E., and Shroder, J.F., Jr., eds., Mountains, witnesses of global changes — Research in the Himalaya and Karakoram — Developments in Earth Surface Processes, v.10, p. 329–340: Amsterdam, Netherlands, Elsevier.

Roohi, Rakhshan, Ashraf, A., Hussain, S., Naz, R., Mool, P.K., and Bajracharya, S.R., 2005, Indus basin, Pakistan, Hindu Kush-Karakoram-Himalaya. Inventory of glaciers and glacier lakes and the identification of potential glacial lake outburst floods (GLOFS) affected by global warming in the mountains of the HKH region: Water Resources Research Institute, National Agricultural Research Centre, Final Report to Government of Pakistan, CD-ROM, unpublished.

Ruddiman, W.F., 1997, Tectonic uplift and climate change: New York, Plenum Press, 535 p.

Schaper, J., Martinec, J., and Seidel, K., 1999, Distributed mapping of snow and glaciers for improved runoff modeling: Journal of Hydrological Processes, v. 13, no. 12–13, p. 2023–2036.

Scott, C.H., 1992, Contemporary sediment transfer in Himalayan glacial systems: Implications for the interpretation of the Quaternary record: unpublished Ph.D. thesis, University of Leicester, U.K., 356 p.

Searle, M.P., 1991, Geology and Tectonics of the Karakoram Mountains: Chichester, John Wiley, 358 p.

Seltzer, G.O., 1993, Late-Quaternary glaciation as a proxy for climate change in the central Andes: Mountain Research and Development, v. 13, p. 129–138.

Shi Yafeng, and Wang Wenying, 1980, Research on snow cover in China and the avalanche phenomena of Batūra Glacier in Pakistan: Journal of Glaciology, v. 26, no. 94, p. 25–30.

Shipton, Eric, 1938, The Shaksgam expedition, 1937: Geographical Journal, v. 91, no. 4, p. 313–339.

Shipton, Eric, 1940, Karakoram, 1939: Geographical Journal, v. 95, no. 6, p. 409–427.

Shroder, J.F., Jr., 1980, Special problems of glacier inventory in Afghanistan, in World Glacier Inventory, Proceedings of the Riederalp workshop: IAHS–AISH Publication No. 126, p. 149–154.

Shroder, J.F., Jr., 1989, Hazards of the Himalaya: American Scientist, v. 77, no. 6, p. 564–573.

Shroder, J.F., Jr., and Bishop, M.P., 1998, Mass movement in the Himalaya — New insights and research directions, in Shroder, J.F, Jr., ed., Special volume on "Mass Movement in the Himalaya": Geomorphology, v. 26, p. 13–35.

Shroder, J.F., Jr., and Bishop, M.P., 2000, Unroofing the Nanga Parbat Himalaya, in Khan, M.A., Treloar, P.J., Searle, M.P., and Jan, M.Q., eds., Tectonics of the Western Himalaya and Karakoram: Geological Society of London Special Publication 170, p. 163–179.

Shroder, J.F., Jr., and Bishop, M.P., 2009, Glaciers of Afghanistan, in Williams, R.S., Jr., and Ferrigno, J.G., eds., Asia, U.S. Geological Survey Professional Paper, 1386–F, p. F167–F199.

Shroder, J.F., Jr., Bishop, M.P., Copland, L., and Sloan, V., 2000, Debris-covered glaciers and rock glaciers in the Nanga Parbat Himalaya, Pakistan: Geografiska Annaler, v. 82A, no. 1, p. 17–31.

Shroder, J.F., Jr., DiMarzio, C.M., Bussom, D.E., and Braslau, David, 1978, Remote sensing of Afghanistan: Afghanistan Journal, v. 5, no. 4, p. 123–128.

Shroder, J.F., Johnson, R., Saqib Khan, M., and Spencer, M., 1984, Batūra Glacier terminus, 1984, Karakoram Himalaya: University Peshawar (Pakistan), Geological Bulletin, v. 17, p. 119–126.

Shroder, J.F., Jr., Kahn, M.S., Lawrence, R.D., Madin, I.P., and Higgins, S.M., 1989, Quaternary glacial chronology and neotectonics in the Himalaya of northern Pakistan, in Malinconico, L.L., Jr., and Lillie, R.J., eds., Tectonics of the western Himalaya: Geological Society of America Special Paper, v. 232, p. 275–294.

Shroder, J.F., Jr., Scheppy, R.A., and Bishop, M.P., 1999, Denudation of small alpine basins, Nanga Parbat Himalaya, Pakistan: Arctic, Antarctic, and Alpine Research, v. 31, no. 2, p. 121–127.

Taylor, P.J., and Mitchell, W.A., 2000, Late Quaternary glacial history of the Zanskar Range, northwest Indian Himalaya: Quaternary International, v. 65, no. 5, p. 81–99.

Thompson, L.G., Yao T., Davis, M.E., Henderson, K.A., Mosley-Thompson, E., Lin, P.N., Beer, J., Synal, H.A., Cole-Dai, J., and Bolzan, J.F., 1997, Tropical climate instability — The last glacial cycle from a Qinghai-Tibetan ice core: Science, v. 276, no. 5320, p. 1821–1825.

Tomkin, J.H., and Braun, J., 2002, The influence of alpine glaciation on the relief of tectonically active mountain belts: American Journal of Science, v. 302, no. 3, p. 169–190.

Tsvetkov, D.G., Osipova, G.B., Zichu, X., Zhongtai, W., Ageta, Y., and Baast, P., 1998, Glaciers in Asia, in Haeberli, W., Hoelzle, M., and Suter, S., eds., Into the second century of worldwide glacier monitoring — Prospects and strategies: Paris, UNESCO, p. 177–196.

Visser, P.C., and Visser-Hooft, J., eds., 1938, Wissenschaftliche Ergebnisse der Neiderlandischen Expeditionen in den Karakoram und die angrenzenden Gebiete in den Jahren 1922, 1925, 1929/30 und 1939, v. 2, Glaziologie: Leiden, E.J. Brill, 216 p.

Wake, C.P., and Searle, M.P., 1993, Rapid advance of Pumarikish Glacier, Hispar Glacier basin, Karakoram Himalaya: Journal of Glaciology, v. 39, no. 131, p. 204–206.

Wang, W., Huang, M., and Chen, J., 1984, A surging advance of Balt Bare Glacier, *in* Miller, K.J., ed., The International Karakoram Project, v.1, p. 76–83.

Washburn, A.L., 1939, Karakoram glaciology: American Journal of Science, v. 237, no. 2, p. 138–146.

Whipple, K.X., Kirby, E., and Brocklehurst, S.H., 1999, Geomorphic limits to climate-induced increases in topographic relief: Nature, v. 401, no. 6748, p. 39–43.

Whipple, K.X., and Tucker, G.E., 1999, Dynamics of the stream-power river incision model — Implications for height limits of mountain ranges, landscape response timescales, and research needs: Journal of Geophysical Research, v. 104, no. B8, p. 17,661–17,674.

Williams, R.S., Hall, D.K., and Benson, C.S., 1991, Analysis of glacier facies using satellite techniques: Journal of Glaciology, v. 37, no. 125, p. 120–128.

Workman, F.B., and Workman, W.H., 1910, The Hispar Glacier: Geographical Journal, v. 35, no. 2, p. 105–132.

Workman, W.H., and Workman, F.B., 1911, The call of the snowy Hispar: New York, Charles Scribner's Sons, 297 p.

Zeitler, P.K., Koons, P.O., Bishop, M.P., Chamberlain, C.P., and others, 2001, Crustal reworking at Nanga Parbat, Pakistan: Metamorphic consequences of thermal-mechanical coupling facilitated by erosion: Tectonics, v. 20, no. 5, p. 712–728.

Zeitler, P.K., Meltzer, A.S., Koons, P.O., Craw, D., and others, 2001, Erosion, Himalayan geodynamics, and the geomorphology of metamorphism: GSA Today, v. 11, p. 4–8.

Glaciers of Asia—

GLACIERS OF INDIA

By Chander P. Vohra

SATELLITE IMAGE ATLAS OF GLACIERS OF THE WORLD

Edited by RICHARD S. WILLIAMS, JR., *and* JANE G. FERRIGNO

U.S. GEOLOGICAL SURVEY PROFESSIONAL PAPER 1386–F–5

CONTENTS

A Brief Overview of the State of Glaciers in the Indian Himalaya in the 1970s
and at the End of the 20th Century, by Chander P. Vohra-------------------------259

Abstract --259

Introduction ---259

Early Studies --260

TABLE 1. Recession of glaciers in the Indian Himalaya----------------------260

Studies in the Late 20th Century---261

Total Number of Glaciers ---262

Glacier Distribution --262

FIGURE 1. Sketch map showing the location of glacierized areas
of India discussed in the text -------------------------------- 262

Nun-Barashigri --263

Gangotri-Chaukhamba--263

FIGURE 2. Landsat 2 MSS image of the Nun-Barashigri area with
some of the many glaciers outlined -------------------------- 264

FIGURE 3. Sketch map showing glacierized area of the Chenab
and Ravi basins -- 265

Satluj River Basin --265

FIGURE 4. Landsat 2 MSS image of the Gangotri-Kāmet-Nanda
Devi areas of the Indian Himalaya -------------------------- 266

Inventory from Landsat Imagery -------------------------------------266

FIGURE 5. Sketch map showing glacierized area of the Ganga basin
showing the Gangotri-Chaukhamba, Kāmet, and Nanda
Devi areas --- 267

TABLE 2. Ice cover determined in the late 1970s from maps of different
scales as a percentage of the basin -------------------------- 267

TABLE 3. Comparison of glacier inventories from maps and Landsat
images-- 268

TABLE 4. Percentage reduction or increase in the ablation, accumulation,
and total area of glaciers ----------------------------------- 268

TABLE 5. Comparison of calculation of ablation, accumulation, and
total area of the Tirung Khad glacier No. 49 between
maps, Landsat images, and field work------------------------ 268

FIGURE 6. Optimum Landsat 1, 2, and 3 images of the glaciers of India---------- 269

TABLE 6. Optimum Landsat 1, 2, and 3 images of the glaciers of India----------- 270

FIGURE 7. Sketch map showing glacierized area of the Satluj River basin --------- 271

Kāmet--271

Nanda Devi --271

Sikkim Himalaya --272

FIGURE 8. Landsat 2 MSS image mosaic of the Sikkim Himalaya ------------------ 272

FIGURE 9. Map and ASTER image of the Zemu Glacier area ----------------------- 273

Eastern Himalaya --274

Surge-Type Glaciers---274

Conclusions--274

A Study of Selected Glaciers under the Changing Climate Regime,
by Syed Iqbal Hasnain, Rajesh Kumar, Safaraz Ahmad, and Shresth Tayal ------ **275**
    Introduction --**275**

    Recent Glacier Studies --**276**

        Gangotri Glacier ---**276**

            FIGURE 10. Location map of Gangotri Glacier and associated glaciers -------------- **277**
            FIGURE 11. Photograph of lake formed on Gangotri Glacier near the
                        junction with *Kirti Bamak Glacier*------------------------------------ **277**
            FIGURE 12. Enlargement of 9 September 2001 ASTER image showing
                        supraglacial lakes on the Gangotri Glacier ------------------------- **278**
            FIGURE 13. Enlargement of 9 September 2001 ASTER image showing
                        position of the terminus of Gangotri Glacier between
                        1780 and 2001 --- **278**
            FIGURE 14. Model showing predicted change in location of the
                        equilibrium line altitude for an increase of 3 °C air
                        temperature in the Gangotri Glacier region ---------------------- **279**

        Chhota Shigri Glacier and Mass-Balance Measurements --------------- **280**

            FIGURE 15. Sketch and location maps and ASTER image of the
                        Chhotta Shigri Glacier -- **280**
            FIGURE 16. Retreat of *Chhota Shigri Glacier* snout------------------------------ **281**
            FIGURE 17. Sketch map and Landsat images showing reduction of
                        the surface area of the *Chhota Shigri Glacier* from 1987
                        to 2000-- **282–283**
            FIGURE 18. Map of the *Chhota Shigri Glacier* showing location of
                        stakes installed in 2002 and the automatic weather
                        station installed during September–October 2003 --------------------- **284**

        Siachen Glacier --**285**

            FIGURE 19. Graphs showing specific net annual mass-balance near
                        measurement points averaged against the elevation on
                        Chhota Shigri Glacier for 2002–03 and 2003–04 ----------------------- **285**
            FIGURE 20. Landsat images showing retreat of the terminus of
                        Siachen Glacier from 1978 to 2001-------------------------------- **286**

        Glacier Lakes--**287**

            FIGURE 21. Landsat and ASTER images showing increase in the number
                        and size of glacier lakes from 1976 to 2001 in the Tista River
                        basin, Sikkim Himalaya--- **288**

    Conclusions---**288**

    References Cited--**289**

GLACIERS OF ASIA—

GLACIERS OF INDIA—A Brief Overview of the State of Glaciers in the Indian Himalaya in the 1970s and at the End of the 20th Century

By Chander P. Vohra[1]

Abstract

Himalaya, "the abode of snow," has many glaciers distributed throughout its higher elevations. The glaciers are commonly clustered in large numbers in areas where many high peaks are located and sufficient precipitation occurs. The first estimate of total ice cover was probably made by von Wissman (1959) who stated that 33,200 km[2] or 17 percent of the Himalaya was ice covered. Few ice-cover estimates for individual river basins were made in the initial years of glacier studies. More recent work by Qin (1999) increased the glacierized area of the Himalaya to about 35,110 km[2]. Based on the preliminary inventory by the Geological Survey of India (Kaul, 1999), about 8,500 km[2] of this ice cover is located in the Indian Himalaya.

Introduction

The study of glaciers in India is very important both from scientific and economical perspectives. The Himalaya contains one of the largest reservoirs of snow and ice outside the Polar Regions. The glaciers are a major source of fresh water, and all the rivers in northern India are nourished by meltwaters of the Himalaya glaciers, thereby affecting the quality of life of millions of people.

As in most parts of the Himalaya, the study of glaciers is handicapped by the remoteness of the area, altitude, topography, and debris cover on the glaciers, as well as the limited availability of large-scale maps, optimum aerial photographs, and high-resolution satellite images.

In 2001, Ravi Shanker, Director General of the Geological Survey of India (GSI), stated that glaciology, the study of snow, ice, and glaciers, is in its nascent stage in India (Srivastava, 2001); however, glacier observations in India actually began in the middle 1800s, and considerable work has been done in the last three decades.

[Editors' note: This manuscript was originally written in 1982 to describe the glaciers of India during the late 1970s and early 1980s, the "benchmark" time period (1972–1981) for the Satellite Image Atlas of Glaciers of the World, U.S. Geological Survey Professional Paper 1386-A–K. Because there were delays in publishing this chapter, the editors updated the manuscript slightly in 2006, using material from the Geological Survey of India, and a supplemental section — A Study of Selected Glaciers under the Changing Climate Regime, by Hasnain and others — was added to provide more current information.]

---

[1]Former Director-General, Geological Survey of India.

Early Studies

Snow and glaciers in the Himalaya attracted the attention of early observers and much was written around 1840–50 about the "line of perpetual snows," beginning with Jack (1844) and von Humbolt (1845). Very few glaciers were visited at that time, but one of the popular ones was Pindari Glacier[2] in the State of Uttar Pradesh (Madden, 1847). Many of the early observations of glaciers were made by geographers and cartographers who were engaged in triangulation and map making of the Himalaya during the 19th century. Godwin-Austin (1862), for example, wrote about the glaciers of Mustakh Range and the upper Indus River basin. Many observers, including geologists, were satisfied with recording the positions of glacier termini and the evidence they observed of more extensive past glaciation. One of the earliest Indian geologists to write about glaciers in the Himalaya was P.N. Bose (1891), who visited *Kabru* and *Pandim Glaciers* in the State of Sikkim. Between 1907 and 1910, the Geological Survey of India initiated a coordinated effort to systematize observations of glacier fluctuations, as part of an international program to study glaciers (Cotter and Brown, 1907; Bridges, 1908–09).

Kanwar Sain (1946) was the first scientist to investigate the role of snow, ice, and glaciers in a comprehensive assessment of the surface water resources of the Himalaya. J.E. Church (1947), a celebrated expert in this field, visited the Himalaya in 1947 to organize a program of snow surveys for hydrological purposes. This program, however, did not flourish (Basu, 1947). In the years between 1910 and 1957, individual studies and observations on some of the glaciers continued (Grinlinton, 1912; Auden, 1937). A second effort at glacier observations was organized during the International Geophysical Year (1957–58), when many glaciers were revisited and a great deal of information regarding glacier fluctuations and their history was gathered (Jangpangi, 1958).

This was followed by further glacier studies during the International Hydrologic Decade (IHD), 1965–74, and the International Hydrologic Programme in 1975. Some results from those observations are listed in table 1. Mercer (1975a, b) prepared a very good review of the status of glaciology in India up to 1975, including references and maps.

TABLE 1.—*Recession of glaciers in the Indian Himalaya*

[Data source: Vohra (1981). Unit: m, meter]

Glacier name	Observational period	Recession (m)
Milam	1849–1957	-1,350
Pindari	1845–1966	-2,840
Shankalpa	1881–1957	-518
Poting	1906–1957	-262
No. 3 in the *Arwa* valley	1932–1956	-24
Gangotri	1935–1976	-600
Zemu	1909–1965	-440
Barashigri	1940–1963	-1,019
Sonapani	1906–1929	-905

[2]The geographic place-names used in this section conform to the usage authorized for foreign names by the U.S. Board on Geographic Names as listed on the GEOnet Names Server (GNS) website: *http://earth-info.nga.mil/gns/html/index.html*. The names not listed on the website are shown in italics.

Studies in the Late 20th Century

Studies begun in the 1970s embraced many aspects of snow and glacier phenomena and included glacier mass balance, the pattern of glacier movement, glacier recession/advance, meltwater discharge, ice thickness in glaciers, crystal fabrics of ice, the radiation balance of snow and ice surfaces, dating of ice by isotopic studies, detailed cartography of glaciers, the study of paleoglaciation, and compilation of glacier inventories. The studies are described by Tewari (1971) and Vohra (1981) and included in unpublished reports of GSI.

The Geological Survey of India organized a glaciology program under the direction of the author, and on 4 January 1974, a Glaciology Division was established at Lucknow. Other national agencies, such as the Survey of India, the Meteorological Department, the Physical Research Laboratory, Indian Space Research Organization, and the Bhakra Beas Management Board collaborated in this program. Some of the early results were as follows:

1. Mass balance on Gara Glacier in the Sutlej River basin was positive from 1974 to 1976, causing previous recession to stop in 1976. [The balance years from 1976 to 1980 were negative, but from 1982 to 1983 were positive (Srivastava, 2001)]. Gangotri Glacier receded between the 1930s and the late 1970s, but the rate of recession kept decreasing. [Between 1977 and 1991, the recession rate increased considerably (Srivastava, 2001)]. Nehnor Glacier in the State of Jammu and Kashmir receded in the 1970s and had only negative mass-balance years. [The negative mass-balance measurements continued from 1976 to 1984 (Srivastava, 2001)].

2. The Himalaya can be divided into four distinct areas on the basis of precipitation patterns: (a) dominant monsoon precipitation (the Mount Everest region and Sikkim), (b) dominant winter (westerly) precipitation (Ladākh and Spiti areas), (c) equal/sub-equal monsoon and winter precipitation (the Ganga basin (equal) and Beas basin (subequal)), and (d) rain-shadow areas (Vohra, 1981). The amount of meltwater produced and mass-balance characteristics of each glacier vary according to the area in which the glacier is found.

3. The ice at the terminus of Nehnor Glacier is less than 400 years old; however, ice at the terminus of Chaugme Khampa Glacier is only 100 years old.

4. Debris cover on the ablation zone of glaciers is ubiquitous throughout the Himalaya. As a consequence, the production of meltwater decreases, instead of increasing, with the decrease in elevation.

5. There are many ice-cored moraines in proglacial areas, and these play significant roles in the water balance of the glacierized basins of the Himalaya.

6. Precipitation in certain basins increases with elevation *within the glaciated* (sic, glacierized) *region* — even during the monsoon season.

Results of later glacier studies from 1974 to 2000 are found in the following description of individual glacier areas, in published and unpublished reports of the Geological Survey of India, and in Kaul (1999), Geological Survey of India (2001), and Srivastava (2001).

Total Number of Glaciers

In the 1980s, the author estimated that about 15,000 glaciers occur in the Himalaya. Considered high, this estimate was based on the average size of glaciers derived from glacier inventories from parts of the Himalaya, and from von Wissman's (1959) estimate of total ice cover. More recent glacier-inventory work by Qin (1999), however, determined that there were 18,065 glaciers in the five drainage basins of the Himalaya, covering 35,110 km²; this number was based on analysis of 1975–78 Landsat MSS band 7 and false-color (bands 4, 5, 7) images, as well as some aerial photographs. So the author's 1980s estimate was actually 20 percent too low. A composite inventory of the glaciers of the Himalaya by Kaul (1999), based on a combination of detailed and regional assessments, yielded a total of only 5,243 glaciers covering an area of about 38,000 km². Kaul (1999) also reported that about 8,500 km² of the glaciers are located in the Indian Himalaya.

Glacier Distribution

Glaciers are found in all geographic areas of the Himalaya which lie above the elevation required to maintain ice. In some areas, they occur in large numbers and attain great lengths — measurable in tens of kilometers. Such mountain areas in India are the *Nun-Barashigri*, Gangotri-*Chaukhamba*, Kāmet group, Nanda Devi group, and Kānchenjunga (fig. 1). The principal glaciers of the Indian Himalaya, including their lengths, are: Gangotri, 30 km; Zemu (Sikkim Himalaya), 28 km; Mīlam (Nanda Devi area), 19 km; and *Kedārnāth* (Gangotri-*Chaukhamba* area), 14.5 km (fig. 1).

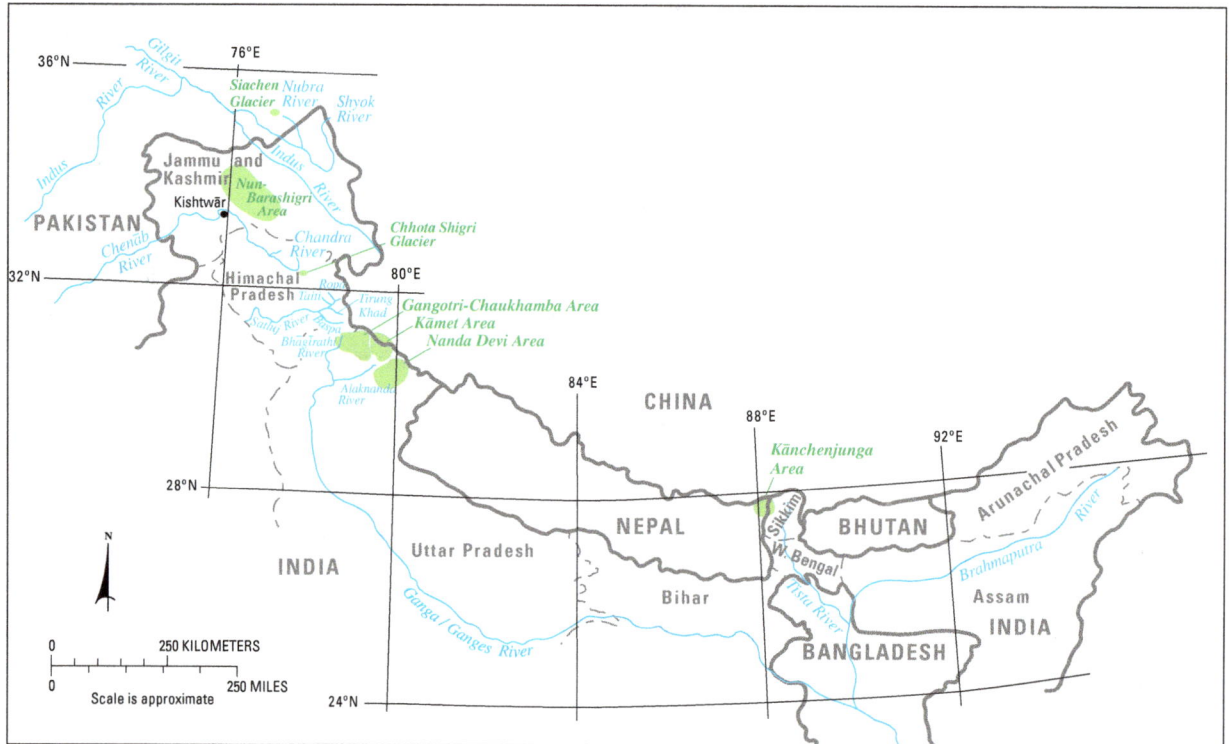

Figure 1.—*Sketch map showing the location of glacierized areas of India discussed in the text.*

Nun-Barashigri

The southern flank of the *Nun-Barashigri* range forms the northern portion of the Chenab River basin, north and east of Kishtwār; it is one of the most highly glacierized areas of the Indian Himalaya (figs. 2, 3). Nearly 1,000 large and small glaciers drain into the Chenab River. The northern flank of the *Nun-Barashigri* range also supports many glaciers, which drain north into tributaries of the Indus River. Several glaciers are greater than 10 km in length; two exceed 20 km, and a third is a little less than 20 km. Several ice fields, drained by outlet glaciers, occur in the general area around Nunkun peak (fig. 2) and also on the mountain itself. Ice fields are formed on mountain summits, or slightly lower down, and the highest summits protrude through the ice. This geomorphic characteristic is unique to this area, and perhaps indicates an earlier erosional surface. The glaciers are fed by both summer (monsoon) and winter (westerly) precipitation. Presently, many cirques and small glacier valleys, especially in the eastern part of the range, are ice free, indicating a post-glacierization phase. Landsat 1 MSS images of the area (Path 158–Rows 37 and 38), acquired on 4 September 1972, show the transient snow line at its highest elevation of the year. Landsat 2 MSS images were acquired on 10 and 11 September 1976 (Path 158, and 159–Rows 36 and 37), following a fresh snowfall (fig. 2). This late monsoon snow covered the highest points of the main Himalaya Range, and was not restricted to a particular elevation as is sometimes thought to occur. Figure 3 shows a map of that area, prepared for the preliminary Inventory of Himalayan Glaciers (Kaul, 1999). Based on the preliminary inventory, it is estimated that there are 989 glaciers covering 2,280 km^2 in the Chenab basin.

Gangotri-*Chaukhamba*

Although the Gangotri-*Chaukhamba*, Kāmet, and Nanda Devi mountain groups are adjacent to each other, they are separated by major river valleys (figs. 4, 5). The Gangotri-*Chaukhamba* area is a cluster of many mountain peaks, 6 to 7 km in elevation, from which many large and small glaciers flow.

Only two glaciers have been studied in this area — the Gangotri and the *Chaurabari*. The most impressive is the Gangotri Glacier, the longest glacier in the Indian Himalaya. This is a "complex glacier" type with large tributaries. Its main trunk is 26 km long, and the overall ice surface of Gangotri Glacier and its tributaries covers an area of a little more than 200 km^2. It is estimated to contain about 20 km^3 of ice. Its meltwater stream becomes a high-volume river after emerging from a sub-glacier tunnel at the glacier terminus — the holy "*Gaumukh*" (cow's mouth) of Hindu mythology. This glacier has been studied since 1817, though only documented since 1886. The glacier has been receding since 1935, but the greatest recession occurred from 1977 to 1990 (Srivastava, 2001). Since 1990, observations were made from 1992 to 1997, and the Gangotri Glacier was in continuous recession (Srivastava, 2001). The river coming from the Gangotri Glacier is called Bhāgīrathi River, and it is considered to be the primary source of the Ganga (Ganges River). All of the glaciers in the Gangotri-*Chaukhamba* area receive nourishment from the monsoon and westerly winter precipitation. Although detailed figures are not available, the total precipitation is only moderate.

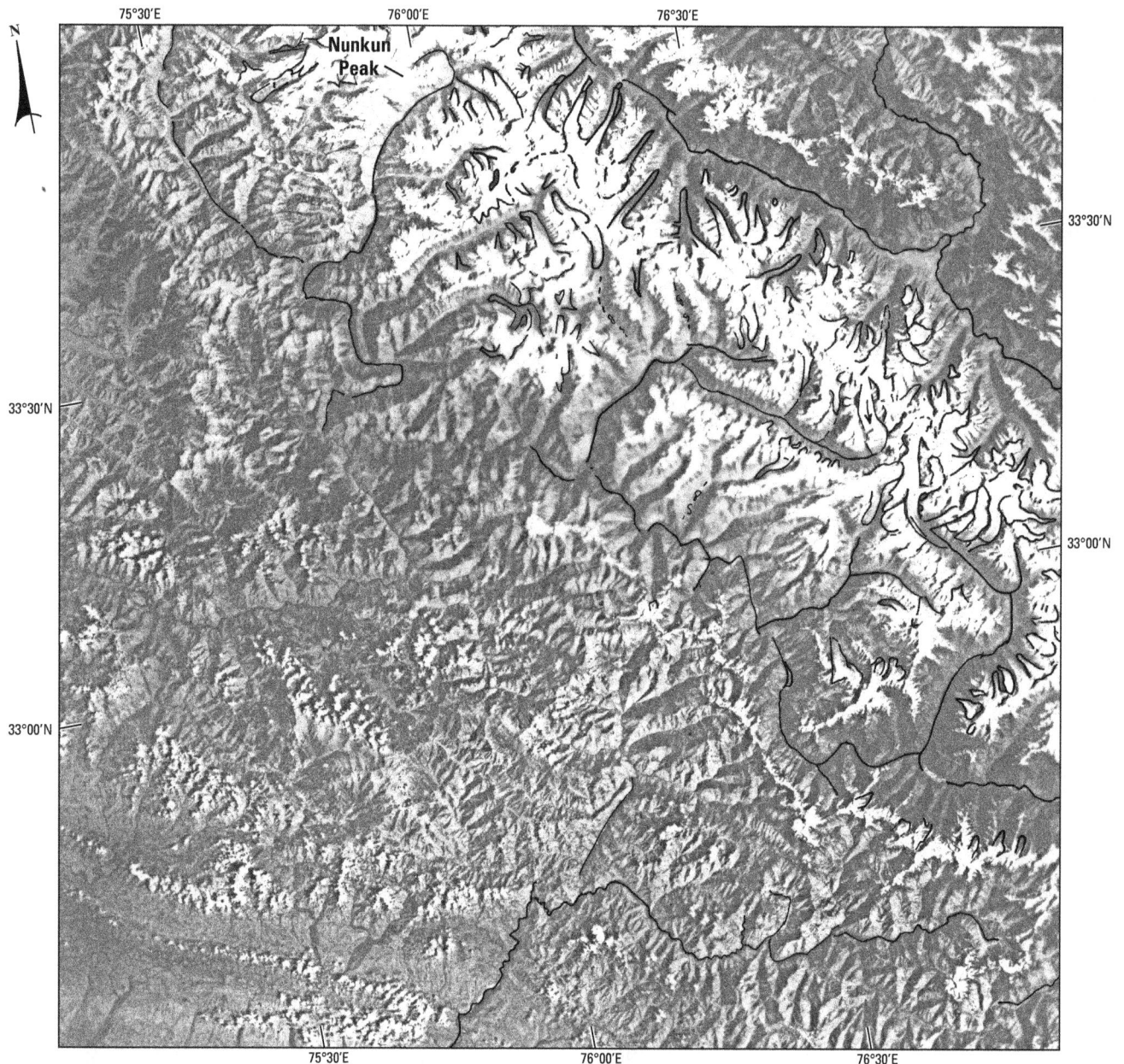

Figure 2.—*Landsat 2 MSS image of the* Nun-Barashigri *area with some of the many glaciers outlined. The image was acquired after a fresh, late monsoon snowfall. The Landsat MSS image (2159037007625590, 11 September 1976; Path 159–Row 37) is from the U.S. Geological Survey EROS Data Center, Sioux Falls, South Dakota.*

The valley below the *Gaumukh* contains abundant evidence of past glacier advances. Seventeen kilometers downstream, at a place named Gangotri, a glacial pavement begins and extends further downstream; a narrow epigenetic gorge, deeply entrenched into the "pavement," also begins at this point. A temple is constructed here to mark the location of the glacier terminus when Bhagirathi Rishi first discovered this glacier. The date of the visit is lost in the maze of mythology, but this previous glacier position may date from the beginning of the Holocene Epoch.

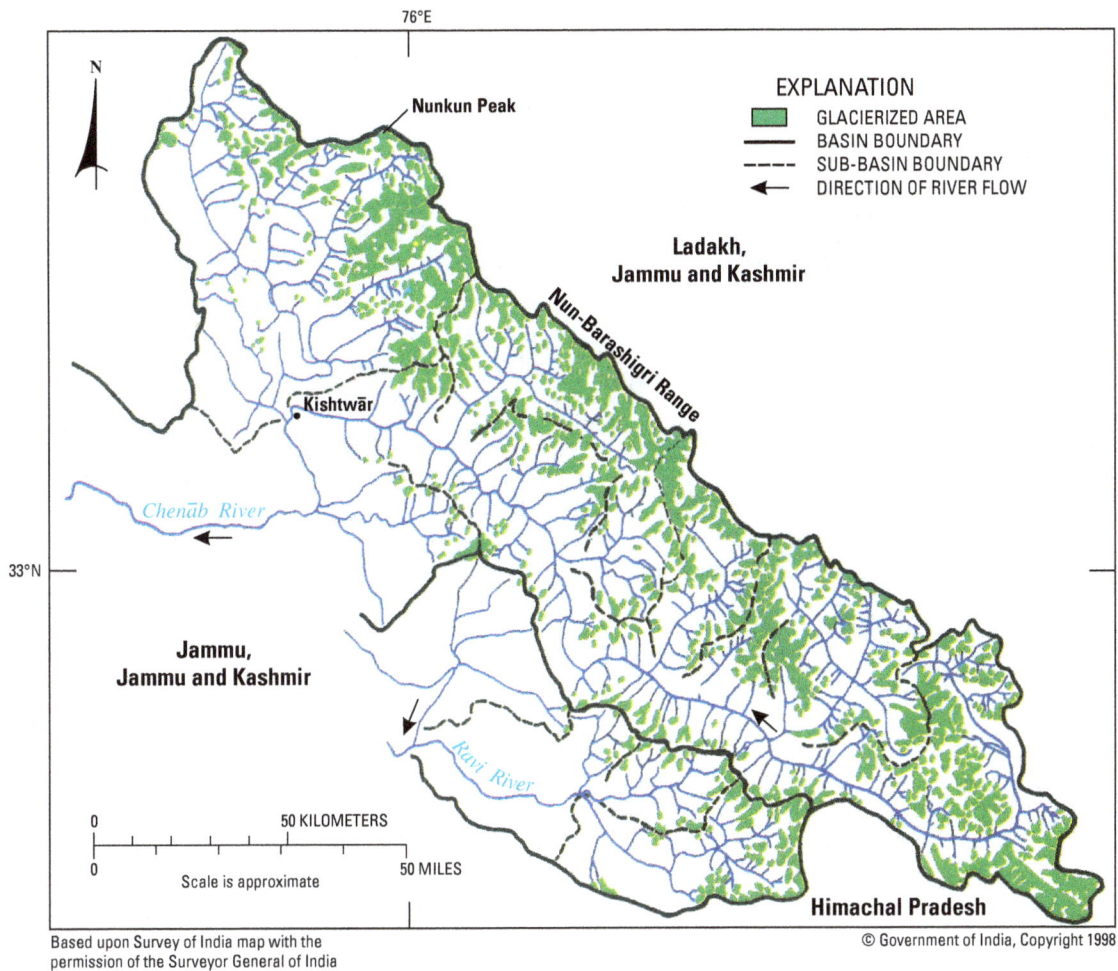

Based upon Survey of India map with the
permission of the Surveyor General of India

© Government of India, Copyright 1998

Figure 3.—*Glacierized area of the Chenab and Ravi basins. From the preliminary "Inventory of the Himalayan Glaciers" by the Geological Survey of India (Kaul, 1999).*

Satluj River Basin

To the north and west of the Gangotri group of peaks flows the Satluj River, which originates in *Mansrovar Lake*. After the Glaciology Division was established as part of the Geological Survey of India in 1974, an initial glacier inventory was planned for the Baspa river basin, the largest glacier-fed tributary of the Satluj basin. The pilot inventory was conducted using the Guidelines of the World Glacier Inventory (Müller and others, 1977, 1978), and was followed by studies of other basins in the Satluj basin. Work was then extended to other parts of the Himalaya (Kaul, 1999). As a result, a thorough survey has been made of part of the Satluj basin and some of the early work and later results are described below.

The total ice cover for the Satluj River basin was estimated in the 1940s to be about 6,390 km^2 or approximately 11 percent of the basin area. The estimates were probably made from the generalized 1:1,000,000-scale maps. There are often major differences in the determination of the area of ice cover when using maps of different scales. When compared with figures from a later glacier inventory in the late 1970s using larger-scale maps, this estimate was found to be far too high. A comparison is given in table 2.

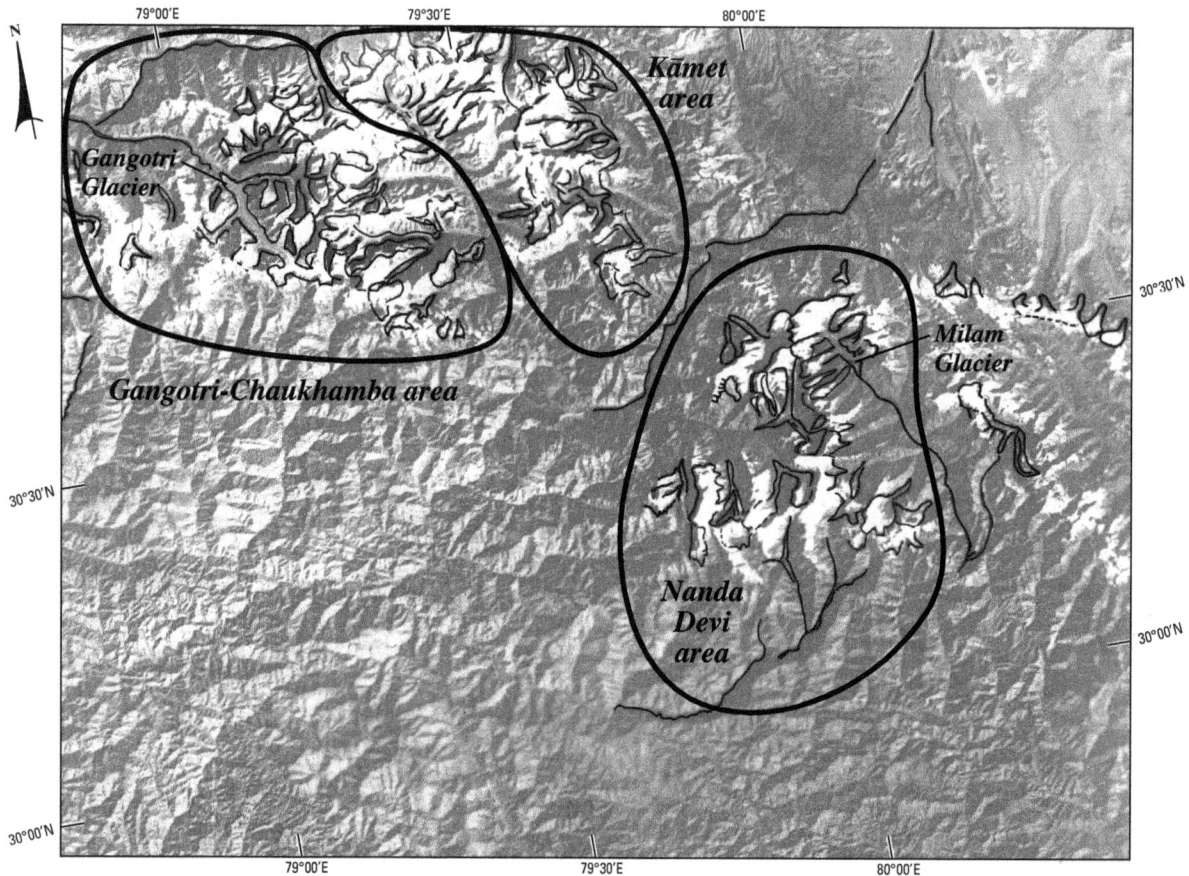

Figure 4.—*Landsat 2 MSS image of the Gangotri-Kāmet-Nanda Devi areas of the Indian Himalaya. The major glaciers are demarcated. This area includes the Gangotri Glacier, largest in India. Its meltwater stream, the Bhāgīrathi River, is considered to be the main source of the Ganga (Ganges River). The Landsat MSS image (2156039007632490, 19 November 1976; Path 156–Row 39) is from the U.S. Geological Survey EROS Data Center, Sioux Falls, South Dakota.*

The glacierization in the Indian part of Satluj River basin is similar to the Chinese part. The four percent figure for ice cover is representative of both parts of the basin. The more glacierized tributaries of the Satluj River (table 2) also indicate the same differences in ice-cover area determined from maps of different scales. Interestingly, the differences resulting from a comparison of ice cover on 1:250,000- and 1:50,000-scale maps are relatively minor, however. The major difference is in calculations based on the 1:1,000,000-scale maps.

Inventory from Landsat Imagery

Landsat Multispectral Scanner (MSS) images from the late 1970s/early 1980s were used for compiling a glacier inventory for a part of the Satluj River basin and for making a glacier map of some parts of this basin. The results were compared with the existing late 1970s inventory for the Baspa Gād and *Tirung Khad*, two tributaries of Satluj River (table 3).

The ice-cover areas match relatively well, the difference being less than 5 percent. The snowline was easy to delineate on the Landsat images, as were the ice margins, except in the case of small ice bodies. Some glaciers, marked as tributaries of larger glaciers on the maps, were easily distinguished on the images as having independent termini.

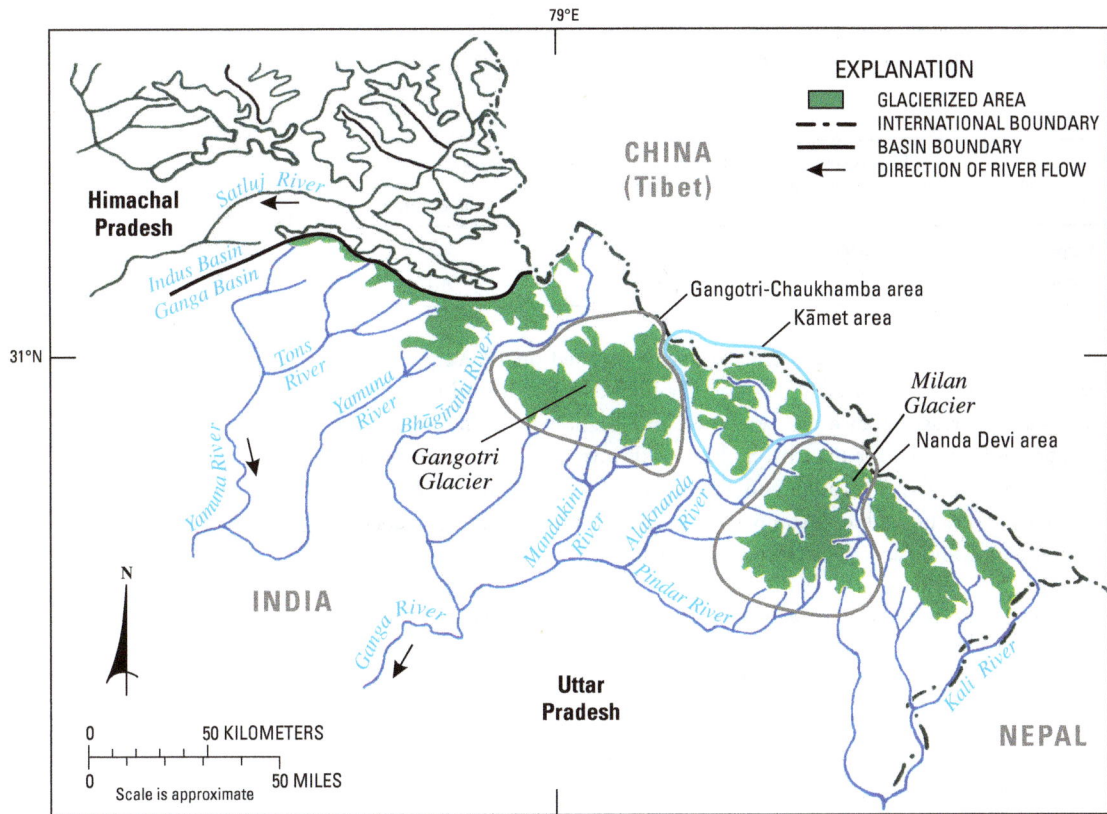

Figure 5.—*Glacierized area of the Ganga basin showing the Gangotri-Chaukhamba, Kāmet, and Nanda Devi glacierized areas based on the preliminary "Inventory of the Himalayan Glaciers" by the Geological Survey of India. Modified from Kaul (1999).*

TABLE 2.—*Ice cover determined in the late 1970s from maps of different scales as a percentage of the basin*

[—, no data]

Basin name	Percent ice cover		
	1:1,000,000-scale map	1:250,000-scale map	1:50,000-scale map
Satluj River	11 (Total basin)	4 (Indian part of basin)	—
Satluj River tributaries:			
Ropa Gād	32	9.3	—
Taiti Gād	37.7	9.7	—
Baspa Gād	—	19.3	22.2
Tirung Khad	—	13.1	14.7

TABLE 3.—*Comparison of glacier inventories from maps and Landsat images in the glacierized drainage basins, prepared in the late 1970s/early 1980s*

[Unit: km², square kilometer]

Basin	Total ice cover delineated from maps (km²)	Total ice cover delineated from Landsat images (km²)	Percentage difference, as seen on the Landsat images
Baspa Gād	186.97	178.46	4.5
Tirung Khad	69.98	67.08	4.1

The snowline on a Landsat 2 MSS image acquired on 17 August 1977 (Path 157–Row 38) was used to delineate the boundary between ablation and accumulation areas. Because this was lower in elevation than the computed equilibrium lines used for the existing late 1970s inventory, some differences in the respective glacier ablation and accumulation areas calculated from the two sources were inevitable. A comparison is given in table 4; only glaciers on which the snowline was well marked are included.

TABLE 4.—*Percentage reduction or increase in the ablation, accumulation, and total area of glaciers, determined by comparing late 1970s Landsat 2 MSS images with the already existing late 1970s inventory*

[Abbreviations: MSS, multispectral scanner; km², square kilometer; %, percent reduction or increase]

River basin	Ablation area, in km²	Accumulation area, in km²	Total area, in km²
Baspa Gād	-15.59 (32.1%)	+13.43 (16.8%)	-2.16 (1.7%)
Tirung Khad	-16.69 (53.4%)	+11.27 (49.4%)	-5.42 (10%)

The difference in respective accumulation and ablation areas because of the lower snowline on the 17 August 1977 Landsat image is obvious. The reduction in the delineation of the ablation area using Landsat imagery is marginal in the case of Baspa Gād basin, but it is significant in the case of *Tirung Khad* basin. In all probability, it is a reflection of the erroneous inclusion of proglacial moraines within the ablation area of the glaciers in the existing late 1970s inventory which was compiled primarily from maps.

Another example from the *Tirung Khad* glacier No. 49 where field work was done is given in table 5. The results from Landsat images and field work are similar, except for some differences due to the lower snowline on the Landsat

TABLE 5.—*Comparison of calculation of ablation, accumulation, and total area of the* Tirung Khad *glacier No. 49 between maps (existing late 1970s glacier inventory), Landsat images (late 1970s/early 1980s), and field work*

[Unit: km², square kilometer]

Source	Ablation area (km²)	Accumulation area (km²)	Total (km²)
Existing glacier inventory	[1]4.91	3.00	[1]7.91
Landsat images	1.63	4.31	5.94
Actual field work	2.42	3.09	5.50

[1]These figures are probably too high, because proglacial moraines could not be separated from the debris-covered termini when topographic maps were prepared.

images. Modern maps were not available when the inventory was compiled for the Spiti River basin. Therefore, the information derived from the Landsat images considerably improved the inventory in that area.

In conclusion, Landsat imagery (fig. 6, table 6) and other higher resolution satellite images are very useful for the Himalayan region where modern maps are not available, and where aerial photography acquired at the end of the ablation season is sparse. Also, it is less time consuming to make an inventory from Landsat images, although the imagery does not provide elevation information.

More recent results from the intensive glacier studies in the Satluj River basin can be seen in figure 7, taken from Kaul (1999). Other recent unpublished and published glacier reports covering this basin were prepared by GSI (Geological Survey of India, 2001; and Srivastava, 2001).

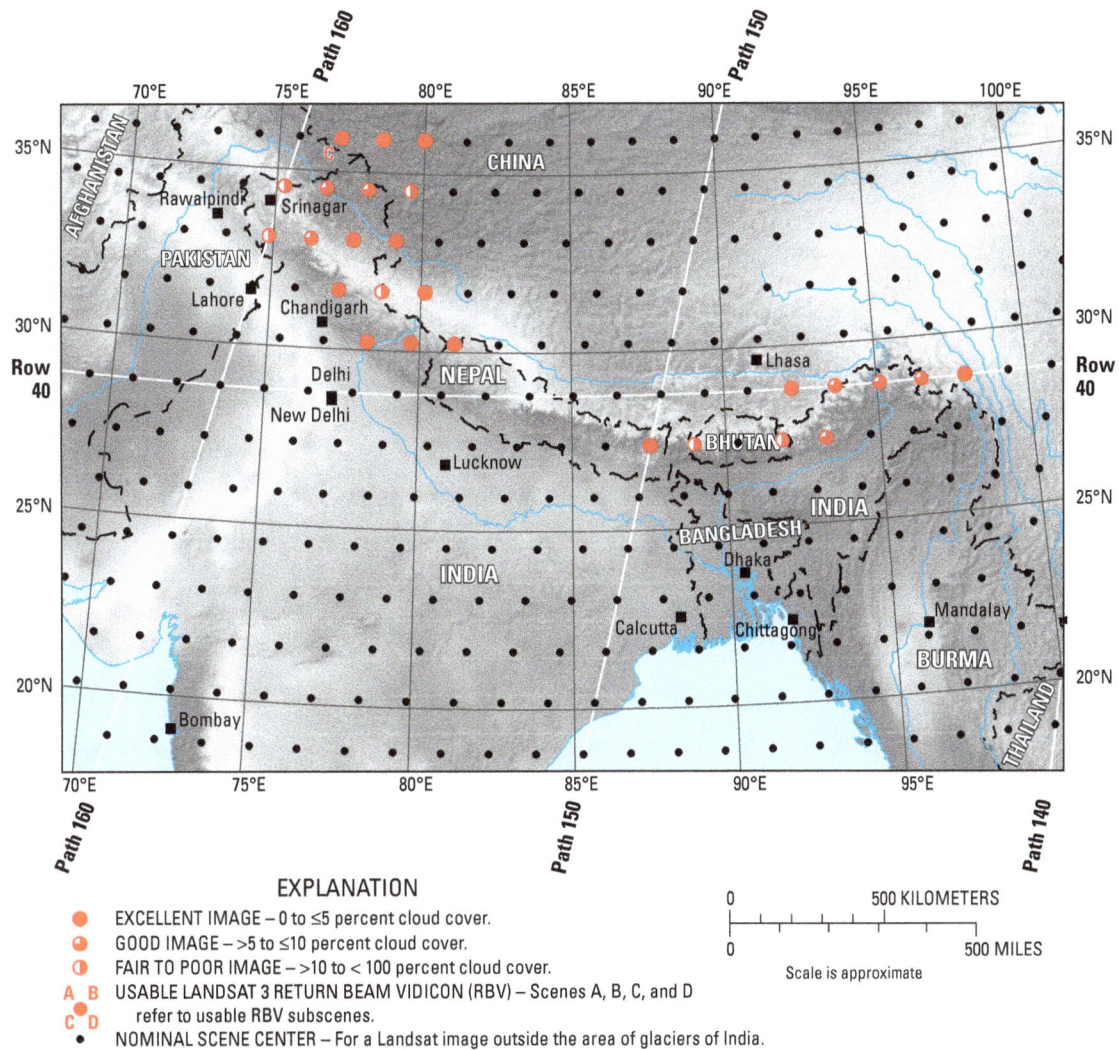

EXPLANATION

- ● EXCELLENT IMAGE – 0 to ≤5 percent cloud cover.
- ◕ GOOD IMAGE – >5 to ≤10 percent cloud cover.
- ◑ FAIR TO POOR IMAGE – >10 to < 100 percent cloud cover.
- A B USABLE LANDSAT 3 RETURN BEAM VIDICON (RBV) – Scenes A, B, C, and D
  C D   refer to usable RBV subscenes.
- • NOMINAL SCENE CENTER – For a Landsat image outside the area of glaciers of India.

Figure 6.—Optimum Landsat 1, 2, and 3 images of the glaciers of India. Compare with table 6.

TABLE 6.—*Optimum Landsat 1, 2, and 3 images of the glaciers of India*

[EROS, Earth Resources Observation and Science.. Code column indicates usability for glacier studies; see figure 6 for explanation]

Path-Row	Nominal scene center latitude and longitude		Entity number (USGS EROS data center)	Date	Code	Cloud cover (percent)	Remarks
143–40	28°49′N.	97°57′E.	2143040007631190	06 Nov 76	●	0	
144–40	28°49′N.	96°31′E.	2144040007631290	07 Nov 76	◕	10	State of Arunachal, Pradesh, eastern Himalaya. Line drops
145–40	28°49′N.	95°05′E.	1145040007335590	21 Dec 73	◕	10	State of Arunachal, Pradesh, eastern Himalaya
146–40	28°49′N.	93°39′E.	2146040007635090	15 Dec 76	◕	10	State of Arunachal, Pradesh, eastern Himalaya
146–41	27°23′N.	93°15′E.	2146041007635090	15 Dec 76	◖	20	State of Arunachal, Pradesh, eastern Himalaya
147–40	28°49′N.	92°13′E.	2147040007635190	16 Dec 76	●	5	
147–41	27°23′N.	91°49′E.	2147041007700390	03 Jan 77	◗	20	
149–41	27°23′N.	88°57′E.	2149041007635390	18 Dec 76	◗	15	Sikkim Himalaya glacier area. Image used for figure 8
150–41	27°23′N.	87°31′E.	2150041007635490	19 Dec 76	●	5	Sikkim Himalaya glacier area. Image used for figure 8
155–39	30°15′N.	81°09′E.	2155039007634190	06 Dec 76	●	5	
156–38	31°41′N.	80°09′E.	2156038007632490	19 Nov 76	●	0	
156–39	30°15′N.	79°43′E.	2156039007632490	19 Nov 76	●	0	Gangotri-Kemet-Nanda-Devi glacier areas; image used for figure 4
157–35	35°58′N.	80°00′E.	2157035007632590	20 Nov 76	●	5	
157–36	34°32′N.	79°34′E.	2157036007726590	22 Sep 77	◐	40	Line drops
157–37	33°07′N.	79°08′E.	2157037007632590	20 Nov 76	●	5	
157–38	31°41′N.	78°42′E.	2157038007632590	20 Nov 76	◖	25	
157–39	30°15′N.	78°17′E.	2157039007630790	02 Nov 76	●	0	A few line drops
158–35	35°58′N.	78°34′E.	2158035007630890	03 Nov 76	●	5	A few line drops
158–36	34°32′N.	78°08′E.	2158036007630890	03 Nov 76	◕	10	Eastern Karakoram Range. A few line drops
158–37	33°07′N.	77°42′E.	2158037007630890	03 Nov 76	●	5	Eastern *Nun-Barashagri* Range
158–38	31°41′N.	77°16′E.	2158038007630890	03 Nov 76	●	5	A few line drops
159–35	35°58′N.	77°08′E.	2159035007630990	04 Nov 76	●	5	
159–36	34°32′N.	76°42′E.	2159036007630990	04 Nov 76	◕	10	A few line drops
159–37	33°07′N.	76°16′E.	2159037007630990	04 Nov 76	◗	20	*Nun-Barashagri* glacier area. Line drops
159–37	33°07′N.	76°16′E.	2159037007625590	11 Sep 76	◕	10	*Nun-Barashagri* glacier area; image used for figure 2
160–36	34°32′N.	75°16′E.	2160036007631090	05 Nov 76	◗	20	A few line drops
160–37	33°07′N.	74°49′E.	2160037007631090	05 Nov 76	◗	50	

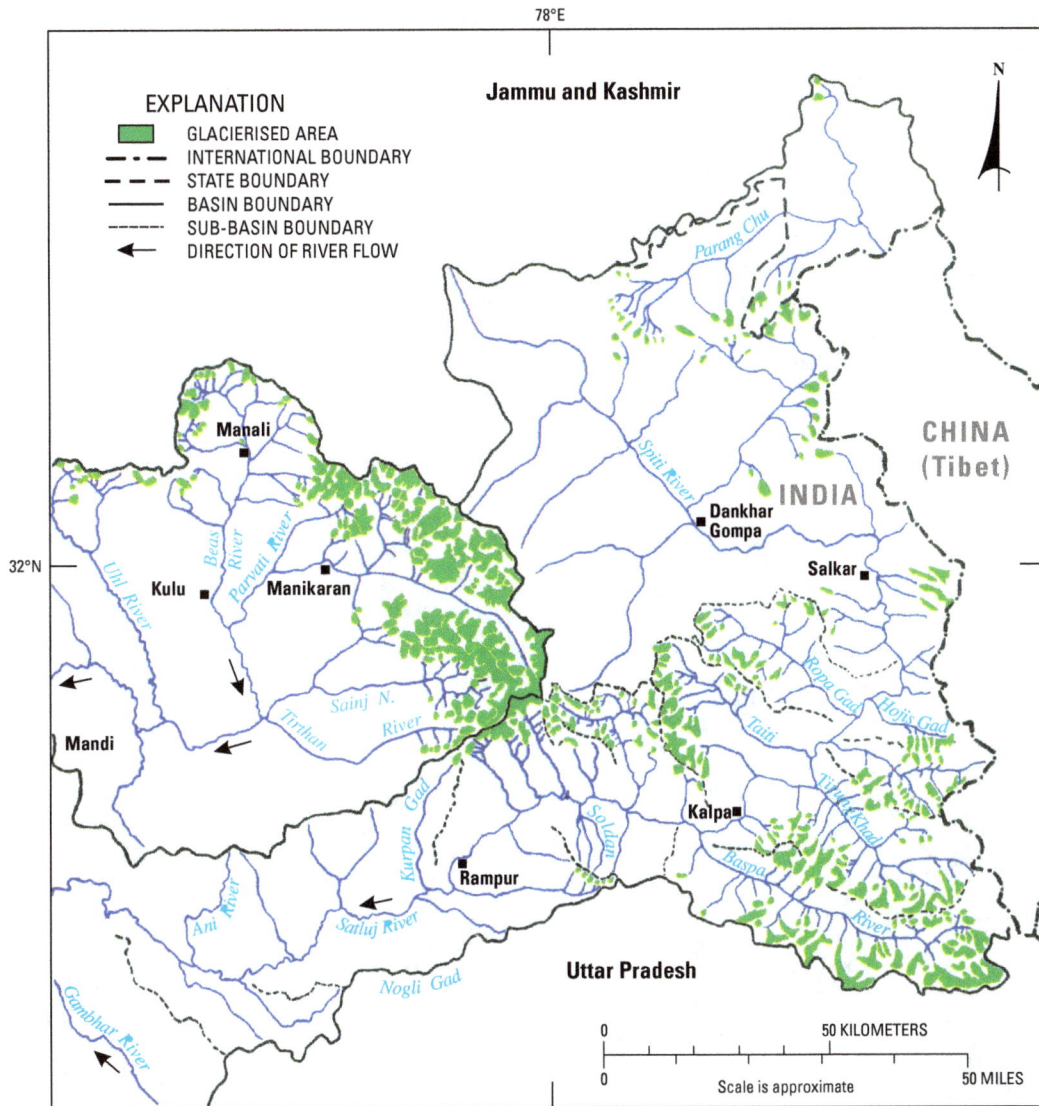

Figure 7.—*Glacierized area of the Satluj River basin. From the preliminary "Inventory of the Himalayan Glaciers" by the Geological Survey of India (Kaul, 1999).*

Kāmet

Kāmet peak is the highest (7,759 m) of the Kāmet group. It supports dozens of major glaciers and many smaller ones (figs. 4, 5). All of the glaciers drain into the Alaknanda River or other tributaries of the Ganga.

Nanda Devi

The Nanda Devi group is dominated by Nanda Devi peak (7,820 m). Several high mountain peaks here form a rim of a large amphitheater — the fabled "sanctuary" of the Himalayan mountaineers — drained by *Rishi Ganga* through an almost impassable gorge. There are large glaciers within the "sanctuary," and the Milam Glacier drains outward from the northeast part into Dhaulīganga River (figs. 4, 5). In this area, Milam, *Poting* and *Shankalpa*

Glaciers (*Shankalpa* is east of Dhaulīganga) have been visited and observed by scientists of the Geological Survey of India since the 1970s. The most recent findings of GSI are published in Kaul (1999), Geological Survey of India (2001), and Srivistava (2001).

Sikkim Himalaya

Sikkim Himalaya, to the east of Nepal, is dominated by the massif of Kānchenjunga (8,591 m), the third highest peak in the Himalaya (fig. 8). It is located in the northwestern part of the State of Sikkim, along the border with Nepal. Zemu Glacier, 28 km long and 41.2 km² in total surface area (Kaul, 1999), is the main glacier descending from the massif (fig. 9). The massif is highly glacierized, and many other major and minor glaciers are located on it.

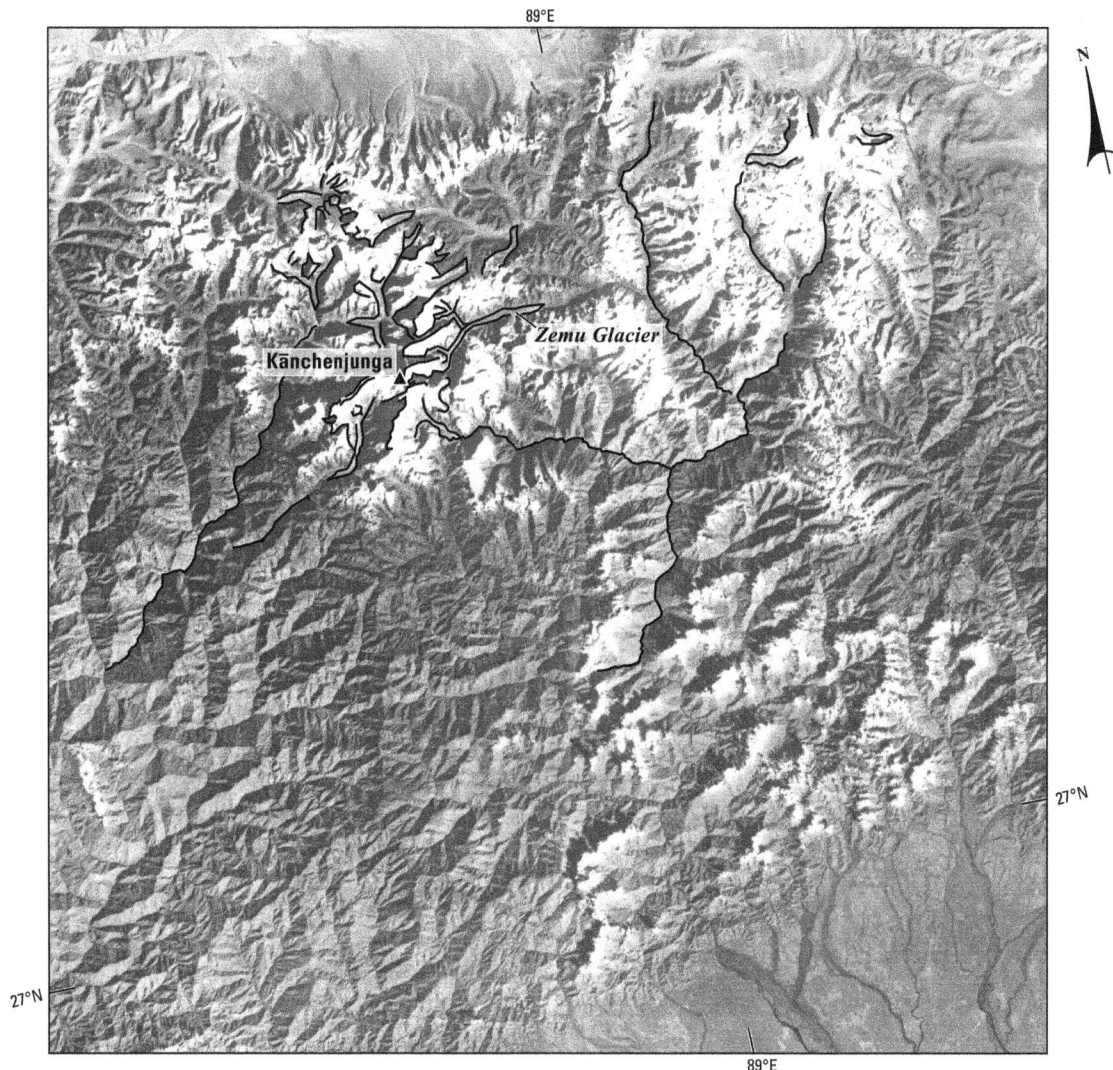

Figure 8.—*Landsat 2 MSS image mosaic of the Sikkim Himalaya dominated by the Kānchenjunga massif, the third highest peak in the Himalaya. Many large glaciers are located here, including the 28 km-long Zemu Glacier. The Landsat MSS images (2149041007635390, 18 December 1976; Path 149–Row 41 (right); and 2150041007635490, 19 December 1976; Path 150–Row 41 (left)) are from the U.S. Geological Survey EROS Data Center, Sioux Falls, South Dakota.*

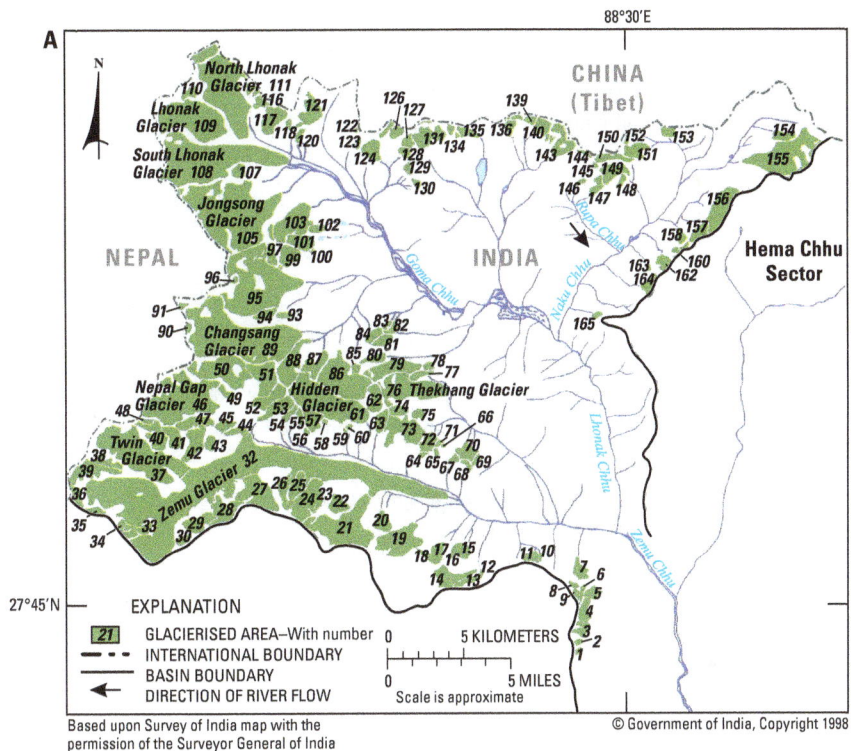

Figure 9.—Map and ASTER image of the Zemu Glacier area. A, Glacier map of the Zemu Basin of the Sikkim Himalaya. From the preliminary "Inventory of the Himalayan Glaciers" by the Geological Survey of India (Kaul, 1999). Zemu Glacier is in the lower left. The numbers on the glaciers relate to the glacier inventory data in Kaul (1999, p. 149). B, 27 November 2001 ASTER image of Zemu Glacier (lat 27°45'N., long 88°16'E.). The image (AST_L1B_00311272001045729_2 0060817160447_7392 hdf) is courtesy of Adina Racoviteanu, Department of Geography and Institute of Arctic and Alpine Research, University of Colorado at Boulder.

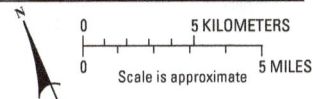

This area apparently receives more precipitation than the adjacent Mount Everest region to the west, whose glaciers are smaller. The northern and northeastern mountains in the State of Sikkim support relatively few major glaciers, although small glaciers do exist. Zemu Glacier and a small glacier in northern Sikkim have been studied by the Geological Survey of India and other Indian agencies.

Eastern Himalaya

In the Eastern Himalaya, a few large glaciers could be delineated on the Landsat 2 MSS image acquired on 3 January 1977 (table 6). The area covers the northern part of the State of Arunachal Pradesh. Smaller glaciers are also present in this area, although only scant information is available about them. Perhaps the strong monsoon, prevalent in the State of Assam, penetrates to this area and provides the requisite nourishment for glacier presence.

Surge-Type Glaciers

Surge-type glaciers are easy to identify on aerial photographs and on Landsat images. Glacier surges often leave their characteristic imprint on glaciers; contorted medial moraines are the most impressive and useful of such characteristics. A search through the Landsat MSS images indicates that surge-type glaciers are rare in the Himalaya, but common in the Karakoram Range. Some Landsat images of the Indian Karakoram show a concentration of surge-type glaciers, including glaciers that have surged in the past. The mountains between the Shyok River (up to Saser) and Gilgit River include many surge-type glaciers.

Glaciers in the Shyok valley (for example, *Chong Kumdan* and *Kichik-Kumdan Glaciers*) are well-known for repeated surge events in this region. During their advances, the Shyok River becomes dammed, causing a large amount of water to accumulate behind the ephemeral dam, which, in time, bursts through the ice dam, producing an outburst flood. The ensuing floods ravage the areas downstream. A huge flood occurred in 1929. More recently, in a Landsat 2 MSS image of July 1978, two glaciers appear to have advanced into the Shyok River valley but have not yet blocked it. Two large glaciers at the head of the river also show surge-type features. Many of the glaciers descending from the northwestern part of the high divide between Shyok and Nubra drainage basins, including those draining into *Kole-Ching Ha* (in China), also seem to exhibit surge-type characteristics.

Conclusions

The study of glaciers is important in India, both scientifically and economically. The meltwater of Indian glaciers affects the quality of life of millions of people. The observation of glaciers in India started in the middle 1800s, and gradually became more prevalent. In the last 25 to 30 years, many intensive glacier studies have been completed.

A Study of Selected Glaciers under the Changing Climate Regime

By Syed Iqbal Hasnain,[3] Rajesh Kumar,[4] Safaraz Ahmad,[5] and Shresth Tayal[3]

Introduction

Low-latitude mountain environments are highly sensitive indicators of climate change (Thompson and others, 1995; Broecker, 1997; Beniston and Fox, 1996). The retreat of glaciers corresponding to global trends has been significant since the middle of the 19th century (Hastenrath, 1995; Kaser, 1999); therefore, glaciers have more recently become the subject of intensive observation. Several studies in the Himalayan region found that the glaciers have retreated considerably during the last two decades (Fujita and others, 1997, 2001; Kadota and others, 1997, 2000; Ageta and others, 2001; Naithani and others, 2001). Several analyses have shown that it is not only increased temperature and/or decreased precipitation that are responsible for recession of glaciers in lower latitudes, but also changes in humidity (Kaser and Noggler, 1991; Hastenrath and Kruss, 1992; Kaser, Georges, and Ames, 1996; Kaser and Georges, 1997; Kaser, 1999; Wagnon and others, 1999).

The glaciers of the high mountains which are located in low latitudes cover a total area estimated at about 2.5×10^3 km^2, which corresponds to 0.16 percent of the total glacier-ice cover of the world (Haeberli and others, 1989). Although negligible in area (if all of these glaciers melted tomorrow, eustatic sea level would not rise more than about 1 cm), they are nevertheless known to be very sensitive components of the environment and deserve attention in the context of both global climate change (Kaser, Hastenrath, and Ames, 1996) and local and regional water supply.

Mass-balance data are useful in glacier-climate interaction studies. The conventional method for mass-balance measurement is normally applied to benchmark glaciers having International Commission on Snow and Ice (ICSI)-approved attributes (Kaser and others, 2002). However, in rugged terrain in remote areas like the Himalaya, remote-sensing techniques are useful for understanding regional climate change and its impact on glaciers because more areal coverage is available with less effort on the ground. A rough estimate of the glacier mass balance can be derived from the accumulation area ratio (AAR) (Kulkarni, 1992; Sapiano and others, 1998; Ramussen and others, 1999). The AAR of a glacier is defined as the ratio between the accumulation area and the area of the entire glacier, and is determined by using the equilibrium line altitude (ELA) that separates the ablation and accumulation areas on a glacier surface.

[3]The Energy and Resources Institute, New Delhi 110003, India

[4]B.M. Birla Science and Technology Center, Statue Circle, Prithviraj Road, Jaipur, 302001, Rajasthan India

[5]Aligarh Muslim University, Aligarh 202002 India

The demand for global fresh water has increased four-fold since 1940, due to a growing population, expansion of irrigation of croplands, increasing urbanization, and industrialization. In the Indian subcontinent, snow and glacier ice of the Hindu-Kush Himalaya provide a very large percentage of the low-land, dry-season flow of the Indus, Ganges (Ganga), and Brahmaputra Rivers. The discharge from glaciers helps to maintain year-round continuity in water availability to surrounding ecosystems and living habitats. Considering Himalayan glaciers as water reserves and effectively managing these mountain water resources could help avoid a future water crisis in coming decades. The glacierized basins in the Himalaya are the primary source of freshwater and perennial river flow in the Indian subcontinent.

Recent Glacier Studies

In recent years, new technologies have been applied to the study of glaciers in India, in addition to continued field investigations. The studies described in this section have provided valuable information about the effects of changing climate on the glaciers.

Gangotri Glacier

The Gangotri Glacier originates at an elevation of 7,143 m ASL in the *Chaukhamba* group of peaks and extends between 30°42'N. and 31°01'N. latitude and between 78°54'E. and 79°18'E. longitude (fig. 10). It terminates in a snout about 25 km below at *Gaumukh* at an elevation of about 4,000 m. The area of ice cover of the Gangotri Glacier is about 250 km². Gangotri Glacier feeds the Bhāgīrathi River and is situated at the transitional zone between the eastern and western Himalaya, reflecting the properties of both. But this glacier is also vulnerable to the threats posed by climate change and associated phenomena.

A detailed field observation of the Gangotri Glacier, carried out during 2002 and 2003, revealed that huge amounts of debris were being deposited on the glacier surface from adjacent mountains because of an increase in glacier erosion. Near the snout, numerous transverse crevasses had developed in the glacier due to the steep slope and unloading of stresses. Material was deposited near the snout because of rapid, continuous deglaciation during the past few decades. A few ice-dammed lakes were located on the supraglacial surface (figs. 11, 12), about 9 km upglacier, near the junction with the tributary glacier, *Kirti Bamak*. The lakes ranged in size from 50 m² to 150 m² (estimated during the field excursion in July 2002, led by Dr. Rajesh Kumar). The existence of ice-dammed lakes may be due to the low gradient (8–9°) in front of *Kirti Bamak*.

The Terra satellite's Advanced Spaceborne Thermal Emission Reflectance Radiometer (ASTER) composite image (fig. 13), produced by Jesse Allen and available on the U.S. National Air and Space Administration's (NASA) Earth Observatory website [*http://earthobservatory.nasa.gov*], shows how the Gangotri Glacier terminus has receded since 1780. The lines showing the location of the terminus are approximate. This satellite image reflects a much enhanced increase in recession of the terminus of the Gangotri Glacier during the last 30 years. Calculations show a recession rate of 3.77 m a⁻¹ for the period 1780 to 1849, 6.53 m a⁻¹ for the period 1849 to1900, 9.38 m a⁻¹ for the period 1900 to 1971, and 27.66 m a⁻¹ for the period 1971 to 2001.

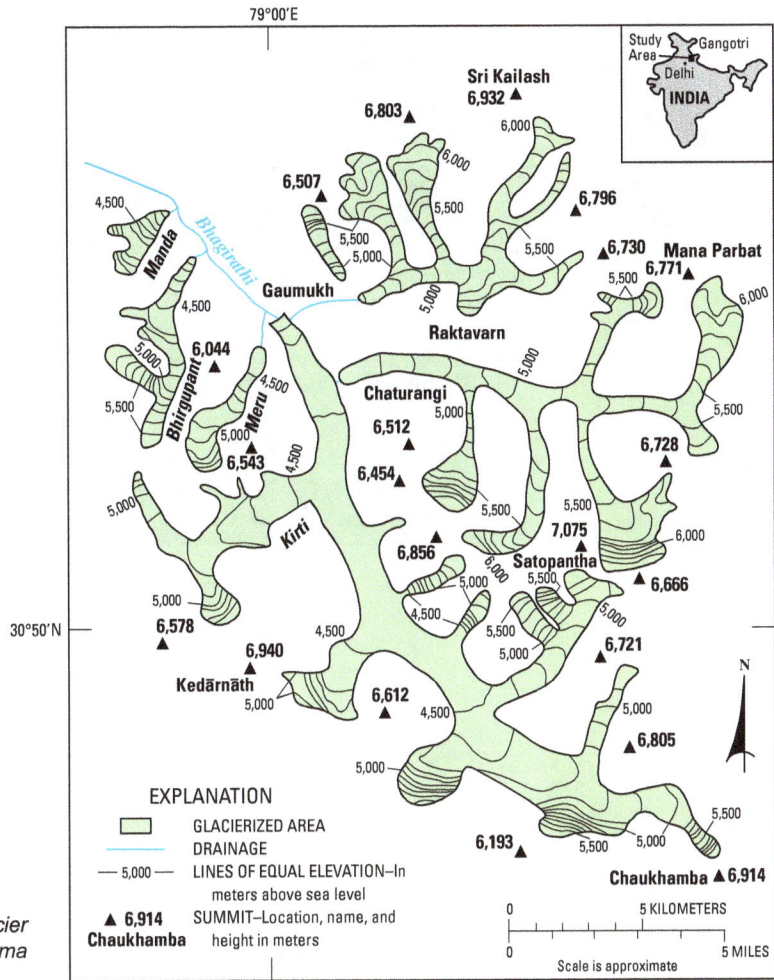

Figure 10.—Location map of Gangotri Glacier and associated glaciers. Modified from Sharma and Owen (1996).

Figure 11.—Lake formed on Gangotri Glacier (approximate lat 31°N., long 79°E.) near the junction with Kirti Bamak Glacier. The background peak is Kedārnāth. Photograph by Dr. Rajesh Kumar, July 2002.

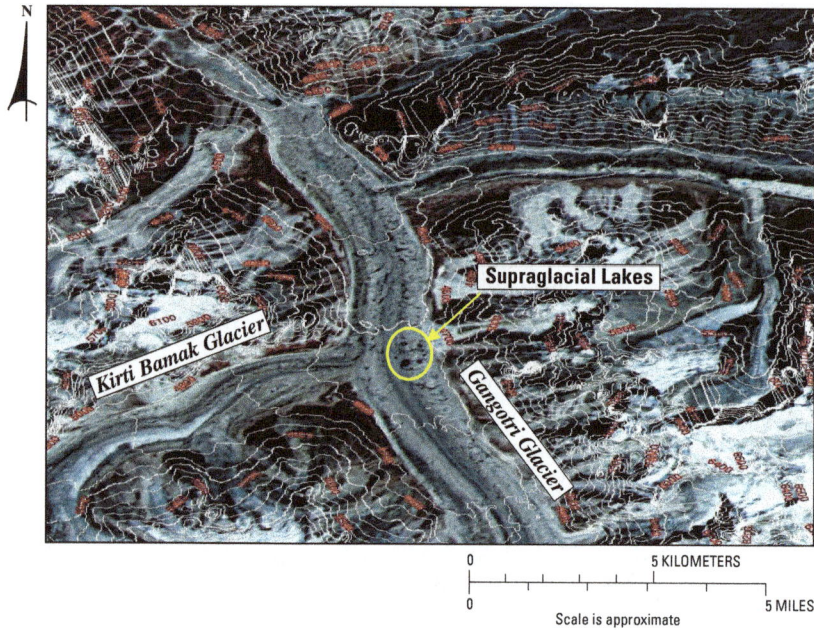

Figure 12.— Enlargement of 9 September 2001 ASTER image showing supraglacial lakes on the Gangotri Glacier (approximate lat 31°N., long 79°E.). Image I.D. AST_LIA:2004126903.

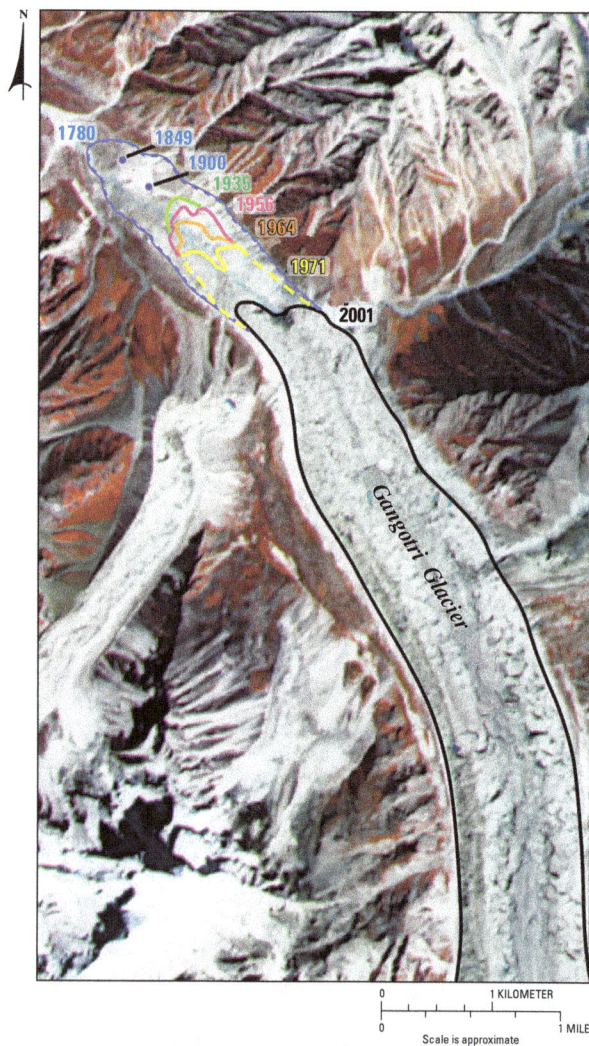

Figure 13.—Enlargement of 9 September 2001 ASTER image showing position of the terminus of Gangotri Glacier between 1780 and 2001 (approximate lat 31°N., long 79°E.). Image from Jesse Allen, NASA's Earth Observatory, based on data provided by the ASTER Science Team. Glacier-retreat boundaries courtesy of the U.S. Land Processes Distributed Active Archive Center (LP DAAC). Image I.D AST_LIA:2004126903.

Shrinkage of the glaciers is linked to climate change due to decrease in winter precipitation, increase in summer precipitation, and atmospheric temperature change. Investigations carried out by the Intergovernmental Panel on Climate Change (IPCC) (Watson and others, 1998) concluded that the Earth's average temperature has increased by $0.6 \pm 0.2°C$ during the 20th century. The increase in air temperature is projected to be about $+1.4°C$ to $+5.8°C$ by the end of 21st century. The projected increase will impact hydrometeorological processes more vigorously than during the 20th century. The impact of increased temperature and liquid precipitation on glaciers and snowfields in the Himalaya is likely to be profound. Recession of snowfields and glaciers will impact the long-term, seasonal pattern and the annual availability of freshwater and the hydropower capacity (Jóhannesson, 1997; Benn and others, 2000). Under the projected scenarios of IPCC, the change in ELA of the Gangotri Glacier has been computed by Hasnain and others (2004) (fig. 14). With an increase in temperature of $+3°C$ the ELA will rise and the AAR will decrease from 0.40 to 0.15, indicating a more negative mass balance of the Gangotri Glacier during the 21st century.

Using an Indian Remote Sensing (IRS) panchromatic satellite 1D image acquired in September 2001, the area of the Gangotri Glacier at the headwaters of the main source of the Ganges river main stem was measured to be about 77 km². This area was 87 km² on the 1:150,000-scale topographic map published by the Survey of India in 1985, implying that the glacier lost about 12 percent of its area during the intervening 16 years. The average rate of the snout recession at *Gaumukh* was also computed by comparing the snout position on the 1985 topographic map and the 2001 panchromatic image. The average rate of recession for this period in this location was about 23 m a⁻¹ (Hasnain and others, 2004).

Figure 14.—*Predicted change in location of the equilibrium line altitude (ELA) for an increase of 3°C air temperature in the Gangotri Glacier region.*

Chhota Shigri Glacier and Mass-Balance Measurements

    Chhota Shigri Glacier, located between 32°11'–32°17'N. and 77°30'–77°32'E., extends from 4,000 m ASL to more than 5,600 m ASL (fig. 15). It is a valley-type glacier, debris covered in the lower ablation zone and lies in the Chandra-Bhāga river basin on the northern ridge of Pir Panjāl Range in the Lāhul-Spiti valley of Himachal Pradesh. The snout was located at 32°17'N., 77°32'E. in 2003. The glacier is in the monsoon-arid transition zone; therefore, it is considered to be a potential indicator of the northern limits of the intensity of the monsoon. It is influenced by both the Asian monsoon in the summer and the westerlies in the winter. From its snout to the accumulation zone near

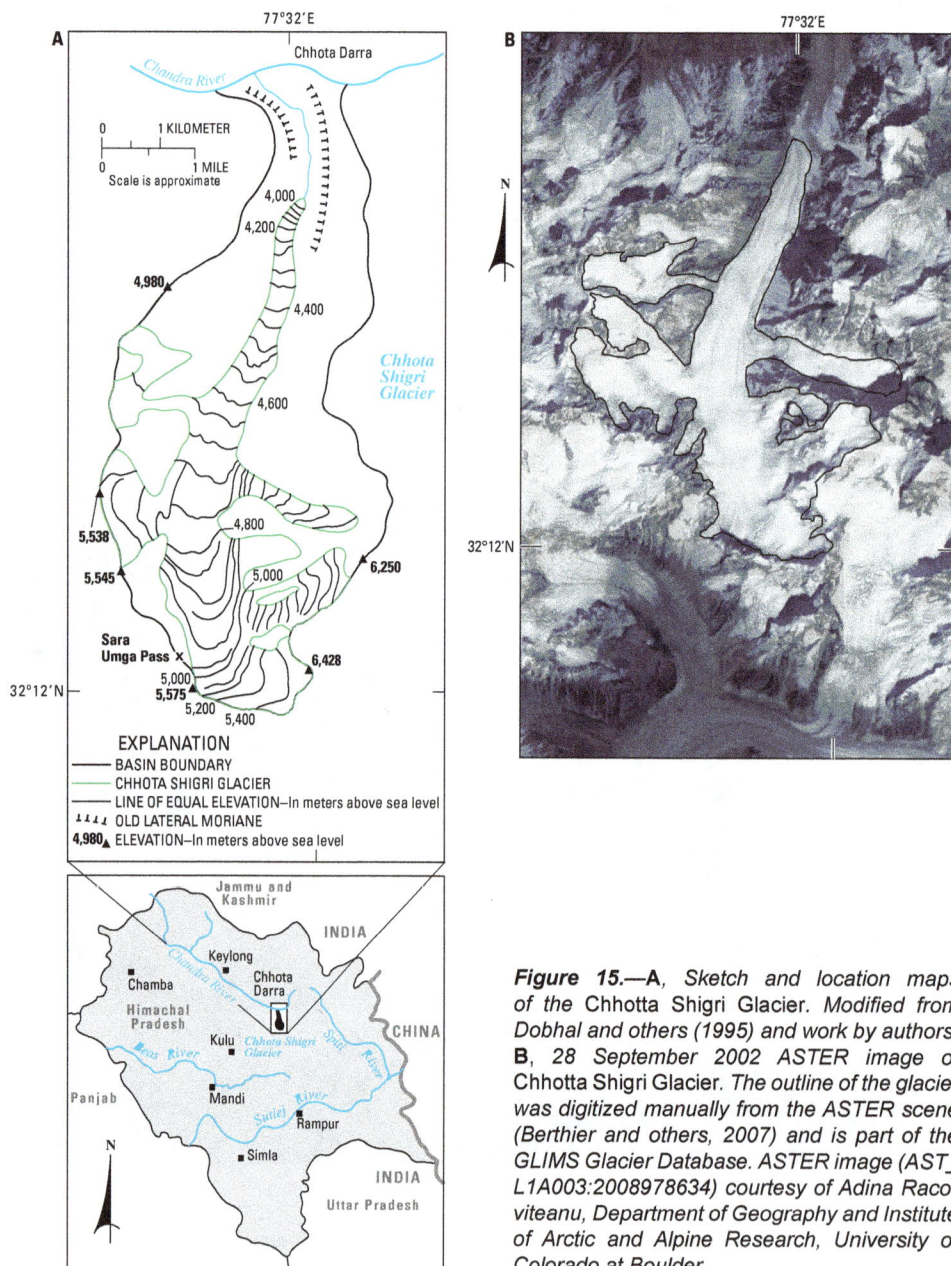

Figure 15.—A, *Sketch and location maps of the* Chhotta Shigri Glacier. *Modified from Dobhal and others (1995) and work by authors.* **B**, *28 September 2002 ASTER image of* Chhotta Shigri Glacier. *The outline of the glacier was digitized manually from the ASTER scene (Berthier and others, 2007) and is part of the GLIMS Glacier Database. ASTER image (AST_L1A003:2008978634) courtesy of Adina Raco-viteanu, Department of Geography and Institute of Arctic and Alpine Research, University of Colorado at Boulder.*

Sara Umga Pass (4,900 m ASL), *Chhota Shigri Glacier* is 9 km long and its width varies from 0.5 to 1.5 km in the ablation zone to about 4.5 km above the equilibrium line (Kumar, 1988). Field studies determined that the equilibrium line was at 5,170 m ASL in 2002–03 and in 2003–04 (Wagnon and others, 2007) — an upward shift since the 1987–89 ELAs, when the equilibrium line was at 4,650–4,840 m (Dobhal and others, 1995). Fluctuations in the width of the ablation zone are between 0.3 and 1.5 km, and are between 1.5 and 3 km in the accumulation zone (Hasnain and others, 1989). The *Chhota Shigri Glacier* drains into the Chandra River. The total drainage area of the *Chhota Shigri Glacier* basin is about 45 km², and the glacier occupies about 20 percent of the drainage area (Dobhal and others, 1995). Several supraglacial streams have formed in the ablation zone; most of them terminate in moulins or crevasses.

The *Chhota Shigri Glacier* snout has been retreating (fig. 16). The average rate of the *Chhota Shigri Glacier* snout recession was calculated by Kumar and Dobhal (1994) to be -7.5 m a^{-1} from 1970 to 1989. However, this rate has increased to -27 m a^{-1} during the period 1989 to 2000.

The areal coverage of the *Chhota Shigri Glacier* has been compared on various satellite images (Landsat 5 TM (1989), Satellite Pour l'Observation de la Terre (SPOT) (1994), Landsat 7 ETM+ (2000)), and on 1:50,000-scale

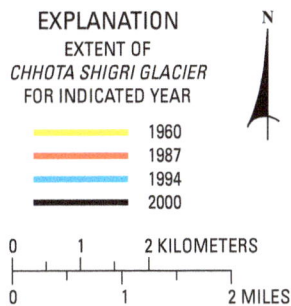

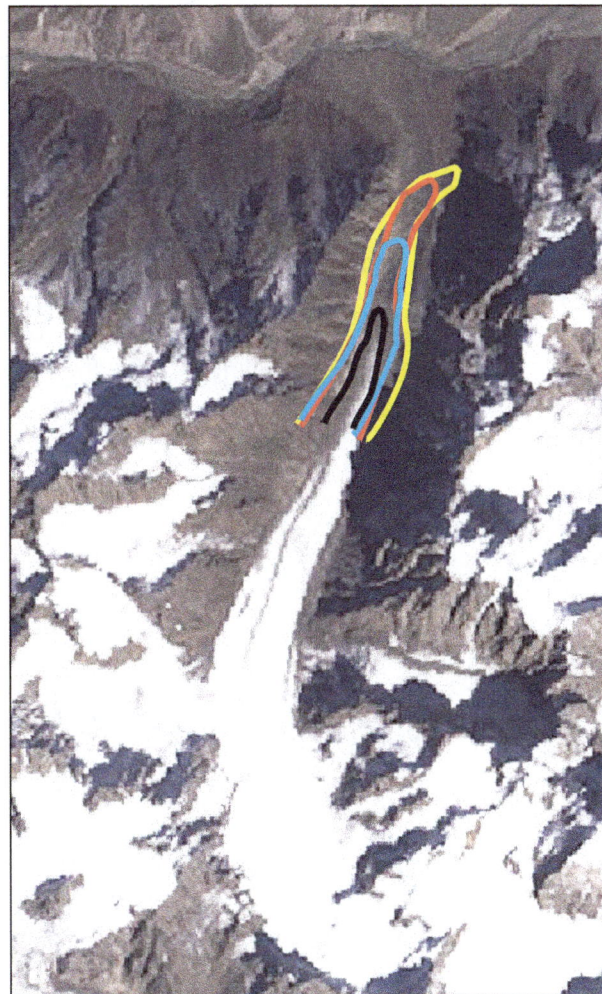

EXPLANATION

EXTENT OF
CHHOTA SHIGRI GLACIER
FOR INDICATED YEAR

1960
1987
1994
2000

Figure 16.—*Retreat of* Chhota Shigri Glacier *snout (approximate lat 32°15'N., long 77°30'E.) in the last 40 years mapped from topographic maps and satellite images. The 1960 and 1987 terminus positions are from topographic maps, the 1994 position from a SPOT image, and the 2000 position from a Landsat 7 ETM+ image.*

topographic maps published in 1960 and 1988. A digital elevation model (DEM) of the *Chhota Shigri Glacier* was prepared with the contour value derived from the topographic map of 1988. A comparative analysis of the planimetric cover of the *Chhota Shigri Glacier* on the geomorphological map of 1988 (surveyed in 1987) and the 2000 Landsat image shows that the areal glacier coverage decreased about 12 percent in the 13-year interval (fig. 17).

Under the projected scenarios of Watson and others (1998), the ELA of the *Chhota Shigri Glacier* has been computed for a temperature increase of +3 °C in the region. The projected AAR of the glacier has also been computed, and shows a decrease from 0.40 to 0.10 for a temperature increase of +3 °C. This implies that the glacier will be about 80 to 90 percent more vulnerable to melting with an increase in temperature of +3 °C.

The International Commission on Snow and Ice (ICSI) selected *Chhota Shigri Glacier* as a benchmark glacier in the Himalayan region in 2002. An international training course on glacier mass balance was held from 24 September

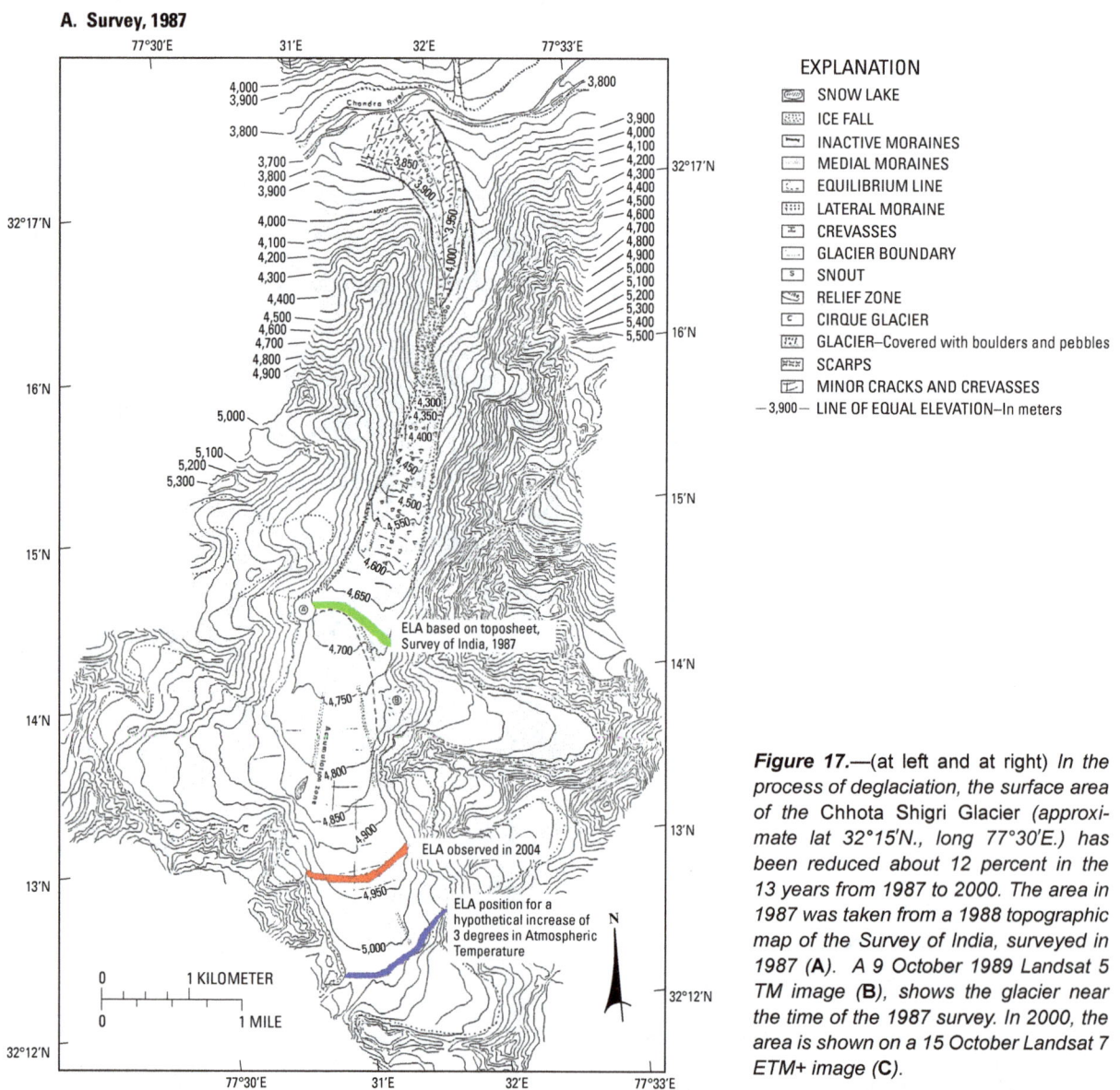

EXPLANATION

⬚	SNOW LAKE
⬚	ICE FALL
⬚	INACTIVE MORAINES
⬚	MEDIAL MORAINES
⬚	EQUILIBRIUM LINE
⬚	LATERAL MORAINE
⬚	CREVASSES
⬚	GLACIER BOUNDARY
⬚	SNOUT
⬚	RELIEF ZONE
⬚	CIRQUE GLACIER
⬚	GLACIER–Covered with boulders and pebbles
⬚	SCARPS
⬚	MINOR CRACKS AND CREVASSES
—3,900—	LINE OF EQUAL ELEVATION–In meters

A. Survey, 1987

Figure 17.—(at left and at right) *In the process of deglaciation, the surface area of the Chhota Shigri Glacier (approximate lat 32°15'N., long 77°30'E.) has been reduced about 12 percent in the 13 years from 1987 to 2000. The area in 1987 was taken from a 1988 topographic map of the Survey of India, surveyed in 1987 (A). A 9 October 1989 Landsat 5 TM image (B), shows the glacier near the time of the 1987 survey. In 2000, the area is shown on a 15 October Landsat 7 ETM+ image (C).*

B. Landsat, 1989

C. Landsat, 2000

to 10 October 2002. It was organized by ICSI, with the Glacier Research Group, Jawaharlal Nehru University (JNU), as local organizers, and technical support from Institut de Recherche pour le Développement (IRD) (France), Institute of Geography (Austria), Geological Survey of Denmark and Greenland (Denmark), Department of Physical Geography and Quaternary Geology and Glaciology (Sweden), and Nagoya University (Japan). As part of the training course, stakes were emplaced in the glacier with a steam drill (fig. 18). The glacier was revisited in 2003 and 2004 and measurements were carried out on the glacier using the stakes as reference points. Also, snow/firn pits were dug in the accumulation area of *Chhota Shigri Glacier* to obtain information on the yearly accumulation of snow. The only recognizable changes in the snow stratigraphy between 2002 and 2004 were a few slight differences in the size of the firn grains and the addition of a few thin ice layers. The ice layers were clear in some places, but contained dirt at other places. Very little change in the density profile has been observed at two pits located at elevations of 5,200 and 5,405 m ASL. A third pit, at an elevation of 5,500 m ASL, showed a large variation in the density profile; density increased with depth and reached 0.86 g cm^{-3} (glacier ice) at a depth of 300–338 cm (Kumar and others, 2005).

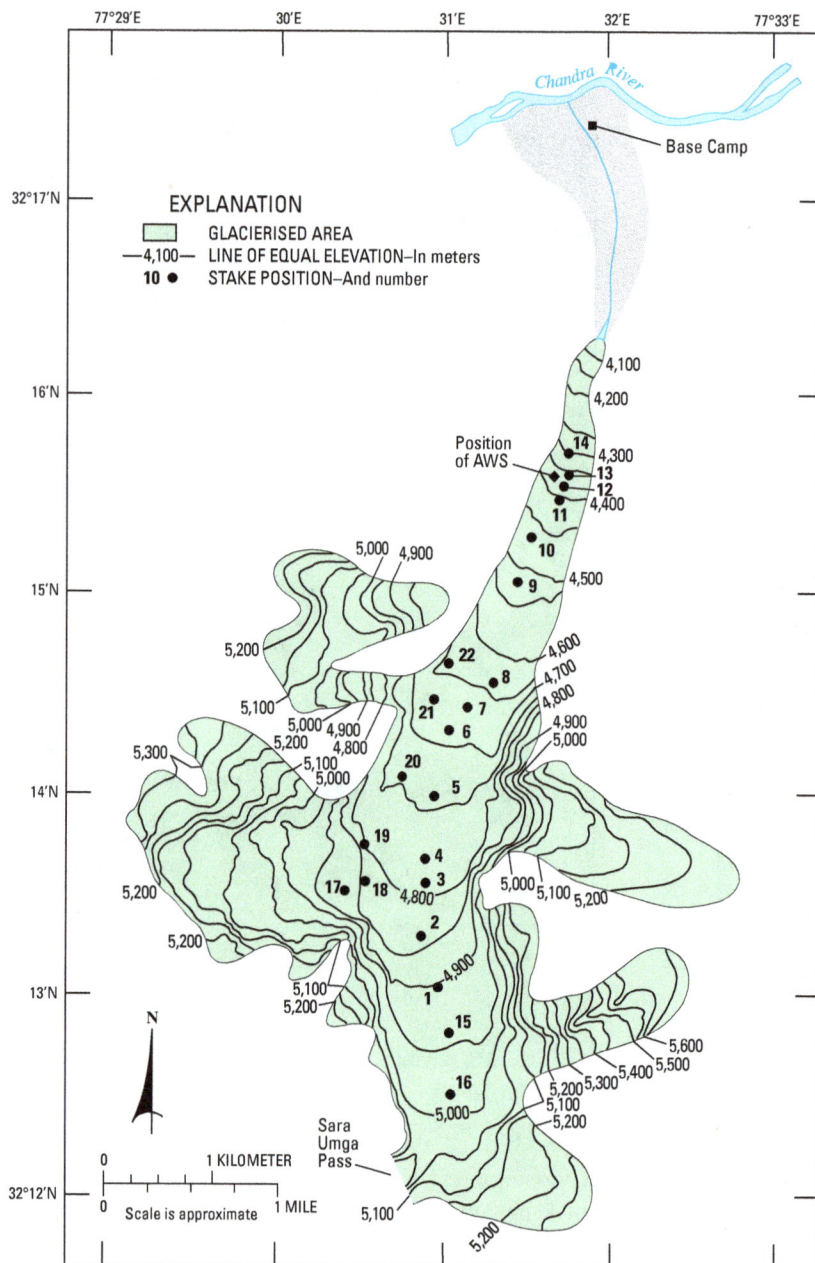

Figure 18.—*Map of the* Chhota Shigri Glacier *showing location of stakes installed in 2002 and the automatic weather station (AWS) installed during September–October 2003. Modified from Kumar and others (2004).*

The various positions of the terminus between 1984 and 1986 (campaign of Department of Science and Technology, New Delhi) were indicative of the annual climatic variation on the glacier, and showed three main episodes of advance and retreat. Fluctuations of the equilibrium line, observed during the same period, support the above observations. The snout of the glacier continued to recede at a rate of 18.7 m a^{-1} during 1986–88. This retreat was accompanied by a negative mass balance of 1.55×10^6 m^3 a^{-1}, observed during 1987–88 (Nijampurkar and Rao, 1992).

Our study, based on two years of observations (2002–03 and 2003–04), shows a negative mass balance of 5.03×10^6 m^3 a^{-1} during 2002–03, and 10.37×10^6 m^3 a^{-1} during 2003–04 (figs. 19A, B) (Kumar and others, 2005). The negative mass balance nearly doubled in two years and increased more than 3- to 6- fold compared with the value in 1987–88, only 15 years before. The glacier is becoming considerably thinner at lower altitudes, based on the increasing negative net mass balance.

Siachen Glacier

Siachen Glacier (approximately 35°11′N., 77°12′E.), is 78 km long and lies in the area claimed by both Pakistan and India, between the Saltoro ridge-line to the west and the main Karakoram Range to the east. The Saltoro ridge originates from the Sia Kangri peak, in the Karakoram Range, and the elevation ranges from 5,500 m to 7,350 m ASL. The major passes on this ridge are Sia La at 6,000 m and Bilafond La at 5,800 m ASL. The Siachen Glacier occupies the great Himalayan watershed that demarcates central Asia from the Indian sub-continent and that separates Pakistan from China in this region.

A comparative analysis of Landsat images of the Siachen glacier for the years 1978, 1989, and 2001 (fig. 20) clearly reflects a retreat in the terminus of the glacier. But a complete determination of the rate of retreat of the glacier with proper field validation is still to be carried out. Although a field study should be done because of the importance of the glacier and its known retreat, the area is under security restrictions and field work is prohibited.

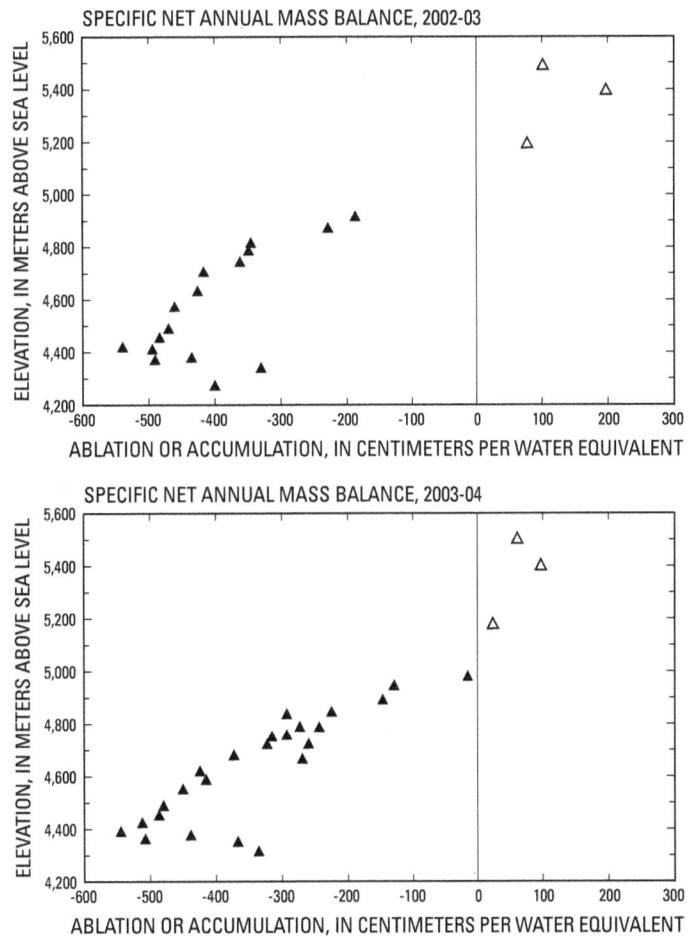

Figure 19.—**A**, *Specific net annual mass-balance (2002–03) near measurement points averaged against the elevation on Chhota Shigri Glacier;* **B**, *Specific net annual mass-balance (2003–04) near measurement points averaged against the elevation on Chhota Shigri Glacier. we = water equivalent.*

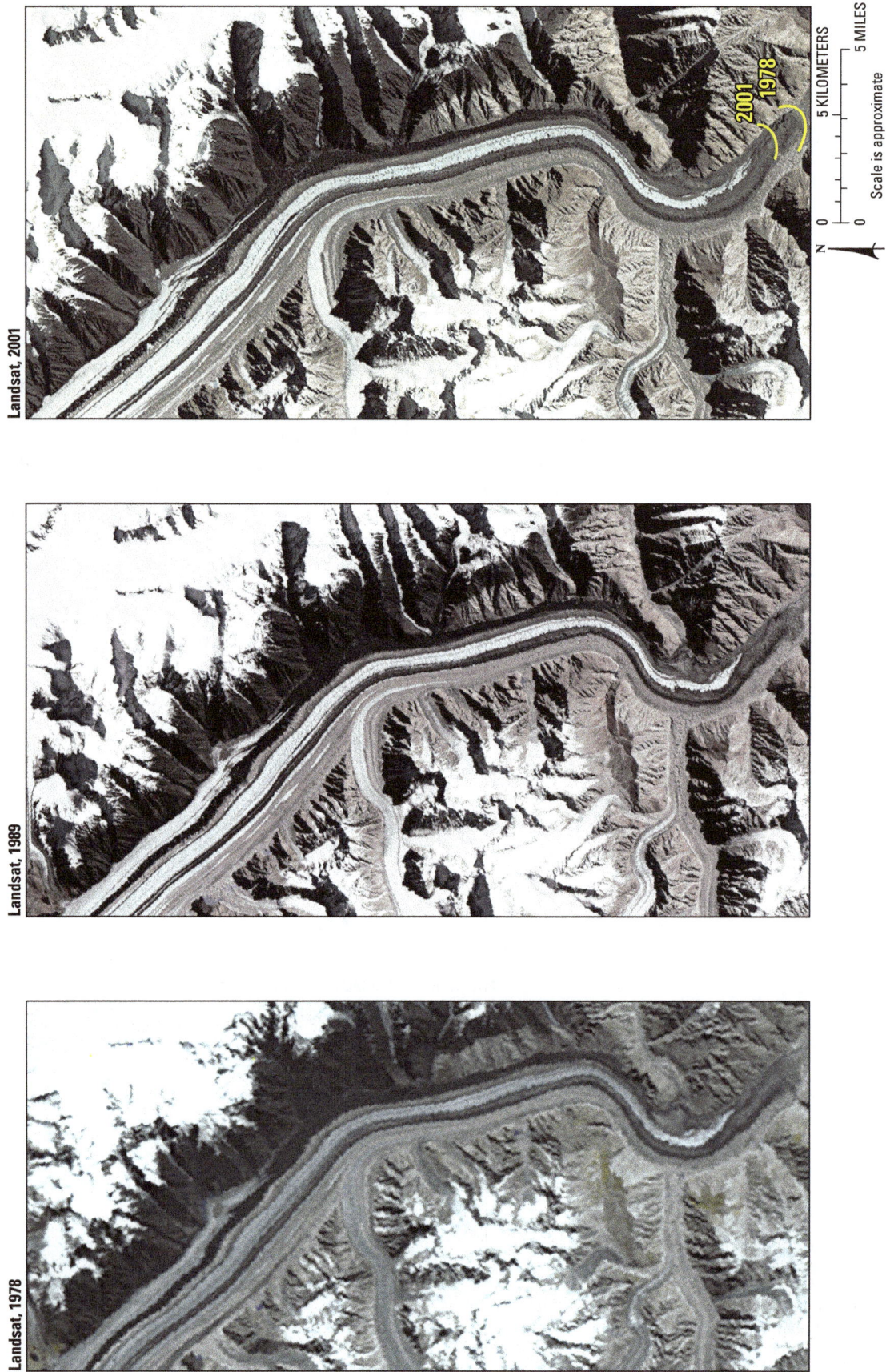

Landsat, 2001

2001
1978

N

5 KILOMETERS
0

5 MILES
0

Scale is approximate

Landsat, 1989

Landsat, 1978

Figure 20.—*Retreat of the terminus of Siachen Glacier (approximate lat 35°11'N., long 77°12'E.) from 1978 to 2001. The 1978, 1989, and 2001 terminus positions are shown on Landsat images: 18 July 1978 Landsat 3 MSS image (Path 159–Row 35), 9 October 1989 Landsat 5 TM image (Path 147–Row 36), and the 31 October 2001 Landsat 7 ETM+ image (Path 147–Row 36).*

Glacier Lakes

Glacier movement and interaction with debris cover on glacier surfaces results in various glacier-surface features. With high rates of melting, these surface features form glacier lakes, both supraglacial and ice-marginal. The formation and disappearance of new supraglacial lakes is a natural phenomenon. But an increase in the number and area of supraglacial lakes on the glacier surface can be linked with high rates of ice melting or excess downwasting of the glaciers. The coalescing of the small lakes results in large glacier lakes that store huge quantities of water and sediment. These glacier lakes can be bounded by the terminal moraine of the glacier, forming moraine-dammed or ice-margin lakes.

The glacial moraine boundary consists of soft and loose material. Steep lateral moraines with highly unstable slopes lie against the steep glacier tongues. Even minor seismic activity, a landslide, ice calving, or snow and rock avalanches can result in the sudden release of huge amounts of water, which cause flash floods downstream. These glacier lake outburst floods (GLOFs) (jökulhlaups) can cause major damage to inhabitants and their infrastructure, and to the ecosystem and environment throughout the Himalayan region

Many glacier lakes have developed during the last half century in the Himalaya, and their numbers have increased in recent years, likely visual evidence of global warming. If the current glacier downwasting trend continues, more potentially dangerous moraine-dammed lakes can be expected to develop. Therefore, it is important to identify the potential GLOF sites in the Himalaya so that necessary preventative action can be taken. For this purpose, an inventory of the glaciers and glacier lakes of Nepal and Bhutan was prepared by the International Centre for Integrated Mountain Development (ICIMOD) (Mool, Bajracharya, and Joshi, 2001; Mool, Wangda, and others, 2001). Aerial photographs and other remote-sensing data were used to identify these sites, and the temporal change in area of these lakes was studied using sequential images. In 1998, engineering construction was begun to reduce the water level in Tso Rolpa, one of the potentially dangerous lakes (see also Morales Arnao, 1998, p. I67–I71, for mitigation of hazards from moraine-dammed lakes in the Cordillera Blanca, Perú).

Similarly, a number of glacier lake studies have been conducted in the Indian part of the Himalaya. The results showed GLOF-potential sites in the Dhaulīganga River in the Ganga headwaters. The area of the largest glacial lake in 1989 was about 0.15 m²; the dimensions of the other small glacier lakes ranged from 0.025 to 0.075 km². Recently, a new assessment of two glacier lakes in the Ganga headwater area was conducted using Landsat 1978, Landsat 1990, and Landsat 2001 images. The results showed that, since 1978, the big glacier lake area in Ganga headwaters has increased by about 40 percent, while the area of the smaller lakes has increased by only about 13 percent. The study indicates that both hydrodynamics and calving are major processes that control the glacier-lake expansion (Chikita and others, 1999). In the Chandra River basin, the area of a glacier lake was 0.359 m² in 1972 and increased to 1.156 m² in 1996 (Kulkarni, 2003). The remapping of this lake, using Landsat 1990 and 2001 images, indicated that the area of the lake has increased about 22 percent since 1990. Evolution of the glacier lakes in the Tista River headwaters are shown in Landsat 1976, Landsat 1990, and ASTER 2001 images (fig. 21). These images clearly show an increase in the number and area of the lakes.

All of the above pilot studies suggest the number and areal cover of the glacier lakes increased in the last 30–40 years. This is a clear sign of global warming and of its impact on glacier lakes. Routine monitoring of the areal coverage of potential GLOF lakes is necessary to prevent the destruction of downstream property and population. Remotely-sensed data, particularly ASTER, can be utilized to monitor these lakes regularly.

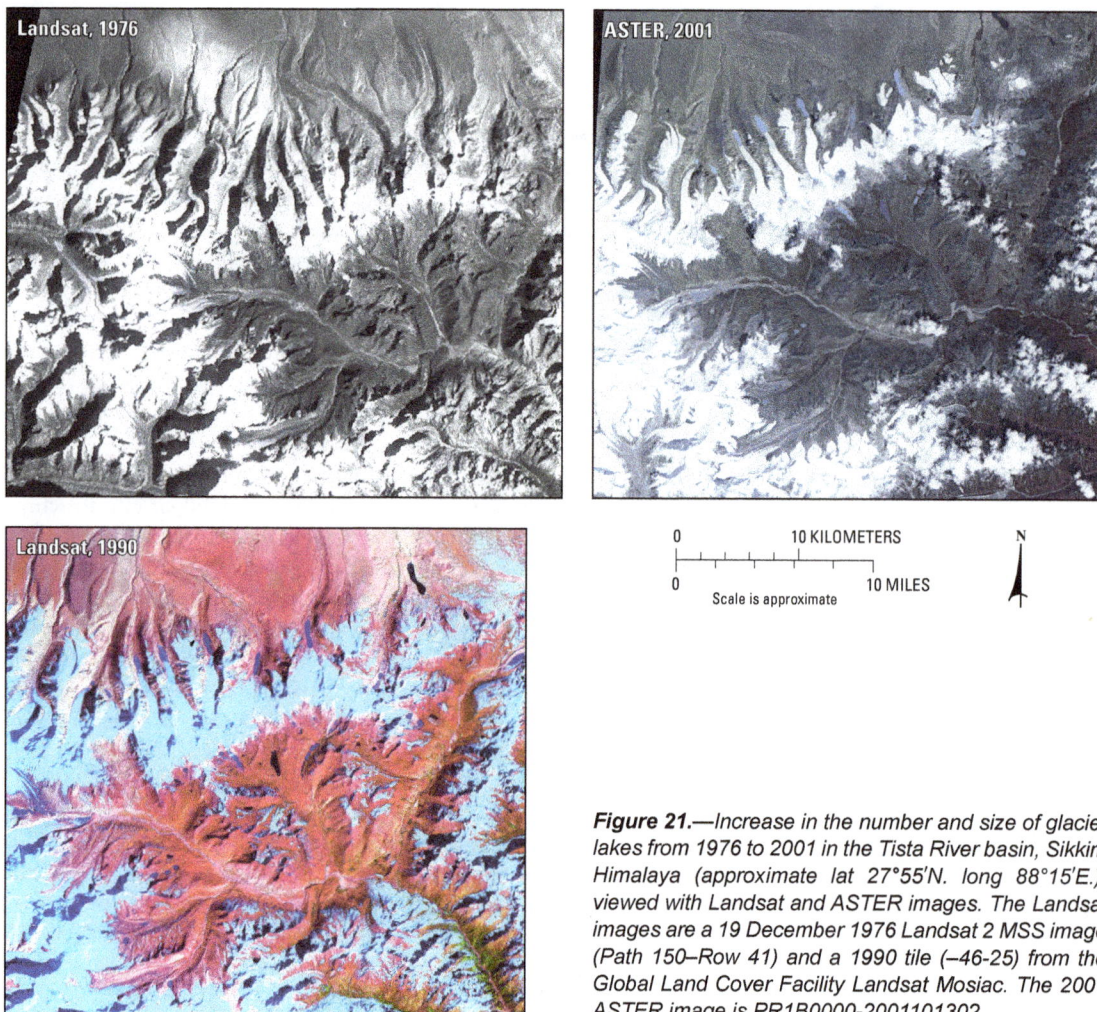

Figure 21.—Increase in the number and size of glacier lakes from 1976 to 2001 in the Tista River basin, Sikkim Himalaya (approximate lat 27°55'N. long 88°15'E.), viewed with Landsat and ASTER images. The Landsat images are a 19 December 1976 Landsat 2 MSS image (Path 150–Row 41) and a 1990 tile (−46-25) from the Global Land Cover Facility Landsat Mosiac. The 2001 ASTER image is PR1B0000-2001101302.

Conclusions

Remote-sensing techniques are very useful for cataloguing changes in glaciers and understanding climate change phenomena. Time-series analysis suggests that the summer temperature at an elevation of 4,000 m ASL has increased in the last forty years in headwaters of the Chenab and Ganga Rivers. Increased air temperatures have resulted in the shrinkage of the *Chhota Shigri Glacier* by about 12 percent in the last 13 years. In addition, 12 percent shrinkage of the main stem of the Gangotri Glacier has occurred in the last 16 years. This implies a rapid rate of shrinkage of many of the glaciers in the Himalaya in recent years. From model results, it was determined that, for an increase of +3 °C air temperature, the ELA will move upward about 400 m, and the AAR of the *Chhota Shigri* and Gangotri Glaciers will decrease from 0.4 to about 0.10 and 0.15, respectively. The model thus suggests that, if there is an increase of +3 °C air temperature during the summer, 80 to 90 percent of the surface area of *Chhota Shigri* and Gangotri Glaciers will be in the ablation area and therefore in the process of melting. This will further increase the meltwater discharges, rapid shrinkage, and rapid snout recession already seen at the end of the last century. Increases in the size and number of glacier lakes in the Himalaya clearly reflect the increasing influence of global climatic changes in the region.

References Cited

Ageta, Y., Naito, N., Nakawo, M., Fujita, K., and others, 2001, Study project on the recent rapid shrinkage of summer-accumulation type glaciers in the Himalayas, 1997–1999: Bulletin of Glaciological Research 18, p. 45–49.

Auden, J.B., 1937, The snout of the Gangotri Glacier, Tehri Garhwal: Geological Survey of India Records, v. 72, pt. 2, p. 135–140.

Basu, Bikash, 1947, Some problems of snow survey in the eastern Himalayas: Journal of the Royal Asiatic Society of Bengal, v. 13, no. 1, p. 43–56.

Beniston, M., Fox, D.G., (and several other lead and contributing authors), 1996, Impacts of climate change on mountain regions, in Watson, R.T., Zinyowera, M.C., and Moss, R.H., eds., Climate change, 1995, Impacts, adaptions and mitigation of climate change — Scientific-technical analyses: Contribution of Working Group II to the second assessment report of the Intergovernmental Panel on Climate Change: Cambridge, U.K., Cambridge University Press, p. 191–213.

Benn, D.I., Wiseman, S., and Warren, C.R., 2000, Rapid growth of a supraglacial lake, Ngozumpa Glacier, Khumbu Himal, Nepal, in Nakawo, M., Raymond, C.F., and Fountain, A., eds, Debris-covered Glaciers: International Association of Hydrological Sciences (IAHS) Publication No. 264, p. 177–185.

Bose, P.N., 1891, Extracts from the journal of a trip to the glaciers of the Kapru, Pandim, etc.: Geological Survey of India Records, v. 24, p. 46–48.

Bridges, H.F., 1908–09, Note on the Hunza and Nagar glaciers: Geological Survey of India Records, v. 37, pt. 2, p. 221.

Broecker, W.S., 1997, Mountain glaciers — Records of atmospheric water vapour content?: Global Biochemical Cycles, v, 11, no. 4, p. 589–597.

Chikita, K., Jha, J., and Yamada, T., 1999, Hydrodynamics of a supraglacial lake and its effect on the basin expansion — Tsho Rolpa, Rolwaling Valley, Nepal Himalaya: Arctic, Antarctic and Alpine Research, v. 31, no. 1, p. 58–70.

Church, J.E., 1947, Snow seeking in the Himalaya (Home of snow): Science and Culture, v. 13, no. 3, p. 82–86.

Cotter, G. de P., and Brown, J.C., 1907, Notes on certain glaciers in Kumaun: Geological Survey of India Records, v. 35, pt. 4, p. 148–157.

Dobhal, D.P., Kumar, S., and Mundepi, A.K., 1995, Morphology and glacier dynamics studies in monsoon-arid transition zone — An example from Chhota Shigri Glacier, Himachal-Himalaya, India: Current Science, v. 68, no. 9, p. 936–944.

Fujita K., Kadota, T., Rana, B., Kayastha, R.B., and Ageta, Y., 2001, Shrinkage of Glacier AX010 in Shorong region, Nepal Himalayas in 1990s: Bulletin of Glaciological Research, v. 18, p 51–54.

Fujita, K., Nakawo, M., Fujii, Y., and Paudyal, P., 1997, Changes in glaciers in Hidden Valley, Mukut Himal, Nepal Himalayas, from 1974 to 1994: Journal of Glaciology, v. 43, no. 145, p. 583–588.

Geological Survey of India, 2001, Symposium on snow, ice and glaciers — A Himalayan perspective, Lucknow, India, 9–11 March 1999, Proceedings: Geological Survey of India Special Publication No. 53, 419 p.

Godwin-Austin, H.H., 1862, On the glaciers of the Mustakh Range: Royal Geographical Society Journal, v. 34, p. 19–56.

Grinlinton, J.L., 1912, Notes on the Poting Glacier, Kamaon Himalaya, June 1911: Geological Survey of India Records, v. 42, pt. 2, p. 102–126.

Haeberli, W., Bösch, H., Scherler, K., Østrem, G., and Wallén, C.C., 1989, World Glacier Inventory, Status 1988: Switzerland, International Association of Hydrological Sciences (International Commission on Snow and Ice), United Nations Environment Programme, United Nations Educational, Scientific, and Cultural Organization (IAHS (ICSI)-UNEP-UNESCO), 430 p.

Hasnain, S.I., Ahmad, S., and Kumar, Rajesh, 2004, Impact of climate change on Chhota Shigri Glacier, Chenab basin, Gangotri Glacier and Ganga headwater in the Himalaya, in Gosnain, A.K., Sharma, S.K., and others, eds., Proceedings, Workshop on vulnerability assessment and adaptation due to climate change on Indian water resources, coastal zones and human health, Delhi, 27–28 June 2003: Government of India, Ministry of Environment and Forest, p. 1–7.

Hasnain, S.I., Subramanian, V., and Dhanpal, K., 1989, Chemical characteristics and suspended sediment load of melt waters from a Himalayan glacier in India: Journal of Hydrology, v. 106, nos. 1/2, p. 99–108.

Hastenrath, S., 1995, Glacier recession on Mount Kenya in the context of the global tropics: Bulletin de l'Institut Français d'Études Andines, v. 24, no. 3, p. 633–638.

Hastenrath, S., and Kruss, P.D., 1992, The dramatic retreat of Mount Kenya's glaciers between 1963 and 1987: Greenhouse forcing: Annals of Glaciology, v. 16, p. 127–133.

Jack, Alexander, 1844, Snow on the Himalayas: Calcutta, Journal of Natural History, v. 4, p. 455–458.

Jangpangi, B.S., 1958, Study of some of the Central Himalayan glaciers: Journal of Scientific and Industrial Research 17A, no. 12 (supplement), p. 91–93.

Jóhannesson, T., 1997, The response of two Icelandic glaciers to climatic warming computed with a degree-day glacier mass-balance model coupled to a dynamic glacier model: Journal of Glaciology, v. 43, no. 144, p. 321–327.

Kadota, T., Fujita, K., Seko, K., Kayastha, R.B., and Ageta, Y., 1997, Monitoring and prediction of shrinkage of a small glacier in the Nepal Himalaya: Annals of Glaciology, v. 24, p. 90–94.

Kadota, T., Seko, K., Aoki, T., Iwata, S., and Yamaguchi, S., 2000, Shrinkage of the Khumbu Glacier, east Nepal from 1978 to 1995, in Nakawo, M., Raymond, C.F., and Fountain, A., eds., Debris-covered glaciers: International Association of Scientific Hydrology (IAHS) Publication. No. 264, p. 235–243.

Kaser, G., 1999, A review of the modern fluctuations of tropical glaciers: Global and Planetary Change, v. 22, nos. 1–4, p. 93–103.

Kaser, G., Fountain, A., and Jansson, P., 2002, A manual for monitoring the mass balance of mountain glaciers, with particular attention to low latitude characteristics: Paris, UNESCO, Technical Documents in Hydrology, No. 59, 137 p. [also available at *http://unesdoc.unesco.org/images/0012/001295/129593e.pdf*]

Kaser, G., and Georges, C., 1997, Changes of the equilibrium-line altitude in the tropical Cordillera Blanca, Peru, 1930–50, and their spatial variations: Annals of Glaciology, v. 24, p. 344–349.

Kaser, G., Georges, C., and Ames, A., 1996, Modern glacier fluctuations in the Huascarán-Chopicalqui Massif of the Cordillera Blanca, Perú: Zeitschrift für Gletscherkunde und Glazialgeologie, v. 32, p. 91–99.

Kaser, G., Hastenrath, S., Ames, A., 1996, Mass balance profiles on tropical glaciers: Zeitschrift für Gletscherkunde und Glazialgeologie, v. 32, p. 75–81.

Kaser, G., and Noggler, B., 1991, Observations on Speke Glacier, Ruwenzori Range, Uganda: Journal of Glaciology, v. 37, no. 127, p. 313–318.

Kaul, M.K., ed., 1999, Inventory of the Himalayan Glaciers: Geological Survey of India Special Publication No. 34, 165 p.

Kulkarni, A.V., 1992, Mass balance of Himalayan glaciers using AAR and ELA methods: Journal of Glaciology, v. 38, no. 128, p. 101–104.

Kulkarni, A.V., 2003, Effect of climatic variations on Himalayan glaciers: Training programme on Remote Sensing for Glaciological Studies: SAC/RESA/MWRG/ESHD/TR-13/2003.

Kumar, Rajesh, Hasnain, S.I., Wagnon, P., Arnaud, Y., and Chevallier, P., 2005, Climate change signals detected through mass balance measurements on benchmark glacier, Himachal Pradesh, India: Presented at the International Seminar on Climate and Anthropogenic Impacts on the variability of Water Resources, Montpellier, France, 22–24 November 2005.

Kumar, Rajesh, Linda, A., Chevallier, P., and Wagnon, P., 2004, Chhota Shigri Glacier mass balance monitoring, September 28–October 12, 2003: [Originally published on the web but now withdrawn]

Kumar, S., 1988, Bed rock control on the quick fluctuations in the Chhota Shigri Glacier: Government of India, Department of Science and Technology, Ministry of Science, Technical Report No. 2, Compiled by A.K. Mundepi and B.R.S. Rawat.

Kumar, S., and Dobhal, D.P., 1994, Snout fluctuation study of Chhota Shigri Glacier, Lahaul and Spiti District, Himachal Pradesh: Journal of the Geological Society of India, v. 44, no. 5, p. 581–585.

Madden, E., 1847, Notes on an excursion to the Pindaree Glacier in September 1846: Journal of the Asiatic Society of Bengal, v. 17, p. 340–450.

Mercer, J.H., 1975a, Glaciers of the Himalaya, *in* Field, W.O., ed., Mountain glaciers of the Northern Hemisphere: Hanover, N.H., U.S. Army Cold Regions Research and Engineering Laboratory, p. 411–447.

Mercer, J.H., 1975b, Glaciers of the Karakoram, *in* Field, W.O., ed., Mountain glaciers of the Northern Hemisphere: Hanover, N.H., U.S. Army Cold Regions Research and Engineering Laboratory, p. 371–409.

Mool, P.K., Bajracharya, S.R., and Joshi, S.P., 2001, Inventory of glaciers, glacial lakes and glacial lake outburst floods, Nepal: Kathmandu, Nepal, International Centre for Integrated Mountain Development, 363 p.

Mool, P.K., Wangda, D., Bajracharya, S.R., and others, 2001, Inventory of glaciers, glacial lakes and glacial lake outburst floods, Bhutan: Kathmandu, Nepal, International Centre for Integrated Mountain Development, 227 p.

Morales Arnao, B., 1998, Glaciers of Perú, with sections on the Cordillera Blanca imagery and Quelccaya ice cap, by Hastenrath, S.L. (I4), *in* Williams, R.S., Jr., and Ferrigno, J.G., eds., Satellite image atlas of glaciers of the world: U.S. Geological Survey Professional Paper 1386–I (South America), p. I51–I79. [*http://pubs.usgs.gov/prof/p1386i*]

Müller, Fritz, Caflisch, T., and Müller, G., 1977, Instructions for the compilation and assemblage of data for a world glacier inventory: Zürich, Swiss Federal Institute of Technology, Temporary Technical Secretariat for World Glacier Inventory, International Commission on Snow and Ice, 28 p.

Müller, Fritz, Caflisch, T., and Müller, G., 1978, Instructions for the compilation and assemblage of data for a world glacier inventory — Supplement; identification/glacier number: Zürich, Swiss Federal Institute of Technology, Temporary Technical Secretariat for World Glacier Inventory, 7 p., plus app.

Naithani, A.K., Nainwal, H.V., Sati, K.K., and Prasad, C., 2001, Geomorphological evidences of retreat of the Gangotri Glacier and its characteristics: Current Science, v. 80, no. 1, p. 87–94.

Nijampurkar, V.N., and Rao, D.K., 1992, Accumulation and flow rates of ice on Chhota Shigri Glacier, central Himalaya, using radioactive and stable isotopes: Journal of Glaciology, v. 38, no. 128, p. 43–50.

Qin Dahe, ed., 1999, Map of glacier resources in the Himalayas; 1st ed.: Beijing, Science Press, scale 1:500,000, 7 sheets. [Compilation of map based on analysis of 1975–1978 Landsat MSS band 7 and false-color images and some aerial photographs]

Rasmussen, L.A., and Krimmel, R.M., 1999, Using vertical aerial photography to estimate mass balance at a point: Geografiska Annaler, v. 81A, no. 4, p. 725–733.

Sain, Kanwar, 1946, Role of glaciers and snow on hydrology of Punjab rivers: Simla, India, Central Board of Irrigation Publication 26, 44 p.

Sapiano, J.J., Harrison, W.D., and Echelmeyer, K.A., 1998, Elevation, volume and terminus changes of nine glaciers in North America: Journal of Glaciology, v. 44, no. 146, p. 119–135.

Sharma, M.C., and Owen, L.A., 1996, Quaternary glacial history of NW Garhwal, Central Himalayas: Quaternary Science Reviews, v. 15, iss.4, p. 335–365.

Srivastava, Deepak, ed., 2001, Glaciology of Indian Himalaya: Geological Survey of India Special Publication No. 63, 213 p.

Tewari, A.P., 1971, A short report on glacier studies in the Himalaya Mountains by the Geological Survey of India: Studia Geomorphologica Carpatho-Balcanica, v. 5, p. 173–181.

Thompson, L.G., Mosley-Thompson, E., Davis, M.E., and others, 1995, Late glacial stage and Holocene tropical ice core records from Huascarán, Peru: Science, v. 269, no. 5520, p. 46–50.

Vohra, C.P., 1981, Himalayan glaciers, *in* Lall, J.S., and Moddie, A.D., eds., The Himalaya, aspects of change: Delhi, Oxford University Press, p. 138–151.

von Humbolt, Alexander, 1845, Remarks on the height of the snowline in the Himalaya: Cosmos, v. 358, p. 483–484.

von Wissman, Herman, 1959, Die heutige vergletscherung und schneegrenze in Hochasien mit hinweisen auf die vergletscherung der letzten Eiszeit [Modern glaciers and the snowline in High Asia with reference to the glaciers of the last ice age]: Abhandlungen Akadamie der Wissenschaften und der Literatur der Mathematisch-Naturwissenschaftlichen Klasse (Mainz), Jahrgang 1959, no. 14, p. 1101–1407.

Wagnon, P., Kumar, R., Arnaud, Y., Linda, A., and others, 2007, Four years of mass balance on Chhota Shigri Glacier, Himachal Pradesh, India, a new benchmark glacier in the the western Himalaya: Journal of Glaciology, v. 53, no. 183, p. 603–611.

Wagnon, P., Ribstein, P., Kaser, G., and Berton, P., 1999, Energy balance and runoff seasonality of a Bolivian glacier: Global and Planetary Change, v. 22, nos. 1–4, p. 49–58.

Watson, R.T., Zinyowera, M.C., and Moss, R.H., eds., 1998, The regional impacts of climate change — An assessment of vulnerability: A special report of Intergovernmental Panel on Climate Change Working Group II: Cambridge, U.K., Cambridge University Press, 517 p.

Glaciers of Asia—

GLACIERS OF NEPAL—Glacier Distribution in the Nepal Himalaya with Comparisons to the Karakoram Range

By Keiji Higuchi, Okitsugu Watanabe, Hiroji Fushimi, Shuhei Takenaka, *and* Akio Nagoshi

SATELLITE IMAGE ATLAS OF GLACIERS OF THE WORLD

Edited by RICHARD S. WILLIAMS, JR., *and* JANE G. FERRIGNO

U.S. GEOLOGICAL SURVEY PROFESSIONAL PAPER 1386–F–6

CONTENTS

Glaciers of Nepal — Glacier Distribution in the Nepal Himalaya with Comparisons to the Karakoram Range, *by* Keiji Higuchi, Okitsugu Watanabe, Hiroji Fushimi, Shuhei Takenaka, *and* Akio Nagoshi--**293**

    Introduction --**293**

        Use of Landsat Images in Glacier Studies ---------------------------------**293**

            FIGURE   1. Map showing location of the Nepal Himalaya and Karokoram Range in Southern Asia-- **294**

            FIGURE   2. Map showing glacier distribution of the Nepal Himalaya and its surrounding regions --- **295**

            FIGURE   3. Map showing glacier distribution of the Karakoram Range ------------ **296**

        A Brief History of Glacier Investigations ------------------------------------**297**

        Procedures for Mapping Glacier Distribution from Landsat Images ---------**298**

            FIGURE   4. Index map of the glaciers of Nepal showing coverage by Landsat 1, 2, and 3 MSS images ----------------------------------- **299**

            FIGURE   5. Index map of the glaciers of the Karakoram Range showing coverage by Landsat 1, 2, and 3 MSS images --------------------- **300**

            TABLE   1. Optimum Landsat 1, 2, and 3 MSS images of the glaciers of Nepal --- **301**

            FIGURE   6. Index map of optimum Landsat 1, 2, and 3 MSS images of the glaciers of Nepal --- **301**

        Glaciers in the Nepal Himalaya and the Karakoram Range--------------------**302**

            Glacier Inventory Based on a Map Compiled from Landsat Images ------**302**

            TABLE   2. Glaciers in the Nepal Himalaya and surrounding regions------- **302–306**

            TABLE   3. Glaciers in the Karakoram Range ------------------------------- **307**

            FIGURE   7. Map showing location of individual glaciers in the Nepal Himalaya and its surrounding regions, identified by glacier number ------------ **308**

            FIGURE   8. Map showing location of individual glaciers in the Karakoram Range identified by glacier number ------------------------------------- **309**

            Comparison of Nepal Himalaya and Karakoram Range Glacier Distribution --**310**

            TABLE   4. List of glaciological characteristics of the Nepal Himalaya and the Karakoram Range ----------------------------------- **310**

        Glacier Inventory from Ground Surveys in the Dudh Kosi Region, Nepal, and Comparison with Landsat Image Analyses------**311**

        Comparison of Results for the Khumbu Glacier, Nepal, from Landsat Image Analysis and Ground Surveys --------------------------**311**

            FIGURE   9. Landsat image, glacier inventory, and glacier distribution in the Dudh Kosi region, east Nepal------------------------------------- **312**

            FIGURE 10. Oblique aerial photograph of the Khumbu Glacier, taken on 11 December 1978 --- **313**

            FIGURE 11. Distribution of supraglacial debris, bedrock, and other features and surface characteristics of the Khumbu Glacier ---------- **314**

        Landsat Image Analysis for Monitoring Glacier Disasters ---------------------**315**

            FIGURE 12. Oblique aerial photograph of the supraglacial lake of the *Imja Glacier*, taken on 30 November 1975 ------------------------- **316**

    Conclusions--**316**

    Supplement to "Glaciers of Nepal," by Yutaka Ageta------------------------------**317**

    Introduction --**317**

    References Cited---**319**

GLACIERS OF ASIA—

GLACIERS OF NEPAL— Glacier Distribution in the Nepal Himalaya with Comparisons to the Karakoram Range[1]

By Keiji Higuchi, Okitsugu Watanabe, Hiroji Fushimi, Shuhei Takenaka, *and* Akio Nagoshi[2]

Introduction

The Nepal Himalaya and the Karakoram Range are the highest mountainous regions on Earth, with many peaks higher than 6,000–8,000 m in elevation. As a result, attempts to study glaciers in these ranges using field and airborne surveys have presented many difficulties historically. The availability of Landsat images in the 1970s, however, offered the opportunity to study the distribution and morphological characteristics of glaciers in the Nepal Himalaya, and to compare them to characteristics of glaciers in the Karakoram Range. The images were particularly valuable because no large-scale official maps were available. Additionally, a series of field investigations of the Nepal Himalaya glaciers, organized by Nagoya and Kyoto Universities, Japan, and conducted over six years (1973–1978), provided ground truth which could be used in resolving inherent Landsat image interpretation issues. This was the first attempt at glacier surveys in these two regions using Landsat images.

The Himalaya dominates the central and northern portions of Nepal (figs. 1 and 2). The Karakoram Range lies in northern India and Pakistan, along the border with China (figs. 1 and 3). Although these ranges are geographically near one another, they are separated by the Indus, Gilgit, and Shyok Rivers and by approximately 150 km of lower elevation terrain.

Use of Landsat Images in Glacier Studies

Because of the many logistical problems in conducting field and airborne surveys, Landsat images are more effective tools for obtaining preliminary information about glaciers in relatively unexplored areas such as the Nepal Himalaya and the Karakoram Range. However, there are several problems in the application of Landsat images to glaciological studies. First, the resolution

---

[1]This section was written in the early 1980s, and describes the glaciers of Nepal during the 1970s, the baseline period for the Satellite Image Atlas of Glaciers of the World, U.S. Geological Survey Professional Paper 1386. Because there was a delay in publication, a supplement by Yuteka Ageta has been added at the end of this section to describe more recent research.

[2]The authors were with the Water Research Institute, Nagoya University, when this section was written.

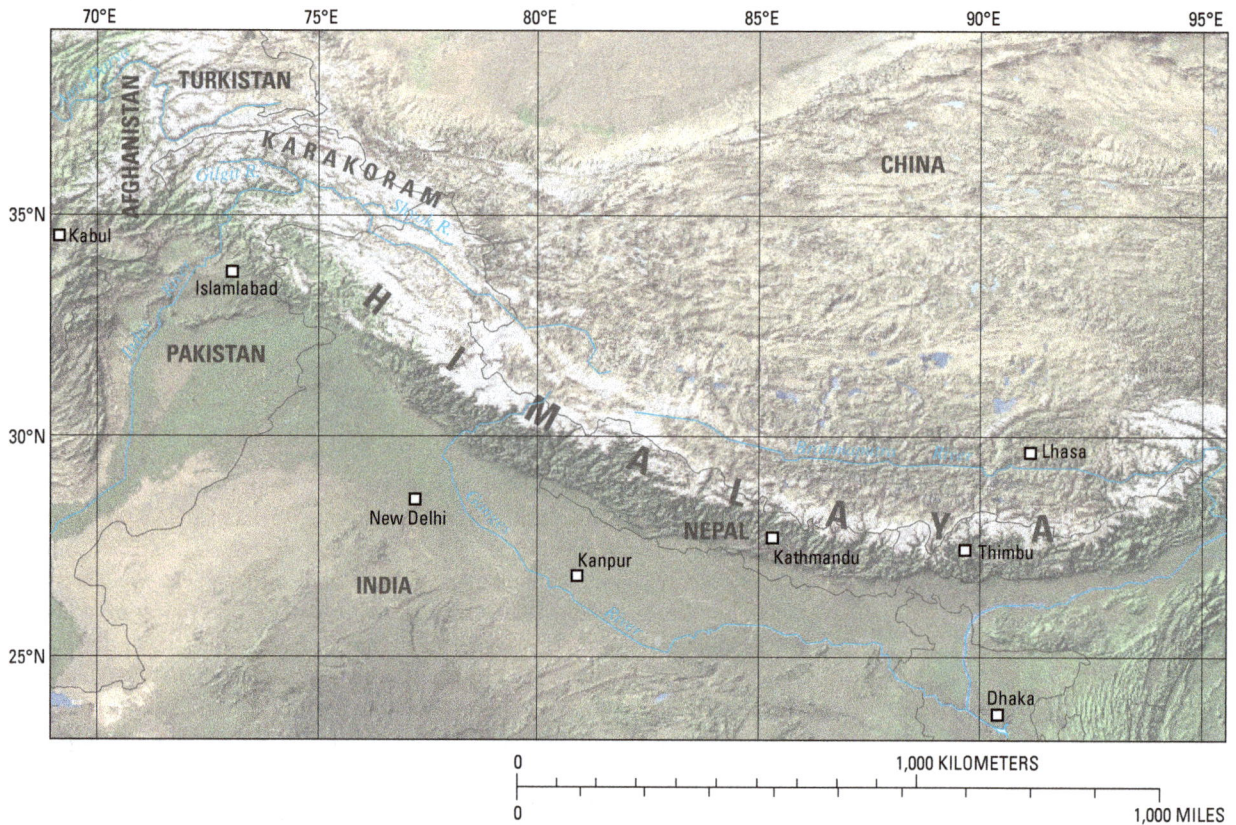

Figure 1.— *Location of the Nepal Himalaya and Karokoram Range in Southern Asia.*

of the image may not be adequate to identify smaller glaciers. For example, small cirque glaciers (less than 100 m in width) are difficult to recognize on Landsat multispectral scanner (MSS) images (Higuchi, 1975). Second, some glacier surfaces are covered by a combination of materials such as soil, ice, and rock that are not only hard to see but difficult to differentiate, so that snow facies and glacier structures are not distinguishable in the images. Third, observations with sensors in the visible part of the spectrum are impossible to make during the monsoon season, and are difficult at other times of the year (particularly in the afternoon) because of cloud cover. In addition, the orbital parameters of the Landsat satellite result in the images for this area being acquired in the morning hours which makes it difficult to see details on the western slopes of mountain ridges because of shadows. Therefore, results of the ground investigations conducted during the same period as this study provided a means of checking questionable Landsat interpretations.

In this paper, the results of a preliminary application of Landsat images to the analysis of glaciological characteristics in the Nepal Himalaya and the Karakoram Range are presented. Ground investigation results from detailed studies in the Dudh Kosi region of Nepal Himalaya (Higuchi and others, 1978, 1980) were used to improve the Landsat interpretations.

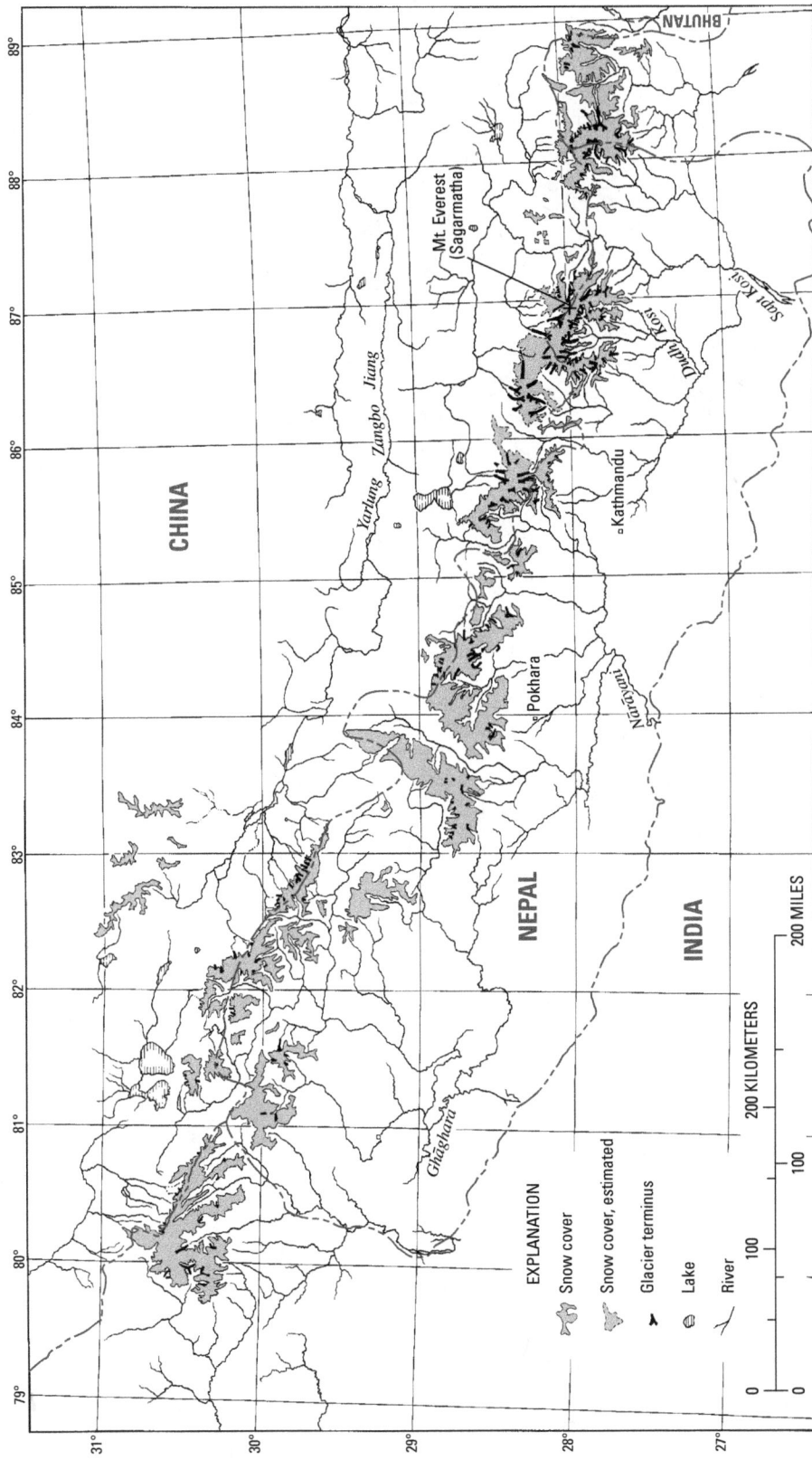

Figure 2.— *Glacier distribution of the Nepal Himalaya and its surrounding regions. Some political boundaries are uncertain.*

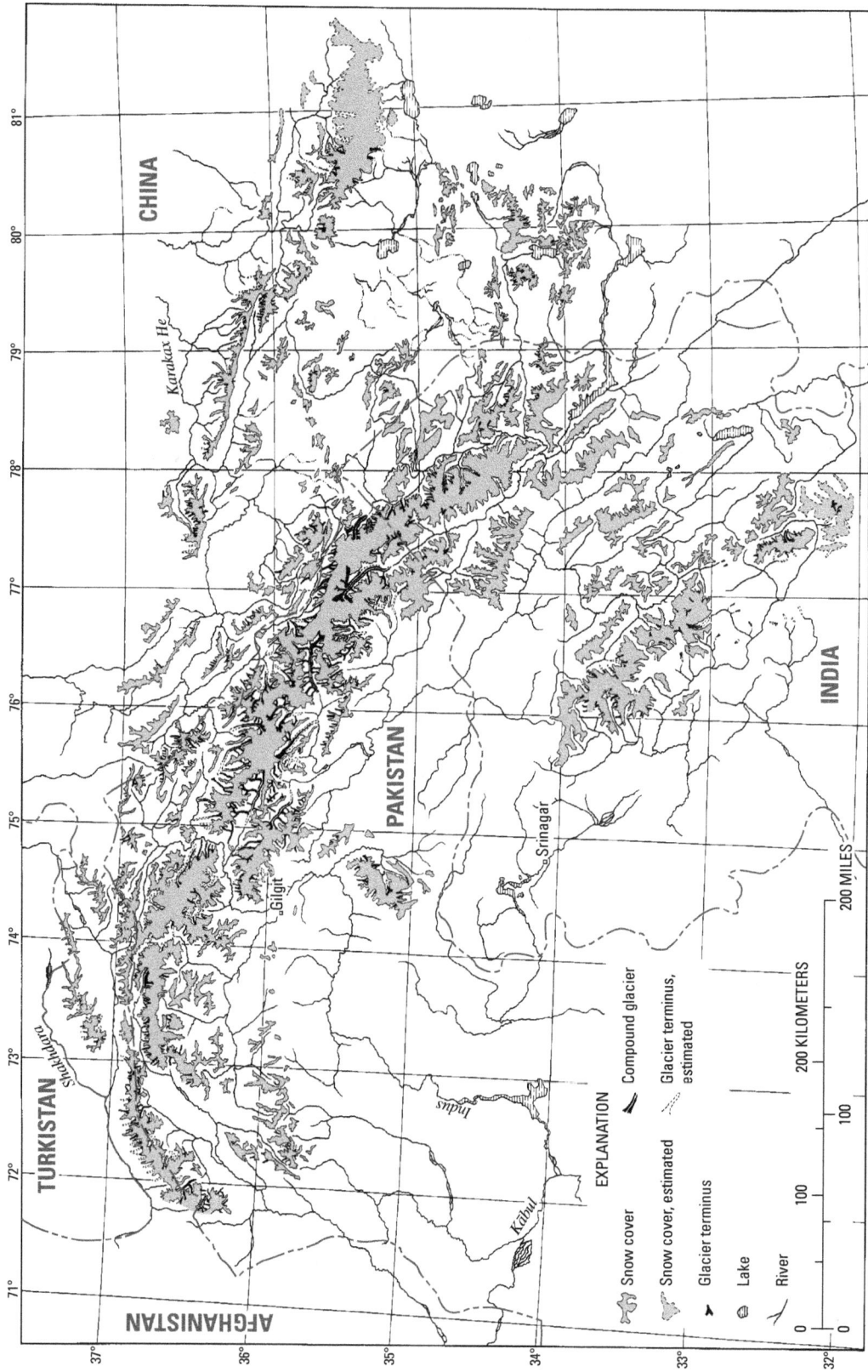

Figure 3.— *Glacier distribution of the Karakoram Range. Some political boundaries are uncertain.*

A Brief History of Glacier Investigations

Since 1949, foreign mountaineers and scientists have repeatedly visited the Nepal Himalaya. The first glaciologist to visit Nepal was Fritz Müller who was a participant in the Swiss Everest Expedition of 1956. He studied the Khumbu Glacier, East Nepal (Müller, 1958). During the following years, the number of mountaineering and scientific expeditions has gradually increased, improving our knowledge of the Nepal Himalayan region and glaciers.

The *Takpu Glacier*,[3] situated in the northwesternmost part of the Nepal Himalaya, was visited by a Japanese expedition in 1963. After additional Japanese glaciological work in 1965 on the *Gustang Glacier* in central Nepal, Watanabe (1976) classified the glaciers of the Nepal Himalaya into two types, based on climatic characteristics of glacierization and surface morphological features: (1) Nepal-type glaciers and, (2) Tibet-type glaciers. The former type has relatively wet-maritime characteristics and the latter type has relatively dry-continental characteristics.

In 1963, Miller and others (1965), as scientific members of the American Everest Expedition, did glaciological studies on the Khumbu Glacier of eastern Nepal. They studied the accumulation processes in the Khumbu Glacier by tritium dating ice layers above the equilibrium line in the *West Cwm*, at the head of the Khumbu Glacier. An inventory of glaciers in the Mt. Everest (Sagarmatha) region was first attempted by Müller (1970) during the International Hydrological Decade (IHD) as a pilot study for the Himalayan glacier inventory.

The concept of using a representative glacier as an index of the regional characteristics of the Nepal Himalayan glaciers was proposed by Watanabe (1976). He selected 80 glaciers from the entire region encompassed by the Nepal Himalaya, and described their topographical features from the perspective of the regional characteristics of glaciers. His study is useful for monitoring long-term change in glaciers.

The first systematic investigation of the Nepal Himalayan glaciers was organized by Nagoya and Kyoto Universities, Japan. The Glaciological Expedition of Nepal (GEN), led by Higuchi (1976, 1977, 1978, 1980), carried out a series of field studies in the Khumbu, *Shorong, Hink*, and *Hunk* regions in eastern Nepal, and the Mukut Himal region in central Nepal from 1973 to 1978. Year-round meteorological observations were made at *Lhajung* Station (4,420 m), Khumbu region, East Nepal, which operated from 1973 to 1976. Also, precise inventory work was performed on the glaciers in the Dudh Kosi region, East Nepal, as part of this research program (Higuchi and others, 1978, 1980). Detailed results of this work are described in this report.

Although the Karakoram glaciers are much larger than those of the Nepal Himalaya, fewer glaciological studies have been performed there. For a discussion of some of the earlier work on the Karakoram glaciers, see the section in this chapter on the Glaciers of Pakistan, as well as the work of Visser (1928), Untersteiner (1955), and von Wissmann (1959).

---

[3] U.S. Government publications require that official geographic place-names for foreign countries be used to the greatest extent possible. In the Glaciers of Nepal section, the use of geographic place-names is based on the U.S. Board on Geographic Names (BGN) Website *http://earth-info.nga.mil/gns/html/index.html*. Names not listed on the BGN Website are shown in italics.

Procedures for Mapping Glacier Distribution from Landsat Images

Landsat images are available for most of the Nepal Himalaya and Karakoram Range. Images acquired between 17 September 1972 and 6 December 1976, were used for the analysis of the Nepal Himalaya glaciers (fig. 4); images acquired between 19 September 1973 and 27 September 1977, were used for the analysis of the Karakoram Range glaciers (fig. 5). The optimum Landsat 1, 2, and 3 images for glacier studies in the 1970s time frame are shown in table 1. An accompanying index map is shown in figure 6.

Landsat image analysis is the most effective remote sensing technique available for analyzing the seasonal and secular variations of cryospheric phenomena. The areal distribution of glaciers and seasonal snow cover are easily discernible on the images. However, it is difficult to map the glaciers using Landsat MSS images because of the low spatial resolution.

Other problems include how to ascertain the extent of individual glaciers and how to distinguish the firn limit from the temporal snow cover in the upper part of the glacier. In the ablation area, drawing a clear boundary between the area of the supraglacial debris cover and that of the moraine field on glaciated ground is also a serious problem in analyzing the images. These difficulties in interpretation directly affect the accuracy of maps made from the images.

To solve these problems, comparison studies were conducted using objective criteria from ground observations to validate the interpretations from the images. The following procedures were used in making the maps.

1. An area was selected where ground survey and aerial photography data were available and much data had been collected on the glaciers.

2. Ground observations were plotted and image data related to the glacial features on a geographic base map of these areas.

3. Glacier features were classified, and compared to information from Landsat images and ground observations.

4. Criteria were established for determining the areal extent of a glacier and its morphological surface characteristics.

After some trial and error, the following criteria were established.

1. All snow and/or ice surface areas above the firn limit would be included in the "cryospheric zone." However, there were severe difficulties in deciding whether these features were glacierized or not. The limit of perennial snow/ice cover becomes clear just before the summer monsoon season, April or May in the Himalaya, so a satellite image taken during that time period should be used, if available.

2. According to results obtained from preliminary studies in the area, it was not difficult to distinguish between the areas of supraglacial debris cover and the areas of moraine which had no subsurface ice body (at the image scale). This was because, after comparing Landsat images with ground observations, the supraglacial debris-covered area did not make much difference in the determination of the glacier extent.

These criteria can only be applied to glaciers larger than a certain limit—approximately several kilometers in length and several hundred meters in width. Therefore, glaciers smaller than this limit were not mapped as individual glaciers but were included in the total cryospheric area. It should be noted that debris-covered areas with stagnant and/or fossil ice, being remnants of past expansion, were not mapped as active glacier areas.

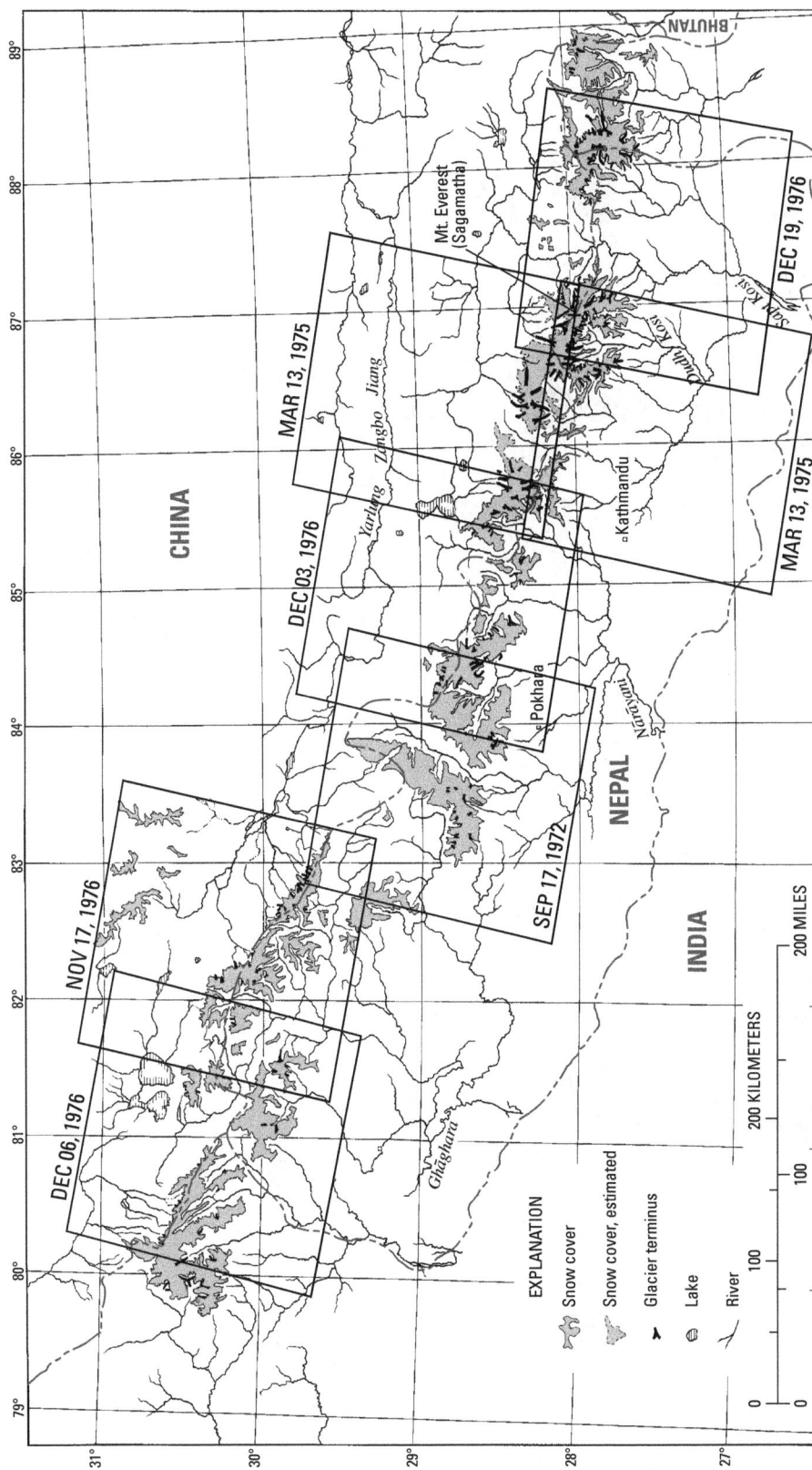

Figure 4.— Index map of the glaciers of Nepal showing coverage by Landsat 1, 2, and 3 MSS images.

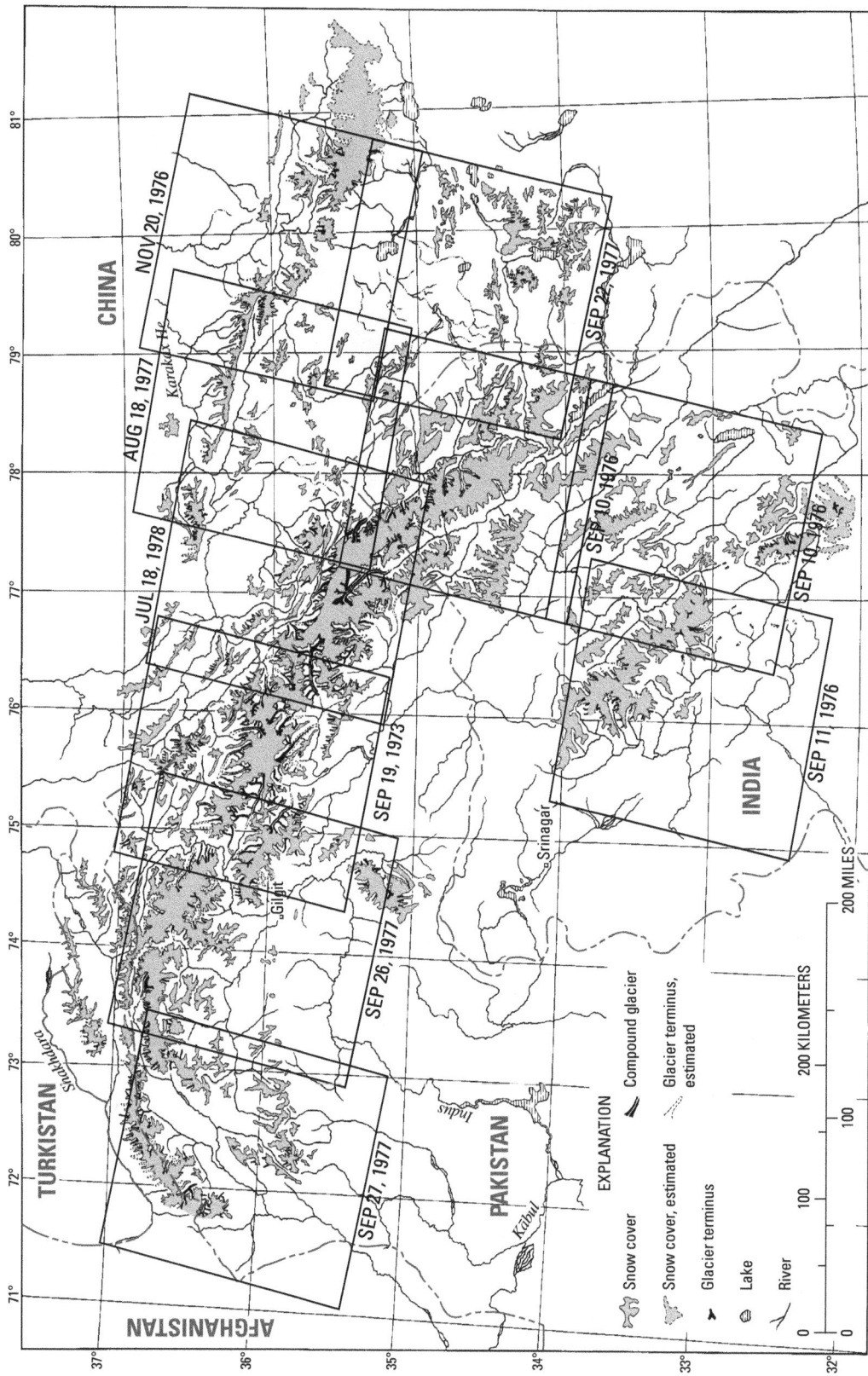

Figure 5.— *Index map of the glaciers of the Karkoram Range showing coverage by Landsat 1, 2, and 3 MSS images.*

TABLE 1.—*Optimum Landsat 1, 2, and 3 MSS images of the glaciers of Nepal*

Path-Row	Nominal scene center latitude and longitude		Landsat identification number	Solar elevation angle (degrees)	Date	Code	Cloud cover (percent)	Remarks
150–41	27°23'N.	87°31'E.	2150041007635490	28	19 Dec 76	●	5	Dudh-Kosi, Mt. Everest region
151–40	28°49'N.	86°29'E.	2151040007507290	43	13 Mar 75	◐	10	
151–40	28°49'N.	86°29'E.	2151040007635590	27	20 Dec 76	●	5	
151–41	27°23'N.	86°05'E.	2151041007507290	44	13 Mar 75	●	5	
151–41	27°23'N.	86°05'E.	2151041007635590	28	20 Dec 76	◑	10	
152–40	28°49'N.	85°03'E.	215204007633890	30	03 Dec 76	●	5	
153–40	28°49'N.	83°37'E.	1153040007226190	~45	17 Sep 72	◑	20	Line drops
153–40	28°49'N.	83°37'E.	2153040007700990	27	09 Jan 77	◑	10	
154–39	30°15'N.	82°35'E.	2154039007632290	32	17 Nov 76	◐	10	
154–40	28°49'N.	82°11'E.	2154040007632290	33	17 Nov 76	◑	30	
155–39	30°15'N.	81°09'E.	2155039007634190	28	06 Dec 76	●	5	

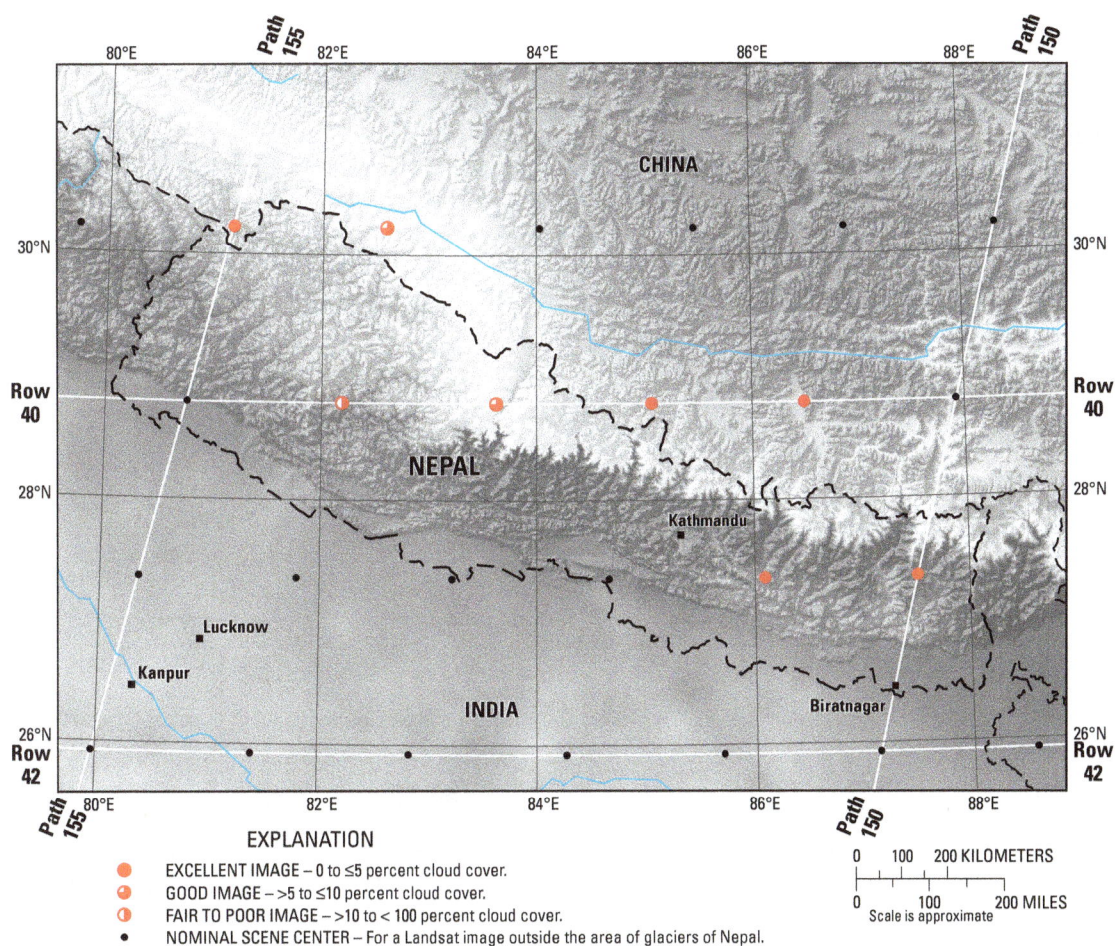

Figure 6.—*Index map of optimum Landsat 1, 2, and 3 MSS images of the glaciers of Nepal.*

Glaciers in the Nepal Himalaya and the Karakoram Range

Based on the Landsat image analysis, a glacier inventory was developed for the Nepal Himalaya and the Karakoram, and the glacier distribution and characteristics were compared between the two areas. A ground survey of glaciers in the Dudh Kosi region of Nepal was conducted and results were compared to results from a Landsat image analysis by Müller (1970). A comprehensive comparison of results from Landsat image and ground investigations was conducted on the Khumbu Glacier, in Nepal.

Glacier Inventory Based on a Map Compiled from Landsat Images

Tables 2 and 3 list the glaciers which were identified using Landsat images, based on the established criteria, for the Nepal Himalaya and Karakoram Range, respectively. The locations of these glaciers are shown on figures 7 and 8. Italicized glacier names in the tables are tentative designations assigned for practical convenience, because official geographic place-names have not yet been established for many of the glaciers in these two regions. The elevation of the terminus and the length and width of each glacier in horizontal projection are estimated from various published papers and maps, and from direct measurements on the Landsat images.

The "Description" column in tables 2 and 3 includes specific surface features and other characteristics of the glacier that were identified directly from the images. Surface conditions of a glacier, such as the distribution of supraglacial debris and the occurrence of medial moraines, glacier lakes, and other glacial features were easily recognized using Landsat images. This information contributed effectively to our knowledge of the present conditions of individual glaciers. Additionally, the monitoring of glacier lake conditions and surge-type glaciers by means of satellite images appears to be an effective means of obtaining advanced warning of potential glaciological catastrophes.

TABLE 2.—*Glaciers in the Nepal Himalaya and surrounding regions*

[Description: D, debris-covered glacier; S, lateral moraine; C, clean glacier; M, medial moraine; Dp, partly debris-covered glacier; P, pond; L, lake around glacier; ?, unknown whether it is covered by snow or debris. References: 1, Higuchi and others (1978, 1980); 2, Schneider (1963); 3, Yoshizawa (1977). Units: m, meter; km, kilometer]

Glacier number	Glacier name	Location latitude and longitude		Lowest glacier elevation (m), when known	Length/width (km), when known	Description	Reference
ALAKNANDA BASIN							
001	*Bagani Bank*	30°45'N.	80°00'E.			D, S	3
002	*Dunagiri Bank*	30°45'N.	80°00'E.			D	3
003	*Uttari Rishi*	30°30'N.	80°00'E.			C, S	3
004	*Uttari Nanda Devi*	30°30'N.	80°00'E.			D, S	3
005	*Dakkhni Nanda Devi*	30°30'N.	80°00'E.			D	3
006	*South Rishi*	30°30'N.	80°00'E.			D, S	3
007	*Trisal*	30°30'N.	79°45'E.			D, S	3
KALI BASIN							
101	*Shalang*	30°15'N.	80°15'E.			D, M, S	3
102	Milam	30°30'N.	80°15'E.			D, M, S	3
103	*Kalabaland*	30°30'N.	80°30'E.			C, S	3
104	*Ngalaphu*	30°15'N.	80°30'E.			D, M, S	3

TABLE 2.—*Glaciers in the Nepal Himalaya and surrounding regions*—Continued

[Description: D, debris-covered glacier; S, lateral moraine; C, clean glacier; M, medial moraine; Dp, partly debris-covered glacier; P, pond; L, lake around glacier; ?, unknown whether it is covered by snow or debris. References: 1, Higuchi and others (1978, 1980); 2, Schneider (1963); 3, Yoshizawa (1977). Units: m, meter; km, kilometer]

Glacier number	Glacier name	Location latitude and longitude	Lowest glacier elevation (m), when known	Length/width (km), when known	Description	Reference
105	*Rama*	30°15′N. 80°45′E.			D	3
106	,	30°15′N. 80°45′E.			Dp	
KARNALI BASIN						
201		30°00′N. 81°00′E.	4,298	6.8/0.7	D, M	3
202		30°00′N. 81°00′E.	4,480	4.9/0.7	D	3
203	*Saipal*	30°00′N. 81°30′E.	3,721	5.0/0.4	Dp, S	3
204	*Ghat*	30°00′N. 81°30′E.	4,419	1.6/	D, S	3
205		30°00′N. 81°30′E.	3,505	7.0/1.4	D, S	3
206	*N. Saipal*	30°00′N. 81°30′E.	4,282	7.6/0.4	Dp, M	3
207		30°30′N. 80°45′E.			C, S	
208		30°30′N. 81°15′E.			C	
209		30°30′N. 81°15′E.			D, M, S	
210		30°15′N. 81°15′E.			Dp	
211		30°15′N. 81°45′E.			C, S	
212		30°15′N. 81°45′E.			C	
213		30°15′N. 82°15′E.			D, S	
214		30°00′N. 82°15′E.			Dp, S	
215		30°00′N. 82°15′E.	4,267	5.1/0.6	D, M, P	3
216		30°00′N. 82°15′E.			Dp, S	
217		29°45′N. 82°45′E.			Dp, S	
218		29°45′N. 82°45′E.	5,029		Dp, M, S	3
219		28°45′N. 83°15′E.	4,115	4.6/0.4	?	3
220		28°45′N. 83°15′E.	4,892	4.3/	?	3
221		28°45′N. 83°15′E.	4,304	5.4/0.4	?	3
222		28°45′N. 83°15′E.	4,877	10.0/0.6	D, S	3
223		28°45′N. 83°15′E.	5,264	4.7/0.3	?	3
224		28°45′N. 83°15′E.	4,724	3.7/0.7	D, S	3
225		28°45′N. 83°15′E.	4,816	5.2/0.3	?	3
226	*Kaphe*	28°45′N. 83°15′E.	4,206	8.3/0.5	?	3
KALI GANDAKI BASIN						
301	*Konabon*	28°45′N. 83°30′E.	4,389	8.5/	?	3
302	*Chhonbarban*	28°45′N. 83°30′E.	3,770	14.6/0.4	?, S	3
303		28°45′N. 83°30′E.	4,282	4.8	?	
304		28°45′N. 83°30′E.	5,014	3.8/0.4	?	3
305		28°45′N. 83°30′E.	5,608	3.2/	?	3
306		28°45′N. 83°30′E.	5,303	4.8/0.6	?	3
307	*S. Annapurna*	28°30′N. 84°00′E.	3,566	13.7/0.5	D, S	3
308	*W. Annapurna*	28°45′N. 84°00′E.	3,789	7.8/0.3	D	3
309	*E. Annapurna*	28°30′N. 84°00′E.		6.9/0.3	D	3
310		28°45′N. 84°15′E.			D	
311		28°45′N. 84°15′E.			D, S, P	
312		28°45′N. 84°30′E.			D, S	
313		28°45′N. 84°30′E.			D, M, S, P	
314		28°45′N. 84°30′E.			D, M, S, P	
315		28°45′N. 84°30′E.			D, M, S, P	
316		28°30′N. 84°30′E.			D, P	
317	*Tulangi*	28°30′N. 84°30′E.	4,023	12.4/0.4	D, P	3
318	*Chuling*	28°30′N. 84°45′E.	3,017	14.0/0.7	D, M, S	3
319	*Lidanda*	28°30′N. 84°30′E.	3,139	14.0/0.9	D, S	3
320	*Pungeu*	28°30′N. 84°30′E.	3,840	10.2/0.7	D, S	3
321	*Manaslu*	28°30′N. 84°30′E.	3,597	7.5/0.4	Dp	3

TABLE 2.—*Glaciers in the Nepal Himalaya and surrounding regions*—Continued

[Description: D, debris-covered glacier; S, lateral moraine; C, clean glacier; M, medial moraine; Dp, partly debris-covered glacier; P, pond; L, lake around glacier; ?, unknown whether it is covered by snow or debris. References: 1, Higuchi and others (1978, 1980); 2, Schneider (1963); 3, Yoshizawa (1977). Units: m, meter; km, kilometer]

Glacier number	Glacier name	Location latitude and longitude		Lowest glacier elevation (m), when known	Length/width (km), when known	Description	Reference
322		28°45′N.	84°30′E.	4,602	6.1/0.5	D	3
323	*Jarkya*	28°45′N.	84°30′E.	4,419	12.4/0.7	Dp, S	3
324		28°45′N.	84°30′E.			?	
325	*Torogompa*	28°30′N.	85°00′E.			D, S	3
326		28°30′N.	85°00′E.			D, S	
327	*Sangjing*	28°30′N.	85°15′E.			D, S	3
328		28°45′N.	85°30′E.			D, S	
329		28°45′N.	85°30′E.			D, S	
330		28°45′N.	85°30′E.			?	
331		28°45′N.	85°30′E.			D, S	
332	*Chinkyong*	28°45′N.	85°30′E.			D, S	3
333	*Lalaga*	28°30′N.	85°30′E.			D, S	3
334	*Phurephu*	28°30′N.	85°30′E.			D, S	3
335	*Chasmudo*	28°15′N.	85°30′E.			D, S	3
336	*Lirang*	28°15′N.	85°30′E.	3,962		D, S	3
337	*Shalbachum*	28°15′N.	85°30′E.	4,114		D	3
338		28°15′N.	85°45′E.	4,663		D, S	3
339	*Langtang*	28°15′N.	85°45′E.	4,550		D, S	3

SAPT KOSI BASIN

401		28°15′N.	85°45′E.	4,267	4.6/0.6	D	3
402		28°15′N.	85°45′E.	4,115	5.5/0.6	D, S	3
403	*Phurbi Cha Chu*	28°15′N.	85°45′E.	3,840	7.8/0.4	D, S	3
404		28°15′N.	85°45′E.	3,846	4.8/0.4	D, S	3
405	*Nyanamphu*	28°30′N.	85°45′E.			?	
406		28°30′N.	85°45′E.			D, S	
407		28°30′N.	85°45′E.			?	
408		28°30′N.	86°15′E.			?	
409		28°15′N.	86°15′E.			?	
410		28°15′N.	86°15′E.			?	
411		28°15′N.	86°30′E.			?	
412		28°00′N.	86°30′E.			D, S, P	
413		28°00′N.	86°30′E.			D	
414		28°00′N.	86°30′E.			D, P	
415		28°00′N.	86°30′E.			D, M, S, P	
416	*Drogpa Nagtsang*	28°00′N.	86°30′E.		10.5/0.8	D, M, S, P	2
417	*Ripimu*	28°00′N.	86°30′E.		10.5/0.7	D, S	2
418	*Tolambau (Drolambao)*	28°00′N.	86°30′E.		14.5/0.8	D, P	2, 3
419	*(Likhu Nup)*	27°45′N.	86°30′E.		4.2/0.5	D	2
420	*(Zurmoche)*	27°45′N.	86°30′E.		7.3/0.5	D, S	2
421	*Dudh Kund*	27°45′N.	86°30′E.	4,400	6.6/0.5	D, S	1, 3
422	*Lumding*	27°45′N.	86°30′E.	4,800	5.3/0.5	D, M, P	1, 3
423	*Thengpo*	28°00′N.	86°30′E.	4,300	2.2/0.3	D	1, 3
424	*Chhule*	28°00′N.	86°30′E.	4,680	4.7/0.4	D, S	1, 3
425	*Dingjung*	28°00′N.	86°30′E.	4,600	7.8/0.7	D, S, P	1, 3
426	*Pangbug*	28°00′N.	86°30′E.	4,880	8.6/0.7	D, S, P	1, 3
427	*Lunak*	28°00′N.	86°30′E.	4,600	18.2/0.8	D, S	1, 3
428	*Nangpa*	28°00′N.	86°30′E.	4,600	18.2/0.8	D, S	1, 3
429	*Sumna*	28°00′N.	86°30′E.	4,880	9.4/0.7	D, S, P	1, 3
430	*Ngojumba*	28°00′N.	86°45′E.	4,680	22.4/1.2	D, S, P	1, 3
431	*Gyubanare*	28°00′N.	86°45′E.	4,680	22.4/1.2	D, S, P	1, 3
432	*Changri (Nup)*	28°00′N.	86°45′E.	5,180	7.2/0.6	D	1, 2, 3

TABLE 2.—*Glaciers in the Nepal Himalaya and surrounding regions*—Continued

[Description: D, debris-covered glacier; S, lateral moraine; C, clean glacier; M, medial moraine; Dp, partly debris-covered glacier; P, pond; L, lake around glacier; ?, unknown whether it is covered by snow or debris. References: 1, Higuchi and others (1978, 1980); 2, Schneider (1963); 3, Yoshizawa (1977). Units: m, meter; km, kilometer]

Glacier number	Glacier name	Location latitude and longitude		Lowest glacier elevation (m), when known	Length/width (km), when known	Description	Reference
433	East Changri (Changri Shar)	28°00′N.	86°45′E.	5,180	5.2/0.7	D	1, 2, 3
434	Khumbu	28°00′N.	87°00′E.	4,920	16.5/0.7	D, M, S, P	1, 3
435	West Cwm	28°00′N.	87°00′E.	4,920	16.5/0.7	D, M, S, P	1, 3
436	Nuptse	28°00′N.	87°00′E.	4,960	6.3/0.4	D, S	1, 3
437	West Lhotse (Lotse Nup)	28°00′N.	87°00′E.	4,980	4.2/0.6	D, M, S	1, 2, 3
438	Lhotse	28°00′N.	87°00′E.	4,960	6.4/1.1	D, S	1, 3
439	Imja	28°00′N.	87°00′E.	5,030	9.8/0.7	D, M, S, P	1, 3
440	Ama Dablam	28°00′N.	87°00′E.	4,850	4.8/0.5	D, S	1, 3
441	Hink (Shar Nup)	27°45′N.	87°00′E.		7.5/0.4	D, S	2, 3
442	(Hink Teng)	27°45′N.	87°15′E.		5.0/1.0	D, P	2
443	Hongu	27°45′N.	87°00′E.	5,580	1.0/	D, P	1, 3
444	Iswa	27°45′N.	87°00′E.			D, S	3
445	Lower Barun	27°45′N.	87°00′E.		14.5/0.9	D, S, L	3
446	Barun	27°45′N.	87°00′E.		15.5/0.5	D, L	2, 3
447		27°45′N.	87°15′E.			D, S	
448	Sakyetong	28°00′N.	87°15′E.			D, S	
449	Kangdoshung	28°00′N.	87°00′E.			D, S	
450	Chomo Lonzo	28°00′N.	87°00′E.		6.0/0.5	D	2, 3
451	Kangshung	28°00′N.	87°00′E.		16.5/1.3	D, M, S, P	2, 3
452		28°00′N.	87°15′E.			D	
453		28°00′N.	87°00′E.			?	
454		28°00′N.	87°00′E.			?	
455		28°00′N.	87°00′E.			?	
456	Kharta	28°00′N.	87°00′E.			?	3
457	East Rongbuk (Ronphu Shar)	28°00′N.	86°45′E.			?	2, 3
458	Rongbuk (Ronphu)	28°00′N.	86°45′E.		15/1.1	?	2, 3
459	Rongta	28°15′N.	86°45′E.			?	3
	Gyachung	28°15′N.	86°45′E.			?	3
460		28°15′N.	86°30′E.			?	
461	Kyetrak (Gyabrag)	28°15′N.	86°30′E.			?	2, 3
462		28°30′N.	86°30′E.			?	
463		28°30′N.	86°15′E.			?	
464		28°30′N.	86°15′E.			?	
465		28°30′N.	85°45′E.			?	
466		28°30′N.	85°45′E.			D	
467	Lashar	28°00′N.	88°00′E.			D, M, S	3
468	Lashar	28°00′N.	88°00′E.			D, M, S	3
469		28°00′N.	87°45′E.			?, S, P	
470		28°00′N.	87°45′E.			?	
471		28°00′N.	87°45′E.			?	
472	Hidden	28°00′N.	87°45′E.			?, S	3
473	Phuchong	27°45′N.	88°00′E.			?, S	3
474	Pamdro	27°45′N.	88°00′E.			?, S	3
475	Tsisima	27°45′N.	88°00′E.			?, S	3
476	Broken	27°45′N.	88°00′E.			?	3
477	Kangchenjanga	27°45′N.	88°00′E.			?, S	3
478	Ramtang	27°45′N.	88°00′E.			?, S	3
479	Jannu	27°45′N.	88°00′E.			?	3
480	Yomatari	27°30′N.	88°00′E.			Dp, S	3
481	Yalung	27°30′N.	88°00′E.			D, S	3
482	Yalung	27°30′N.	88°00′E.			D, S	3

[Description: D, debris-covered glacier; S, lateral moraine; C, clean glacier; M, medial moraine; Dp, partly debris-covered glacier; P, pond; L, lake around glacier; ?, unknown whether it is covered by snow or debris. References: 1, Higuchi and others (1978, 1980); 2, Schneider (1963); 3, Yoshizawa (1977). Units: m, meter; km, kilometer]

Glacier number	Glacier name	Location latitude and longitude		Lowest glacier elevation (m), when known	Length/width (km), when known	Description	Reference
TISTA BASIN							
501		27°30′N.	88°15′E.			D	
502		27°30′N.	88°15′E.			D, M	
503		27°45′N.	88°15′E.			?, S	
504		27°45′N.	88°15′E.			D, S, P	
505		27°45′N.	88°15′E.			D, M, S	
506		27°45′N.	88°15′E.			D, S, P	
507		27°45′N.	88°15′E.			D, S, P	
508	*Jongsan*	28°00′N.	88°00′E.			D, S, P	3
509	*Lhonak*	28°00′N.	88°00′E.			D, M, S	3
510		28°00′N.	88°45′E.			D	
511		28°00′N.	88°45′E.			D	
512		28°00′N.	88°45′E.			D	
513		28°00′N.	88°45′E.			?	
514		27°45′N.	89°00′E.			?	
515		27°45′N.	89°00′E.			D	
YARLUNG ZANGBO JIANG							
601		28°30′N.	85°45′E.			D, M, S	
602		29°00′N.	84°30′E.			?	
603		29°00′N.	84°30′E.			?	
604		29°00′N.	84°30′E.			?	
605		29°00′N.	84°30′E.			?	
606		29°00′N.	84°15′E.			?	
607		29°00′N.	84°15′E.			?	
608		29°45′N.	83°00′E.			C, P	
609		29°45′N.	83°00′E.			C, S	
610		29°45′N.	83°00′E.			C, S	
611		29°45′N.	82°45′E.			C	
612		29°45′N.	82°45′E.			C	
613		29°45′N.	82°45′E.			C, S	
614		29°45′N.	82°45′E.			C	
615		29°45′N.	82°45′E.			C, P	
616		30°00′N.	82°45′E.			?	
617		30°00′N.	82°45′E.			C, S	
618		30°00′N.	82°45′E.			C, P	
619		30°00′N.	82°45′E.			C, S	
620		30°00′N.	82°45′E.			C, L	
621		30°00′N.	82°45′E.			C, P	
622		30°15′N.	82°15′E.			C, S, P	
623		30°15′N.	82°15′E.			C, S	
624		30°15′N.	82°15′E.			C	
625		30°15′N.	82°15′E.			C, S, L	
SUTLEJ BASIN							
701		30°30′N.	81°15′E.			C, S	
702		30°30′N.	80°30′E.			Dp	
703		30°30′N.	80°30′E.			Dp, S	
704		30°30′N.	80°15′E.			Dp, S	
705		30°30′N.	80°15′E.			Dp, S	
706		30°30′N.	80°15′E.			?	

TABLE 3.—*Glaciers in the Karakoram Range*

[Description: Dp, partly debris-covered glacier; M, medial moraine; C, clean glacier; D, debris-covered glacier; S, lateral moraine; L, lake around glacier; ?, unknown whether it is covered by snow or debris. Unit: km, kilometer]

Glacier number	Glacier name	Location latitude and longitude		Length (km)	Number of constituent glaciers	Description
1	Upper Tirich	36°26'N.	71°51'E.	17	17	Dp, M
2	Atark	36°30'N.	72°00'E.	16	3	C
3	Niroghi	36°30'N.	72°10'E.	12	2	D
4	Chiantar	36°50'N.	73°40'E.	30	4	Dp, S, M
5	Yashkuk Yaz	36°45'N.	74°20'E.	16	3	Dp, M
6	Batura	36°30'N.	74°40'E.	51	4	Dp, S, M
7	*Pasu*	36°30'N.	74°45'E.	21	1	Dp, S
8	*Ghulkin*	36°25'N.	74°50'E.	16	1	D
9	*Gulmit*	36°25'N.	74°50'E.	8	1	Dp, M
10	*Lupghur Yaz*	36°25'N.	75°00'E.	13	1	D
11	Momhil	36°25'N.	75°05'E.	20	1	Dp, S, M
12	Mulungutti	36°25'N.	75°15'E.	16	1	Dp, S, M
13	Yazghil	36°20'N.	75°25'E.	26	2	C, S
14	Khurdopin	36°15'N.	75°30'E.	35	2	Dp, S, M
15	Virjerab	36°15'N.	75°40'E.	37	2	Dp, M
16	Braldu	36°10'N.	75°50'E.	31	1	D, M
17	*Skamri*	36°05'N.	76°10'E.	35	3	Dp, S, M
18	*Sarpo Laggo*	35°55'N.	76°20'E.	27	3	Dp, S, M
19	K2	36°00'N.	76°30'E.	19	2	D, S, M
20	*Gasherbrum*	35°50'N.	76°40'E.	23	2	C, S, M
21	*Urdok*	35°45'N.	76°45'E.	21	2	D, M
22	*Staghar*	35°45'N.	76°50'E.	16	1	C, S
23	*Singhi*	35°40'N.	77°00'E.	?	2	Dp, S, M, L
24	*Kyagar*	35°35'N.	77°15'E.	17	3	C, S, M, L
25	Rimo	35°25'N.	77°30'E.	38	2	Dp, S, L
26	*S. Rimo*	35°20'N.	77°30'E.	23	2	Dp, L
27	*N. Shukpa Kunchang*	34°50'N.	77°50'E.	?	3	C, S, M
28	*S. Shukpa Kunchang*	34°40'N.	77°50'E.	?	3	D, S, M
29	*S. Terong*	35°10'N.	77°25'E.	24	2	D
30	*Teram Shehr*	35°30'N.	77°10'E.	28	1	Dp, S, M, L
31	Siachen	35°25'N.	77°05'E.	70	4	Dp, S, M
32	Bilafond	35°20'N.	76°55'E.	20	2	Dp, S, M
33	*Sherpi Gang*	35°20'N.	76°45'E.	16	1	Dp, S
34	Kaberi	35°30'N.	76°35'E.	25	2	Dp, S, M
35	*Chogolisa*	35°25'N.	76°30'E.	17	2	D, L
36	*Ghondokoro*	35°35'N.	76°25'E.	17	2	Dp, S, M
37	Godwin Austin	35°50'N.	76°35'E.	23	2	Dp, S, M
38	Baltoro	35°45'N.	76°20'E.	61	11	Dp, S, M, L
39	*Panmah*	35°55'N.	76°00'E.	41	4	Dp, S M
40	Biafo	35°50'N.	75°45'E.	52	3	D, S, M
41	*Kero Lungma*	36°00'N.	75°15'E.	15	3	D
42	Chogo Lungma	35°55'N.	75°10'E.	42	5	Dp, S, M
43	*Khotia*	35°50'N.	75°00'E.	12	3	D, S
44	*Mani*	35°55'N.	74°55'E.	14	2	D
45	Bualtar	36°10'N.	74°45'E.	19	2	D, M
46	Barpu	36°10'N.	74°50'E.	22	2	Dp, S
47	*Kham Basa*	36°10'N.	75°25'E.	13	2	D, S, M
48	*Julmau*	36°10'N.	75°20'E.	15	3	D, S, M
49	*Khingyang*	36°15'N.	75°10'E.	20	3	D, S
50	Hispar	36°05'N.	75°20'E.	47	4	D, S, M, L
51	*Triror (Gharesa)*	36°15'N.	74°55'E.	20	2	Dp, S

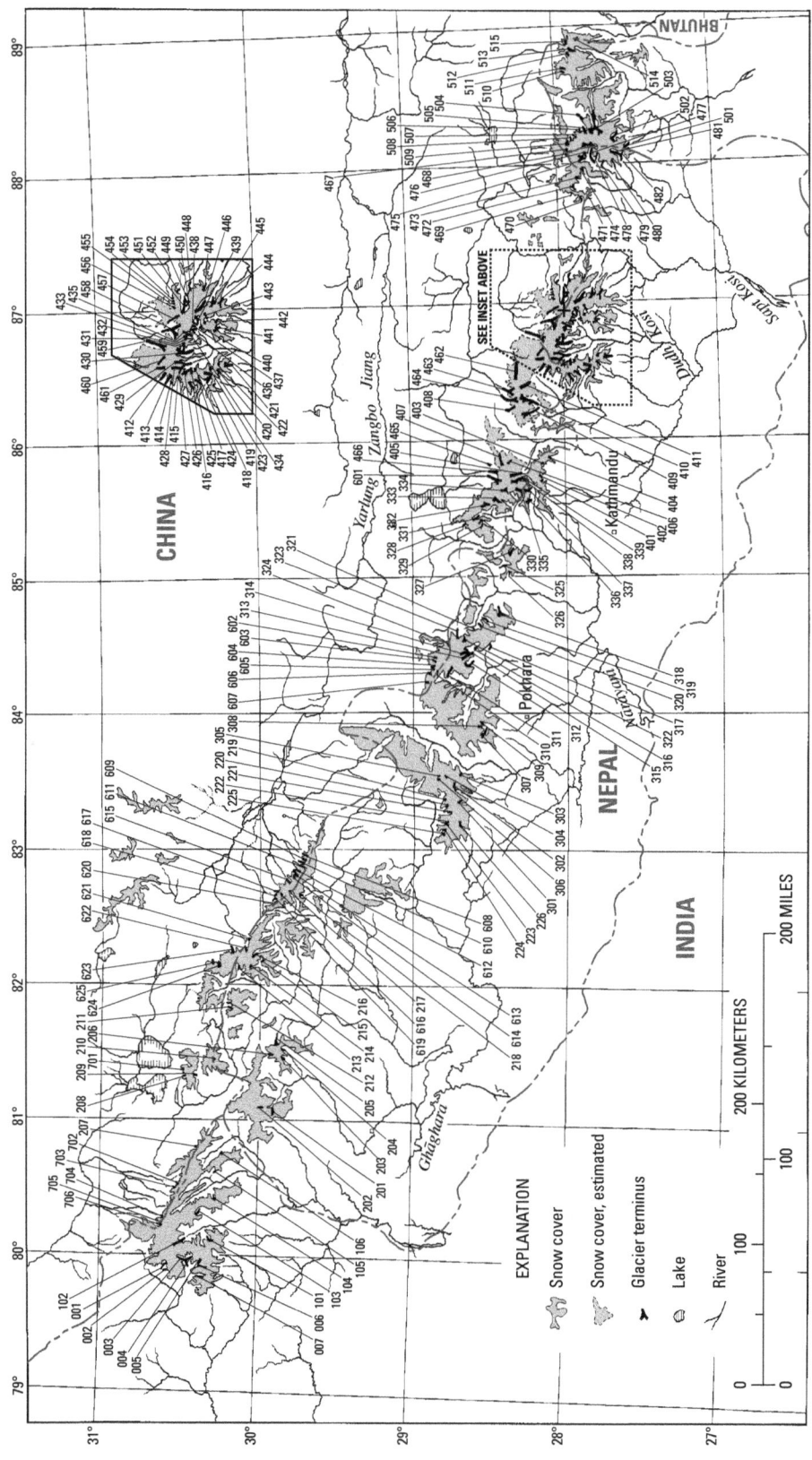

Figure 7.— Location of individual glaciers in the Nepal Himalaya and its surrounding regions, identified by glacier number (see table 2).

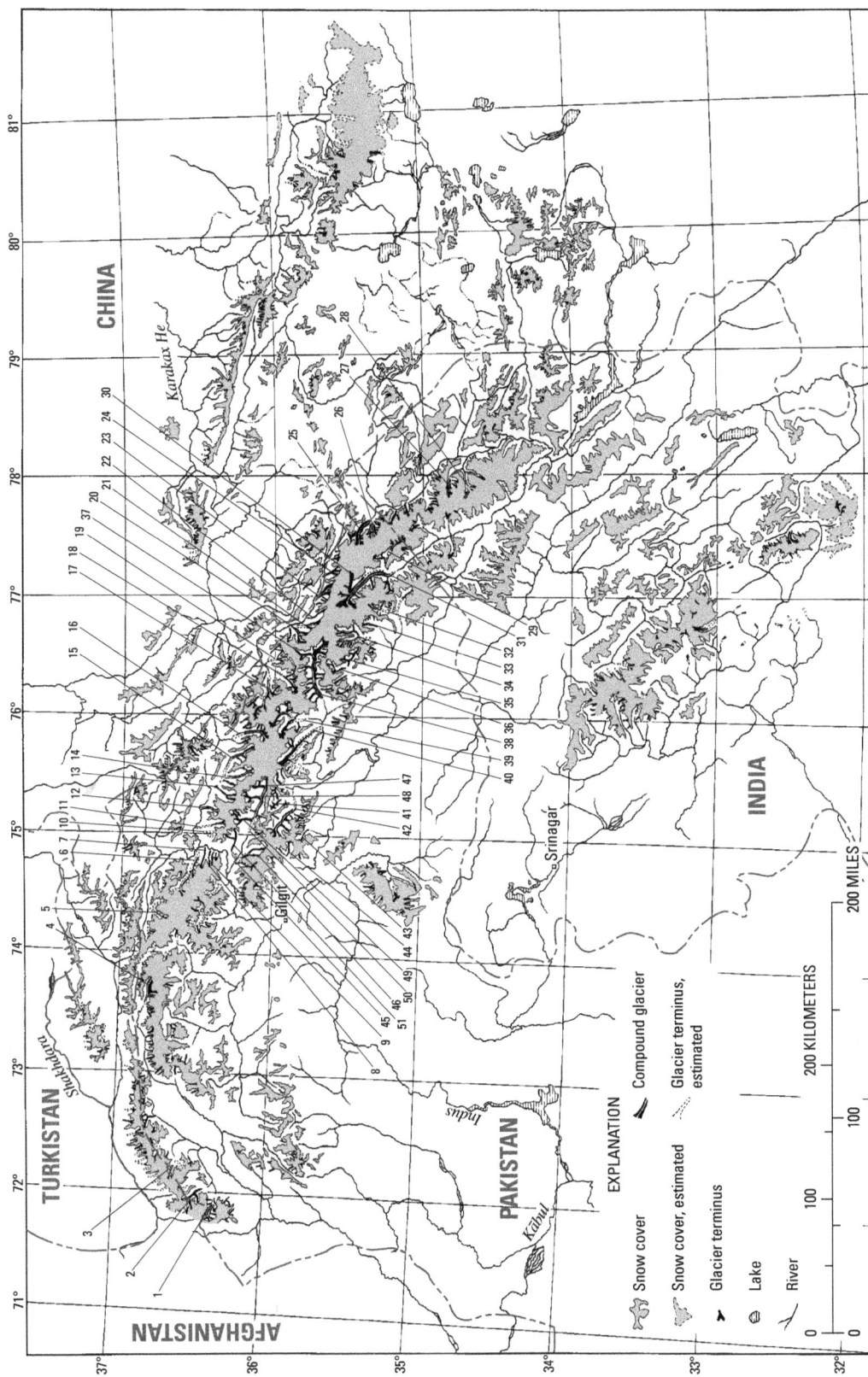

Figure 8.— *Location of individual glaciers in the Karakoram Range identified by glacier number (see table 3).*

Comparison of Nepal Himalaya and Karakoram Range Glacier Distribution

The Nepal Himalaya extend from about 80° to 87°E. in longitude and from about 27.5° to 30.5°N. in latitude (fig. 2). The central region of the Karakoram Range extends from 72° to 79°E. and from 34° to 37°N. (fig. 3). Both of these regions are nearly equivalent in areal extent (7 degrees of longitude and 3 degrees of latitude), and both have a northwest-southeast trending crest with several mountains over 8,000 m in elevation.

The topographical and climatic characteristics of these two regions are very different, however, and because of this, the distribution and features of the glaciers in these two regions are very different. The glacierized area of the Karakoram is much larger than that of the Nepal Himalaya in horizontal scale and is distributed as large, connected, glacier systems, whereas that of the Nepal Himalaya is distributed as individual glaciers.

Figures 2 and 3 show that the obvious contrast in cryospheric area between the Nepal Himalaya and the Karakoram is also reflected in the shapes and sizes of the individual glaciers. The total area of the glacierized region and the average length of glaciers of both mountain ranges, and the mean elevation of glacier termini for the Nepal Himalaya, are shown in table 4. In general, the accumulation basins of the Karakoram glaciers tend to concentrate the ice, and the river system is well controlled by the regional geological structure, so that compound glaciers with long lengths are well developed. In regard to the geomorphology, mountains in the Karakoram Range are massive, and the rivers are wholly parallel to the northwest/southeast crest line. Conversely, the Nepal Himalaya are composed of mountain ranges that are more separated; and although the mountains have a northwest-southeast trend, rivers cut the main range perpendicularly in a north-south direction. Additionally, climatic influence of the summer monsoon is different between regions — the Karakoram Range is less influenced by the summer monsoon than the Nepal Himalaya.

TABLE 4.—*List of glaciological characteristics of the Nepal Himalaya and the Karakoram Range*

[Abbreviations: km, kilometer; m, meter; ?, unknown]

Characteristic	Nepal Himalaya and surrounding regions	Karakoram Range
Total area of the glacierized region including snow cover	24,900 km²	65,300 km²
Average length of glacier	All glaciers: 8.1 km Karnāli Basin: 5.6 km Kāli Gandaki Basin: 9.0 km Sapt Kosi Basin: 9.6 km	25.4 km
Mean elevation of glacier terminus	All glaciers: 4,480 m Karnāli Basin: 4,450 m Kāli Gandaki Basin: 4,190 m Sapt Kosi Basin: 4,710 m	?

Glacier Inventory from Ground Surveys in the Dudh Kosi Region, Nepal, and Comparison with Landsat Image Analyses

Detailed studies on glaciers in the Dudh Kosi region, Khumbu Himal, were carried out from 1973 to 1978 as part of the Glaciological Expedition of Nepal (GEN) by Japanese scientists. The *Lhajung* station, located near *Periche* village where the upper *Imja Khola* joins the tributary originating from the Khumbu Glacier, was operated for three and a half years to collect meteorological data and to make glaciological observations on many glaciers of various sizes.

As a result of these ground studies, a glacier inventory was prepared for the Dudh Kosi region, using data from ground surveys and aerial photographs (Higuchi and others, 1978). In the glacier inventory, glaciers were divided into several categories according to the surface topography: (1) ice body, (2) ice body covered by debris, (3) estimated ice body, and (4) rock glacier. The glacier distributions were plotted on the "Mount Everest Regions" 1:100,000-scale map of the Royal Geographical Society (1975), and glaciers were denoted with numbers according to the above-mentioned categories.

The Dudh Kosi region was divided into several sub-basins identified by the letters AX to HX, and each glacier was assigned a three-digit number. Glacier numbers from Müller's pilot study (Müller, 1970) were shown as well as glacier names that were collected from various other sources. In the Dudh Kosi region, there are 664 glaciers recorded in this ground surveyed glacier inventory by GEN; Müller (1970) identified 166 glaciers in the *Nagpo Tsaugpo* and *Imja Khola* basins of the northernmost Dudh Kosi region.

Figure 9A is a Landsat MSS image of the Dudh Kosi region. The glacier distribution in this region as determined from the GEN ground survey is shown in figure 9B, and the glacier distribution based on the Landsat images is shown in figure 9C. When using the Landsat multispectral scanner (MSS) images, the minimum width and length of glaciers that are discernable is 0.3 km and 1.0 km, respectively; these can be seen without difficulty by simple visual analyses.

The smaller glaciers are considered as perennial ice-and-snow areas when analyzing glacier distribution on the Landsat image. It is possible to discern the smaller glaciers from the images only when meteorological conditions are good, and it is necessary for them to be checked by ground surveys.

Comparison of Results for the Khumbu Glacier, Nepal, from Landsat Image Analysis and Ground Surveys

The Khumbu Glacier (fig. 10) is one of the best studied glaciers in the Nepal Himalaya. The glacier is composed of an upper part, the *Western Cwm*, that has an accumulation area that receives ice from the steep cliff of Sagarmatha (Mount Everest, 8,848 m) and an ice fall ranging from 5,400 to 6,000 m, and an ablation area with rich supraglacial debris in its lower part. The equilibrium line altitude is estimated to be at about 5,600 m and the terminus is at 4,920 m.

Below the ice fall, there are longitudinal rows of ice pinnacles which disappear in the lower part where the supraglacial debris occurs. Based on the flow measurements and surface topography, the present active terminus is thought to be located where the *Changri Glacier* joins from the west (fig. 11A). Below the present active terminus, it was thought there was stagnant ice and fossil ice from past glacier expansion that occurred in the 16th century (Fushimi, 1978). [Editors' note: However, it was later found that most of the glacier was active and flow velocities were measured (Nakawo and others, 1999).]

A

C

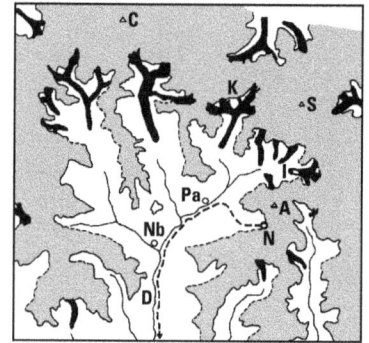

EXPLANATION

▨	Snow cover
---▶	Path of outburst flood
A	*Mt. Ama Dablam*
C	Cho Oyu peak, (8,153 meters above sea level)
D	Dudh Kosi River
I	*Imja Glacier*
K	Khumbu Glacier
Nb	Namche settlement
N	*Nare Glacier Lake*
Pa	Pangboche
S	Sagarmatha peak (Mt. Everest, 8,848 meters above sea level)

B

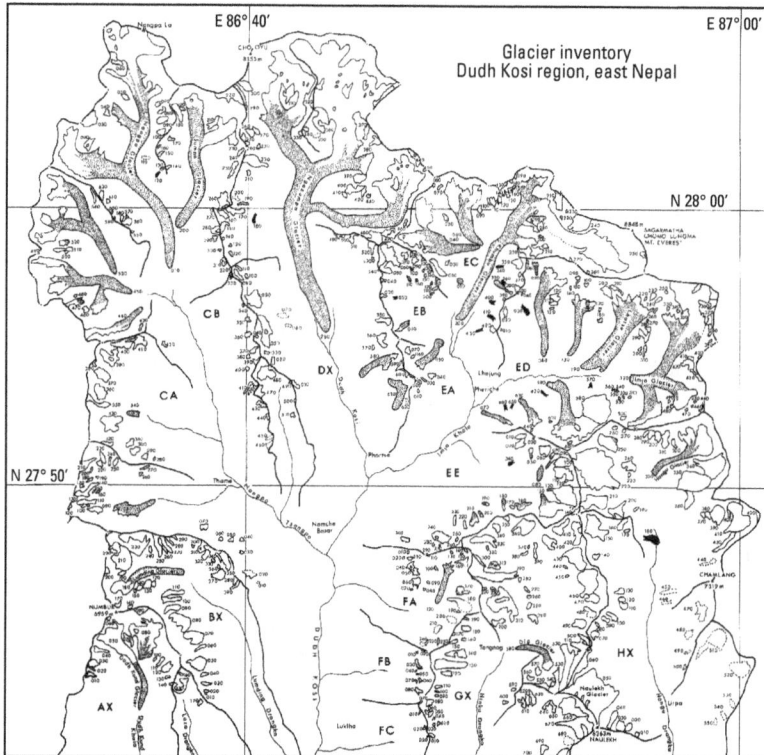

Figure 9.— **A**, *Landsat image;* **B**, *glacier inventory based on ground surveys; and* **C**, *glacier distribution mapped from Landsat image, in the Dudh Kosi region, east Nepal.*

Figure 10.— *Oblique aerial photograph of the Khumbu Glacier, Khumbu region, east Nepal, taken on 11 December 1978.*

Thus, the Khumbu Glacier is broadly divided into a debris-free area and a debris-covered area, with ice pinnacles along its boundary. There are two types of rock in the supraglacial debris: dark-colored metasediments of generally black schists and hornfels, and light-colored plutonic rocks such as granite and diorite. Both types come from weathering of bedrock in the upper basins. The distribution of supraglacial debris gives information about the flow and ablation processes of compound glaciers.

Detailed topographical surveys and ground investigations of the debris-cover in the ablation area of Khumbu Glacier were carried out in 1978 to clarify the origin of the debris-cover, the heat balance characteristics, and the effects of debris cover on the glacier variations (Watanabe and others, 1980; Iwata and others, 1980; Fushimi and others, 1980; Inoue and Yoshida, 1980). As a result, a map of the debris-covered area was completed (fig. 11*B*). The map shows the ablation area of the glacier up to 5.5 km from the terminus, and contains 125-m grid squares which identify the coverage of debris-covered ridges, ice cliffs, lakes, and streams in each square.

A   Distribution of supraglacial debris, bedrock, and other features of the Khumbu Glacier, Khumbu Region, east Nepal

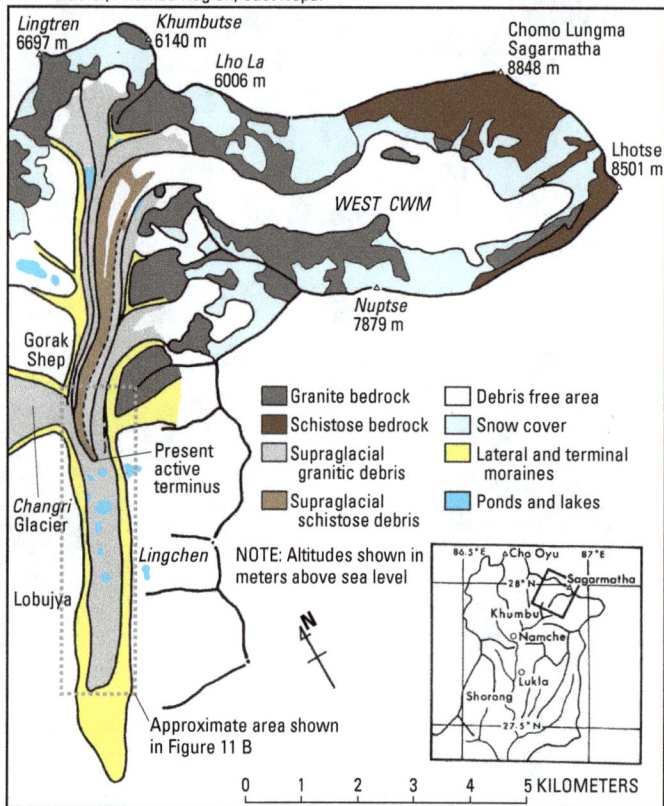

Lingtren
6697 m

Khumbutse
6140 m

Lho La
6006 m

Chomo Lungma
Sagarmatha
8848 m

Lhotse
8501 m

WEST CWM

Nuptse
7879 m

Gorak
Shep

Present
active
terminus

Changri
Glacier

Lingchen

Lobujya

■ Granite bedrock		□ Debris free area	
■ Schistose bedrock		■ Snow cover	
■ Supraglacial granitic debris		■ Lateral and terminal moraines	
■ Supraglacial schistose debris		■ Ponds and lakes	

NOTE: Altitudes shown in meters above sea level

Approximate area shown in Figure 11 B

0   1   2   3   4   5 KILOMETERS

B

■ Lakes
✕ Ridges
Ice cliffs

Lateral moraine ridge

0   1   2 KILOMETERS

Distribution of debris-covered ridges: length, in meters
□ 0
□ 1-100
■ 101-200
■ 201-300
■ 301-400
■ 401-500

Distribution of ice cliffs: coverage, in percent
□ 0
□ 1-10
■ 11-20
■ 21-30
■ 31-40
■ 41-50

Distribution of lakes: coverage, in percent
□ 0
□ 1-20
■ 21-40
■ 41-60
■ 61-80
■ 81-100

Distribution of streams: length, in meters
□ 0
□ 1-80
■ 81-160
■ 161-240
■ 241-320
■ 321-400

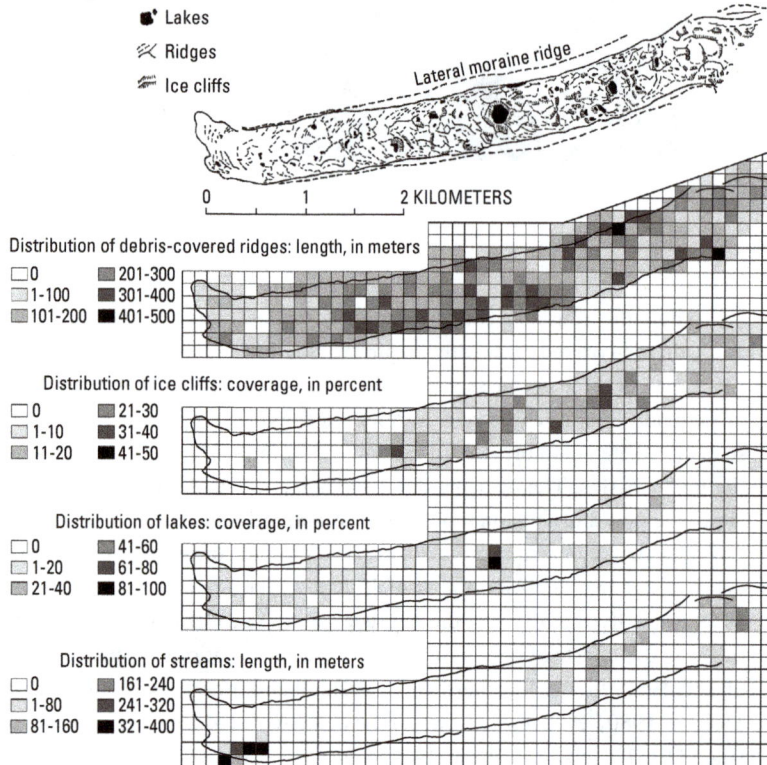

Figure 11.— A, Distribution of supraglacial debris, bedrock, and other features of the Khumbu Glacier, Khumbu region, east Nepal. B, Surface characteristics of a 5.5 km length of the ablation area of the Khumbu Glacier, mapped in 125 m grid squares.

During the same period, Rundquist and Samson (1980) examined the Khumbu Glacier using digital Landsat images. The aim of their study was to evaluate the utility of digital Landsat data for the analysis of alpine glaciers. They assumed the following six land-cover classes and analyzed their distribution from digital data: (1) water body (lakes, ponds, and streams), (2) barren rock, (3) moraine (unconsolidated deposits such as lateral and terminal moraines), (4) supraglacial moraine (light deposits termed "debris covered ice"), (5) clean ice, and (6) firn and snow. They concluded that their method provided a relatively accurate assessment of the surface characteristics of the Nepalese glacier.

Results of these simultaneous studies, the digital analysis of the Landsat image, and ground surveys of the surface features and topography in the ablation area of the Khumbu Glacier, were compared. Though the comparison indicated discrepancies in the interpretation of granitic debris and bare ice, generally good agreement was obtained between the surveyed topography and the map compiled from digital analysis of Landsat images. Fundamental topographic features such as the area of the active glacier, debris-covered area, the moraine distribution, and structures of the compound glacier flow are almost identical.

Landsat Image Analysis for Monitoring Glacier Disasters

Two types of glacier disaster may occur in Nepal— the extraordinary advances of surge-type glaciers and glacier outburst floods (GLOFs) (jökulhlaups) caused by collapse of moraine- or ice-dammed proglacial lakes. These happen simultaneously in some cases and independently in others. In the Khumbu region, it is still a controversial point whether glacier surges happened in the past or not; however, there are oral traditions among local people that advances of glaciers destroyed pastures, farm lands, and mountain paths of the Great Himalayas in the past.

In September 1977, an outburst flood occurred south of *Mt. Ama Dablam* (6,856 m) when the morainic dam of the *Nare Glacier* lake collapsed and caused a flood along the Dudh Kosi river. The flood destroyed river-side houses, camping tents, and killed several people.

This outburst flood was caused by rapid drainage from the ice-cored, moraine-dammed lake. The size of the lake before the flood was 200×300 m, and its depth was 25 m. The flood level reached up to 10 m, and flood sediments were deposited along the Dudh Kosi river. The location of the *Nare Glacier* lake, the course of the flood, and the affected areas are indicated by a dotted line on figure 9C.

It is difficult to predict glacier disasters, but it is possible, as a future application of repeated analysis of sequential Landsat images, to observe the expansion or partial drainage of glacier lakes, and extraordinary glacier advance (surge-type glaciers) by periodic monitoring for many years. Figure 12 shows the supraglacial lake of the *Imja Glacier* (fig. 7) which should be monitored in the Khumbu region.

It is clear from published maps and photographs that the supraglacial lake of Imja Glacier did not show any indication of expansion in the 1960s. However, in the 1970s, the area covered by the glacial lake was estimated from various sources and showed a large expansion in total area.

[Editors' note: Several recent studies have been published about the glacier hazards in Nepal. See, for example, the following supplement by Ageta, publications by Young, 1993; Young and Neupane, 1996; Upreti, 2000; Mool

Figure 12.— *Oblique aerial photograph of the supraglacial lake of the* Imja Glacier, *Khumbu region, east Nepal taken on 30 November 1975. Mt. Makalu (8,481 m) is seen in the background.*

and others, 2001; the Nepal Case Study from the NAPA (National Adaption Program of Action) Workshop, Thimphu, Bhutan, September 2003, and the website of the International Centre for Integrated Mountain Development (ICIMOD), *http://www.icimod.org*]

Conclusions

It is clear that Landsat and other high-resolution satellite images are an invaluable tool for glaciological studies in Nepal. In comparisons with glacial distribution and features identified from ground investigations, Landsat images revealed many of the same features. Also, the distribution and features of the glaciers in the Nepal Himalaya and Karakoram Range could be compared using Landsat images. These images can be used for inventory, monitoring, mapping, and hazard prediction and detection. Perhaps the most important use is the latter — to help mitigate the disasters caused by potential glacial lake outburst floods. The images can be used to determine changes in water supply, to monitor lake size and flood threat potential, to monitor glacier termini positions which may indicate increased glacial melt, and to show other vegetation and land cover changes that may signal climate change.

Supplement to "Glaciers of Nepal"

By Yutaka Ageta[4]

Introduction

The preceding section, "Glaciers of Nepal" by Higuchi and others, was prepared in the early 1980s to describe the status of glaciers during the 1970s as analyzed by Landsat images and localized ground investigations. Here I briefly note supplementary information acquired after that time.

Glaciers in Nepal have been observed mainly by Japanese glaciologists since the 1970s. Their studies were reported by Higuchi (1993), Nakawo and others (1997), and Ageta and others (2001). Studies of Khumbu Glacier and other debris-covered glaciers which related to analyses in the preceding section were published by Seko and others (1998) and Nakawo and others (2000); glacier lakes and their outburst floods in Nepal were reported by Yamada (1998).

Another publication describing the glaciers of the Nepal Himalaya is a review of glacier distribution, characteristics, and glacier inventories in Asia by Tsvetkov and others (1998). Also, recently, Karma and others (2003) analyzed the distribution of the highest and lowest glaciers, and estimated equilibrium line altitudes of glaciers over most of the Himalayan Range, based on several glacier inventories.

Higuchi and others, in table 4 of the preceding section, calculated the total area of glaciers in Nepal to be 24,900 km^2. Their value is much larger than later results because they included the regions surrounding Nepal and the area of temporal snow cover. When referring to their work, it is also important to know that for several glaciers in the glacier list of table 2, one glacier was counted as two glaciers (for example, glacier numbers 430 and 431, 434 and 435). After their analysis, further estimates were made. The total glacierized surface area in Nepal was estimated at 6,000 km^2 from a preliminary glacier inventory based on Landsat image maps (Haeberli and others, 1989). [Editors' note: A map compiled by Qin Dahe and others (1999) reported a total of 3,466 glaciers in Nepal covering an area of 7,929 km^2 based on an analysis of 1975–78 Landsat MSS images.] An updated glacier inventory of Nepal was compiled by Mool and others (2001), utilizing Landsat images and other sources. They calculated the area of glaciers in Nepal to be 5,324 km^2.

[Editors' note: In addition to the excellent glaciological studies made by Japanese scientists in Nepal, other scientists from Nepal, the U.S., the U.K., Switzerland, Russia, and Germany, among others, have also made valuable contributions to all aspects of the glaciological research in Nepal in the last few decades.]

---

[4]Graduate School of Environmental Studies, Nagoya University, Nagoya 464-8601, Japan (Retired)

Vertical aerial photographs were taken in 1992 and 1996 by the Survey Department of the Nepalese Government in cooperation with the Government of Finland. In the last several years, a topographical map series at 1:50,000 scale has been produced based on the photographs. Compilation of a more detailed and accurate glacier inventory of the whole country of Nepal using these aerial photographs and maps is now possible. Some work has already been done using these materials to analyze glacier variations by comparison with other earlier maps, photographs and satellite images and more is expected (Asahi, 2001).

References Cited

Ageta, Yutaka, Naito, N., Nakawo, M., Fujita, K., Shankar, K., Pokhrel, A.P., and Wangda, D., 2001, Study project on the recent rapid shrinkage of summer-accumulation type glaciers in the Himalayas, 1997–1999: Bulletin of Glaciological Research, v. 18, p. 45–49.

Asahi, K., 2001, Inventory and recent variations of glaciers in the eastern Nepal Himalayas: Seppyo, v. 63, p. 159–169. (In Japanese)

Fushimi, Hiroji, 1978, Glaciations in the Khumbu Himal (2): Seppyo, v. 40, special iss., p. 71–77.

Fushimi, Hiroji, Yoshida, M., Watanabe, O., and Upadhyay, B.P., 1980, Distributions and grain sizes of supraglacial debris on the Khumbu Glacier, Khumbu Region, East Nepal: Seppyo, v. 41, special iss., p. 18–25.

Haeberli, Wilfried, Bösch, H., Scherler, K., Østrem, G., and Wallén, C.C., eds., 1989, World glacier inventory status 1988: Nairobi, IAHS (ICSI)-UNEP-UNESCO, 448 p.

Higuchi, Keiji, 1975, Evaluation of ERTS–1 imagery for inventory work of perennial snow patches in central Japan: Journal of Glaciology, v. 15, no. 73, p. 474.

Higuchi, Keiji, ed., 1976, Glaciers and climates of Nepal Himalayas-Report of the Glaciological Expedition to Nepal: Seppyo, v. 38, special iss., 130 p.

Higuchi, Keiji, ed., 1977, Glaciers and climates of Nepal Himalayas-Report of the Glaciological Expedition of Nepal-Pt. 2: Seppyo, v. 39, special iss., 67 p.

Higuchi, Keiji, ed., 1978, Glaciers and climates of Nepal Himalayas-Report of the Glaciological Expedition of Nepal-Pt. 3: Seppyo, v. 40, special iss., 84 p.

Higuchi, Keiji, ed., 1980, Glaciers and climates of Nepal Himalayas-Report of the Glaciological Expedition of Nepal-Pt. 4: Seppyo, v. 41, special iss., 111 p.

Higuchi, Keiji, 1993, Nepal-Japan cooperation in research on glaciers and climates of the Nepal Himalaya, in Snow and Glacier Hydrology, Proceedings of the Kathmandu Symposium, 16–21 November 1992: International Association of Hydrological Sciences (IAHS) Publication No. 218, p. 29–36.

Higuchi, Keiji, Fushimi, H., Ohata, T., Iwata, S., Yokoyama, K., and others, 1978, Preliminary report on glacier inventory in the Dudh Kosi region: Seppyo, v. 40, special iss., p. 78–83.

Higuchi, Keiji, Fushimi, H., Ohata, T., Takenaka, S., Iwata, S., and others, 1980, Glacier inventory in the Dudh Kosi region, East Nepal, in World Glacier Inventory, Proceedings of the workshop at Reideralp, Switzerland, 17–22 September 1978: International Association of Hydrological Sciences-Association Internationale de Sciences Hydrologiques (IASH-AISH) Publication No. 126, p. 95–103.

Inoue, Jiro, and Yoshida, M., 1980, Ablation and heat exchange over the Khumbu Glacier: Seppyo, v. 41, special iss., p. 26–33.

Iwata, Shuji, Watanabe, O., and Fushimi, H., 1980, Surface morphology in the ablation area of the Khumbu Glacier: Seppyo, v. 41, special iss., p. 9–17.

Karma, Ageta, Y., Naito, N., Iwata, S., and Yabuki, H., 2003, Glacier distribution in the Himalayas and glacier shrinkage from 1963 to 1993 in the Bhutan Himalayas: Bulletin of Glaciological Research, v. 20, p. 29–40.

Miller, M.M., Leventhal, J.S., and Libby, W.F., 1965, Tritium in Mount Everest ice — Annual glacier accumulation and climatology at great equatorial altitudes: Journal of Geophysics Research, v. 70, no. 16, p. 3885–3888.

Mool, B.K., Bajracharya, S.R., and Joshi, S.P., 2001, Inventory of glaciers, glacial lakes, and glacial lake outburst floods — Nepal: Kathmandu, International Center for Integrated Mountain Development (ICIMOD), 363 p.

Müller, Fritz, 1958, Eight months of glacier and soil research in the Everest region, in Swiss Foundation for Alpine Research, The Mountain World, 1958/59: New York, Harper, p. 191–208.

Müller, Fritz, 1970, A pilot study for an inventory of the glaciers in the eastern Himalayas, in Perennial ice and snow masses: Paris, UNESCO/IASH, p. 47–59.

Nakawo, M., Fujita, K., Ageta, Y., Shankar, K., Pokhrel, A.P., and Yao T., 1997, Basic studies for assessing the impacts of the global warming on the Himalayan cryosphere, 1994–1996: Bulletin of Glacier Research, v. 15, p. 53–58.

Nakawo, M., Raymond, C.F., and Fountain A., ed., 2000, Debris-covered glaciers, Proceedings of a workshop held at Seattle, Washington, 13–15 September 2000: International Association of Hydrological Sciences (IAHS) Publication No. 264, 288 p.

Nakawo, M., Yabuki, H., and Sakai, A., 1999, Characteristics of Khumbu Glacier, Nepal Himalaya—Recent changes in the debris-covered area: Annals of Glaciology, v. 28, p. 118–122.

Qin Dahe, chief ed., 1999, Map of glacier resources in the Himalayas: Beijing, Science Press, 7 sheets, scale 1:50,000.

Royal Geographical Society, 1975, Mount Everest region: London, Cook, Hammond, and Kell Ltd., 1:100,000-scale map.

Rundquist, D.C., and Samson, S.A., 1980, A Landsat digital examination of Khumbu Glacier, Nepal: Remote Sensing Quarterly, v. 2, no. 1, p. 4–15.

Schneider, Erwin, 1963, Khumbu Himal, (Nepal): Vienna, Freytag-Berndt, and Artaria, 1:50,000-scale map (also published in Hellmick, Walter, ed., Ergebnisse des Forschungsunternehmens Nepal Himalaya, v. 1, iss. 5: München).

Seko, K., Yabuki, H., Nakawo, M., Sakai, A., Kadota, T., and Yamada, Y., 1998, Changing surface features of Khumbu Glacier, Nepal Himalayas revealed by SPOT images: Bulletin of Glacier Research, v. 16, p. 33–41.

Tsvetkov, D.G., Osipova, G.B., Xie, Z., Wang, Z., Ageta, Y., and Baast, P., 1998, Glaciers in Asia, in Haeberli, W., Hoelzle, M., and Suter, S., eds., Into the second century of worldwide glacier monitoring — Prospects and strategies: Paris, UNESCO Publishing, Studies and Reports in Hydrology 56, p. 177–196.

Untersteiner, N., 1955, Some observations on the banding of glacier ice: Journal of Glaciology, v. 2, no. 17, p. 502–506.

Upreti, B.N., convenor, 2000, Proceedings of the international symposium on engineering geology, hydrogeology, and natural disasters with emphasis on Asia: Journal of the Nepal Geological Society, v. 22.

Visser, P.C., 1928, Von den Gletschern am obersten Indus [Glaciers in the uppermost Indus]: Zeitschrift für Gletscherkunde, v. 16, no. 3/4, p. 169–229.

von Wissmann, H., 1959, Die heutige Vergletsherung und Schneegrenze in Hochasian [Modern glaciers and the snowline in High Asia]: Wiesbaden, Akademie der Wissenschaften und der Literatur in Mainz in Kommission bei Franz Steiner Verlag GMBH, 307 p.

Yamada, T., 1998, Glacier lake and its outburst flood in the Nepal Himalaya: Data Center for Glacier Research, Japanese Society of Snow and Ice, Monograph No. 1, 96 p.

Watanabe, Okitsugu, 1976, On the types of glaciers in the Nepal Himalayas and their characteristics: Seppyo, v. 38, special iss., p. 10–16.

Watanabe, Okitsugu, Fushimi, H., Inoue, J., Iwata, S., Ikegami, K., Tanaka, Y., and others, 1980, Outline of debris cover project in Khumbu Glacier: Seppyo, v. 41, special iss., p. 5–8.

Yoshizawa, I., ed., 1977, Mountaineering maps of the world — Himalayas: Tokyo, Gakken Publication Inc., 330 p. (In Japanese)

Young, G.J., ed., 1993, Snow and glacier hydrology, Proceedings of an international symposium held at Kathmandu, Nepal, 16–21 November 1992: International Association of Hydrological Sciences, IAHS Publication No. 218, 410 p.

Young, G.J., and Neupane, B., 1996, Bibliography on the hydrology of the Himalaya-Karahoram region: Glaciological Data Report GD–29, 122 p.

Glaciers of Asia—

GLACIERS OF BHUTAN—An Overview

By Shuji Iwata

SATELLITE IMAGE ATLAS OF GLACIERS OF THE WORLD

Edited by RICHARD S. WILLIAMS, JR., *and* JANE G. FERRIGNO

U.S. GEOLOGICAL SURVEY PROFESSIONAL PAPER 1386–F–7

CONTENTS

Introduction --- 321

Glacier Distribution and Types of Glaciers --- 322

    FIGURE   1. Map showing river basins and glacier distribution of the
                Bhutan Himalaya -- 323

    TABLE   1. Total number, area, volume, and average thickness of each
                glacier type in Bhutan --- 323

    FIGURE   2. Photograph showing small debris-free plateau glacier with glacial
                lakes at the *Gangrinchemzoe La* in 1998 --------------------------- 324

    FIGURE   3. Graph showing distribution of the lowest elevations of debris-free
                glaciers and debris-mantled glaciers in the Bhutan Himalaya ----------- 325

Glacier Equilibrium Line Altitude (ELA) and Climate --------------------------------- 325

    FIGURE   4. Diagram showing variation of annual precipitation along
                the Himalaya --- 326

Glacier Variations -- 327

    FIGURE   5. Diagram showing horizontal distance of glacier-front retreat
                in the Bhutan Himalaya compared with latitude between
                1963 and 1993 --- 327

    FIGURE   6. Map of *Gangrinchemzoe-Gophu La Plateau* area showing
                debris-free and debris-covered glaciers and small glacial lakes ----------- 328

    FIGURE   7. Geomorphological map of the area around *Thanza* village
                in the eastern part of Lunana, northern Bhutan, showing
                debris-mantled glaciers, glacial lakes, and moraines ---------------------- 329

Glacial Lakes -- 330

    FIGURE   8. Photograph of *Lugge Tsho* lake from the western end ------------------- 331

    FIGURE   9. Photograph of *Raphsthreng Tsho* lake and erosion on the toe
                of the lateral moraine caused by the 1994 glacier lake outburst
                flood from *Lugge Tsho* --- 331

    FIGURE 10. Photograph of *Chubda Tsho* supraglacial lake on the debris-
                mantled ablation area of *Chubda Glacier*, upstream in the
                Chamkhar Chhu basin, north-central Bhutan ----------------------- 331

    FIGURE 11. Map of the optimum Landsat 1, 2, and 3 MSS images of the glaciers
                of Bhutan --- 332

    TABLE   2. Optimum Landsat 1, 2, and 3 MSS images of the glaciers of Bhutan ------ 332

References Cited -- 333

GLACIERS OF ASIA—

GLACIERS OF BHUTAN—An Overview

By Shuji Iwata[1]

Introduction

The Kingdom of Bhutan is a little-known mountainous and heavily forested country located in the eastern Himalaya. Having a total area of 46,500 km^2, it extends about 300 km east to west, and 170 km north to south. The total area covered by glaciers in Bhutan is 1,317 km^2 (Mool and others, 2001).

Officials, explorers, and naturalists who traveled to Bhutan in the past provided no descriptions of glaciers in the Bhutan Himalaya; therefore, not much historical information is available. The first modern descriptions of the glaciers of Bhutan were undertaken in the 1970s by Augusto Gansser, who carried out intensive geological surveys and developed a chronology of past glaciations in northern Bhutan (Gansser, 1970, 1983). In addition, Gansser (1970) warned of the potential for glacier lake outburst floods [GLOFs (or jökulhlaups)] in the Lunana area and documented his report with excellent photographs of glaciers and glacial lakes.

The report by Gansser (1970) prompted interest in the glaciers and glacial lakes of Bhutan, and in 1984 and 1986, the Geological Surveys of Bhutan and India carried out joint surveys of glaciers and glacial lakes for an assessment of the hazard potential in the Lunana area (Sharma and others, 1986). Their report concluded that there was no danger of outburst floods from *Raphsthreng Tsho* (called Lunana Lake in the report). Unfortunately, on 7 October 1994, a GLOF from *Lugge Tsho* glacial lake killed more than 20 people in the town of Punākha in the lower reaches of the drainage basin (Watanabe and Rothacher, 1996; Geological Survey of Bhutan, 1999).

After the GLOF disaster, several field studies focused on glaciers and glacial lakes in the Lunana area (Indo-Bhutan Expedition, 1995; Royal Government of Bhutan (RGOB) Expedition, 1995; Water and Power Consultancy Services (WAPCOS), 1997; Haeusler and Leber, 1998, Ageta and Iwata, 1999). At the same time, some of the field teams visited areas outside Lunana to observe glaciers and glacial lakes, conducting GLOF-hazard assessments (RGOB Expedition, 1995, Ageta and Iwata, 1999; Karma and Tamang, 1999; Karma and others, 1999, 2004; Ageta and others, 2000; Iwata, Ageta, and others, 2002; Iwata, Narama, and Karma, 2001), and compiling inventories of glaciers and glacial lakes (Division of Geology and Mines, 1996; Geological Survey of Bhutan, 1999; Mool and others, 2001). Bhutan and Japan initiated cooperative glaciological studies of large debris-mantled glaciers and small debris-free glaciers (Yamada and Naito, 2003; Ageta and Kohshima, 2004).

[1]Department of Geography, Tokyo Metropolitan University, Minami-Osawa, Hachiochi, Tokyo 192–0397, Japan [*iwata-s@comp.metro-u.ac.jp*]

Intensive hazard-assessment and mitigation efforts have continued in Lunana (Leber and others, 2000, 2002, 2003). Based on the findings of these studies, this paper provides brief descriptions of glaciers and related phenomena in the Bhutan Himalaya.

Glacier Distribution and Types of Glaciers

Glacier distribution in a region, including exact numbers and areas of glaciers, is shown by a glacier inventory. The first glacier inventory in Bhutan was compiled in 1996 by the Geological Survey of Bhutan (Division of Geology and Mines, 1996), and was subsequently published in a revised form with 83 pages of text and 13 map sheets (Geological Survey of Bhutan, 1999). A more complete inventory was published by the International Center for Integrated Mountain Development (ICIMOD) in Katmandu, Nepal (Mool and others, 2001). These inventories were compiled from analysis of 1993 Satellite Pour l'Observation de la Terre (SPOT) imagery, and 1:50,000-scale topographic maps made by the Geological Survey of India from vertical aerial photographs taken in the 1960s. Mool and others (2001) counted 677 glaciers in Bhutan covering a total area of 1,317 km². [Editors' note: An inventory of glacier resources by Qin (1999), compiled from analysis of 1975–1978 Landsat MSS band 7 and false-color infrared images and some aerial photographs, documented 649 glaciers in Bhutan, covering a total area of 1,304 km², and including a total volume of 150 km³.]

Glaciers are common on peaks and mountain flanks along the main topographic divide of the Bhutan Himalaya, and on plateaus and ridges that stretch to the south from the main Himalayan divide (fig. 1). Most glaciers are located in river basins, where the rivers flow from north to south. However, Bhutan also includes the north flank of the main divide in the northernmost part of the country, an area called the *Northern Basin*,[2] where rivers flow from Bhutan northward, toward the Autonomous Region of Tibet, China.

The total number, area, estimated volume, and average thickness of each glacier type in Bhutan is listed in table 1. Valley glaciers on the southern side of the main divide have narrow and steep upper basins, whereas glaciers in the *Northern Basin* have large, gently sloping accumulation areas. The longest glacier in Bhutan is 20.1 km — *Wachey Glacier* in Pho Chhu (river) basin (Mool and others, 2001). The largest single glacier basin, with a total area of 99.7 km², is located on the north slope of *Table Mountain* in the *Northern Basin*, northeast of Lunana (Mool and others, 2001). The accumulation areas of glaciers in the *Northern Basin* form an ice field that constitutes the north slopes of *Table Mountain*.

Large glaciers within valleys are characterized by debris-mantled snouts. On the south flank of the Bhutan Himalaya, slopes are steeper than those on the north slopes because topography on the south is formed by headward erosion of the rivers. As a result, rock cliffs occupy the heads of glacierized valleys. Headwalls of glacierized basins have ice and/or snow avalanches and rockfalls, processes enhanced by high precipitation from the summer monsoon.

---

[2]U.S. Government publications require that official geographic place-names for foreign countries be used to the greatest extent possible. In the Glaciers of Bhutan section, the use of geographic place-names is based on the U.S. Board on Geographic Names as listed on the GEOnet Names Server (GNS) website: *http://earth-info.nga.mil/gns/html/index.html*. Names not listed in the BGN website are shown in italics.

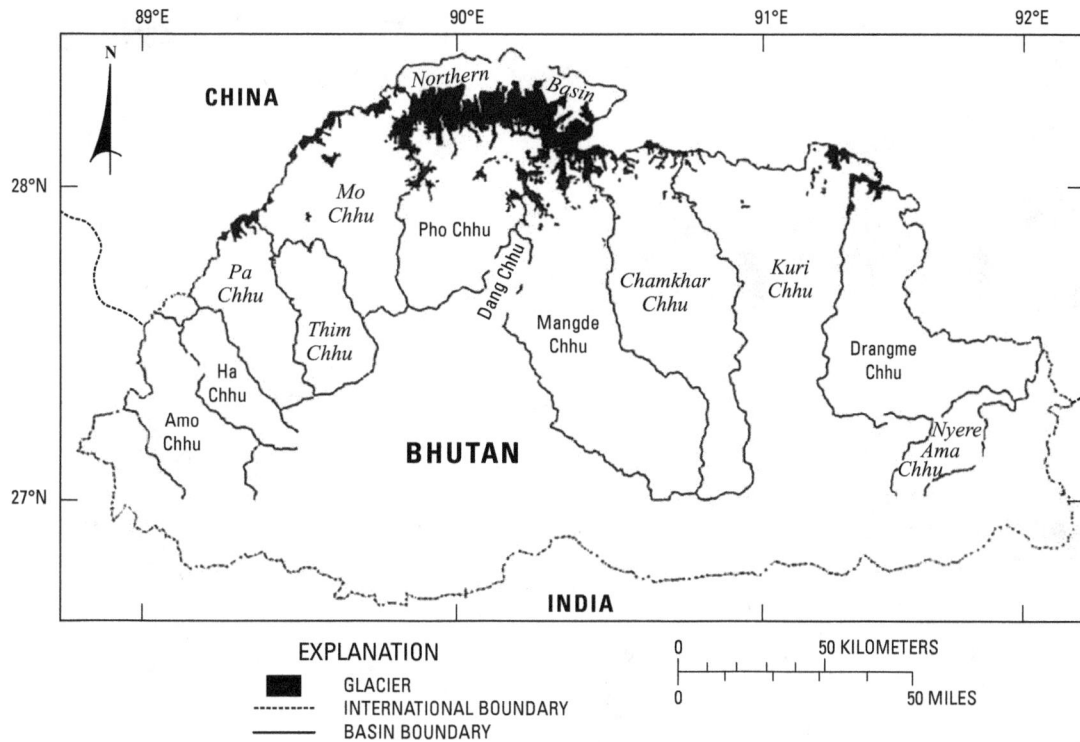

Figure 1.—River basins and glacier distribution of the Bhutan Himalaya. After Mool and others (2001).

TABLE 1.—*Total number, area, volume, and average thickness of each glacier type in Bhutan*

[Data sources: Mool and others (2001), and Karma and others (2003). Units: km, kilometer; m, meter]

Glacier type[1]	Number	Area (km²)	Estimated volume (km³)	Average thickness (m)
Valley glacier	51	691.76	90.222	130
Mountain glacier	453	579.02	35.510	61
Ice apron	94	33.40	1.197	36
Ice cap	16	5.19	0.182	35
Niche glacier	51	5.57	0.105	19
Cirque glacier	12	1.79	0.035	20
Total	677	1,316.73	127.251	97

[1]Glacier types are adapted from Müller and others (1997). Mountain glaciers are the most common, and their profile shows a hanging type.

Figure 2.—*Small debris-free plateau glacier with glacial lakes at the* Gangrinchemzoe La *(pass is at 5,200 m) in 1998. The lake started to form around the mid-1980s.*

The large amount of supraglacial debris in accumulation areas produces extensive debris-mantles in ablation areas downglacier. Large debris-mantled glaciers develop in Lunana, the upper part of the Pho Chhu basin, and in the *Northern Basin,* where there are many steep, high mountains.

Numerous mountain glaciers, especially small hanging-type glaciers, occur on steep slopes of high peaks and on back walls of valley glaciers. Cirque glaciers, small ice caps, ice aprons, and niche glaciers are common on plateaus and ridges that stretch to the south from the main Himalayan divide (fig. 2). These small glaciers are virtually debris-free.

According to Mool and others (2001), the highest elevation of glacier basins is at about 7,500 m, near *Gankerphuensum* (7,570 m) in the Mangde Chhu basin, and the second highest is at about 7,300 m, near Chomo Lhāri (7,314 m) in the *Pa Chhu* basin. Although no measurements were conducted for the ICIMOD inventory, glacier basins at higher elevations may exist on Kula Kangri (*Kunla Khari*, 7,538 m). Values in Mool and others (2001) show that the lowest elevations of glacier termini are found slightly above 4,000 m. Glacier termini between longitudes 90.2°E. and 90.9°E. (the Mangde Chhu and *Chamkhar Chhu* basins) are somewhat higher, above 4,300 m. Glaciers in this area cannot flow down to lower elevations because of plateau-like landforms and high valley floors. A group of small mountain glaciers in the *Kuri Chhu* basin, located east of long 90.9°E., and the Drangme Chhu basin farther east, have very low maximum (below 5,000 m) and minimum (nearly 4,000 m) elevations.

The Geological Survey of Bhutan inventory (1999) lists a smaller number of glaciers than do Mool and others (2001), but glaciers on the list can be classified into debris-mantled and debris-free from the associated maps. Karma and others (2003) plotted frontal elevations of both debris-mantled and debris-free glaciers (fig. 3). The graph clearly shows that the termini of debris-mantled glaciers extend to lower elevations (4,200 m) than those of debris-free ones (4,700 m).

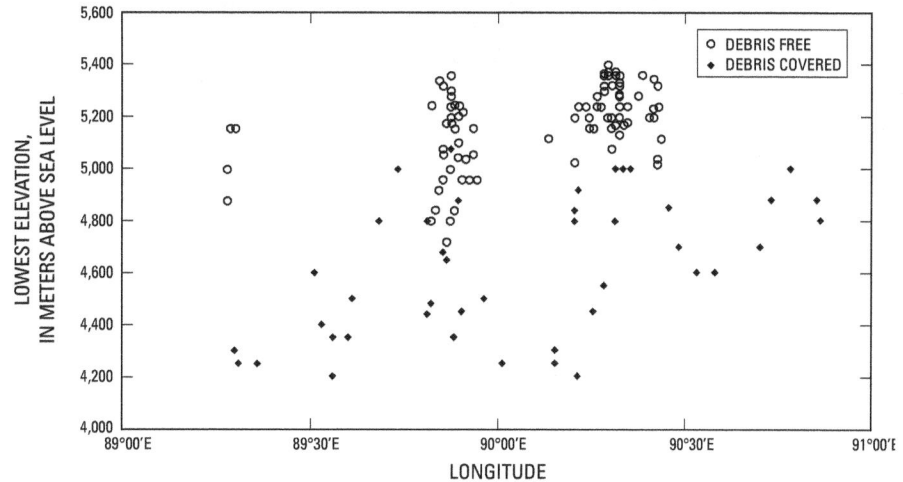

Figure 3.—Distribution of the lowest elevations (glacier termini) of debris-free glaciers (open circles) and debris-mantled glaciers (solid diamonds) in the Bhutan Himalaya. Karma and others (2003).

Glacier Equilibrium Line Altitude (ELA) and Climate

Bhutan has a monsoonal climate and most precipitation occurs in summer. Because monsoonal precipitation is much greater in the eastern and southern parts of the Himalaya than in the western and northern parts (Eguchi, 1991; fig. 4), glaciers have the characteristics of "summer-accumulation types." Glaciers receive most of their annual accumulation in the summer when ablation occurs simultaneously. The higher summer air temperature provides the following negative effects on mass balance of the summer-accumulation type glaciers (Ageta and Higuchi, 1984; Ageta and Kadota, 1992): (1) increased percentage of total precipitation in the form of rainfall reduces accumulation by snowfall, (2) decreased albedo due to less snowfall enhances ablation by insolation, (3) higher temperatures increase ablation. As a result, the variations of monsoon-type glaciers are very sensitive to the summer air temperature.

The equilibrium line altitude (ELA) is the most important parameter used to analyze glacier response to climate. Unfortunately, ELA information on glaciers in Bhutan is not available from the existing inventories. The ELA obtained from the author's observation in the field in September and October 1998 was 5,200–5,300 m on glaciers on the south side of the Himalaya in the northwestern and northern parts of Bhutan (Iwata, Naito, and others, 2003). This value is the lowest elevation of the ELA throughout the central and eastern Himalaya (Matsuoka, 2003). The intensity of the monsoon apparently produces the low ELA.

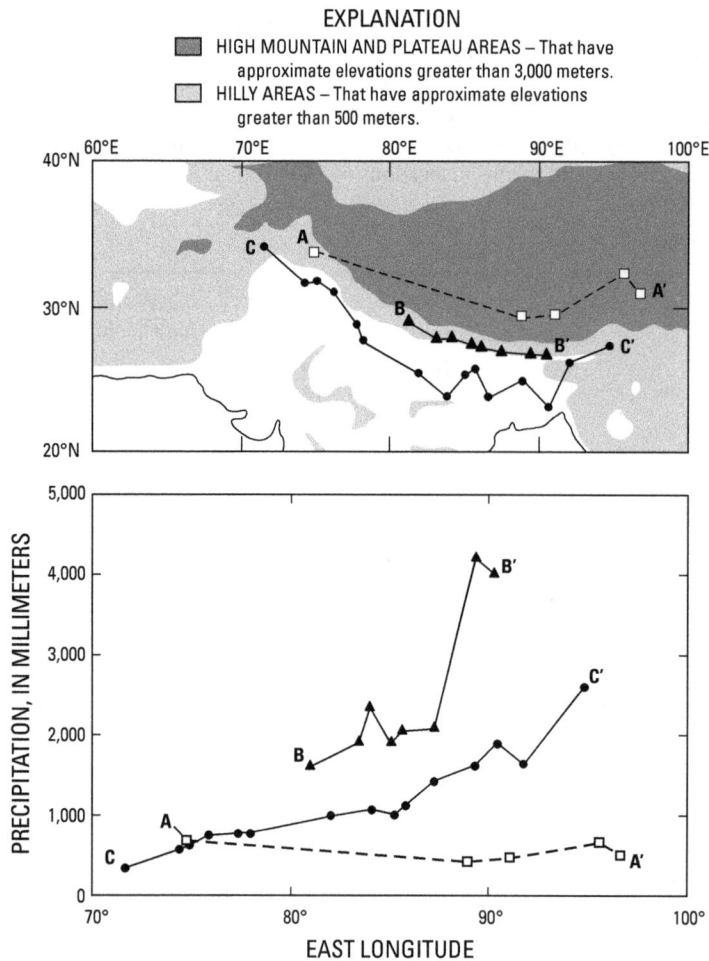

EXPLANATION

▨ HIGH MOUNTAIN AND PLATEAU AREAS – That have
approximate elevations greater than 3,000 meters.
▢ HILLY AREAS – That have approximate elevations
greater than 500 meters.

Figure 4.—*Variation of annual precipitation along the Himalaya.*
A, *North of the Himalaya,* **B**, *Southern slopes of the Himalaya,*
C, *Indian plain. From Eguchi (1991).*

Rock glaciers exist in many places along the *Snowman Trekking Route*
in northwestern and northern Bhutan. Iwata, Gurung, and Komori (2003)
reported on the lower limit of permafrost based on the distribution of active
periglacial rock glaciers. The lower limit of discontinuous permafrost is at
4,800 m on north-facing slopes and at 5,000 m on south- and east-facing
slopes. Results of automatic weather-station measurements between 1998
and 1999 suggest that mean annual air temperatures at the lower limits of the
permafrost zone are –1.1 °C at 5,000 m and +0.1 °C at 4,800 m. These values
mean that the relative height difference between the ELA and the lower limit
of permafrost in the Bhutan Himalaya is the smallest in the Asian Continent
(Matsuoka, 2003).

Isacks and others (1995) discussed the ELA depression during the Last
Glacial Maximum (LGM) based on a determination of the maximum extent
of glaciation using satellite imagery. Although their attempt was noteworthy,
their conclusions are not supported by any field observations, and no glacial
landforms have been dated using morphostratigraphic or cosmogenic-isotope
dating methods.

Glacier Variations

Karma and others (2003) identified termini variations of debris-free glaciers in the 30-year period between 1963 and 1993 by analyzing published inventories (Geological Survey of Bhutan, 1999; Mool and others, 2001), topographic maps (1:50,000 scale, showing the early 1960s positions), and satellite images (1993 SPOT images). A total of 103 debris-free glaciers were selected for measurement, because these types of glaciers are sensitive to climatic change, and most debris-mantled glaciers show no changes of their frontal positions. The debris-free glaciers selected for analysis were small and had similar lengths. Karma and others (2003) determined that 90 glaciers (87.3 percent) are retreating, 13 (12.7 percent) are stationary, and no glacier was advancing. The horizontal distance of the frontal retreat of each glacier was plotted versus its latitude (fig. 5). Results indicate that the magnitude of glacier retreat is larger in the south and smaller in the north because of the higher sensitivity of glacier mass balance to the warmer temperature and greater precipitation in the south. Karma and others (2003) concluded that the average retreat rates of glacier termini in Bhutan are higher than those in Nepal.

Areal shrinkage of 66 glaciers selected from the 103 debris-free glaciers was measured on the maps with a planimeter (Karma and others, 2003). The 66 glaciers had a total area of 146.9 km^2 in 1963, and the area decreased to 134.9 km^2 in 1993 — an 8.1 percent shrinkage in 30 years. The smaller glaciers show higher shrinkage rates than larger glaciers, and three small glaciers with areas of 0.1, 0.15, and 0.2 km^2 in 1963 had completely disappeared by 1993 (Karma and others, 2003).

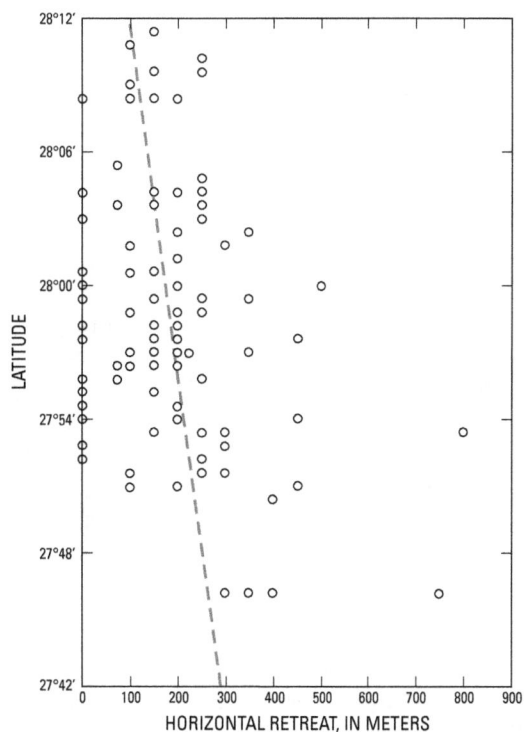

Figure 5.—Horizontal distance of glacier-front retreat in the Bhutan Himalaya compared with latitude between 1963 and 1993. Dashed line shows the linear regression. From Karma and others (2003).

Frontal variations of several debris-free glaciers on *Gangrinchemzoe-Gophu La Plateau* were observed in the field by Dr. Nozomu Naito in 1999 and compared with photographs taken in 1983–1984 by the Kyoto University Expedition (Yamada and Naito, 2003). *Gangrinchemzoe-Gophu La Plateau*, located to the southeast of Lunana, is 5,000–5,500 m in elevation, with many small glaciers and small glacial lakes (fig. 6). Observed glaciers have been shrinking markedly in recent decades as Karma and others (2003) reported.

Glacier variations since the late Pleistocene were mapped by Gansser (1983). He classified moraines and erosional landforms into six glacial stages in the entire area of northwestern and northern Bhutan. Gansser's glacial stages, however, were illustrated on the map but without explanation, and, in some cases, landforms with similar morphostratigraphic characteristics were assigned to two or more glacial stages.

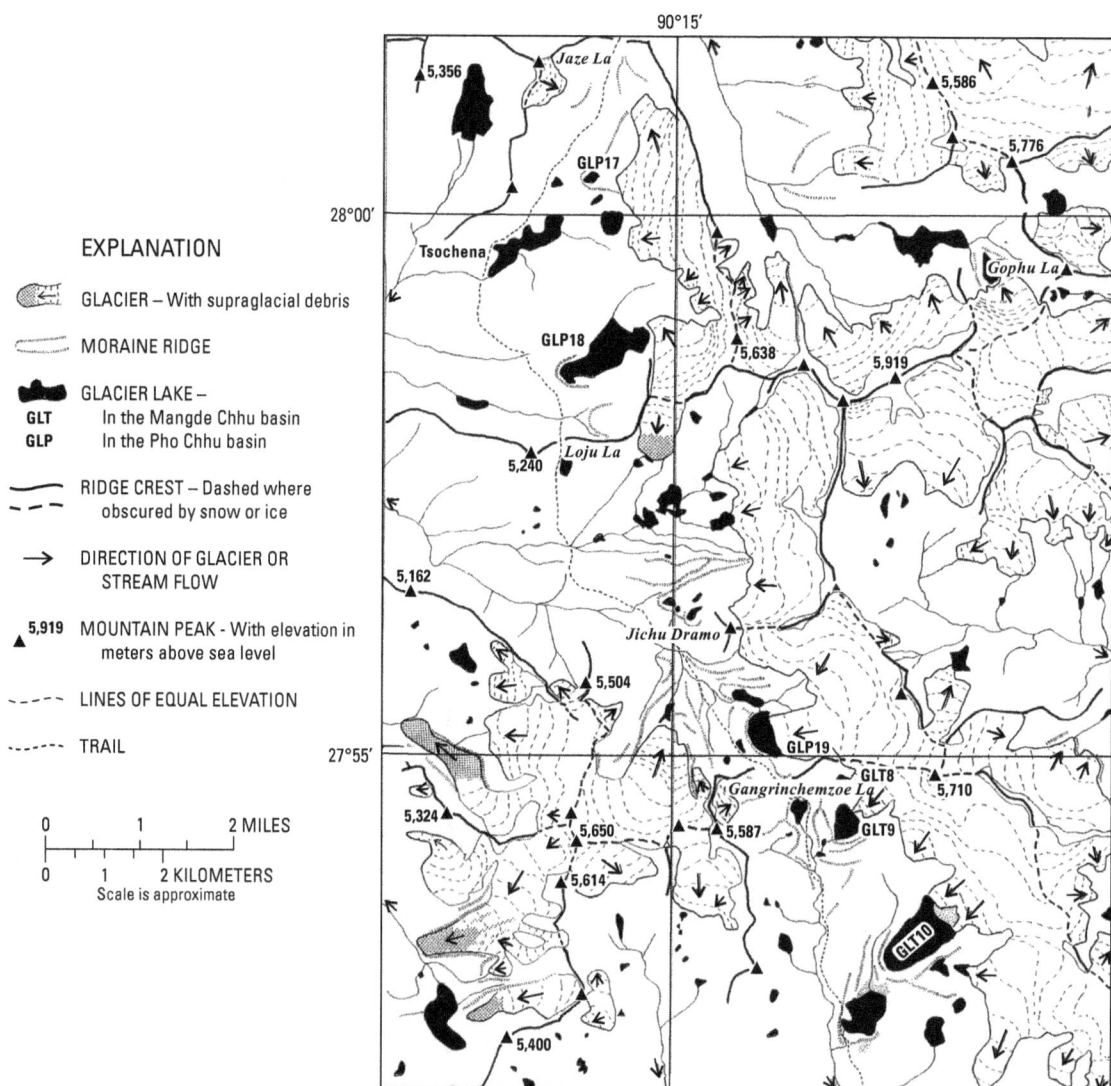

Figure 6.—Gangrinchemzoe-Gophu La Plateau *area showing debris-free and debris-covered glaciers and small glacial lakes. After Yamada and Naito (2003); revised from Ageta and Iwata (1999), based on interpretation of 1993 SPOT image.*

Iwata and his colleagues carried out geomorphological research in the Lingshi area, northwestern Bhutan, and Lunana in northern Bhutan (Iwata, Narama, and Karma, 2002). Moraines formed by valley glaciers in the Lingshi area constitute three distinct stages, with differences in spatial situations, volume of moraines, and surface features, such as clasts, soils, and vegetation. Moraines in the *Thanza* village area in Lunana also show three stages (fig. 7), with characteristics similar to those in the Lingshi area. Based on morphostratigraphic criteria (Iwata, 1976), these three stages correlate with some glacial stages in the Khumbu Himal, eastern Nepal. Dates of moraines in the Khumbu Himal (Richards and others, 2000) provide a tentative chronological sequence of these three stages. Valley glaciers in Bhutan expanded during the following periods: (1) the Takaphu and Raphsthreng Stages that may correlate to the "Little Ice Age" and late Holocene Epoch glacial advances, (2) the Jykhuje and Tenchey Stages that may represent the late Pleistocene Epoch and/or early Holocene glacial advances, and (3) the Lingshi Dzong and Lunana Stages that may coincide with the "Late glacial stage" of the Pleistocene Epoch.

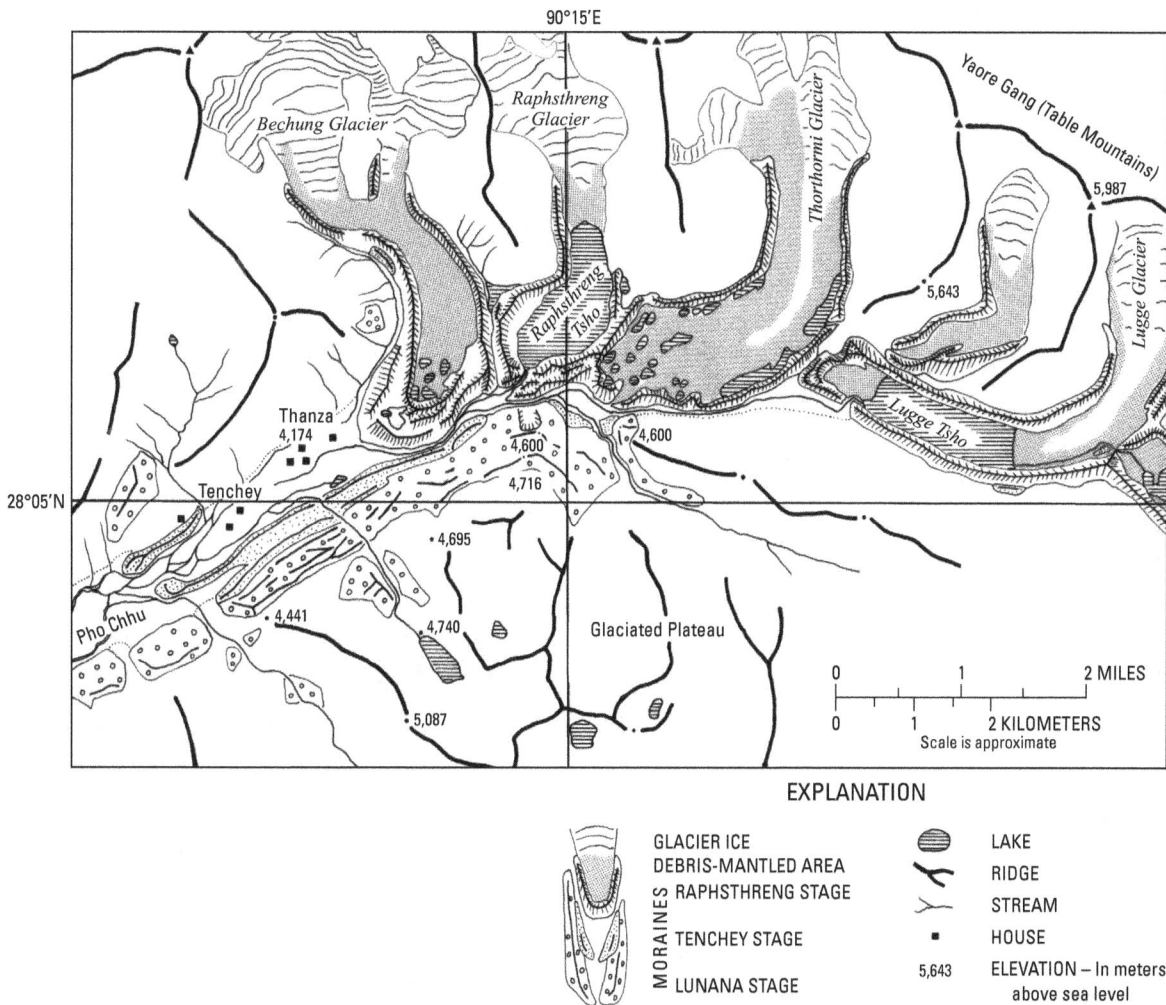

Figure 7.—*Geomorphological map of the area around* Thanza *village (4,170 m) in the eastern part of Lunana, northern Bhutan, showing debris-mantled glaciers, glacial lakes, and moraines. From Iwata, Narama, and Karma (2002).*

Glacial Lakes

Glacial lake outburst flood (GLOF) events have occurred in the high-mountain areas of the Bhutan Himalaya during the past 50 years and have increased in recent years. After the 1994 GLOF from *Lugge Tsho* glacial lake (fig. 8), inventories of glacial lakes were compiled based on 1989 and 1990 SPOT images to identify potentially hazardous lakes (Division of Geology and Mines, 1996; revised version: Geological Survey of Bhutan, 1999). A total of 62 glacial lakes were listed with information, such as code name, types of lake, elevation, dimensions, geographic coordinates, and location. The remarks noted that 37 glacial lakes should be field checked and/or monitored.

The inventory of glacial lakes made by ICIMOD (Mool and others, 2001) was compiled from topographic maps made by the Survey of India (1960s), aerial photographs, and Landsat Thematic Mapper (TM) (1993–1999), Indian Remote Sensing (IRS) 1D Linear Self Scanning (LIS) S3 (1999), and SPOT (1994) satellite images. A total of 2,674 lakes were listed with a total area of 106,776 km². The number of lakes is very large, because it includes all lakes in glaciated and glacierized areas, including small tarns in empty cirques. Mool and others (2001) identified 24 potentially dangerous glacial lakes concentrated in the Pho Chhu (Lunana) and Mangde Chhu basins; lakes in the *Northern Basin* (north of the main divide) were excluded from the assessment. Potentially dangerous glacial lakes are mostly located in or adjacent to the ablation areas of large debris-mantled glaciers. Field observations and analyses of maps and satellite images indicate that most supraglacial lakes tend to connect with each other and can enlarge rapidly. Since the 1970s, some moraine-dammed lakes formed and have rapidly expanded in area, caused by retreating and/or melting of glaciers (Ageta and Iwata, 1999; Ageta and others, 2000; Iwata, Ageta, and others, 2002).

After the 1994 GLOF from the *Lugge Tsho* glacial lake, many research groups visited the Lunana area to carry out risk assessments, risk mitigation, and disaster prevention efforts (Indo-Bhutan Expedition, 1995; RGOB Expedition, 1995; WAPCOS, 1997; Haeusler and Leber, 1998; Leber and others, 2000, 2002, 2003). The Bhutan Government carried out hydraulic-engineering projects, and between 1996 and 1998, they excavated a drainage channel at the outlet of *Raphsthreng Tsho* (figs. 7 and 9) using hand labor. As a result, the water level of the lake was lowered by 4 m. [Editors' note: see figure 14 (p. I71) and the Glacier Hazards section (p. I68–I71) in Morales Arnao (1998) for a discussion of similar hazard-mitigation work in Perú.]

A Bhutan-Japan joint team, which included the author, visited and assessed the risk potential of glacial lakes in areas along *Snowman Trekking Route* through northwestern and northern Bhutan (Ageta and Iwata, 1999). Analysis of the occurrence of trigger mechanisms, impact on the lakes, and stability of moraine-dams suggests that there are at least five potentially dangerous glacial lakes of the 30 lakes studied in northwestern and northern Bhutan. Three of these in eastern Lunana, including *Lugge Tsho, Raphsthreng Tsho,* and *Thorthomi Tsho,* still contain a large volume of water, are connected to each other, and interact sensitively through water flux and erosion (Iwata, Ageta, and others, 2002).

The inventory by Mool and others (2001) identified a large and dangerous glacial lake called *Chubda Tsho* in the upper part of *Chamkhar Chhu* basin, Bumthang, central Bhutan. In 2002, this author carried out a short field study at the lake. The *Chubda Tsho* is not a moraine-dammed lake, but a large supraglacial lake (fig. 10). No definite terminal moraine dams the lake, but at the lower end of the lake there is only a gentle threshold composed of the debris-mantled glacier. The lake tends to expand more toward the upglacier end than the downglacier end. However, the expansion rate is not high, most

Figure 8.—Lugge Tsho *lake from the western end. The ice cliff of* Lugge Glacier *is at the eastern end of the lake. A large volume of water still remains after the 1994 glacier lake outburst flood (GLOF) (1998 photograph by S. Iwata).*

Figure 9.—Raphsthreng Tsho *lake and erosion on the toe of the lateral moraine caused by the 1994 glacier lake outburst flood (GLOF) from* Lugge Tsho. *The excavation made at the outlet of* Raphsthreng Tsho *lowered the water level 4 m. To the left is the debris-mantled* Bechung Glacier *(1998 photograph by S. Iwata).*

Figure 10.—Chubda Tsho *supraglacial lake on the debris-mantled ablation area of Chubda Glacier, upstream in the* Chamkhar Chhu *basin, north-central Bhutan (2002 photograph by S. Iwata).*

likely because expansion toward the upglacier end may be counter balanced by glacier advance. Although this glacial lake does not pose an urgent danger of GLOF, the lake threshold itself may fail and so detailed inspections of the threshold by glaciologists with knowledge of debris-covered glaciers are necessary (Iwata, Gurung, and Komori, 2003; Komori and others, 2004).

The inventories of glacial lakes from satellite images indicate that many large and dangerous glacial lakes exist in headwater areas of *Kuri Chhu* in the Autonomous Region of Tibet, China. *Kuri Chhu* flows from the Autonomous Region of Tibet, China, to Bhutan, so that, if a GLOF from these lakes occurs, areas in the lower reaches of *Kuri Chhu* in Bhutan could sustain heavy losses.

In analyzing SPOT images, Dr. Jiro Komori found evidence of a recent GLOF and warned of the risk of future GLOFs in the region (Ageta and Kohshima, 2004). The Bhutanese Government is concerned about this danger, but cannot take preventative action because the lakes are located in China.

Constant and regular monitoring of glaciers and glacial lakes is urgently required to develop necessary hazard-mitigation strategies (Wangda, 2000). Monitoring of satellite imagery is the most effective tool for the purpose. Table 2 and figure 11 show optimum Landsat 1, 2, and 3 imagery for conducting historic studies of the glaciers of Bhutan.

TABLE 2.—*Optimum Landsat 1, 2, and 3 MSS images of the glaciers of Bhutan*

[Code column indicates usability for glacier studies; see figure 11 for explanation]

Path-Row	Nominal scene center latitude and longitude		Landsat identification number	Date	Code	Cloud cover (percent)
147–41	27°23′N.	91°49′E.	2147040007635190	16 Dec 76	●	5
148–40	28°49′N.	90°47′E.	2148040007635290	17 Dec 76	●	5
148–41	27°23′N.	90°23′E.	2148041007635290	17 Dec 76	◕	10
149–40	28°49′N.	89°21′E.	2149040007635390	18 Dec 76	◕	10
149–41	27°23′N.	88°57′E.	2149041007635390	18 Dec 76	◑	15

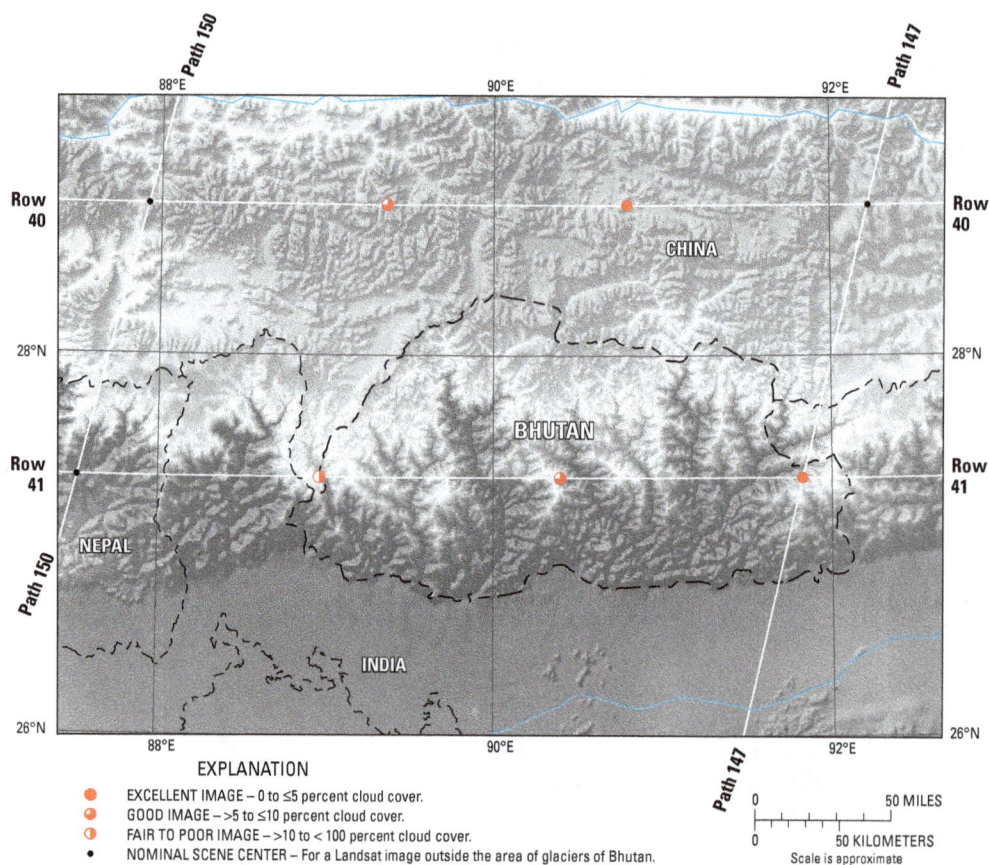

EXPLANATION

● EXCELLENT IMAGE – 0 to ≤5 percent cloud cover.
◕ GOOD IMAGE – >5 to ≤10 percent cloud cover.
◑ FAIR TO POOR IMAGE – >10 to < 100 percent cloud cover.
• NOMINAL SCENE CENTER – For a Landsat image outside the area of glaciers of Bhutan.

Figure 11.—*Map of the optimum Landsat 1, 2, and 3 MSS images of the glaciers of Bhutan. Compare with table 2.*

References Cited

Ageta, Y., and Higuchi, K., 1984, Estimation of mass balance components of a summer-accumulation type glacier in the Nepal Himalaya: Geografiska Annaler, v. 66A, no. 3, p. 249–255.

Ageta, Y., and Iwata, S., eds., 1999, Report of Japan-Bhutan joint research 1998 on the assessment of glacier lake outburst flood (GLOF) in Bhutan: Nagoya University, Institute for Hydrospheric-Atmospheric Science (not officially published; available for inspection in the Geological Survey of Bhutan, Thimphu), 161 p.

Ageta, Y., Iwata, S., Yabuki, H., Naito, N., Sakai, A., and others, 2000, Expansion of glacier lakes in recent decades in the Bhutan Himalayas, in Nakawo, M., Raymond, C.F., and Fountain, A., eds., Debris-covered glaciers, Proceedings of a workshop held at Seattle, Washington, USA, 13–15 September 2000: IAHS Publ. no. 264, p. 165–175.

Ageta, Y., and Kadota, T., 1992, Predictions of changes of glacier mass balance in the Nepal Himalaya and Tibetan Plateau — A case study of air temperature increase for three glaciers: Annals of Glaciology, v. 16, p. 89–94.

Ageta, Y., and Kohshima, S., eds., 2004, Report of Japan-Bhutan joint research 2003 on the assessment of glacier lake outburst flood (GLOF) in Bhutan: Nagoya University, Graduate School of Environmental Studies; Tokyo Institute of Technology, Graduate School of Bioscience and Biotechnology; Geological Survey of Bhutan, 64 p. (not officially published; available for inspection in the Geological Survey of Bhutan, Thimphu).

Division of Geology and Mines, 1996, Glaciers and glacier lakes in the headwaters of major river basins of Bhutan: Thimphu, Ministry of Trade and Industry, 55 p. (not officially published; available for inspection in the Geological Survey of Bhutan, Thimphu).

Eguchi, T., 1991, Regional and temporal variations in precipitation in the eastern part of the Himalayas: Kochi, Japan, Kochi University, Faculty of Humanities and Economics.

Gansser, A., 1970, Lunana: The peaks, glaciers and lakes of northern Bhutan: The Mountain World 1968/69, p. 117–131.

Gansser, A., 1983, Geology of the Bhutan Himalaya: Basel, Birkhauser Verlag, 181 p.

Geological Survey of Bhutan, 1999, Glaciers and glacier lakes in Bhutan: Thimphu, Ministry of Trade and Industry, Geological Survey of Bhutan, 2 vols., 83 p., and 13 map sheets.

Haeusler, H., and Leber, D., 1998, Final report of the Raphstreng Tsho outburst flood mitigation project: (Lunana, Northwestern Bhutan), Phase I: University of Vienna, Institute of Geology, 33 p. (not officially published; available for inspection in the Geological Survey of Bhutan, Thimphu).

Indo-Bhutan Expedition, 1995, Geology, environmental hazards and remedial measures of the Lunana area, Gasa Dzongkha — Report of 1995 Indo-Bhutan Expedition: Geological Survey of India, 40 p., and maps (not officially published; available for inspection in the Geological Survey of Bhutan, Thimphu).

Isacks, B.L., Duncan, C.C., and Klein, A.G., 1995, Modern and LGM glacial extents from satellite images and digital topography — Implications for LGM climate in the Bhutan Himalayas [abs.]: Abstracts-International Himalayan/Tibetan Plateau Paleoclimate Workshop, Kathmandu, p. 24.

Iwata, S., 1976, Late Pleistocene and Holocene moraines in the Sagarmatha (Everest) region, Khumbu Himal: Seppyō (Japanese Journal of Snow and Ice) v. 38, special iss., p. 109–114.

Iwata, S., Ageta, Y., Naito, N., Sakai, A., Narama, C., and Karma, 2002, Glacial lakes and their outburst flood assessment in the Bhutan Himalaya: Global Environmental Research [Tokyo], v. 6, no. 1, p. 3–17.

Iwata, S., Gurung, Deo Raj, and Komori, J., 2003, Report of Chubda Tsho in the head waters of Chamkar Chhu, Bumthang, 2002, in Yamada, T., and Naito, N., eds., Report of Japanese - Bhutan joint research 2002 on the assessment of glacier lake outburst flood (GLOF) in Bhutan: Hokkaido University, p. 54–66.

Iwata, S., Naito, N., Narama, C., and Karma, 2003, Rock glaciers and the lower limit of mountain permafrost in the Bhutan Himalayas: Zeitschrift für Geomorphologie, N.F. Supplementband 130, p. 129–143.

Iwata, S., Narama, C., and Karma, 2002, Three Holocene and late Pleistocene glacial stages inferred from moraines in the Lingshi and Thanza village areas, Bhutan: Quaternary International v. 97/98, p. 69–78.

Karma, Ageta, Y., Naito, N., Iwata, S., and Yabuki, H., 2003, Glacier distribution in the Himalayas and glacier shrinkage from 1963 to 1993 in the Bhutan Himalayas: Bulletin of Glaciological Research v. 20, p. 29–40.

Karma, and Tamang, K.B., 1999, Preliminary report on Tsokar Tsho in the head waters of Chamkhar Chhu, Bhumthang: Geological Survey of Bhutan, (not officially published; available for inspection in the Geological Survey of Bhutan, Thimphu).

Karma, Thapa, T., and Ghalley, K.S., 1999, Preliminary report on Chubda Tsho in the headwaters of Chamkhar Chhu, Bhumthang: Geological Survey of Bhutan, 11 p., and tables, figures, and maps (not officially published; available for inspection in the Geological Survey of Bhutan, Thimphu).

Komori, J., Gurung, D.R., Iwata, S., and Yabuki, H., 2004, Variation and lake expansion of Chubda Glacier, Bhutan Himalayas, during the last 35 years: Bulletin of Glaciological Research, v. 21, p. 49–55.

Leber, D., Haeusler, H., Brauner, M., Payer, T., and Wangda, D., 2003, Final report of the glacier lake outburst flood (GLOF) mitigation project (2002-2003) — Pho Chhu-Western branch (Tarina and Tang Chhu Watershed), Bhutan: University of Vienna, Department of Geological Sciences; Thimpu, Bhutan, Ministry of Trade and Industry, Department of Geology and Mines, 97 p., and maps (not officially published; available for inspection in the Geological Survey of Bhutan, Thimphu).

Leber, D., Haeusler, H., Brauner, M., and Skuk, S., 2000, Technical Report — Luggye Mitigation Project, Lunana, Bhutan: University of Vienna, Institute of Geology, 71 p. (not officially published; available for inspection in the Geological Survey of Bhutan, Thimphu).

Leber, D., Haeusler, H., Brauner, M., and Wangda, D., 2002, Final report of the glacier lake outburst flood (GLOF) mitigation project — Pho Chhu-Eastern branch (Thanza-Lhedi; 2000–2002),

Lunana, Bhutan: University of Vienna, Institute of Geology; Thimpu, Bhutan, Ministry of Trade and Industry, Department of Geology and Mines, 189 p., and maps (not officially published; available for inspection in the Geological Survey of Bhutan, Thimphu).

Matsuoka, N., 2003, Contemporary permafrost and periglaciation in Asian high mountains — An overview: Zeitschrift für Geomorphologie, N.F. Supplement Bd. 130, p. 145–166.

Mool, P.K., Wangda, D., Bajracharya, S.R., Kunzang, K., Gurung, D.R., and Joshi, S.P., 2001, Inventory of glaciers, glacial lakes and glacial lake ouburst floods — Monitoring and early warning systems in the Hindu Kush-Himalayan region, Bhutan: Kathmandu, International Centre for Integrated Mountain Development (ICIMOD), 227 p.

Morales Arnao, B., 1998, Glaciers of Perú (I-4), with a section on Quelccaya Ice Cap, by Hasternrath, S., in Williams, R.S., Jr., and Ferrigno, J.G., eds, Satellite image atlas of glaciers of the world: U.S. Geological Survey Profesional Paper 1386–I (South America), p. I51–I79. [http://pubs.usgs.gov/prof/p1386i/]

Müller, F., Caflish, T., and Müller, G., 1977, Instruction for compilation and assemblage of data for a World Glacier Inventory: Zürich, Temporary Technical Secretariat for World Glacier Inventory, Swiss Federal Institute of Technology; International Commission on Snow and Ice, International Association of Hydrological Sciences, United Nations Environment Programme, United Nations Educational, Scientific and Cultural Organization, 28 p.

Qin Dahe, chief ed., 1999, Map of glacier resources in the Himalayas; 1st ed.: Beijing, Science Press, scale 1:500,000, 7 sheets. [Compilation of map based on analysis of 1975–1978 Landsat MSS 7 and false-color images and some aerial photographs]

Richards, B.W.M., Benn, D.I., Owen, L.A., Rhodes, E.J., Spencer, J.Q., 2000, Timing of late Quaternary glaciations south of Mount Everest in the Khumbu Himal, Nepal: Geological Society of America, Bulletin, v. 112, no. 10, p. 1621–1632.

Royal Government of Bhutan (RGOB) Expedition, 1995, Preliminary report on the investigation of glacial lakes at Pho Chhu source and the assessment of flood affected areas in Lunana: Royal Government of Bhutan Expedition (not officially published; available for inspection in the Geological Survey of Bhutan, Thimphu).

Sharma, A.R., Ghosh, D.K., and Norbu, P., 1986, Report on Lunana Lake expedition —1986: Geological Survey of India (not officially published; available for inspection in the Geological Survey of Bhutan, Thimphu, Bhutan).

Wangda, D., 2000, The need to monitor glaciers and glacier lakes in the headwaters of major river basins of Bhutan: Bhutan Geology News Letter SL No. 2, p. 17–19.

Watanabe, T., and Rothacher, D., 1996, The 1994 Lugge Tsho glacial lake outburst flood, Bhutan Himalaya: Mountain Research and Development, v. 16, no. 1, p. 77–81.

Water and Power Consultancy Services (WAPCOS), 1997, Raphstreng Lake, Lunana (Bhutan) — Report on flood mitigatory measures (Phase 1—1996): Water and Power Consultancy Services (India) Limited.

Yamada, T., and Naito, N., eds., 2003, Report of Japan-Bhutan joint research 2002 on the assessment of glacier lake outburst flood (GLOF) in Bhutan: Hokkaido University, Institute of Low Temperature Science; Nagoya University, Graduate School of Environmental Studies; Geological Survey of Bhutan, 106 p. (not officially published; available for inspection in the Geological Survey of Bhutan, Thimphu, Bhutan).

Glaciers of Asia—

THE PALEOENVIRONMENTAL RECORD PRESERVED IN MIDDLE LATITUDE, HIGH-MOUNTAIN GLACIERS—AN OVERVIEW OF THE U.S. GEOLOGICAL SURVEY EXPERIENCE IN CENTRAL ASIA AND THE UNITED STATES

By L. DeWayne Cecil, David L. Naftz, Paul F. Schuster, David D. Susong, *and* Jaromy R. Green

SATELLITE IMAGE ATLAS OF GLACIERS OF THE WORLD

Edited by RICHARD S. WILLIAMS, JR., *and* JANE G. FERRIGNO

U.S. GEOLOGICAL SURVEY PROFESSIONAL PAPER 1386–F–8

CONTENTS

Abstract --- **335**

Introduction --- **335**

    FIGURE   1. Map showing locations of the *Nangpai Gosum Glacier*, Lednik
               Engil'chek (Inylchek Glacier), *Belukha Glacier*, Upper
               Fremont Glacier, and the Idaho National Laboratory -------------------- **338**

  Study Sites -- **339**

    Nangpai Gosum Glacier **339**

    Lednik Engil'chek **339**

    Belukha Glacier **339**

    Upper Fremont Glacier **339**

Selected Results and Discussion -- **340**

  Nangpai Gosum Glacier -- **340**

    FIGURE   2. Graph showing chlorine-36 and cesium-137 concentrations
               in the ice core collected from the *Nangpai Gosum Glacier* in 1998 ------ **340**

    FIGURE   3. Graphs showing concentrations of tritium, delta oxygen-18,
               and chlorine-36 in ice-core samples from the 1991 Upper
               Fremont Glacier expedition --- **341**

  Lednik Engil'chek -- **341**

  Upper Fremont Glacier -- **342**

    FIGURE   4. Graph showing refined age-to-depth profile for the 1991 Upper
               Fremont Glacier ice core -- **344**

    FIGURE   5. Graph showing profile of historic concentrations of mercury
               measured in selected ice-core sections from the Upper
               Fremont Glacier -- **345**

Applications --- **345**

  Paleoclimatic Records -- **345**

  Environmental Concentrations of Mercury ------------------------------------- **346**

Summary --- **347**

References Cited -- **348**

GLACIERS OF ASIA—

The Paleoenvironmental Record Preserved in Middle-Latitude, High-Mountain Glaciers—An Overview of the U.S. Geological Survey Experience in Central Asia and the United States

By L. DeWayne Cecil,[1] David L. Naftz,[2] Paul F. Schuster,[3] David D. Susong,[2] *and* Jaromy R. Green[4]

Abstract

The U.S. Geological Survey (USGS) is conducting a research program to study the geochemical and isotopic content of precipitation, snow, ice, and runoff samples from high-mountain glaciers in the middle latitudes of central Asia and the United States. Related topics of research, such as the reconstruction of paleoclimate records, the description of anthropogenic input of chemicals to the environment, and computer modeling of global climate change, are important to the well being of the Earth's human population. The collection and chemical analysis of snow and glacier ice cores can improve our knowledge of all of these topics. Nearly all of the chemical constituents that compose snow and ice-core samples contribute important scientific information. This research is made critical by the fact that middle-latitude, high-mountain glaciers, and the environmental and other information preserved in them, are rapidly disappearing as a result of global warming. Information collected to date includes the documentation of fallout from nuclear-weapons testing conducted during the 1940s, 1950s, and 1960s, quantification of pre-industrialization levels of mercury in the environment, evidence of rapid regional climate change, and identification of microbial communities entrained in the ice. Researchers from national and international academic institutions and from federal and provincial governmental agencies are collaborating on this project.

Introduction

Global environmental changes have occurred throughout the 4.5 billion-year history of the Earth, and such changes will continue to take place throughout the Earth System in the future. The changes are now recognized to have the potential for immediate as well as long-term consequences for the Earth's varied ecosystems. As a result, the importance of understanding current and potential global environmental change has dramatically increased (Cecil and others, 2004).

Archives of past climatic and environmental conditions are preserved in the Earth's alpine glaciers, ice caps, ice fields, and ice sheets. Ice cores from the polar regions have provided the scientific community with an unprecedented view of past environmental change through analysis of chemical, isotopic, and stratigraphic data. These records extend back in time for 110,000 years in ice

---

[1]U.S. Geological Survey, Boise, Idaho 83702 U.S.A.

[2]U.S. Geological Survey, Salt Lake City, Utah 84104 U.S.A.

[3]U.S. Geological Survey, Boulder, Colorado 80303 U.S.A.

[4]Garden City Community College, Garden City, Kansas 67846 U.S.A., formerly with the U.S. Geological Survey

cores from the Greenland ice sheet (Alley, 2000) and for more than 800,000 years (EPICA, 2004) in the Antarctic ice sheet. However, changes in weather and climate patterns affect high-latitude regions of the world differently than they do middle- to low-latitude regions because of latitudinal differences in global circulation. Because the majority of the Earth's population, at least 85 percent, lives between latitudes 50°N. and 50°S., it is of prime importance to understand potential environmental change in middle- and low-latitude regions; ice cores collected from selected high-mountain glaciers in temperate regions are valuable tools for this effort.

Although some middle-latitude, high-mountain regions of the Earth are glacierized, glaciers in these regions are generally considered to have unusable records of past environmental and climatic changes because of thawing and refreezing, and associated meltwater percolation. Recently, however, it has been shown that carefully selected middle-latitude, high-mountain glaciers accurately preserve isotopic and chemical records, thus providing information about climatic and environmental changes (Naftz and others, 1996, 2004; Cecil and Vogt, 1997; Cecil and others, 1998, 2004; Thompson, 2000, 2004, in press; Schuster and others, 2002, 2004; Green, Cecil, Synal, and others, 2004). Middle-latitude, high-mountain glaciers must have certain characteristics to preserve retrievable environmental records in glacier ice-cores. These include relatively simple ice-flow dynamics, flat to low-angle bedrock topography, limited redistribution of snow from wind and avalanches, minimal snowmelt during the summer season or minimal effects from the snow melt, and sufficiently thick glacier ice for maximum record length. To determine if a glacier has some of these characteristics, snow samples that fall on the glacier surface can be analyzed for lateral variability of chemical and particulate constituents across the glacier surface, caused by processes such as wind, temperature, and elevation gradients.

In addition to recording naturally occurring past environmental changes, middle-latitude, high-mountain glaciers also preserve a record of atmospheric input from human activities. Increased levels of many modern substances are archived in younger glacier ice (less than about 100 years old). Such substances include pollutants from refrigerants, sulfate from acid rain, mercury and other heavy metals from coal-burning power-plants and other industrial emissions, isotopic fallout from nuclear facilities including accidents such as at Chernobyl, Russia, in April 1986, and fallout from above-ground testing of nuclear weapons in the 1940s, 1950s and 1960s.[5] Detonation of nuclear devices by the United States and Great Britain over the Pacific Ocean in the mid-1960s created a significant quantity of radioactive isotopes, many of which were incorporated or injected directly into the upper atmosphere.[6] Chlorine-36 (^{36}Cl), tritium (^{3}H), and cesium-137 (^{137}Cs) with half-lives of 301,000, 12.26, and 30.2 years, respectively, are three isotopes that were spread throughout the atmosphere and deposited around the globe by means of both wet precipitation and dry deposition. During the nuclear-weapons-testing era, small amounts of these isotopes became trapped each year in the glacier ice, thus compiling an atmospheric-based record of the recent past.

---

[5]Editors' note: The first above-ground test of a nuclear weapon was on 16 July 1945; most above-ground tests of nuclear weapons ceased in 1963, after 711 tests in the atmosphere or the ocean. According to Greenpeace, the source of information in this footnote, "the last atmospheric nuclear weapons test occurred on 16 October 1980 in China." *http://archive.greenpeace.org/comms/nukes/ctbt/read9.html*

[6]Editors' note: During 1962, five high-altitude atmospheric tests of nuclear weapons were conducted by the United States in the Pacific Ocean under Operation "Fishbowl." Four of the tests were 10s of kilometers above the Earth. The fifth test, "Starfish Prime," detonated a 1.4 megaton nuclear device 400 km above Johnston Island (exoatmospheric) on 9 July 1962. The tests produced many fallout products that were distributed globally.

With the advent of ultra-sensitive analytical methods such as accelerator mass spectrometry (AMS) and the collaboration of expert scientists with diverse expertise and experience, glaciers worldwide are more accessible for study. The USGS and other institutions (Thompson, in press) are currently conducting research on middle-latitude, high-mountain glaciers in central Asia and North America, applying new scientific methods to an understanding of human influence on global environmental processes. Sites in central Asia are located on the *Nangpai Gosum Glacier*, Nepal; Lednik Engil'chek,[7] Kyrgyzstan/Kazakhstan/China; and *Belukha Glacier*,[8] Russia/Kazakhstan. The site in North America is on the Upper Freemont Glacier, Wyoming (fig. 1). The studies described here are being carried out in collaboration with researchers from several domestic and foreign universities, U.S. national laboratories, and other federal and provincial governmental agencies.

The scientific validity of results from analysis of glacier ice cores is dependent on the following precise protocols for each step: ice-core collection, storage, analysis, and interpretation (Green, Cecil, and Frape, 2004). Mechanical drills were utilized for ice-core collection at the central-Asia sites, and a thermal drill was used for ice-core collection at the site in North America. This difference in drilling technique is a result of the difference in elevation of the sites. The site in North America is the lowest in elevation above sea level; therefore, a thermal drill was utilized to insure maximum ice-core recovery. The mechanical and thermal drills were used to obtain continuous ice-core samples for chemical and physical analyses. At each site, 1-m long ice-core sections were removed from an aluminum core barrel, quickly sealed in polyethylene bags, and placed in lexan core tubes by personnel wearing TyvekTM suits and powder-free latex gloves. The core tubes were stored at a maximum temperature of 0 °C in snow vaults constructed at each collection site. Immediately after drilling was completed, the ice-core containers were packed in dry ice and transported by yak, helicopter, jet aircraft, and/or freezer truck to cold storage in the United States. Temperature recorders were placed in the shipping containers to monitor any changes during the shipping phase. The temperature inside the containers did not exceed 0 °C during shipping. The cores were equilibrated to exam-room temperature, 10 °C to 24 °C, for 24 hours prior to processing for chemical and physical analyses.

Ice and meltwater samples from the glacier ice core were analyzed for their elemental and isotopic content using various laboratory methods and established protocols. After melting the ice in a controlled environment to ensure minimal or no contamination from ambient air, the resultant melt-water was prepared for analysis.

Certain constituents present in ice cores can be used to reconstruct the timing and significance of past environmental and climatic events. For example, radionuclides are useful for establishing time lines in ice-core records. Direct-current electrical-conductivity measurements (ECM), in combination with concentrations of selected anions, are used to document historic volcanic events that also aid in establishing time lines. Changes in the isotopic composition of the oxygen atom in ice molecules can be used to reconstruct changes in air temperature and relative precipitation rates during long time periods. Initial research indicates that microbial populations preserved in glacier ice may have responded to changes in atmospheric circulation patterns, land use, and biogeographical conditions.

---

[7]Although Lednik Engil'chek is the officially approved version of the glacier name, the more commonly known and used versions are Inyl'chek or Inilchek Glacier.

[8] U.S. Government publications require that official place-names for foreign countries be used to the greatest extent possible. In this section, the use of geographic place-names is based on the U.S. Board on Geographic Names (BGN) website, as listed on the GEOnet Names Server (GNS): *http://earth-info.nga. mil/gns/html/index.html*. Names not listed on the BGN website are shown in italics.

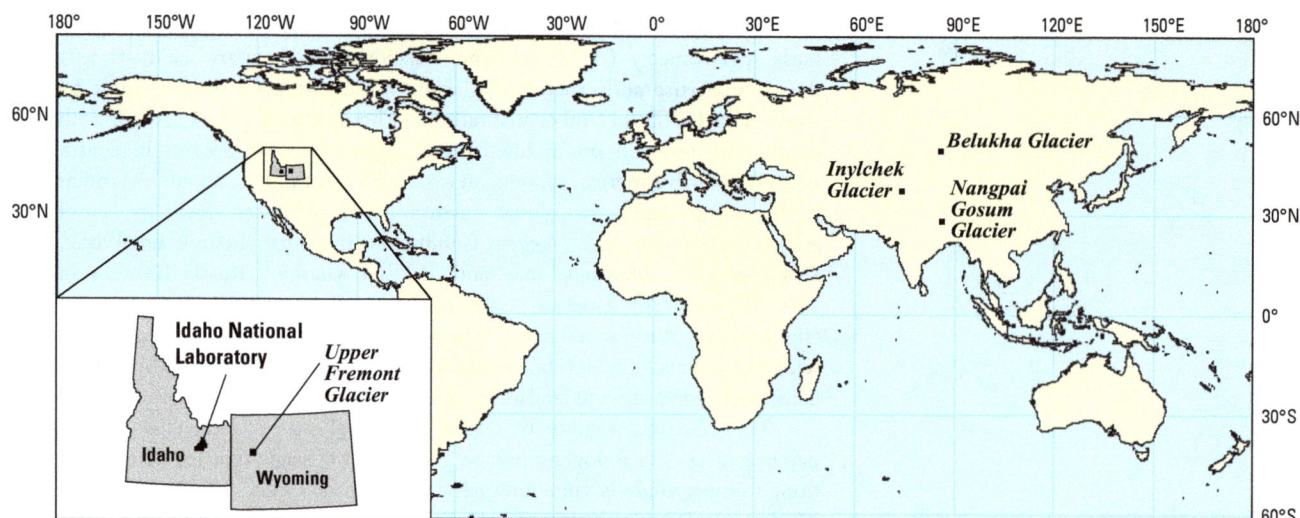

Figure 1.—*Locations of the* Nangpai Gosum Glacier, *Nepal;* Lednik Engil'chek (Inylchek Glacier), *Kyrgyzstan/Kazakhstan/China;* Belukha Glacier, *Russia/Kazakhstan; and the Upper Fremont Glacier, Wyoming, U.S.A.; and the Idaho National Laboratory, U.S.A.*

For the research described here, the radionuclides chlorine-36 (^{36}Cl), cesium-137 (^{137}Cs), and carbon-14 (^{14}C) were analyzed by accelerator mass spectrometry (AMS) methods (Currie and others, 1985; Elmore and Phillips, 1987). Tritium (^{3}H) was analyzed by using electrolytic enrichment (Ostlund and Werner, 1962; Thatcher and others, 1977). Concentrations of chloride (Cl^{-}), nitrate (NO$_3^{-}$), and sulfate (SO$_4^{-2}$) are determined by ion exchange chromatography (Fishman and Friedman, 1989). Concentrations of sodium (Na^{+}), magnesium (Mg^{++}), and calcium (Ca^{++}) can be determined by inductively-coupled, plasma-emission spectroscopy (Garbarino and Taylor, 1979); however, recent advances in ion-exchange-chromatography equipment have permitted the analyses of these constituents, along with ammonium (NH$_4^{+}$) and potassium (K^{+}), at small environmental concentrations. Values for delta oxygen-18 (δ^{18}O) were determined using a gas- or solid-source mass spectrometer (Kendall and Caldwell, 1998). The mercury (Hg) analyses presented here were performed by Dual Amalgamation Cold Vapor Atomic Fluorescence Spectrometry (USEPA Method 1631, 1999).

The electrical-conductivity measurements (ECM) were performed by drawing a pair of electrodes along the entire length of an ice core at a constant velocity (Schuster and others, 2000). For microbiological analyses, total numbers of microbial cells and organic matter were determined by acridine-orange, direct-count microscopy at the U.S. Department of Energy's Idaho National Laboratory (INL), Idaho, USA (M. Delwiche, oral commun., 2005). The glacier ice cores were melted according to protocols described by Naftz (1993) that were modified for microbial analyses. Standard quality assurance and quality control protocols and guidelines were followed for all analyses presented in this paper.

Study Sites

Nangpai Gosum Glacier

Researchers from the University of New Hampshire are studying the *Nangpai Gosum Glacier*, located 25 km west northwest of Mount Everest in the Nepal Himalaya (fig. 1). It is located at lat 28°02'N., long 86°36'E. The ice-core drilling site is situated at 5,700 m above mean sea level. In 1998, a 37-m ice core was extracted from the glacier by Cameron P. Wake of the University of New Hampshire and transported to the University of New Hampshire for analysis. At the request of the USGS study team, radionuclide analyses were performed for ^{36}Cl and ^{137}Cs on selected sections of the ice core by Hans-Arno Synal at the Paul Scherrer Institut (PSI) in Villigen, Switzerland.

Lednik Engil'chek

The Lednik Engil'chek (fig. 1) is being studied in collaboration with researchers from several U.S. and foreign universities. It is located in the Tien Shan of central Asia at lat 42°10'N., long 80°15'E. It is located in parts of three countries: northern China, southern Kazakhstan, and eastern Kyrghyzstan. Altitudes exceed 6,000 m on many parts of the remote, 65-km-long glacier. Its maximum 300-m depth is estimated to contain 1,000 to 5,000 years of accumulation. The elevation of our selected drilling site, 5,300 m, is high enough that minimal or no melting occurs (Aizen and others, 1997). In the summer of 2000, two deep ice cores, 162 m and 165 m in length, fresh snow, snow pit, and crevasse-wall samples were also recovered from this site (Kreutz and others, 2003, 2004; Green, Cecil, Synal, and others, 2004). Ice chips from the uppermost 100 m of one of the deep cores were shipped frozen to PSI where ^{36}Cl analyses were performed with the AMS. The cores provide a high-resolution isotopic and geochemical record that spans approximately the last 200 years.

Belukha Glacier

The *Belukha Glacier* [also called the *Belukha Firn Plateau*] is situated at an elevation of about 4,000 m in the Altay mountains of southern Siberia, at lat 49°48'26"N., long 86°34'43"E. on the border of Kazakhstan and Russia (fig. 1). The glacier's remote location makes it potentially an ideal site for environmental studies. A team of scientists, including Americans, Russians, and Japanese, traveled to this remote glacier in the summer of 2001 to assess the feasibility of studying the glacier and extracting ice cores at the site. A Swiss-Russian team also worked on the glacier in 2001. Research was carried out from 2001 to 2003; glaciological observations were made, and both shallow cores and cores to bedrock were extracted and analyzed (Olivier and others, 2003; Fujita and others, 2004). Based on tritium analysis to date, the deeper cores may contain as much as 3–5,000 years of climatic and environmental records (Aizen, oral commun., 2005)

Upper Fremont Glacier

The Upper Fremont Glacier (fig. 1) is located at lat 43°07'52"N., long 109°36'55"W., in the Wind River Range of Wyoming (figs. 18 and 19 on p. J360 and J361, respectively, in Krimmel, 2002). It is a relatively large middle-latitude glacier with a surface area between 2.5 and 3 km², a maximum altitude of 4,100 m above sea level, and an ice thickness greater than 150 m. It is located nearly 40 km inside the boundary of a designated wilderness area; access to the glacier requires a rigorous 2-day hike from the nearest road as well as the use of pack goats to transport scientific equipment and supplies. Continuous ice cores were collected from the glacier in 1991 and 1998. The site contains an ice-core record that is representative of past climatic and environmental changes, and of precipitation falling in remote, high-altitude environments in this part of North America (Naftz, 1993; Naftz and others 1993, 1996, 2004, 2009; Cecil and Vogt, 1997; Cecil and others, 1998, 1999, 2004; Schuster and others, 2000, 2004).

Selected Results and Discussion

Various constituents from archived inventories of snow-and-ice cores from middle-latitude glaciers have been identified. To date, ice cores from the *Nangpai Gosum Glacier*, Lednik Engil'chek, and the *Belukha Glacier* have only been analyzed for select constituents (for example, $\delta^{18}O$, ^{36}Cl, and ^{137}Cs). The initial work involving researchers from the USGS began on these glaciers in 1998. For the Upper Fremont Glacier, the ice-core and snow dataset is the most comprehensive of its kind for a middle-latitude glacier in the contiguous United States, the result of more than a decade of work on this glacier (Naftz and Miller, 1992; Naftz, 1993; Naftz and others, 1993, 1996, 2004, 2009; Cecil and Vogt, 1997; Cecil and others, 1998, 1999, 2004; Schuster and others, 2000, 2004).

Nangpai Gosum Glacier

Many radioactive isotopes were produced during the 35 years of above-ground nuclear-weapons testing, including ^{137}Cs. As with ^{36}Cl and ^{3}H, a small amount of ^{137}Cs (greater than background concentrations) became trapped in glacier ice during the 1940s, 1950s, and 1960s. Analysis of the 37-m ice core collected from the *Nangpai Gosum Glacier* in 1998 showed a distinct peak of anthropogenic ^{36}Cl and ^{137}Cs concentrations (fig. 2), just as for ^{36}Cl in the Upper Fremont Glacier ice core (fig. 3C). It was determined that the peak nuclear-weapons-tests fallout of ^{36}Cl and ^{137}Cs occurs at a depth of about 31 m in this glacier.

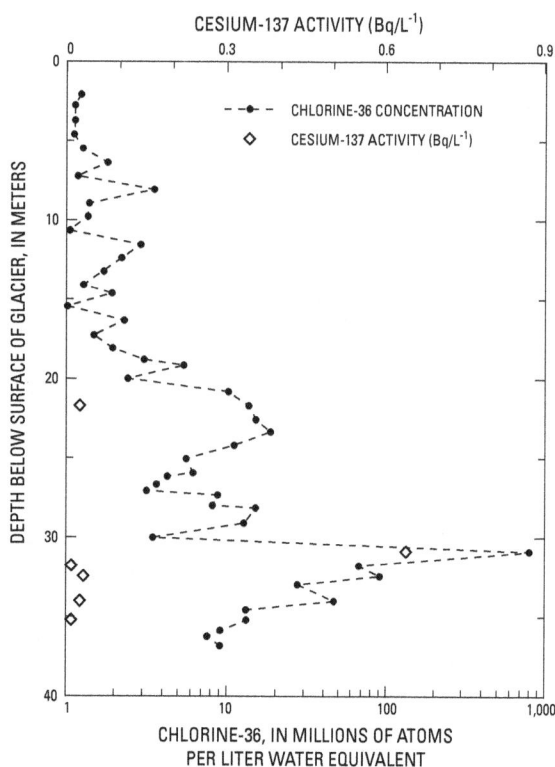

Figure 2.—Chlorine-36 and cesium-137 concentrations in the ice core collected from the Nangpai Gosum Glacier in 1998. BqL^{-1} (Becquerels per liter) is a measure of radioactivity.

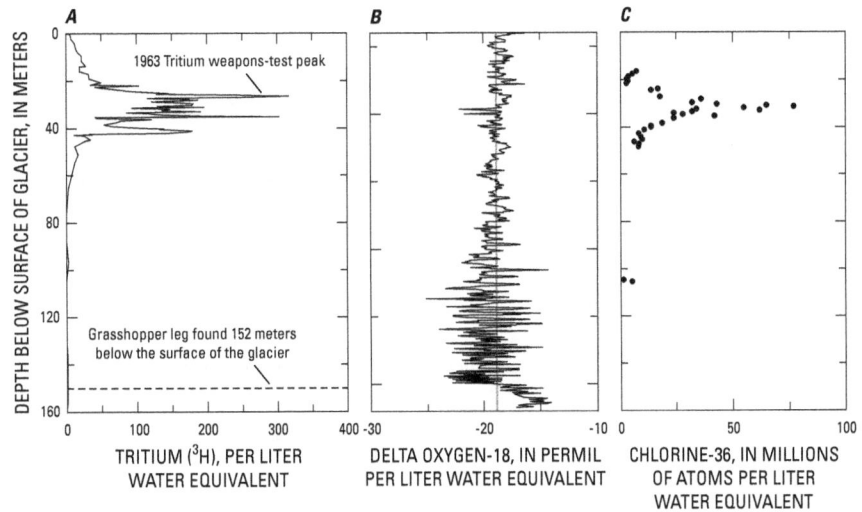

Figure 3.—Concentrations of **A**, Tritium (^{3}H) (data from Naftz and others, 1996), **B**, delta oxygen-18 (δ^{18}O) (data from Naftz and others, 1996) and **C**, chlorine-36 (^{36}Cl) (data from Cecil and others, 1999) in ice-core samples from the 1991 Upper Fremont Glacier expedition.

Lednik Engil'chek

Kreutz and others (2004) documented the limited spatial and elevational dependence that exists in isotopic and major ion data from the fresh snow pack at the Lednik Engil'chek site. There was no apparent trend with increasing elevation for any of the isotopic or geochemical variables measured. These findings lead to the conclusion that snow formation occurs from a horizontal cloud base across the Lednik Engil'chek and vicinity. In addition, based on the estimated accumulation rates at this site (1.6 m a^{-1}), it was reported that the interannual-scale variability in the ice-core profiles is significantly greater than the spatial variability during individual snowfall events. The conclusion from this analysis is that the majority of the down-core isotopic variability may be a reflection of climate changes rather than ice-flow effects or noise in the snow-deposition signal.

These conclusions have implications for further interpretation of the apparently high-resolution isotopic and geochemical records that will be developed from the two deep cores collected in 2000 (Kreutz and others, 2003). The implications are: (1) the high-resolution records may be used to evaluate the time-series, signal-to-noise ratio related to small-scale deposition processes, (2) given the relatively high accumulation rate at this site, it may be possible to reconstruct climate and environmental change on a snowfall-event basis, and (3) given the isotopic signals observed in the fresh snow events and the snow pits, it may be possible to investigate moisture sources and pathways to the Lednik Engil'chek through time on a subannual scale.

For example, Green, Cecil, Synal, and others (2004) reported a high-resolution subannual record of nuclear-weapons-test era ^{36}Cl from an ice core collected at the Lednik Engil'chek in 2000. In spite of the accumulation differences between the Upper Fremont Glacier and Lednik Engil'chek, 0.76 m a^{-1} and 1.6 m a^{-1}, respectively, the peak ^{36}Cl concentrations were remarkably similar in shape and magnitude. However, because of the greater accumulation rate calculated for the Lednik Engil'chek site, it is possible that the isotopic and geochemical record archived in this glacier will provide the opportunity for detailed study of environmental changes during the last 150 years. Each section of the Lednik Engil'chek ice core that was analyzed for ^{36}Cl was representative of as little as one-third of the annual accumulation; the same volume of ice used for ^{36}Cl analyses at the Upper Fremont Glacier may have been representative of more than one year's accumulation. This high-resolution record from Lednik Engil'chek may aid in the study of the documented diminishing water resources that may be one of the results of global warming in heavily-populated central Asia.

Upper Fremont Glacier

Chlorine-36 (^{36}Cl) and ^{3}H have been measured in a series of precipitation, snow, ice, and runoff samples collected in 1991 and 1998 from the Upper Fremont Glacier. The ice core collected from the Upper Fremont Glacier in 1991 shows a profile with depth for ^{3}H and ^{36}Cl that is expected from a minimally disturbed, temperate-glacier environment (fig. 3A, C) (Naftz and others, 1996; Cecil and others, 1999). Tritium (^{3}H) and ^{36}Cl concentrations near the surface of the glacier are similar to concentrations in recent precipitation measured in the Rocky Mountains. In deeper sections of the 1991 core, ^{3}H and ^{36}Cl concentrations are higher, reaching maximums at 29 m below the glacier surface for ^{3}H and 32 m for ^{36}Cl. These "markers" represent the 1963 and 1958 peak productions from atmospheric weapons tests, respectively (Cecil and others, 1999). The ^{3}H and ^{36}Cl concentrations decrease below these depths. Below a depth of 40 m, ^{3}H concentrations are essentially zero and ^{36}Cl concentrations return to pre-nuclear weapons test levels of 10^6 atoms per liter (l^{-1}) water equivalent or less.

In glacier environments, any plant or animal material within the ice potentially can be dated by determining the amount of ^{14}C in the sample (5,730 year half-life). Several assumptions apply when using this method of dating organic materials incorporated into the ice. The assumptions include: (1) the organic material was incorporated into the snow and ice at the actual time of death of the plant or organism; and (2) the initial concentration of ^{14}C in the plant or animal material is well-known and is independent of time, geographic location of the sample, and species of plant or animal. Naftz and others (1996) dated a species of grasshopper contained in the ice core collected from the Upper Fremont Glacier in 1991. The sample was taken from the ice core at a depth of 152 m below the surface (fig. 3A). The ^{14}C content of the grasshopper was determined using AMS (Currie and others, 1985). The ^{14}C age of the sample (A.D. 1,729 ±95 years) indicates that the grasshopper was deposited on the glacier surface sometime in the mid-1600s to the mid-1700s. To build confidence in this ^{14}C age, additional grasshopper parts were extracted from a depth of 146 m in the ice core collected in 1998. The ^{14}C model age for the grasshopper parts was determined to be A.D. 1,680 ±80 years. This information, coupled with the mercury (Hg) profile (discussed below) and the nuclear weapons-testing dates provided by the ^{3}H and ^{36}Cl profiles was instrumental in establishing a detailed glacial ice-core chronology and age-to-depth profiles for the Upper Fremont Glacier.

Major ion analyses of ice cores are essential components of ice-core interpretation. Changes in the concentrations of selected major ions can be used to identify specific events (natural or anthropogenic) that affect the chemistry of precipitation deposited on a glacier. For example, increases in Cl$^-$, NO$_3^-$, and SO$_4^{-2}$ concentrations can provide supporting evidence for volcanic-event horizons used as time markers to develop an ice-core chronology. Colder periods are typically windier and dryer thus facilitating increased deposition of dust layers (rich in Mg^{++} and Ca^{++} in western North America). Increased concentrations of these same constituents can be caused by anthropogenic influences, such as acid rain and bio-mass burning; Na^+, K^+, Mg^{++}, and Ca^{++} can be enriched in seasonal dust layers deposited on the surface of a glacier. Increased concentrations of these cations not only provide a potential mechanism for development of an ice-core chronology through counting annual dust layers, but they also may serve as indicators of paleoclimatic change.

Changes in the $\delta^{18}O$ concentrations in ice cores can be used to reconstruct changes in air temperature from the water-vapor source region to the ice-core site (White and others, 1989). Values of $\delta^{18}O$ change in relation to temperature of formation of and distance from source water, storm track, altitude, and evaporation. In most ice cores, more negative $\delta^{18}O$ values represent cooler air temperatures. Relative changes in air temperature at the Upper Fremont Glacier were reconstructed by determining the $\delta^{18}O$ values from equally spaced samples along the entire length of the 1991 ice core (fig. 3B). Between the depths of 102 m and 150 m, numerous high-amplitude oscillations in the $\delta^{18}O$ values were detected. The mean $\delta^{18}O$ value for this depth range abruptly shifted to more negative values, corresponding to the approximate time interval from the mid-A.D. 1700s to mid-A.D. 1800s. This period of time coincides with the latter part of the "Little Ice Age" (LIA) (Naftz and others, 1996, 2004).

Atmospheric circulation can deposit microbial cells on snowfields and glacier-ice surfaces. Some of these windborne organisms may survive in a preserved state for extended times, and some types may colonize microhabitats within the accumulated snow layers. It is possible that preserved microbial ice populations contain a record of atmospheric circulation patterns, land use, and biogeographical conditions upwind from deposition sites. Upwind sources of heavy metals from mining and smelting operations may have an influence on the microbial communities that can survive and flourish at higher altitudes; where heavy metal fallout is significant at the glacier's surface, metal-reducing microbial types may prevail. In addition, because of the intensity of high-altitude sunlight (for example, more intense UV radiation), radiation-tolerant species may dominate in these middle-latitude glacier sites and may act as monitors to changes in ultraviolet radiation caused by thinning of the ozone layer. Scientists at the INL have begun investigations on microbial organisms found in ice cores from our study sites. The ice samples, which have been in frozen storage since collection, are being used to determine total biomass content and to estimate viability of englacier populations.

The total number of microbial cells and organic matter from nine ice-core samples collected from the Upper Fremont Glacier have been determined (M. Delwiche, personal commun., 2005). After melting, 0.1 milliliter (mL) of core interior meltwater was immediately plated onto rich (R2A) and lean (1 percent PTYG) types of standard solid medium for growth of heterotrophic microorganisms (Gerhardt and others, 1994). Also, 5 mL of each inner core and of each "rinse" from the core exterior, and 10 mL of filtered DI water as a negative control were stained with 0.01 percent acridine orange fluorescent dye. Standard microscopy methods were used for preparation and counting of stained cells on black polycarbonate membrane filters (Hobbie and others, 1977). In all, 8 negative controls, 18 core, and 9 rinse slides were counted. In addition to analyses performed immediately after melting, meltwater was stored and retested after time periods of 12 days and 2 months to test the effects of storage on culturability of contained microbial populations. All plates and meltwater were stored at 12 °C.

Significant numbers of "cell-like" objects were observed in most ice samples. Total numbers were as high as 2.1×10^5 cells mL^{-1} of melted ice with numerous discernable morphologies; clumps of brightly fluorescing rods associated with inorganic particles were common in at least one third of the samples and appeared to be more numerous in the shallower sections of the ice core. There was no perceptible correlation between depth in the core and microbial numbers. Samples also contained numerous arthropod parts and large cellular objects believed to be unicellular algae. The results are consistent with reports in the literature from polar ice cores (Priscu and others, 1998; Karl and others, 1999) and appear to correspond with expectations of subsurface microbial communities (Pedersen, 1993; Phelps and others, 1994).

Direct current electrical conductivity measurements (ECMs), used to determine the acidity of ice cores, can be used to assist in the determination of ice-core chronology. ECM application includes the identification of seasonal/ summer dust layers for layer counting and the identification of volcanic events to be used as time-line markers. Schuster and others (2000, 2004) used the 1991 Upper Fremont Glacier ice core to produce an ECM log (fig. 4). Together with major ion analyses, ^{3}H, ^{36}Cl, and ^{14}C data, two of the largest explosive volcanic eruptions during the last 10,000 years (fig. 5) (Krakatau, Indonesia, on 27 August 1883; and Tambora, Sumbawa, Indonesia, on 10-11 April 1815) (Simkin and Siebert, 1994) were identified in the ECM signal produced from

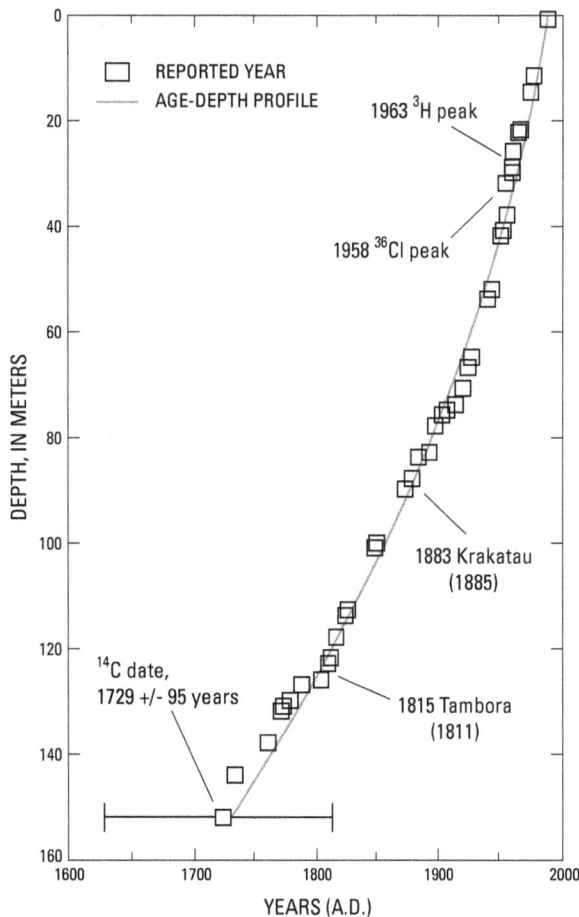

Figure 4.—*Refined age-to-depth profile for the 1991 Upper Fremont Glacier ice core. Modified from Schuster and others (2004). Boxes are the reported year for historic events that are preserved in the ice.*

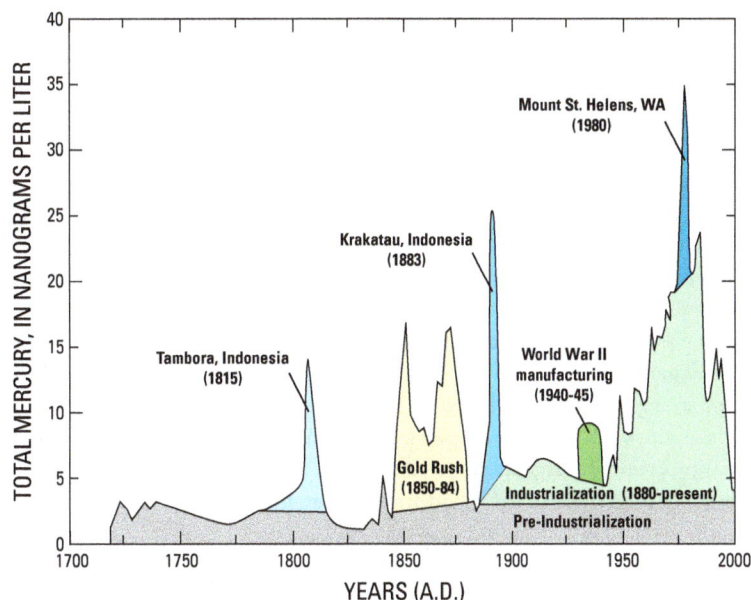

Figure 5.—*Profile of historic concentrations of mercury (Hg) measured in selected ice-core sections from the Upper Fremont Glacier. Modified from Schuster and others (2000).*

the Upper Fremont Glacier ice core along with more than 30 lesser volcanic eruptions. The refined ice-core chronology indicated that the transition from the LIA that was preserved in the ice core occurred around A.D. 1845. Additionally, the transition was abrupt, probably fewer than 10 years, in the alpine regions of the Wind River Range, Wyoming U.S.A. (Naftz and others, 1996, 2004; Schuster and others, 2000, 2004).

Applications

Future information obtained from the study of these glaciers will add significantly to our understanding of atmospheric processes and possibly provide a linkage to other middle-latitude ice cores that record global environmental changes. Middle-latitude glacier research can potentially be applied to numerous climatic and environmental studies worldwide. Applications of the results from the Upper Fremont Glacier study site are discussed below and include reconstructing paleoclimatic records on a global scale and documentation of environmental mercury concentrations during the past 270 years. Similar interpretations and applications of the datasets from the central Asia glacier sites are underway.

Paleoclimatic Records

In addition to the documentation of an abrupt end to the LIA, Naftz and others (2004) developed a transfer function for estimating the change in ambient air temperature over time using the $\delta^{18}O$ profile from the Upper Fremont Glacier ice core and on-site snow-depth sensor data. The transfer function was then used to reconstruct trends in average air temperature during the last 300 years. Based on this transfer function, an increase in air temperature of about +2 °C has occurred at the Upper Fremont Glacier site from the end of the LIA to the present, a relatively rapid air-temperature change.

A consequence of this apparent rapid change in climate was documented in the form of a glacier-outburst flood event (jökulhlaup) (GLOF) that occurred at Grasshopper Glacier in September 2003, 14 km north of the Upper Fremont Glacier (Naftz and others, 2009). A 12-hectare lake at the head of Grasshopper Glacier instantaneously burst releasing approximately 3.2×10^6 m^3 of water, with an associated maximum stream flow of about 35.7 m^3 s^{-1} on Dinwoody Creek downstream from the lake. Annual peak stream flows on this creek generally occur during middle- to late-June as a result of snowmelt runoff and average 26.7 m^3 s^{-1} based on 38 years of record. As a result of the jökulhlaup, there may be an increased flood risk in the upper reaches of the drainage basin because of aggradation of the stream channels. Continued warming and melting of glaciers in the Wind River Range may result in future and perhaps more frequent jökulhlaups.

Environmental Concentrations of Mercury

A record of atmospheric mercury (Hg) deposition has been documented in ice from the Upper Fremont Glacier (Schuster and others, 2000, 2004). Although some polar ice-core studies have provided a limited record of past Hg deposition, polar cores are, at best, proxy indicators of historic Hg deposition in the middle-latitudes, where 80 to 90 percent of the Earth's human population resides. Two ice cores extracted from the Upper Fremont Glacier in 1991 and 1998 (each totaling 160 meters in length) provided a chronology and paleoenvironmental framework that were essential to the interpretation of the Hg deposition record. A total of 97 ice core samples were selected using low-level processing methods and were analyzed to reconstruct a 270-year atmospheric Hg deposition record for this part of North America. Trends in Hg concentration measured in ice from the Upper Fremont Glacier record major releases to the atmosphere of both natural and anthropogenic Hg from regional and global sources (fig. 5).

The record shows that Hg concentrations are significantly, but for relatively short-time intervals, elevated during periods corresponding to explosive volcanic eruptions, an indication that natural volcanic events "punctuate" the record. Anthropogenic activities, such as industrialization (global-scale), gold mining, and wartime manufacturing (regional scale), indicate that chronic levels of elevated Hg emissions have a greater influence on the historical atmospheric deposition record from the Upper Fremont Glacier. In terms of total Hg deposition recorded by the Upper Fremont Glacier during approximately the past 270 years: anthropogenic inputs appeared to contribute 52 percent; volcanic events contributed 6 percent; and pre-industrialization (or background level) accounted for 42 percent of the total input. More significantly, during the last 100 years, anthropogenic sources contributed 70 percent of the total Hg input. A declining trend in Hg concentrations is obvious during the past 20 years. Declining Hg concentrations in the upper section of the ice-core profile (fig. 5) are corroborated by recent declining trends observed in sediment cores (Engstrom and Swain, 1997; Norton and others, 1997; Bindler and others, 2001). This is also verified by similar concentrations in Upper Fremont Glacier snow samples collected in 1999 (P.F. Schuster, written commun., 2005). This decline may be in response to the United States Clean Air Act of 1970.

Summary

Precipitation, snow, ice, and runoff samples from four glaciers, widely distributed geographically in the middle latitudes of the Northern Hemisphere, are currently being analyzed utilizing a wide array of geochemical, physical, and isotopic-measurement techniques. Radioactive fallout from nuclear-weapons testing in the 1940s, 1950s, and 1960s, large explosive volcanic eruptions with regional and global fallout, and radiocarbon dating of bio-organic material, have been identified as time markers in selected ice cores from the *Nangpai Gosum Glacier*, Nepal; Lednik Engil'chek, Kazakhstan/Kyrghyzstan/China; *Belukha Glacier*, Kazakhstan/Russia; and Upper Fremont Glacier, Wyoming, U.S.A., ice cores. These time markers, in conjunction with other types of analyses such as the delta oxygen-18 ($\delta^{18}O$) and mercury (Hg), aid in refining the chronology of the ice cores. The reconstruction of ice-core chronologies provides critical support to other areas of research being undertaken. Linking data from middle-latitude glaciers around the world will aid in the reconstruction of paleoclimatic and paleoenvironmental records on regional and global scales.

References Cited

Aizen, V.B., Aizen, E.M., Dozier, J., Melack, J.M., Sexton, D.D., and Nesterov, V.N., 1997, Glacial regime of the highest Tien Shan mountain, Pobeda-Khan Tengry massif: Journal of Glaciology, v. 43, no. 145, p. 503–512.

Alley, R.B., 2000, The two-mile time machine—Ice cores, abrupt climate change, and our future: Princeton, N.J., Princeton University Press, 229 p.

Bindler, R., Renberg, I., Appleby, P.G., Anderson, N.J., and Rose, N.L., 2001, Mercury accumulation rates and spatial patterns in lake sediments from West Greenland — A coast to ice margin transect: Environmental Science and Technology, v. 35, no. 9, p. 1736–1741.

Cecil, L.D., Green, J.R., and Thompson, L.G., eds., 2004, Earth paleoenvironments—Records preserved in mid- and low-latitude glaciers: Dordrecht, Kluwer Academic Publishers [now Springer], 250 p.

Cecil, L.D., Green, J.R., Vogt, S., Frape, S.K., Davis, S.N., Cottrell, G.L., and Sharma, P., 1999, Chlorine-36 in water, snow, and mid-latitude glacial ice of North America — Meteoric and weapons-tests production in the vicinity of the Idaho National Engineering and Environmental Laboratory, Idaho: U.S. Geological Survey Water-Resources Investigations Report 99–4037, 27 p.

Cecil, L.D., Green J.R., Vogt, S., Michel, R.L., and Cottrell, G., 1998, Isotopic composition of ice cores and meltwater from Upper-Fremont Glacier, and Galena Creek rock glacier, Wyoming: Geografiska Annaler, v. 80, nos. 3–4, p. 287–292.

Cecil, L.D., and Vogt, S., 1997, Identification of bomb-produced chlorine-36 in mid-latitude glacial ice of North America: Nuclear Instruments and Methods in Physics Research, v. B123, nos. 1–4, p. 287–289.

Currie, L.A., Klouda, G.A., Elmore, D., and Gove, H.E., 1985, Radiocarbon dating of microgram samples — Accelerator mass spectrometry and electromagnetic isotope separation: Nuclear Instruments and Methods in Physics Research, v. B12, no. 3, p. 396–401.

Elmore, D., and Phillips, F.M., 1987, Accelerator mass spectrometry for measurement of long-lived radioisotopes: Science, v. 236, no. 4801, p. 543–550.

Engstrom, D.R., and Swain, E.B., 1997, Recent declines in atmospheric mercury deposition in the upper midwest: Environmental Science and Technology, v. 31, no. 4, p. 960–967.

EPICA, 2004, Eight glacial cycles from an Antarctic ice core: Nature, v. 429, no. 6992, p. 623–628.

Fishman, M.J., and Friedman, L.C., 1989, Methods for determination of inorganic substances in water and fluvial sediments: U.S. Geological Survey Techniques of Water-Resources Investigations, 3d ed., book 5, chap. A1, 709 p.

Fujita, K., Takeuchi, N., Aizen, V., and Nikitin, S., 2004, Glaciological observations on the plateau of Belukha Glacier in the Altai Mountains, Russia from 2001 to 2003: Bulletin of Glaciological Research, no. 21, p. 57–64.

Garbarino, J.R., and Taylor, H.E., 1979, An inductive-coupled plasma atomic emission spectrometric method for routine water quality testing: Applied Spectroscopy, v. 33, no. 3, p. 220–225.

Gerhardt, P., and others, eds., 1994, Methods for general and molecular bacteriology: Washington, D.C., American Society for Microbiology, 791 p.

Green, J.R., Cecil, L.D., and Frape, S.K., 2004, Methods of mid- and low-latitude glacial record collection, analysis, and interpretation, in Cecil, L.D., Green, J.R., and Thompson, L.G., eds., Earth paleoenvironments — Records preserved in mid- and low-latitude glaciers: Dordrecht, Kluwer Academic Publishers [now Springer], p. 17–36.

Green, J.R., Cecil, L.D., Synal, H.-A., Santos, J., Kreutz, K.J., and Wake, C.P., 2004, A high resolution record of chlorine-36 nuclear-weapons-tests fallout from central Asia: Nuclear Instruments and Methods in Physics Research, v. B223–224, p. 854–857.

Hobie, J.E., Daley, R.J., and Jasper, S., 1977, Use of nuclepore filters for counting bacteria by fluorescence microscopy: Applied and Environmental Microbiology, v. 33, no. 5, p. 1225–1228.

Karl, D.M., Bird, D.F., Bjorkman, K., Houlihan, T., Shackelford, R., and Tupas L., 1999, Microorganisms in the accreted ice of Lake Vostok, Antarctica: Science, v. 286, no. 5447, p. 2144–2147.

Kendall, C., and Caldwell, E.A., 1998, Fundamentals of isotope geochemistry, in Kendall, C., and McDonnell, J.J., eds., Isotope tracers in catchment hydrology: Amsterdam, Elsevier Science, p. 51–86.

Kreutz, K.J., Wake, C.P., Aizen, V.B., Cecil, L.D., Green, J.R., and Synal, H-A., 2004, Event to decadal-scale glaciochemical variability on the Inilchek Glacier, central Tien Shan, in Cecil, L.D., Green, J.R., and Thompson, L.G., eds., Earth Paleoenvironments — Records preserved in mid- and low-latitude glaciers: Dordrecht, Kluwer Academic Publishers [now Springer], p. 61–79.

Kreutz, K.J., Wake, C.P., Aizen, V.B., Cecil, L.D., and Synal, H.-A., 2003, Seasonal deuterium excess in a Tien Shan ice core — Influence of moisture transport and recycling in Central Asia: Geophysical Research Letters, v. 30, no. 18, 1922, doi:10.1029/2003 GLO 17896.

Krimmel, R.M., 2002, Glaciers of the western United States, with a section on Glacier retreat in Glacier National Park, Montana by Key, C.H., Fagre, D.B., and Menicke, R.K., Glaciers of the conterminous United States (J–2), in Williams, R.S., Jr., and Ferrigno, J.G., eds., Satellite image atlas of glacier of the world: U.S. Geological Survey Professional Paper 1386–J (North America), p. J329–J381.

Naftz, D.L., 1993, Ice-core records of the chemical quality of atmospheric deposition and climate from mid-latitude glaciers, Wind River Range, Wyoming: Golden, Colorado School of Mines, Ph.D. dissertation, 204 p.

Naftz, D.L., Klusman, R.W., Michel, R.L, Schuster, P.F., Reddy, M.M., and others, 1996, Little Ice Age evidence from a south-central North American ice core, U.S.A.: Arctic and Alpine Research, v. 28, no. 1, p. 35–41.

Naftz, D.L., Michel, R.L., and Miller, K.A., 1993, Isotopic indicators of climate in ice cores, Wind River Range, Wyoming, in Swart, P.K., Lohmann, K.C., McKenzie, J., and Savin, S., eds., Climate change in continental isotopic records: Geophysical Monograph 78, p. 55–66.

Naftz, D.L., and Miller, K.A., 1992, USGS collects ice core through alpine glacier: EOS (Transactions, American Geophysical Union), v. 73, no. 3, p. 27.

Naftz, D.L., Oswald, L., Schuster, P.F., and Miller, K., 2009, Ice-core and flood-flow evidence of rapid climate change, Fitzpatrick Wilderness Area, Wind River Range, Wyoming, *in* Wagner, F.H., ed., Climate warming in western North America—Evidence and environmental effects: Salt Lake City, Utah, University of Utah Press, p. 35–48.

Naftz, D.L., Susong, D.D., Cecil, L.D., and Schuster, P.F., 2004, Variations between ^{18}O in recently deposited snow and on-site air temperature, Upper Fremont Glacier, Wyoming, *in* Cecil, L.D., Green, J.R., and Thompson, L.G., eds., Earth paleoenvironments—Records preserved in mid and low latitude glaciers: Dordrecht, Kluwer Academic Publishers [now Springer], p. 217–234.

Norton, S.A., Evans, G.C., and Kahl, J.S., 1997, Comparison of Hg and Pb fluxes to hummocks and hollows of ombrotrophic Big Heath bog and to nearby Sargent Mountain Pond, Maine, USA: Water, Air, and Soil Pollution, v. 100, no. 3–4, p. 271–286.

Olivier, Susanne, and others, 2003, Glaciochemical investigation of an ice core from Belukha Glacier, Siberian Altai: Geophysical Research Letters, v. 30, no. 19, 2019 p., doi:10.1029/2003GL018290.

Ostlund, H.G., and Werner, E., 1962, The electrolytic enrichment of tritium and deuterium for natural tritium measurements, *in* Tritium in the physical and biological sciences, Proceedings of the symposium on the detection and use of tritium in the physical and biological sciences: Vienna, International Atomic Energy Agency, v. 1, p. 95–104.

Pedersen, K., 1993, The deep subterranean biosphere: Earth-Science Reviews, v. 34, no. 4, p. 243–260.

Phelps, T.J., Murphy, E.M., Pfiffner, S.M., and White, D.C., 1994, Comparison between geochemical and biological estimates of subsurface microbial activities: Microbial Ecology, v. 28, no. 3, p. 335–349.

Priscu, J.C., Fritsen, C.H., Adams, E.E., Giovannoni, S.J., Paerl, H.W., and others, 1998, Perennial Antarctic lake ice — An oasis for life in a polar desert: Science, v. 280, no. 5372, p. 2095–2098.

Schuster, P.F., Naftz, D.L., Cecil, L.D., and Green, J.R., 2004, Evidence of abrupt climate change and the development of an historic mercury deposition record using chronological refinement of ice cores at Upper Fremont Glacier, *in* Cecil, L.D., Green, J.R., and Thompson, L.G., eds., Earth paleoenvironments — Records preserved in mid- and low-latitude glaciers: Dordrecht, Kluwer Academic Publishers [now Springer], p. 181–216.

Schuster, P.F., White, D.E., Naftz, D.L., and Cecil, L.D., 2000, Chronological refinement of an ice core record at Upper Fremont Glacier in south central North America: Journal of Geophysical Research, v. 105, no. D4, p. 4657–4666.

Simkin, Tom, and Siebert, Lee, 1994, Volcanoes of the world — A regional directory, gazetteer, and chronology of volcanism during the last 10,000 years, 2d ed.: Tucson, Arizona, Geoscience Press, 349 p.

Thatcher, L.L., Janzer V.J., and Edwards, K.W., 1977, Methods for determination of radioactive substances in water and fluvial sediments: U.S. Geological Survey Techniques of Water-Resources Investigations, book 5, chap. A5, 95 p.

Thompson, L.G., 2000, Ice-core evidence for climate change in the Tropics—Implications for our future: Quaternary Science Reviews, v. 19, no. 1–5, p. 19–35.

Thompson, L.G., 2004, High altitude, mid- and low-latitude ice core records — Implications for our future, *in* Cecil, L.D., Green, J.R., and Thompson, L.G., eds., Earth paleoenvironments — Records preserved in mid- and low-latitude glaciers: Dordrecht, Kluwer Academic Publishers [now Springer], p. 3–15.

Thompson, L.G., in press, Ice cores, high-mountain glaciers, and climate, *in* Williams, R.S., Jr., and Ferrigno, J.G., eds., Satellite image atlas of glaciers of the world: U.S. Geological Survey Professional Paper 1386–A (State of the Earth's cryosphere at the beginning of the 21st century: Glaciers, global snow cover, floating ice, and permafrost and periglacial environments).

U.S. Environmental Protection Agency, 1999, Method 1631-Revision B — Mercury in water by oxidation, purge and trap, and cold vapor atomic fluorescence spectrometry: Washington, D.C., U.S. Environmental Protection Agency, Office of Water, EPA 821-R-99-005, 40 p.

White, J.W.C., Brimblecombe, P., Brühl, C., Davidson, C.I., Delmas, R.J., and others, 1989, How do glaciers record environmental processes and preserve information?, *in* Oeschger, H., and Langway, C.C., Jr., eds., The environmental record in glaciers and ice sheets: Chichester, John Wiley, p. 85–98.

www.ingramcontent.com/pod-product-compliance
Lightning Source LLC
Chambersburg PA
CBHW080704220326
41598CB00033B/5307